Acquired Mitochondropathy – A New Paradigm in Western Medicine Explaining Chronic Diseases

Enno Freye

Acquired Mitochondropathy – A New Paradigm in Western Medicine Explaining Chronic Diseases

The Safety Guide for Prevention and Therapy of Chronic Ailments

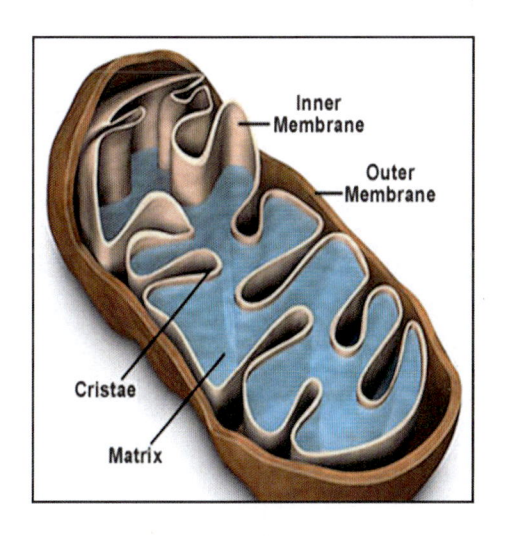

Springer

Enno Freye
University of Düsseldorf
Deichstr. 3 a
41468 Neuss-Uedesheim
Germany
enno.freye@uni-duesseldorf.de

ISBN 978-94-007-2035-0   e-ISBN 978-94-007-2036-7
DOI 10.1007/978-94-007-2036-7
Springer Dordrecht Heidelberg London New York

Library of Congress Control Number: 2011937456

© Springer Science+Business Media B.V. 2012
No part of this work may be reproduced, stored in a retrieval system, or transmitted in any form or by any means, electronic, mechanical, photocopying, microfilming, recording or otherwise, without written permission from the Publisher, with the exception of any material supplied specifically for the purpose of being entered and executed on a computer system, for exclusive use by the purchaser of the work.

Printed on acid-free paper

Springer is part of Springer Science+Business Media (www.springer.com)

Foreword

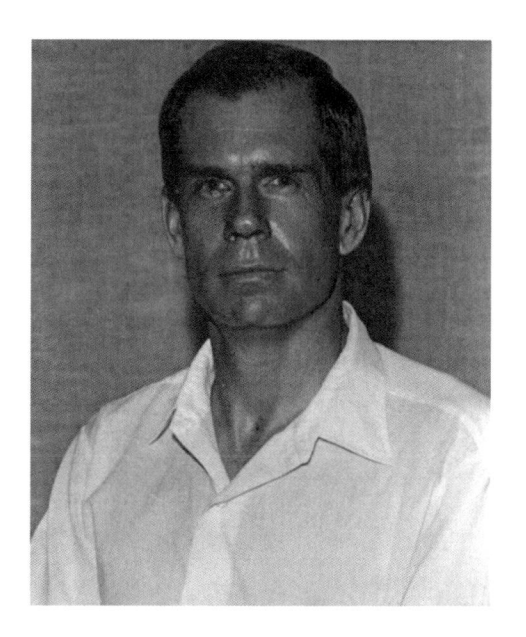

More than 80% of all chronic disease is preventable, but only if people would know how. Learn how the most important parts in cellular function, i.e. mitochondria, are affected by environmental toxins as well as food additives and how the causes for cancer, heart disease, diabetes, depression, hypertonus, Alzheimer, Parkinson and many other neurodegenerative health conditions can be prevented. The book is aimed to provide a detailed explanation for the overall mechanism of mitochondropathy and supply a proposed mechanism for the many symptoms and signs all of which share common features and how each of a specific symptom is generated by a similar basic mechanism. Based on scientifically proven facts from basic science and supported by clinical studies, a comprehensive paradigm for the nitric oxide/ peroxynitrite-cycles and their succeeding effects on mitochondrial activity is outlined. The goal is to increase awareness, bring attention and expand the understanding of many of these mitochondrial-related illnesses, which are caused by the increasingly addition of toxic by-product in the dauly food chain. Alternative ways of treatment in mitochondropathy with micronutrients is outlined in detail explaining the biochemical pathways as to how supplementation with antioxidants exhibit their mode of action at these energy-generating power plants. It is because antioxidants nowadays are considered key-products in the prevention and treatment of chronic ailments of the Western world. Because they are natural components, which interact in the biochemical pathways of cellular function, they are free of any side effects, regularly associated with allopathic medicine. As mitochondria are vital in normal function of practical every organ, the book is also aimed to aid the physician in their decision making as to what kind of complementary, alternative orientated therapy is best for his patients.

Neuss, Germany   Enno Freye, MD, PhD

Contents

1 Introduction 1

2 The Mitochondrion – Essential Part in Cellular Function 3
   2.1 Function of Mitochondria – the Electron Transport Chain 5
      2.1.1 Mitochondrion Complex I 7
      2.1.2 Mitochondrion Complex II 8
      2.1.3 Mitochondrion Complex III 8
      2.1.4 Mitochondrion Complex IV 9
      2.1.5 Mitochondrion Complex V 9
   2.2 Function of Mitochondria – Oxidative Phosphorylation 10
   References 12

3 Definition of Mitochondropathy – Nitrosative Stress 13
   3.1 Significance of T1-Th2 Switch Within the Immune System 19
   References 24

4 Nitric Oxide, a Janus Two-Faced Molecule 27
   4.1 Biological Functions of Nitric Oxide 28
   4.2 Nitric Oxide-Peroxynitrite, Causes for Multisystem Ailments 31
   References 35

5 Representative Ailments with Excess Nitric Oxide Formation 37
   5.1 Chronic Fatigue Syndrome, Symptom of Mitochondrial Failure 39
   5.2 Excess Formation of Nitric Oxide in Inflammation 43
   5.3 Excessive Nitric Oxide Formation Damage Mitochondria 50
   References 52

6 Principles in Nitrosative Stress 55
   6.1 Factors Triggering NO/ONOO Cycle – Inflammation,
       Radical Nitrosative and Oxygen Substrates 56
   6.2 Leptin, Link for Inflammation in NO-Related Diseases 60
   6.3 Tetrahydrobiopterin Impaired by Nitrosative Stress 61
   References 67

Contents

7 Nitrosative Stress in Diverse Multisystem Diseases 71
  7.1 Factors Leading to a Reduction of BH_4 74
      7.1.1 BH_4 Level Restoration Results in Cardiovascular Improvement 77
      7.1.2 Local Malfunction of Mitochondria Related to Multisystem Diseases 79
      7.1.3 Factors Driving Nitrosative Stress 86
  7.2 Exogenic Toxicity Leading to Mitochondrial Function 88
      7.2.1 Pharmaceutical Products 88
      7.2.2 Stress Hormones on Mitochondrial Function 94
      7.2.3 Additives in Food Chain with Mitochondrial Toxicity 95
      7.2.4 Chemical Substances – Originators of Mitochondropathy 97
  7.3 Neurodegeneration Resulting from Blockade in the Energy Pathway 101
      7.3.1 Neurodegenerative Diseases Have a Common Denominator 106
      7.3.2 Mitochondropathy and Neurodegeneration in Different Disease States 115
      7.3.3 Chronic Fatigue Syndrome, Fibromyalgia Caused by Mitochondrial Failure 121
      7.3.4 Chronic Fatigue Caused by Mitochondrial Failure 127
      7.3.5 Hormonal Problem in CFS – Hypothyroidism, Adrenal Insufficiency 135
  7.4 Symptoms of Mitochondropathy of the CNS 146
      7.4.1 General Approach for Diagnosing and Restoring Mental Health 147
      7.4.2 Fibromyalgia Closely Related to Mitochondropathy 149
      7.4.3 Situations Developing into Hypoglycemia 170
  7.5 The Methylation Cycle-Hypothesis for Pathogenesis of Fatigue 187
      7.5.1 Biochemistry of the Methylation Cycle 189
      7.5.2 Testing How Well the Methylation Cycle Works in the Body 190
      7.5.3 Block in the Methylation Cycle – Which Supplements to Take 190
      7.5.4 Problems Arise When Starting Treatment 192
      7.5.5 Creating a Healthy Digestive Tract 193
  References 203

8 Points to Consider in Therapy of Mitochondropathy 217
  8.1 Phospholipids Restoring Cellular Function 217
      8.1.1 Switching Off the Sugar Addiction Gene 221
      8.1.2 Disturbed Sleep, a Common Symptom of Hypoglycemia 222
      8.1.3 Insulin a Stress Hormone Activated by High Blood Glucose Levels 222

	8.1.4	Monitoring Symptoms of Hypoglycemia	223
	8.1.5	Longterm Treatment of Hypoglycemia	223
	8.1.6	Resetting the Recommended Daily Allowance (RDA)	224
8.2	Toxic Stress in Western World (Xenobiotics)	227
	8.2.1	Origin of Xenobiotics Resulting in a Toxic Load	228
	8.2.2	Risk Factor for Alzheimer/Parkinson – Toxins Affecting Mitochondrial Function	230
	8.2.3	Glycation End-Products Impede Supply of Cholesterol and Fats to Neurons	233
	8.2.4	Growing your Own Probiotics	242
	8.2.5	Kefir for Growing Good Bacteria in the Intestinal Tract	242
8.3	Mitochondria Involved in Mutation to Cancer Cells?	244
	8.3.1	Exogenic Factors Rising Nitric Oxide, Blocking Endogenous Synthesis	248
8.4	Avoiding Inflammation, Mediator of Nitrosative Stress	251
	8.4.1	Inflammation via NF-κB Activation	254
	8.4.2	Cryptopyrroluria, Nitrosative Stress and Mitochondrial Disease	256
	8.4.3	Specific Treatment Options in Chronic Fatigue Syndrome	260
8.5	Therapeutics for Normal Mitochondrial Function	265
	8.5.1	Antioxidants for Mitochondrial Protection	268
	8.5.2	The Normal Antioxidant System	269
	8.5.3	Principles/Practice of the "Cave Man Diet"	270
	8.5.4	Unstable Blood Sugar Results in Reactive Oxygen/Nitrosative Stress	273
	8.5.5	Recommendations for Diet in Mitochondropathy	274
	8.5.6	Treatment Package for Failing Mitochondria	280
	8.5.7	Why Is There Fatigue in CFS and FMS Patients	284
	8.5.8	Implications for Treatment of CFS and FMS	286
8.6	Basics in Therapy of Mitochondrial-Related Diseases	293
8.7	Reasoning Supplemental Therapy in Mitochondropathy	297
8.8	Food Used as Medicine in Mitochondropathy	301
	8.8.1	Classification of Nutraceuticals	303
	8.8.2	Oxygen Radical Absorbance Capacity of Food	305
	8.8.3	Rational for Specific Supplements in Mitochondropathy	307
8.9	Newer Therapeutic Options in Mitochondropathy	311
	8.9.1	Rational for Supplementation with Vitamin D_3	311
	8.9.2	Supplementation with Neutraceuticals	313
	8.9.3	NADH Essential Within the Electron Transport Chain	313
	8.9.4	NADH in Parkinson-Clinical Studies Demonstrate Efficacy	319

		8.9.5	Supplementation of NADH as a Powerful Antioxidant	320
		8.9.6	Benefits of Curcumin in Mitochondrial Disease	322
	8.10	Mechanisms of Action of Agents for Therapy in CFS-FMS-Parkinson-Alzheimer		323
		8.10.1	Rational for Taking High dose B_{12} in Mitochondropathy	330
		8.10.2	Tetrahydroxybiopterin (BH_4) – Key Candidate in Mitochondropathy	332
		8.10.3	NADH in Parkinson–de Novo Pathway for the Production of BH_4	333
		8.10.4	Rational for Betain in Mitochondropathy	338
		8.10.5	Dietary Considerations in Mitochondropathy	339
		8.10.6	Ribose – Option for Restoring Energy Depletion	342
		8.10.7	Mitochondria Remodeling with Intermittent Hypoxia Therapy (IHT)	347
		8.10.8	Practice in Applying Intermittent Hypoxic Training	350
		8.10.9	Improving Antioxidant Status with Bicarbonate	352
	8.11	New Avenues in Treatment of Mitochondropathy		355
		8.11.1	Pyrroloquinoline Quinone, Another CoQ10 for Neuroprotection	356
		8.11.2	The Positive Effects of Saturated Fats	357
		8.11.3	Cholesterol, the Culprit for Arteriosclerosis?	358
	References			364
9	**Lab Tests for Mitochondropathy, Nitrosative Stress**			**371**
	9.1	The Extended Lab Tests		372
	9.2	Diagnosing a Possible Allergy		376
		9.2.1	Testing for a Toxic Load in the Organism	376
	References			380

List of Certified Labs in Europe for Diagnosing Mitochondropathy (Not Claimed for Completeness) 381

Index 387

Chapter 1
Introduction

What allows us to live and our bodies to function are billions of chemical reactions in the body which occur every second. They are essential for the production of energy, which drives all the processes of life such as nervous function, movement, heart function, digestion and so on. If all these enzyme reactions invariably occurred perfectly, there would be no need for an antioxidant system. However, even our own enzyme systems make mistakes and the process of producing energy in mitochondria is highly active. When mistakes occur, free radicals are produced. Essentially, a free radical is a molecule with an unpaired electron, it is highly reactive and to stabilize its own structure, it will literally stick on to anything. That "anything" could be a cell membrane, a protein, a fat, a piece of DNA, or whatever. In sticking on to something, it denatures that something so that it has to be replaced. Free radicals come from inside the body (mitochondrial energy production, the P450 detox system in the liver and immune activity from inflammation) and from outside the body (poisoning from pesticides, volatiles organic compounds, heavy metals, radiation, etc), and by having free radicals, this is extremely damaging to the body. Therefore the body has evolved a system to get rid of these free radicals before they have a chance to do such damage and this is called our antioxidant system. In recent years, however, more stress has been placed on our antioxidant system because we are increasingly exposed to internal toxins (modern diets) and external toxins (pollution), which often exert their malign influence by producing free radicals. Therefore, it is even more important than ever to ensure a good antioxidant status, because free radicals effectively accelerate the normal ageing process while antioxidants slow the normal ageing process. The best example that is visible to all, are the effects of smoking. Cigarette smoking produces large amounts of free radicals and people who have smoked for many years have prematurely aged skin. Smokers also die younger from cancer or arterial disease – problems one only expects to see in the elderly. Conversely, people who live and eat the healthy way age more slowly.

Chapter 2
The Mitochondrion – Essential Part in Cellular Function

And while free radicals increasingly built up in modern life, they in return attack an essential cell organelle, the mitochondrion (Fig. 2.1). A mitochondrion is a cytoplasmic organelle that provides most of the energy necessary for a cell to survive and operate. It is present in all eukaryotic cells, and a typical human cell has several 100 mitochondria. For instance, there are 1,000–2,000 mitochondria in a single liver cell, occupying roughly a fifth of its total volume and most mitochondria are found in the muscle cells of the heart and the neuronal cells of the central nervous system.

Presently it is widely accepted that the mitochondrion is a remnant of a prokaryotic organism that had become a vital symbiotic partner to the eukaryotic cell early in evolution. The protozoon *Reclinomonas americana* harbors the most bacteria-like mitochondrial genome, and *Rickettsia prowazekii* the most mitochondria-like eubacterial genome. Another hypothesis on the origin of eukaryotes, a 'revisionist view of eukaryotic evolution', has gained some acceptance within the past years. In this scenario there is no mitochondriate eukaryote, but the fusion of anaerobic archaeobacteria (the 'host') with respiration-competent proteobacteria (the 'symbiont') represents the initiation of the formation of eukaryotes [1, 2]. There is evidence that modern mitochondria eukaryotes (Archeozoa, Giardia) are derived from this initial event but they have lost their mitochondria in the course of time [3]. All the evidence points to a monophyletic origin, i.e. the endosymbiosis has occurred only once. The mitochondrion has its own circular genome containing an identical set of genes in most of the organisms supporting the hypothesis for monophyletic origin [4].

Mitochondria are 0.5–1 µm in size and are bounded by two membranes. The smooth outer membrane contains many copies of a transport protein porin, which forms aqueous channels allowing molecules with a maximal molecular weight of 5,000 Da to penetrate the membrane. The inner membrane is conspicuously folded, forming tubular or lamellar structures called cristae (Fig. 2.1), which are connected to it by narrow tubular structures, cristae junctions. The energy-generating apparatus is composed of five multipolypeptide enzyme complexes and is located in the inner membrane, which surrounds the matrix space of the mitochondrion (Fig. 2.2).

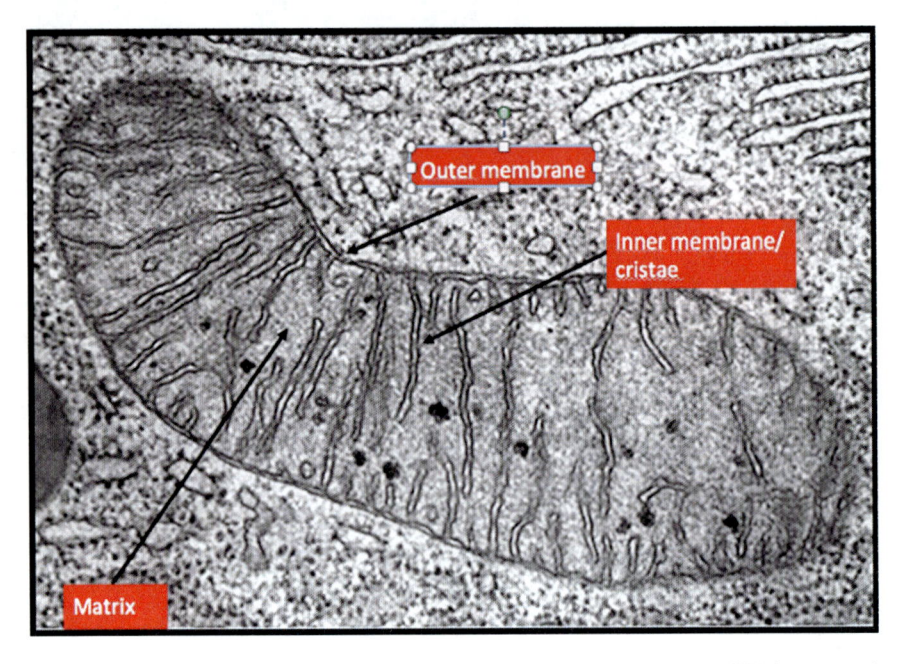

Fig. 2.1 The mitochondrion, an essential part of all living cells where energy (ATP) is generated

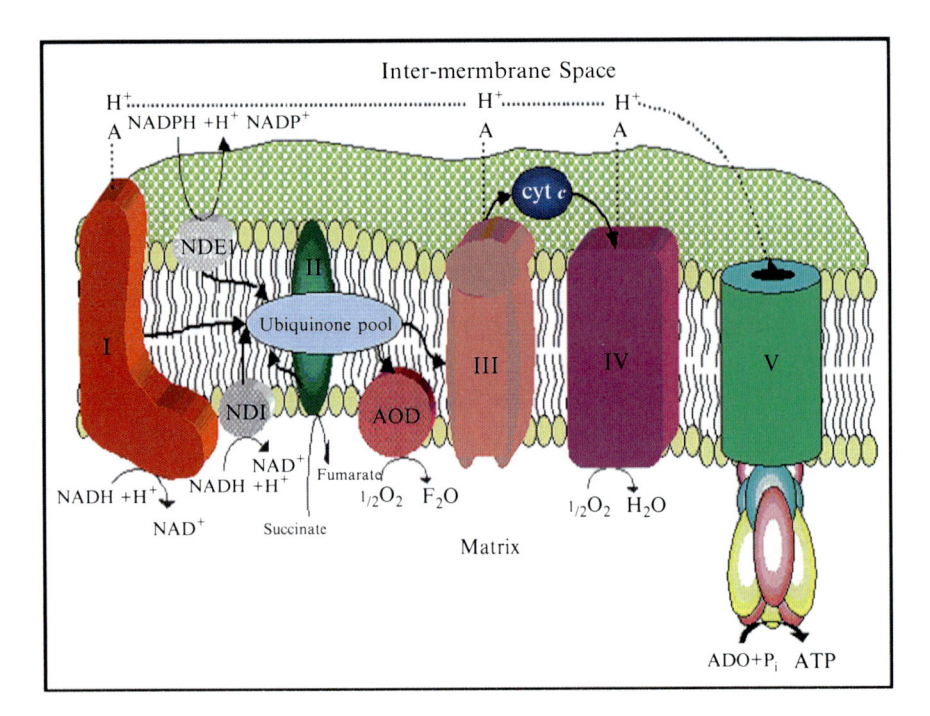

Fig. 2.2 Representation of the five multipolypeptide enzyme complexes, located in the inner membrane of a mitochondrion, necessary part of oxidative phosphorylation for the generation of energy (ATP)

The matrix contains mitochondrial DNA (mtDNA) molecules, ribosomes, tRNAs and various enzymes needed in protein synthesis, the oxidation of pyruvate and fatty acids and the citric acid cycle, for example [5, 6]. There are about 1,000 proteins in a mitochondrion, most of which are nuclear-encoded.

The location of mitochondria in the cytosol is non-random. They can move or be moved according to the needs of the cell [7, 8] and they can change in shape. Some mitochondria in a cell are in contact with the endoplasmic reticulum and others with cytoskeletal structures. The movement of mitochondria depends on microtubules [9].

2.1 Function of Mitochondria – the Electron Transport Chain

The main and vital action of mitochondria is the generation of energy in form of adenosine triphosphate (ATP), which is generated via the electron transport chain (ETC; Fig. 2.3). This chain couples a reaction between an electron donor (such as NADH) and an electron acceptor (such as O_2) to the transfer of H⁺ ions across a membrane, through a set of mediating biochemical reactions. These H⁺ ions are used to produce adenosine triphosphate (ATP), the main energy intermediate in living organisms, as they move back across the membrane. Electron transport chains are used for extracting energy from sunlight (photosynthesis) in plants and from redox reactions such as the oxidation of sugars (oxidative phosphorylation).

While in plants, light drives the conversion of water to oxygen and NADP⁺ to NADPH and a transfer of H⁺ ions. NADPH is used as an electron donor for carbon fixation in chloroplasts. In mitochondria, it is the conversion of oxygen to water, NADH to NAD⁺ and succinate to fumarate that drives the transfer of H⁺ ions (Fig. 2.4). While some bacteria have electron transport chains similar to those in chloroplasts or mitochondria, other bacteria use different electron donors and acceptors. Both the respiratory and photosynthetic electron transport chains are major sites of premature electron leakage to oxygen, thus being major sites of superoxide production and drivers of oxidative stress.

During oxidative phosphorylation, electrons derived from NADH and $FADH_2$ combine with O_2, and the energy released from these oxidation/reduction reactions is used to drive the synthesis of ATP from ADP. The transfer of electrons from NADH to O_2 is a very energy-yielding reaction, with $\Delta G^{\circ\prime} = -52.5$ kcal/mol for each pair of electrons transferred. To be harvested in usable form, this energy must be produced gradually, by the passage of electrons through a series of carriers, which constitute the electron transport chain. These carriers are organized into four complexes in the inner mitochondrial membrane. A fifth protein complex then serves to couple the energy-yielding reactions of electron transport to ATP synthesis. Electrons from NADH enter the electron transport chain in complex I (Fig. 2.4), which consists of nearly 40 polypeptide chains. These electrons are initially

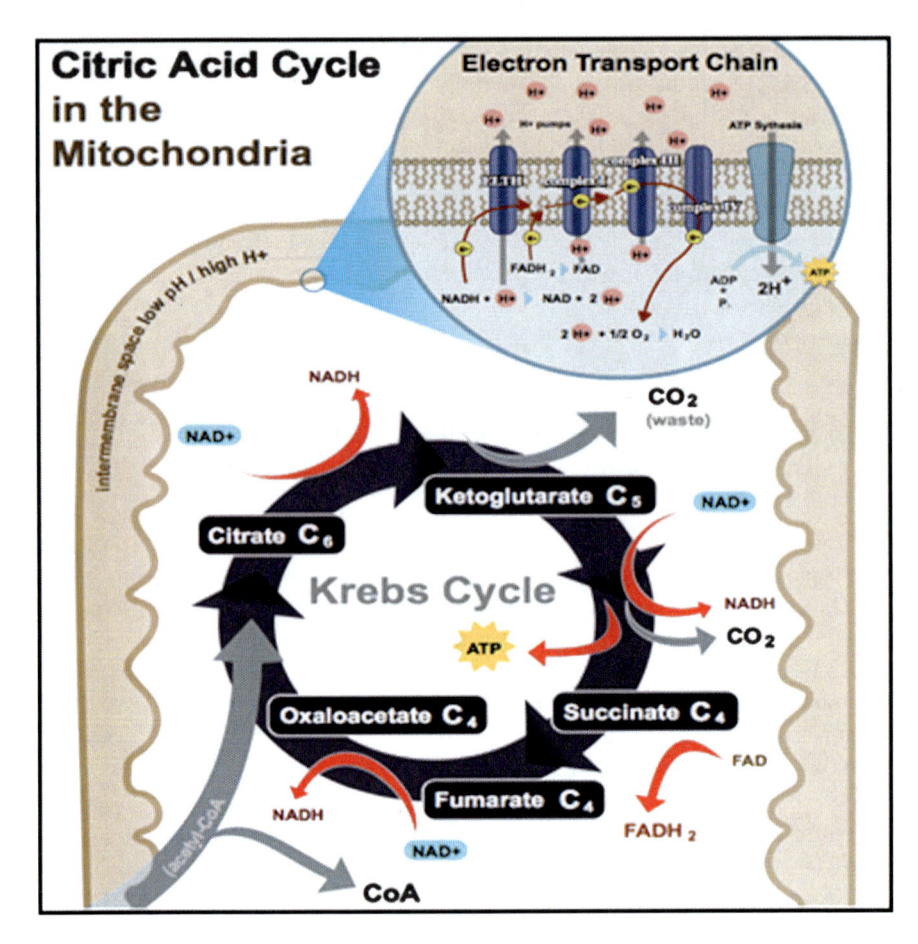

Fig. 2.3 Cellular respiration in a typical eukaryotic cell with the Krebs-cycle in the cytosol and the electron transport chain within the inner lining of mitochondria

transferred from NADH to flavin mononucleotide and then, through an iron-sulfur carrier, to coenzyme Q10 an energy-yielding process with $\Delta G^{o\prime} = -16.6$ kcal/mol. Coenzyme Q10 (also called ubiquinone or Q10) is a small, lipid-soluble molecule that carries electrons from complex I through the membrane to complex III, which consists of about ten polypeptides. In complex III, electrons are transferred from cytochrome *b* to cytochrome *c*—an energy-yielding reaction with $\Delta G^{o\prime} = -10.1$ kcal/mol. **Cytochrome *c***, a peripheral membrane protein bound to the outer face of the inner membrane, then carries electrons to complex IV (cytochrome oxidase), where they are finally transferred to O_2 ($\Delta G^{o\prime} = -25.8$ kcal/mol). The protein complexes I, III, IV, and V also depend upon proteins encodes by nuclear DNA which are synthesized in the cytosol and imported into the mitochondria.

2.1 Function of Mitochondria – the Electron Transport Chain

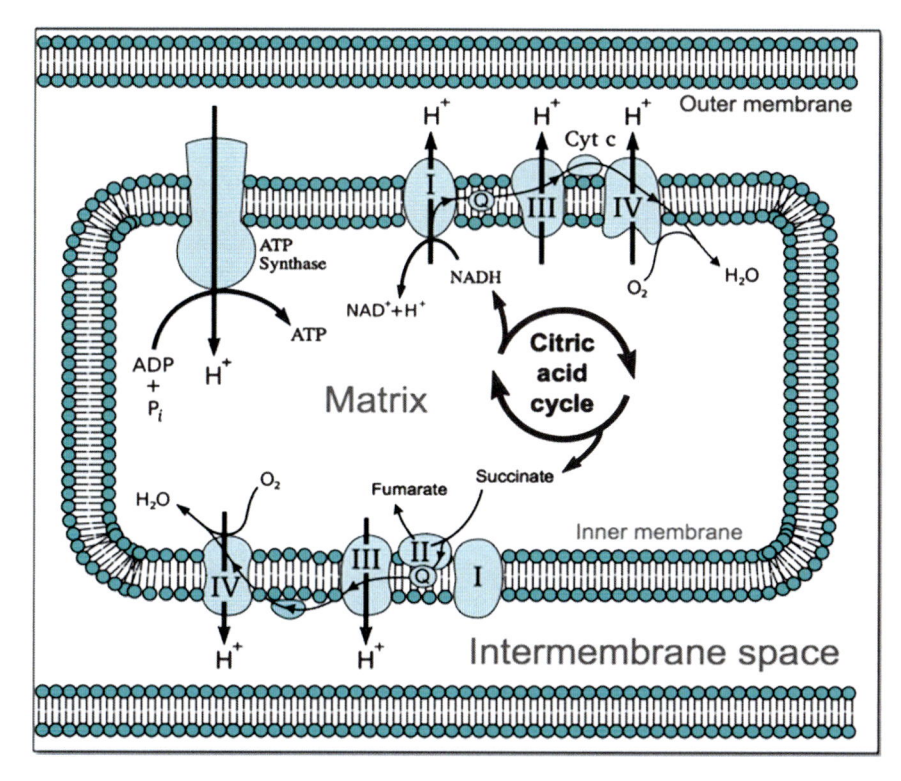

Fig. 2.4 The electron transport chain (ETC) with complex I-IV in the mitochondrion is the site of oxidative phosphorylation in eukaryotes. The NADH and succinate generated in the citric acid cycle are oxidized, providing energy to power ATP synthase

2.1.1 Mitochondrion Complex I

In Complex I (NADH dehydrogenase, also called NADH: ubiquinone oxidoreductase; EC 1.6.5.3) two electrons are removed from NADH and transferred to a lipid-soluble carrier, ubiquinone (Q). The reduced product, ubiquinol (QH_2) freely diffuses within the membrane, and Complex I translocates four protons (H^+) across the membrane, thus producing a proton gradient. Complex I is one of the main sites at which premature electron leakage to oxygen occurs, thus being one of the main sites of production of harmful superoxide. The pathway of electrons occurs as follows: NADH is oxidized to NAD^+, by reducing flavin mononucleotide to $FMNH_2$ in one two-electron step. $FMNH_2$ is then oxidized in two one-electron steps, through a semiquinone intermediate. Each electron thus transfers from the $FMNH_2$ to an Fe-S cluster, from the Fe-S cluster to ubiquinone (Q). Transfer of the first electron results in the free-radical (semiquinone) form of Q, and transfer of the second electron reduces the semiquinone form to the ubiquinol form, QH_2. During this

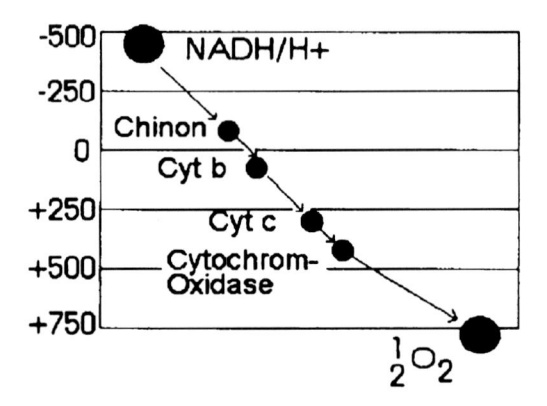

Fig. 2.5 Principle of energy production (i.e. the redoxpotential) within the inner mitochondrial membrane via different coenzymes resulting in a high gradient of positively charged hydrogen ions, which later are used for ATP synthesis via the ATP-synthetase; *Cyt b, c* cytochrome b, c; *Chinon* ubiquinone Q10

process, four protons are translocated from the mitochondrial matrix to the intermembrane space, creating a proton gradient that generates ATP through oxidative phosphorylation.

2.1.2 Mitochondrion Complex II

In *Complex II* (succinate dehydrogenase; EC 1.3.5.1) additional electrons are delivered into the quinone pool (Q) originating from succinate and transferred (via FAD) to Q. Complex II consists of four protein subunits: SDHA, SDHB, SDHC, and SDHD. Other electron donors (e.g., fatty acids and glycerol 3-phosphate) also direct electrons into Q (via FAD).

2.1.3 Mitochondrion Complex III

In *Complex III* (cytochrome bc_1 complex; EC 1.10.2.2) two electrons are removed from QH_2 at the Q_O site and sequentially transferred to two molecules of cytochrome c, a water-soluble electron carrier located within the intermembrane space. The two other electrons sequentially pass across the protein to the Q_i site where the quinone part of ubiquinone is reduced to quinol. A proton gradient is formed by two quinol (4H+4e−) oxidations at the Q_O site to form one quinol (2H+2e−) at the Q_i site. (in total six protons are translocated: two protons reduce quinone to quinol and four protons are released from two ubiquinol molecules). The bc1 complex does not 'pump' protons, but helps to build the proton gradient by an asymmetric absorption/release of protons. When electron transfer is reduced (by a high membrane potential or respiratory inhibitors such as antimycin A), Complex III may leak electrons to molecular oxygen, resulting in superoxide formation (Fig. 2.5).

2.1.4 Mitochondrion Complex IV

In *Complex IV* (cytochrome c oxidase; EC 1.9.3.1) four electrons are removed from four molecules of cytochrome c and transferred to molecular oxygen (O_2), producing two molecules of water. At the same time, four protons are translocated across the membrane, contributing to the proton gradient. The activity of cytochrome c is inhibited by cyanide. In cell division the movements of mitochondria are highly regulated [10], chiefly by inheritance components located either in the cytosol or in the outer membrane of mitochondria, which have been identified in yeasts. Cytosolic proteins, such as Mdm1p, which is essential for both mitochondrial and nuclear inheritance, are likely to be associated with the cytoskeleton [11, 12]. Four proteins located in the outer membrane have been identified to date [11, 13].

2.1.5 Mitochondrion Complex V

In complex V the final step in conversion of ADP to ATP via the ATP synthase is conducted. ATP synthase is a general term for an enzyme that can synthesize adenosine triphosphate (ATP) from adenosine diphosphate (ADP) and inorganic phosphate by using a form of energy. This energy is in the form of protons moving down an electrochemical gradient, such as from the lumen into the stroma of chloroplasts or from the inter-membrane space into the matrix in mitochondria. The overall reaction sequence is:

$$ADP + P_i \rightarrow ATP$$

Enzymes like synthase are of crucial importance in almost all organisms, because ATP is the common "energy currency" of cells. However, they are also a sensitive site to toxins resulting in a deficit or a total malfunction. For instance, the antibiotic oligomycin inhibits the F_O unit of ATP synthase resulting in insufficient ATP production. Like other enzymes, the activity of F_1F_O ATP synthase is reversible. In respiring organism under physiological conditions, ATP synthase generally runs in the direction from the inside to the outside, creating ATP while using the protonmotive force created by the electron transport chain as a source of energy (Fig. 2.6). The overall process of creating energy in this fashion is termed oxidative phosphorylation. This very process takes place in each of the 1,500 mitochondria located within every cell, where ATP synthase is located in the inner mitochondrial membrane, so that F_1-part sticks into mitochondrial matrix, where ATP synthesis takes place.

The transport of ATP and ADP across the inner membrane is mediated by an integral membrane protein, the adenine nucleotide translocator, which transports one molecule of ADP into the mitochondrion in exchange for one molecule of ATP transferred from the mitochondrion to the cytosol. Because ATP carries more negative charge than ADP (−4 compared to −3), this exchange is driven by the voltage component of the electrochemical gradient. Since the proton gradient

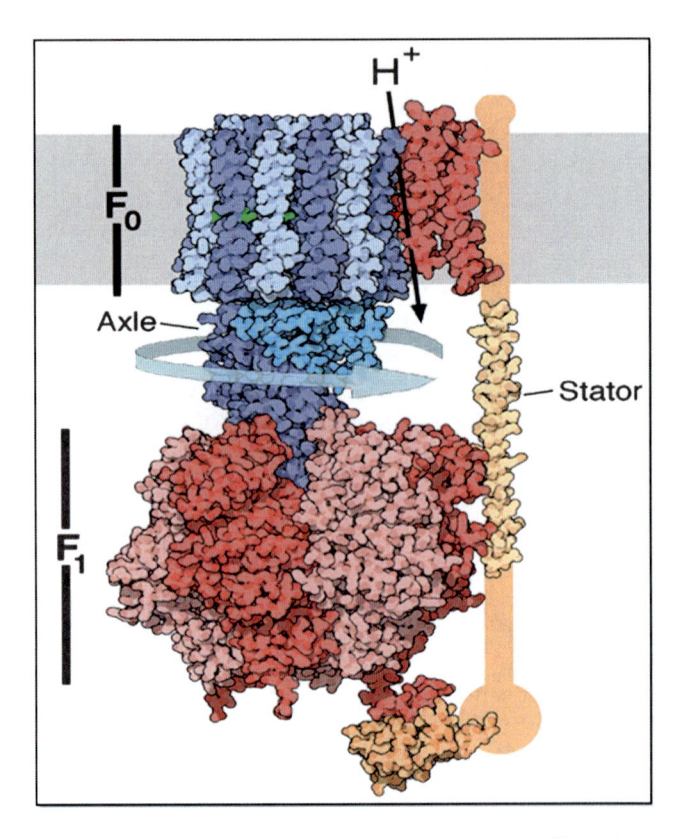

Fig. 2.6 Principal structure of the ATP synthase. The mitochondrial ATP synthase (complex V) consists of two multisubunit components, F_0 and F_1, which are linked by a slender stalk. The F_0 proton channel and rotating stalk are shown in *blue*, the F1 synthase domain in *red* and the membrane in *grey* resulting in a rotation movement, which shifts H⁺-ions from the in- to the outside

establishes a positive charge on the cytosolic side of the membrane, the export of ATP in exchange for ADP is energetically favorable. The synthesis of ATP within the mitochondrion requires phosphate ions (P_i) as well as ADP, so P_i must also be imported from the cytosol. This is mediated by another membrane transport protein, which imports phosphate ($H_2PO_4^-$) and exports hydroxyl ions (OH⁻). This exchange is electrically neutral because both phosphate and hydroxyl ions have a charge of −1. However, the exchange is driven by the proton concentration gradient; the higher pH within mitochondria corresponds to a higher concentration of hydroxyl ions, favoring their translocation to the cytosolic side of the membrane.

2.2 Function of Mitochondria – Oxidative Phosphorylation

The mitochondrion converts energy derived from chemical fuels by an oxidative phosphorylation process that is more efficient than anaerobic glycolysis. In mitochondrion the metabolism of one molecule of glucose produces about 30 molecules

2.2 Function of Mitochondria – Oxidative Phosphorylation

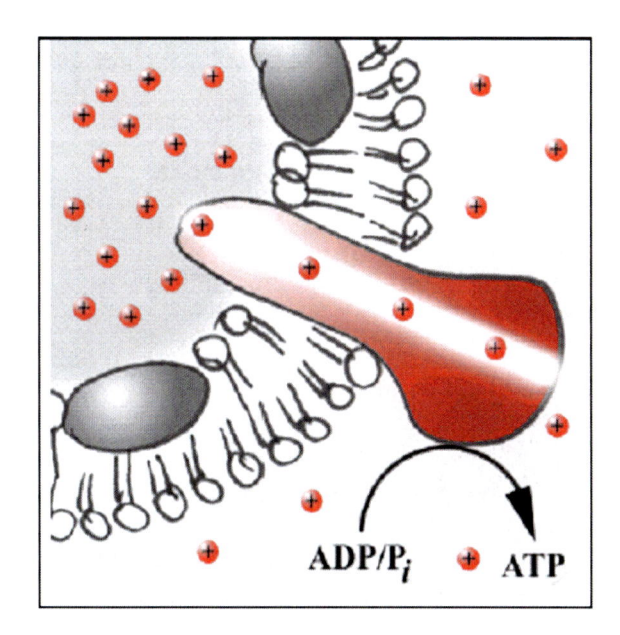

Fig. 2.7 ATP synthase of complex V at the inside membrane of a mitochondrion, exchanges three positively charged protons for conversion of one molecule of ATP to one ATP by attaching one phosphorous group

of ATP (adenosine triphosphate), while only two molecules of ATP are produced by glycolysis alone. This means that organs with a high energy demand are vulnerable to the depletion of mitochondrial energy production. Oxidative metabolism in the mitochondria is fuelled by pyruvate produced from carbohydrates by glycolysis and fatty acids produced from triglycerides. These are selectively imported into the matrix of the mitochondria and broken down into acetyl CoA by the pyruvate dehydrogenase complex or the β-oxidation pathway. The acetyl group then participates in the citric acid cycle, which produces molecules of NADH and FADH. Electrons generated from NADH are passed along a series of carrier molecules called the electron transport chain (ETC), the products of this process being H_2O and energy.

The oxidative phosphorylation pathway (OXPHOS) is composed of the electron transport chain (ETC) and the enzyme ATPase. The whole OXPHOS system embedded in the lipid bilayer of the mitochondrial inner membrane is composed of five multiprotein enzyme complexes (I–V) and two electron carriers – coenzyme Q10 and cytochrome c. The main function of the system is the coordinated transport of electrons and protons and the production of ATP. This passage of electrons releases energy, which is largely stored in the form of a proton gradient across the inner mitochondrial membrane and is used by the last OXPHOS complex (F_1F_O-ATPase; Fig. 2.7) to generate ATP from ADP and inorganic phosphate [14, 15]. Since the inside membrane potential is negative of around 180–200 mV, and a proton gradient of about one unit, this energy is able to drive the synthesis of ATP to fuel the cell's energy dependent processes (Fig. 2.7).

References

1. Gray MW, Burger G, Lang BF (1999) Mitochondrial evolution. Science 283:1476–1481
2. Vellai T, Vida G (1999) The origin of eukaryotes: the difference between prokaryotic and eukaryotic cells. Proc R Soc Lond B: Biol Sci 266:1571–1577
3. Roger AJ et al (1998) A mitochondrial-like chaperonin 60 gene in Giardia lamblia: evidence that diplomonads once harboured an endosymbiont related to the progenitor of mitochondria. Proc Natl Acad Sci USA 95:229–234
4. Scheffler IE (2001) Mitochondria make a come back. Adv Drug Deliv Rev 49:3–16
5. Alberts B et al (1994) Energy conversion: mitochondria and chloroplasts. In: Alberts B et al (eds) Molecular biology of the cell. Garland Publishing, Inc., New York, pp 653–720
6. Wallace DC, Brown MD, Lott MT (1997) Mitochondrial genetics. In: Rimoin DL et al (eds) Emory and Rimoin's principles and practice of medical genetics. Churchill Livingstone, London, pp 277–332
7. Johnson LV, Walsh ML, Chen LB (1980) Localization of mitochondria in living cells with rhodamine 123. Proc Natl Acad Sci USA 77:990–994
8. Bereiter-Hahn J, Voth M (1994) Dynamics of mitochondria in living cells: shape changes, dislocations, fusion, and fission of mitochondria. Microsc Res Tech 27:198–219
9. Overly CC, Rieff HI, Hollenbeck P (1996) Organelle motility and metabolism in axons vs dendrites of cultured hippocampal neurons. J Cell Sci 109:971–980
10. Yaffe MP (1999) The machinery of mitochondrial inheritance and behavior. Science 283:1493–1497
11. Berger KH, Yaffe MP (1996) Mitochondrial distribution and inheritance. Experientia 52:1111–1116
12. Hermann GJ, Shaw JM (1998) Mitochondrial dynamics in yeast. Annu Rev Cell Dev Biol 14:265–303
13. Pelloquin L et al (1998) Identification of a fission yeast dynamin-related protein involved in mitochondrial DNA maintenance. Biochem Biophys Res Commun 251:720–726
14. Hatefi Y (1985) The mitochondrial electron transport and oxidative phosphorylation system. Annu Rev Biochem 54:1015–1069
15. Saraste M (1999) Oxidative phosphorylation at the fin de siècle. Science 283:1488–1493

Chapter 3
Definition of Mitochondropathy – Nitrosative Stress

There are two types of mitochondropathy: The inborn mitochondropathy and the acquired mitochondropathy. The first is a clinically heterogeneous multisystem disease characterized by defects of brain–mitochondrial encephalopathies and/or muscle–mitochondrial myopathies due to alterations in the protein complexes of the electron transport chain of oxidative phosphorylation, including Alper syndrome, Leber's hereditary optic neuropathy, Lowe syndrome, Luft syndrome, Menke's kinky hair syndrome, Zellweger syndrome, MELAS, MERRF, mitochondrial myopathy, rhizomelic chondrodysplasia punctata, and stroke-like episodes. Mutations in mitochondrial genes can lead to a number of mitochondrial disorders. The muscle or the brain are two most commonly affected sites, since they rely so heavily on mitochondria for their energy needs. Mitochondrial DNA (mtDNA) is inherited from the mother, and the mtDNA is present in five to ten copies per mitochondria. When mitochondria divide, the mtDNA copies are divided randomly among the two new mitochondria. If only a few maternally inherited mtDNA copies are defective, mitochondrial division may cause most of the defective mtDNA to reside in just one of the new mitochondria. For this reason, mitochondrial disease often only becomes apparent when the number of affected mitochondria reaches a certain level (threshold experience). Mutations in mtDNA are common because the error checking mechanisms present during nuclear DNA replication are absent. The impact of mitochondrial disease can vary widely, ranging from mild exercise intolerance up to lethal system-wide effects. Some examples of mitochondrial diseases include mitochondrial myopathies, leber hereditary optic neuropathy (LHON), Leigh syndrome, neuropathy/ataxia/retinitis pigmentosa/ptosis (NARP), and myoneurogenic gastrointestinal encephalopathy (MNGIE).

A recently discovered protein plays a central role in translating genetic information into a proteins structure. A biochemical process that early in the origin of life played an important role was discovered by Roland Lill and his group [1]. He discovered that a selective protein is responsible for an essential function, i.e. the production of ribosomes, which play a central role in the formation of proteins to structure DNA. Thus, mitochondria have, at least from now on have to be looked upon as an

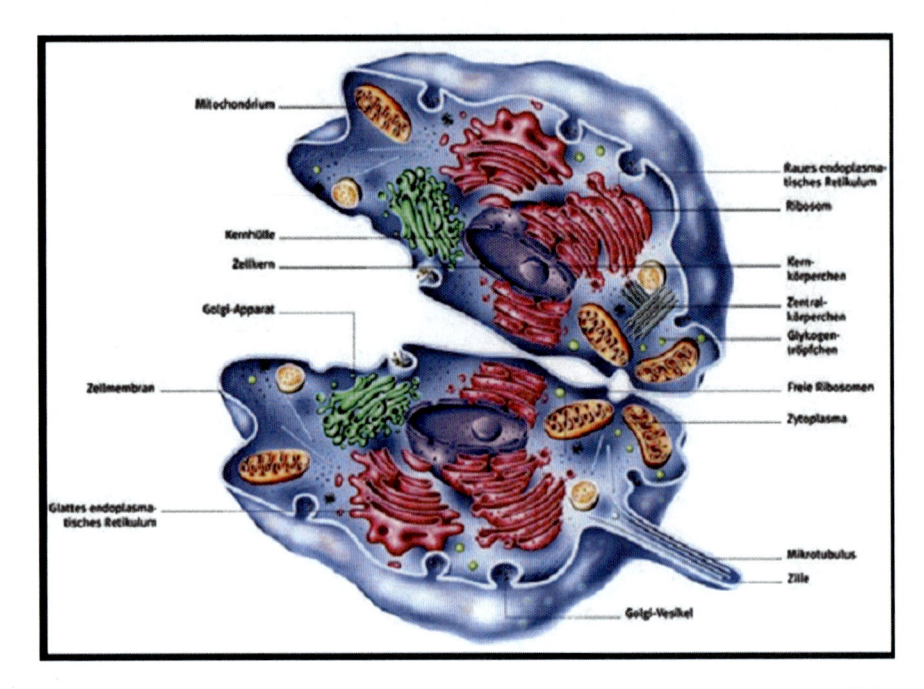

Fig. 3.1 Location of cell organelles mitochondria being an essential part in the regulation of cellular energy production but also of replication

organelle fulfilling important task in replication (Fig. 3.1). Mitochondria control synthesis of proteins in cells. In contrast to previous belief, these researchers now found that synthesis of the iron-sulfur proteins is not a spontaneous, but a complex biochemical process. So far, Lill has discovered with his coworkers, 12 of the previously known 8 protein molecules, mostly mitochondrial proteins, which are important for this process. Once synthesis of iron-sulfur proteins is inhibited, the cell dies. Therefore, these proteins fulfill a vital function. After many decades of mitochondrial research these findings finally resolve the significance of mitochondria in cellular activity and identify these organelles as the most essential components of a cell. However, the actual task of the iron-sulfur protein in cellular action is still left open to speculation, leaving questions like why the formation of these components in mitochondria is essential and what function they fulfill.

To summarize:

1. Mitochondria are the real rulers in the body or the cells.
2. Cancer is a mitochondria-related disease.
3. Mitochondria affect the body's defenses by NO – gas production.
4. Lack of NO – gas increases in the long term, the risk of allergies and autoimmune diseases.

5. NO – gas is absolutely necessary to regulate the blood vessels. If NO is absent for long periods, it can result in an increase of blood pressure.
6. Mitochondria are the main problem in patients with lethargy, fatigue, and people with burn out syndrome, all dealing with an insufficiency of their mitochondria.
7. Cancer therapy must primarily be oriented to regenerate the function of mitochondria.

Being usually regarded as a part of the cell, conducting the biochemical function of cellular respiration, and is the source of energy in the form of adenosine triphosphate (ATP). But even without this cellular respiratory cycle, a cell can live at least for some time when being fed with sufficient glucose. This newly discovered additional fundamental importance of mitochondria is that of synthesis of iron-sulfur containing proteins the prerequisites of ribosomes of the cell, which are used to read the genetic material of a cell. Once there is a pathological impairment in the production of iron-sulfur proteins, seen e.g. in a neurodegenerative disease called Friedreich's ataxia. With an average incidence of one out of 50,000 people, it usually leads to death due to cardiomyopathy-triggered heart failure. The scarcity of this disease is an indicator for the importance of an iron-sulfur protein synthesis in mitochondria, a synthesis which presents a fundamental process in survival so that individual can develop into a higher level in the evolutionary cycle.

But aside from this relevant task, mitochondria are the cell organelles that are responsible for energy production. Each cell of the human being, depending on the organ, contains from 1,500 to 6,000 (liver) mitochondria with a size ranging from 0.8 to 4.9 μm. They have two membranes and are inherited only from the mother to the child, since paternal mitochondria do not enter the sperm during fertilization in the egg. Within the cell they produce energy that is stored in units of ATP using the oxygen as well the glucose, protein and fatty acid molecules from the digested food, Previously it has been thought that mitochondria are subordinate to the DNA of the nucleus. This however, has been proven to be wrong, since mitochondria are the real rulers of the cell.

Mitochondria function is vulnerable to hypoxia, nutrient deficiency, heavy metal poisoning, carbon monoxide and excess nitric oxide

When the mitochondria are exposed to stressors, e.g. hypoxia, nutritional deficit, heavy metal poisoning or a continuous exposure to ROS and NOS, they switch to energy without oxygen utilization, called anaerobic glycolysis. Fatally, the mitochondria after prolonged cellular stress can enter a state resulting in apoptosis (programmed cell death) or even worse, they switch into a state of cell proliferation. This is due to chemical messengers, which are being sent from the cell nucleus. These two mechanisms can be activated when mitochondrial activity drops below 20%. Once, however, when mitochondria stimulate the program of cell proliferation, cancerous cells can arise. Why do mitochondria resort to this option? During the evolution of life they have learned to maintain, if necessary, and activate a cellular ancient program, which is only directed for survive of their own species. This strategy of the so-called Ark of bacteria, or Archaea, has evolved on earth over millions of years and was already observed in the sulphur active volcanic cones of the deep sea. Their resistance is enormous, as they can survive temperatures well above 100°. However, is life in a depth of 1,000 m, where sunlight has never been reached, possible? According to general knowledge, sunlight is a prerequisite of life, and Ark bacteria or Archaea have developed a technique to capture light. This is because active elements of the so-called respiratory chain of mitochondria can absorb light rays emitted from volcanic lava. Similar to many other essential amino acids, mitochondria use sulfur embedded within their protein molecules. With the help of these molecules and the mitochondria complex I to V, the emitted rays of volcanic lava is absorbed through resonance. Such an archaic program is being reactivated in cancerous cells, which however, because of the absence of oxygen, gain their energy by means of fermentation of sugar molecules and not by oxidative phosphorylation. In such circumstances there is an exceptionally high turnover of energy substrates, which is much higher than the usual metabolic turnover with a more efficient respiratory cycle. In addition, sugar molecules are no longer completely metabolized to carbon dioxide and water, which is why there is a built-up of sugar degradation products such as lactic acid. As a compensatory mechanism, energy from proteins and fatty acids is being used, used up for uncontrolled cell growth. And finally, the end product of glucose metabolism, i.e. lactic acid by itself presents an advantage for any cancerous cells. This is because the acidity drives the surrounding healthy cells into programmed cell death, while at the same time it dissolves the protective scaffold that surrounds the cells, resulting in a faster proliferation of cancer cells.

Another important part of the maintenance of cellular activity of mitochondria is the production of the gas nitric oxide (NO). This aggressive gas NO is not only the first defense line to guard the cellular level protecting it from aggressive bacteria, virus and cancer cells, but also it is an important neurotransmitter. These findings are relatively young, since the Nobel Prize for the development and unraveling the significance of NO in normal cell function has only been extensively being studied since 1998. As a neurotransmitter, NO is able to dilate the blood vessels and lower blood pressure. A lack of NO production may thus not only result in an inadequate defense against pathogens as found for example in AIDS, but also results in an increased blood pressure and erection problems in men. In the immune system the poisonous gas NO is being used as a defense against viruses and bacteria (Fig. 3.2).

However, when in former times highly organized organisms were attacked by organized parasites, i.e. multicellular organisms like helminthes, the NO production

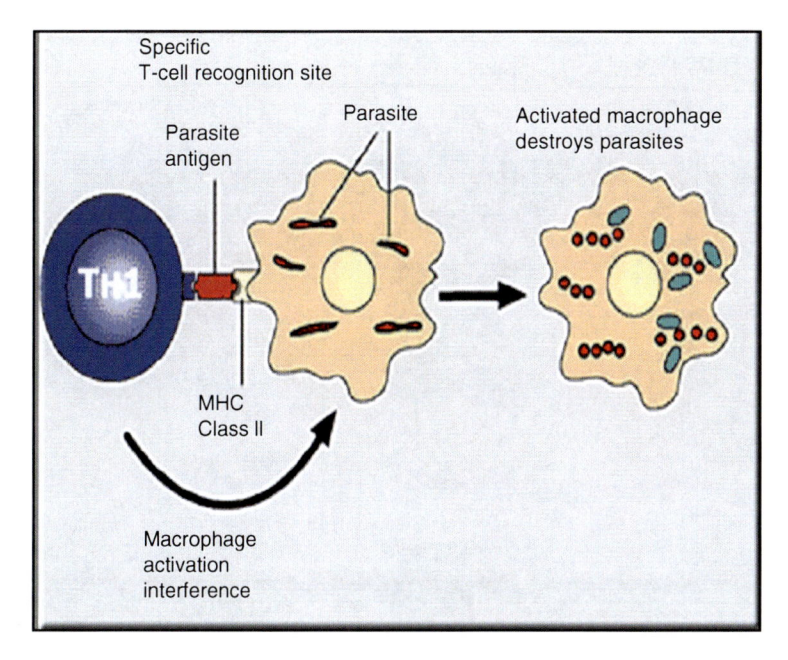

3 Definition of Mitochondropathy – Nitrosative Stress

Fig. 3.2 Antigens of parasites in a macrophage such as leishmania or TBC bacteria are recognized by TH1-cells, releasing cytokines (e.g. interferon-gamma), which activates the production of the gaseous radicals such as NO or O_2^- in macrophages. This activation induces an intracellular decimation of pathogens

were no longer efficient. This is because killing a worm would have needed an excess of NO, a concentration that would have killed the animal himself. This was the time when nature developed a new strategy of defense starting around the appearance of fishes, and resulted in the formation of antibodies. Antibodies against parasite antigens on the tegument of plathelminthes or the epicuticula of nematodes induced adherence of different effector cells like macrophages and granulocytes (Fig. 3.3). Since the oversize target does not permit phagocytosis, the parasite is beige damaged of exocytosis of effector molecules (lysosomal enzymes, basic proteins, nitric oxide). This, in evolutionary terms relatively young defense system, nowadays termed the Th2 immune defense line, is also activated when there is an excess (or a deficient) production of NO (Fig. 3.2). Such activation of TH2 for instance is also found in case of a deficiency in sulfur-containing organic compounds, necessary to synthesize the body's own potent antioxidative glutathione (GSH), which results in insufficient scavenging of ROS and NOS and a shift of the immune balance towards Th2 producing cells.

> **A lack of NO production results in a dominance of TH2 immune cells responsible for antibody response promoting autoimmune diseases.**

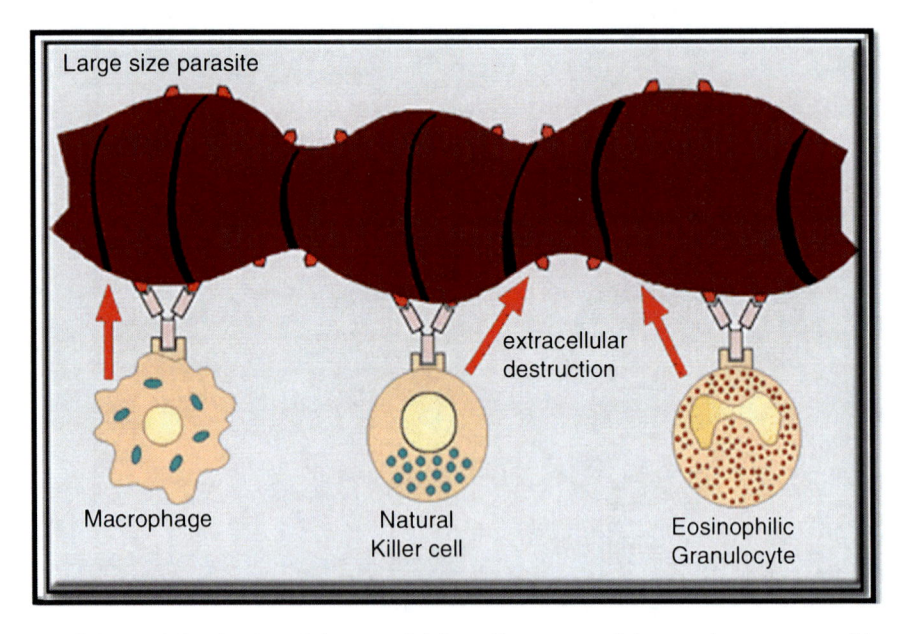

Fig. 3.3 Principle of action of the second defense line system of the organism, represented by macrophages, NK-cells as well as eosinophil granulocytes, all of which after parasites have invaded the body are activated via iNOS producing the poisonous gas NO. Through extracellular binding large parasites are enzymatically destroyed

Any disturbance of NO gas becomes evident, when sulfur-containing organic compounds are missing in the cellular metabolism. These sulfur-containing thiols refresh the NO gas and maintenance of the NO budget is essential. The importance of thiols in the human body already has been well established for a long time being part of a healthy diet, since thiols are found in onions, garlic, mustard, horseradish, eggs, and in other food elements containing a pungent odor.

In contrast, the inherited or acquired mitochondropathy is the net result of outside toxins, radiation, **radical oxygen substrates (ROS) or radical nitric oxygen substrates (RNS)**, all of which affect mitochondria in their productivity to synthesize ATP. It should be borne in mind that ROS are produced constitutively by the mitochondrion, in that 1–4% of oxygen consumed by mitochondria emerges as superoxide radical [2]. Under normal conditions, there is a homeostatic set point where a steady state balance exists between ROS and their scavenging by endogenous antioxidants [3]. The effects of an acquired mitochondrial disease can vary substantially, since distribution of the defective mitochondrial DNA may vary from organ to the other within the body, and each mutation is modulated by other genome variants. Thus, such defect in one individual may cause liver disease while in another person it may cause a brain disorder. The severity of the specific defect may also be large or small depending on the amount of mitochondria being impaired. It is well recognized that if 60% of all mitochondria in one organ are defective, only than clinical signs of malfunction will become visible. Minor defects will only become visible in "exercise intolerance", with no obvious serious illness or disability.

If, however, the function of mitochondria is severely affected in several tissues, it will result in a so-called multi-system disease. Many diseases of the aging organism are caused by such defects in mitochondrial function, since the mitochondria are responsible for processing oxygen and converting parts of the nutrition into energy, which is essential for regular cellular function. Once there is a problem within mitochondrial function, this eventually will lead to a malfunction of the organ, such as Type 2 diabetes, Parkinson's disease, atherosclerosis, heart insufficiency, stroke, Alzheimer's disease, and/or cancer. Certain medications can also injure mitochondrial function resulting in mitochondropathy and as a rule any mitochondrial-related disease gets worse when defective mitochondria are present in the musculature, the central, or the peripheral nervous system [4], because those cells need more energy than most of the other cells in the body.

3.1 Significance of T1-Th2 Switch Within the Immune System

Today it is well established that Th1 controls immune balance and Th2 cells being critical for the protection of the host from pathogenic invasion while its imbalance becomes the cause of various immune disorders including autoimmune diseases. Cytokines, such as IL-12 and IL-4, are critical factors to drive the differentiation of naïve CD4[+] T-cells to Th1- or Th2-cells. In addition to cytokines, steroid hormones have been demonstrated to affect on the control of Th1/Th2 balance and the onset of autoimmune diseases.

Scientists engaged in "HIV"-AIDS-research around 1980 originally detected the Th1/Th2-Switch. They found two subgroups of T4-immune cells: the inflammatory T4-Immune cells of type 1, also termed as TH1-immune cells, und the T4-helper immune cells, also termed TH2-immune cells, which transmit trigger signals to antibody producing immune cells, once they are activated (Fig. 3.4). TH1- and TH2-immune cells can be distinguished by means of the synthesis of different protein transmitters, whereby type 1 cytokines and type 2 cytokines are formed. The latter directly govern the function of T4-immune cells and the interaction with other cellular-mediated activity of the immune network and many other non-immune mediated cellular function.

Synthesis of the different cytokines depends on dendritic immune cells, patrolling throughout the whole body and once they get in contact with pathogens, present this information or part of the pathogens to the cellular surface of T4-immune cells. Activity of these dendritic immune cells depends on their glutathione content, by which a different cytokine pattern activates the T4 immune cells. For instance, a high glutathione level results in the generation of cytokines of Th1 cells, while a low glutathione levels induces release of cytokines of Th2 immune cells (Fig. 3.4). In case of pathogen invasion or in case of a non-specific toxic immune stress, the Th1-cytokines in return release of the poisonous gas nitric oxide (NO) an essential part in eliminating intracellular microbes. In contrast, type 2 cytokines inhibit the synthesis of this poisonous gas NO mobilizing the synthesis of antibodies for the Th1-site (Fig. 3.5). Therefore, long-term depletion of glutathione (or an immune

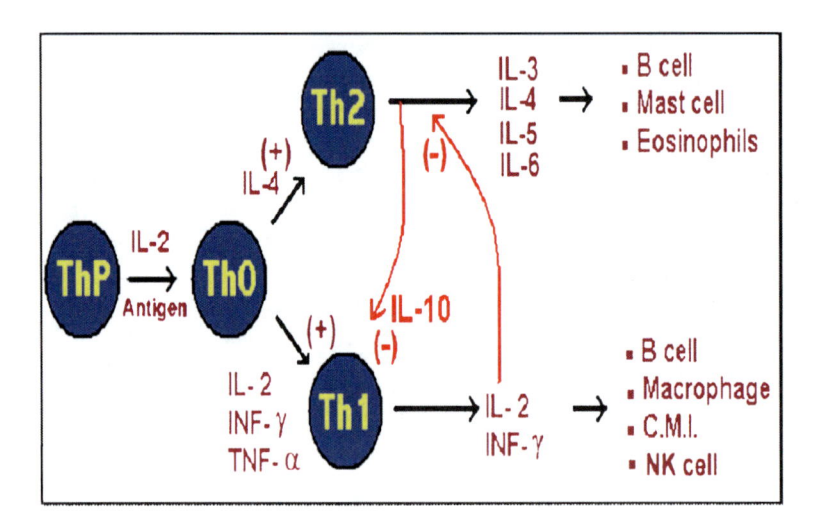

Fig. 3.4 Humoral (Th2) and intracellular (Th1) immune defense system being in balance with each other. Release of cytokoines (inflammatory mediators) following contact with an antigen (i.e. virus, bacteria) of the native ThP (parental) cell, which later differentiates into a Th1 or a Th2 cell when being confronted with the antigen-presenting cell (APC). Once confrontation is with cytokine IL-12, progression is directed to Th1-cells, however when confronted with the IL-4 cytokine, Th-2 cell will develop, one opposing the other and resulting in a delicate balance

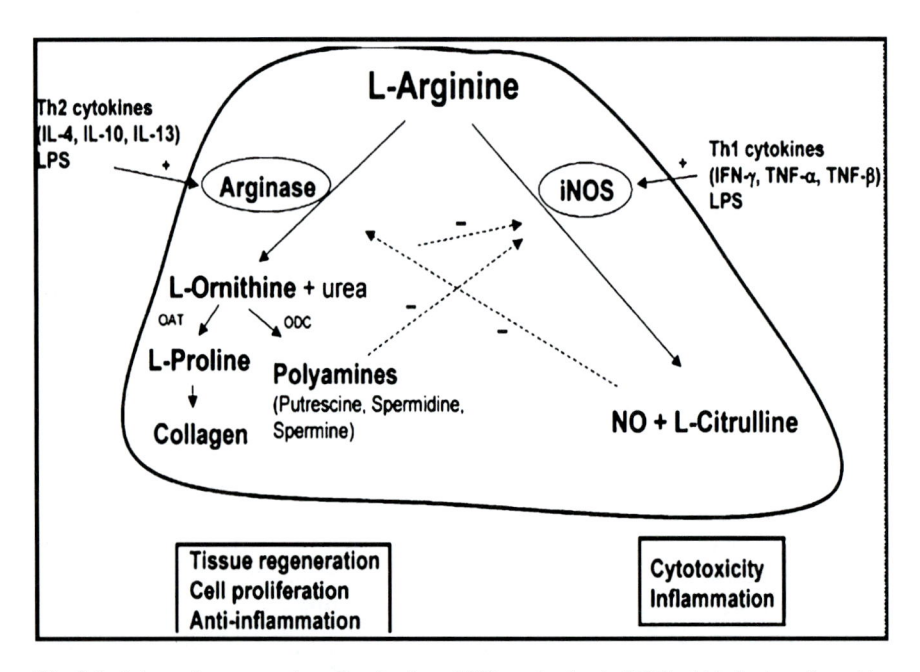

Fig. 3.5 Schematic presentation of activation of NO synthesis via iNOS within body cells and its activation through cytokines from Th1 cells

depression by toxins) of dendritic immune cells, results in a subsequent Th2 activation with a switch to Th2 cytokine release with insufficient production of cytokines from Th1-cells. As a result there is insufficient production of NO from Th1-cells, necessary for destruction of intracellular pathogens. Thus, early contamination with pathogens has a long-lasting effect on adaptive immune response. For instance, in the mouse leishmania tropica model, the dominant immune response decides over life and death. C57BL-mice with a dominant Th1-immune response are able to build up a protective immune response. In contrast Balb/c-mice with a dominant Th2 immune response are not able to fight the mycobacterium lepra and will die from disseminated leishmaniosis. Similar effect can be observed in human leprosy where the immune response effects clinical symptoms of the disease with a dominant Th1 response resulting in slow progression and a dominant Th2 response with antibody formation found in cutaneous leprosy, which rapidly may end up in death.

As a result of such a shift towards Th2 immunological reaction there is an increase in allergic reactions (e.g. neurodermitis, thyreoiditis, rheumatic disease, colitis, etc.), which can be diagnosed by using the lymphocyte transformation test (LTT norm 2:1). Because T- and B-lymphocytes are not able to intracellularly invade the cell, the body already tries to inactivate pathogens in the blood stream, before they have entered the cell. This is because once these pathogens (e.g. protozoans, viruses, fungi and also tuberculosis bacteria) then undergo uninhibited multiplication. As a result those 1,500 so called power plants (i.e. mitochondria) in each cell will be damaged, with diminished energy production and the use of oxygen within the electron chain reaction, resulting in an inhibition of intracellular respiration and the generation of ATP.

On the tegument of plathelminthes or the epikutikula of nematodes, which act like antigens, antibodies against the parasite induce an adherence of different effector cells such as macrophages and eosinophilic granulocytes (Fig. 3.4). Since a large size target object does not permit phagocytosis through the release of exocytosis of effector molecules (e.g. lysosomal enzymes, basic proteins, and most of all of nitric oxide) the parasite indirectly is damaged permanently. Therefore, when antigens of parasites such as leishmania and tuberculosis are being recorgnised by Th1 cells they then release cytokines (i.e. inerferon-gamma), which activate macrophages in order to produce the poisonous gas nitric oxide (NO) and oxygen radicals (O_2^-) for intracellular decimation.

The illustration in Fig. 3.6 shows how complex interactions between the pathogen and host lymphocytes and antigen-presenting cells may drive immune responses, with regulatory T cells having an important role in the balance between protection and immunopathology. Host factors involved in the immune response, such as interferon-gamma (IFN-γ), tumor-necrosis factor alpha (TNF-α), nitric oxide (NO) and the interleukins IL-12 and IL-23, all contribute to both immunopathology and suppression of bacterial persistence. A change in intestinal microflora has been implicated, in association with genetic factors, as a putative mechanism responsible for the initiation and persistence of inflammation in intestinal bowel disease (IBS). Indeed, it has been suggested that the failure to maintain immunologic tolerance toward the indigenous microflora leads to a disease-associated dysregulation of the

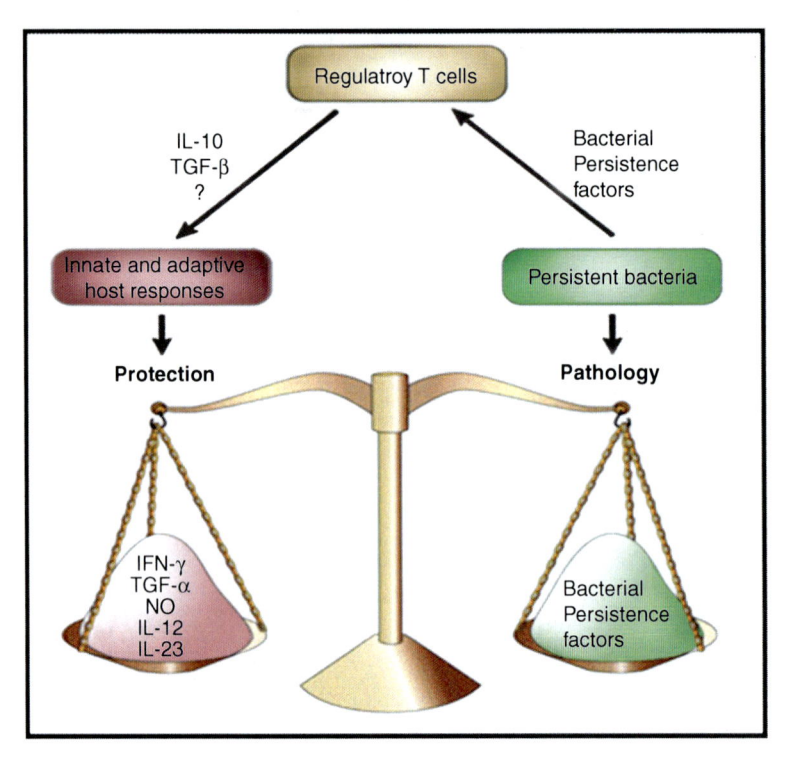

Fig. 3.6 The balance between the immune response and infection during persistent infections is important for both the host and the pathogen, with the host switching off the immune response when it becomes more harmful than the presence of the pathogen. This is depicted by the horizontal axis between protection and pathology on either side of the balance

gut-associated immune system. Direct and indirect evidence of altered flora of the large and small intestine has been reported in IBS patients. Bacterial persistence is mediated by specific bacterial persistence factors and potentially dysregulated immune responses that facilitate persistence by contributing to chronic inflammation and low-level tissue injury that facilitates bacterial survival. The causes for such a switch are diverse; while there is a natural switch towards Th2 in persons over 40 years of age, environmental factors can accelerate this switch remarkably. This includes smoke, smut, ultraviolet rays, oxidative stress, long lasting psychogenic stress, long-term medication (especially chemotherapy, antifungals, antivirals, heavy metal intoxication, and antibiotics), vaccination, protein deficit, intoxications with addictive agents, food with preserversatives, artificial coloring, flavor enhancer etc; for more information see Sect. 7.2 resulting in the outbreak of a number of ailments all of which are able to induce a Th2 switch and at the same time reducing Th1 activity (Table 3.1). Thus, any therapy is aimed to reduce such shift regaining balance in the Th1/Th2 immune response with probiotics playing an active role in controlling the intestinal immune responses.

Table 3.1 Summary of different ailments resulting from the dominance of the Th1- or the Th2-immune system

Th1 dominance (infrequent)	Th2 dominance (frequent)
Autoimmune diseases	Food allergy, IgE and IgGI type
Hashimoto thyreoiditis	Tuberculosis (miliar type)
Goiter	HIV, combined with Th1 deficit
Multiple sclerosis	Cancer
Sjörgen syndrome	Leaky gut syndrome
Schistosomiasis	Allergic reaction with excess histamine release
Psoriasis	Gulf war syndrome
Depression	Chronic fatigue syndrome
Sarkoidosis	Multiple chemical sensitivity
Tuberculosis	Fibromyalgia
Dermatitis in Ni-allergy	Hepatitis B & C
Depression, endogenous	Lupus erythematodes
	Arteriosclerosis, systemic
	Bronchial asthma, allergy type
	Allergic rhinitis
	Atopic dermatitis
	Mycosis

And lastly, vitamin D deficiency has been linked to allergic reactions, gut microbiota, asthma and obesity. For instance, it has been found that vitamin D deficiency has been associated with early-life wheeze, reduced asthma control [5] and allergic diseases [5] and increased body mass index [6]. In recent review researchers have identified both gut microbiota and vitamin D as potential common early life exposures for asthma and obesity [7]. It is unknown whether vitamin D deficiency affects the composition of the intestinal microbiota, decreased vitamin D intake is correlated with differences in fecal microbiota composition. Given the governmental role of vitamin D in T-regulatory and dendritic cell development and function [8, 9] it is possible that the host's vitamin D status could modify the effect of the intestinal microbiota on the immune system. For example, mice that lack the vitamin D receptor (VDR) have chronic, low-grade inflammation in the gastrointestinal tract [10]. Furthermore, the absence of the VDR leads to decreased homing of T cells to the gut, resulting in further inflammation in response to normally nonpathogenic bacterial flora [10]. Intestinal VDR has also been shown to be directly involved in suppression of bacteria-induced NF-κB activation [11]. Wu and colleagues also showed that commensally bacterial colonization affects both the distribution and expression of VDR in intestinal epithelial cells, suggesting a dynamic interplay between these bacteria and the receptor [11]. Thus, emerging evidence suggest that the vitamin D pathway is a potentially important modifier of the effects of intestinal flora on inflammatory disorders.

Since Bacteroides and Lactobacillus show decreased representation in the intestines of many intestinal bowel disease patients, it is intriguing to speculate

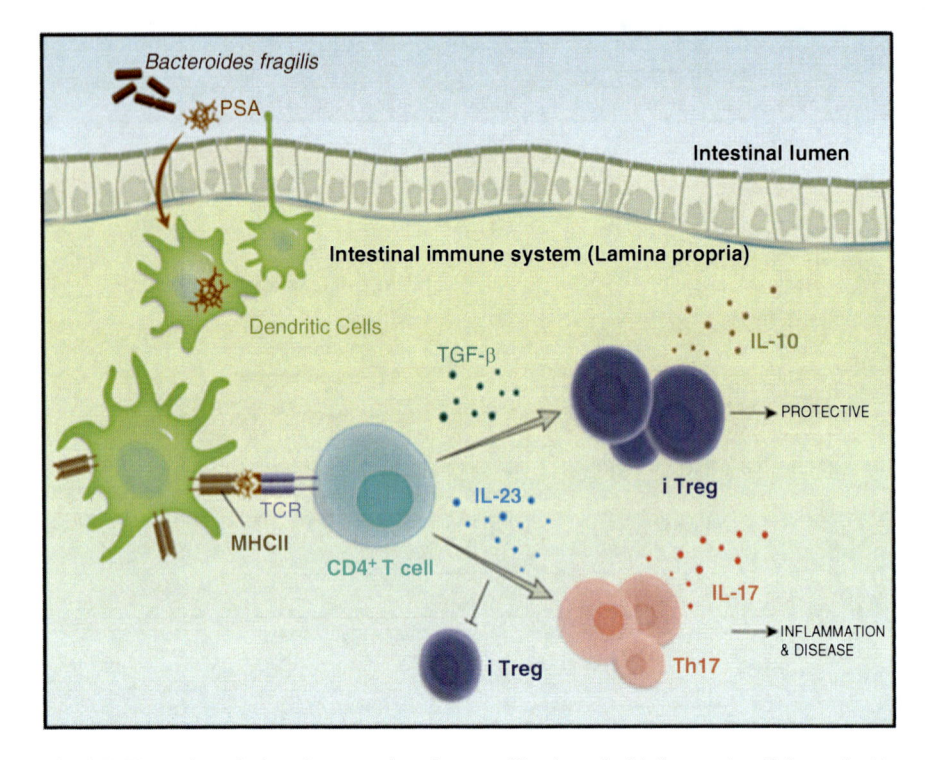

Fig. 3.7 The active role for microorganisms in controlling intestinal inflammation. Polysaccharide A (PSA) is taken up by lamina propria dendritic cells, processed, and presented to naïve CD4+ T cells. In the presence of activated TGF-β, these cells can become induced regulatory T cells (iTreg). Production of IL-10 by these and other T-lineage cells promotes control of immune activation. IL-23 inhibits control by Treg, and promotes expansion of inflammatory Th17 cells. For simplicity, many other pro- and anti-inflammatory mechanisms present in the intestines are not shown (Adapted from [12])

that symbiotic colonization may be deficient in these patients. Therefore PSA (polysaccharides) expressed by other probiotics play an active role in controlling intestinal immune responses (Fig. 3.7).

References

1. Lill R (2009) Function and biogenesis of iron-sulphur proteins. Nature 460:831–838
2. Kowaltowski AJ, Vercesi AE (1999) Mitochondrial damage induced by conditions of oxidative stress. Free Radic Biol Med 26:463–471
3. Inoue M et al (2003) Cross talk of nitric oxide, oxygen radicals, and superoxide dismutase regulates the energy metabolism and cell death and determines the fates of aerobic life. Antioxid Redox Signal 5:475–484
4. Finsterer J (2007) Hematological manifestations of primary mitochondrial disorders. Acta Haematol 118:88–98

References

5. Brehm JM et al (2009) Serum vitamin D levels and markers of severity of childhood asthma in Costa Rica. Am J Respir Crit Care Med 179:765–771
6. Black PN, Scragg R (1995) Relationship between serum 25-hydroxyvitamin D and pulmonary function in the third national health and nutrition examination survey. Chest 126:3792–3798
7. Ly NP et al (2011) Gut microbiota, probiotics, and vitamin D: interrelated exposures influencing allergy, asthma, and obesity? J Allergy Clin Immunol 127(5):1087–94
8. Griffin MD, Xing N, Kumar R (2003) Vitamin D and its analogs as regulators of immune activation and antigen presentation. Annu Rev Nutr 23:117–145
9. Adorini L, Penna G (2009) Dendritic cell tolerogenicity: a key mechanism in immunomodulation by vitamin D receptor agonists. Hum Immunol 70:345–352
10. Yu S et al (2008) Failure of T cell homing, reduced CD4/CD8 alpha intraepithelial lymphocytes, and inflammation in the gut of vitamin D receptor KO mice. Proc Natl Acad Sci USA 105:20834–20839
11. Wu S et al (2010) Vitamin D receptor negatively regulates bacterial-stimulated NF-kappaB activity in intestine. Am J Pathol 177:686–697
12. Lee YK, Mazmanian SK (2010) Has the microbiota played a critical role in the evolution of the adaptive immune system? Science 330:1768–1773

Chapter 4
Nitric Oxide, a Janus Two-Faced Molecule

Nitric oxide (common name) or nitrogen monoxide (systematic name) is a chemical compound with chemical formula NO. This diatomic gas is an important cell signaling molecule in mammals, including humans, and is an extremely important intermediate in the chemical industry. It is also an air pollutant produced by combustion of substances in air, like in automobile engines and fossil fuel power plants. NO is an important cellular messenger molecule involved in many physiological and pathological processes within the mammalian body both beneficial and detrimental [1]. Appropriate levels of NO production are important in protecting an organ such as the liver from ischemic damage. Since the original demonstration of the vasodilatory properties of nitric oxide nearly 30 years ago, it is now known that nitric oxide also regulates other aspects of cardiovascular physiology as well as playing a role in inflammatory and immune responses, reproductive functions, bronchodilation, bone formation, insulin sensitivity, and gastrointestinal protection [2].

Nitric oxide is a remarkable molecule with an important role in many physiologic processes that maintain human health. Although simple in chemical structure, the biological actions of Nitric Oxide are complex. In the late 1970s and early 1980s, Robert Furchgott described a key biologic function of compound, called endothelium-derived relaxing factor (EDRF), that was produced by blood vessels with intact endothelium in response to acetylcholine [3]. It was discovered that EDRF is produced by vascular endothelial cells and acts on underlying vascular smooth muscle cells to produce vasodilation. In the same period, Ferid Murad was studying the activation of soluble guanylyl cyclase (sGC) by nitrogen-containing compounds and the role of cyclic guanosine monophosphate (cGMP) in the relaxation of arterial smooth muscle [4]. Later in 1987, Salvador Moncada and Louis Ignarro separately proved that EDRF was, in fact, NO which acts intracellularly through the activation of cGMP [5, 6].

Nitric oxide should not be confused with nitrous oxide (N_2O), an anesthetic and greenhouse gas, or with nitrogen dioxide (NO_2), a brown toxic gas and a major air pollutant. However, nitric oxide is rapidly oxidized in air to nitrogen dioxide.

4.1 Biological Functions of Nitric Oxide

NO is one of the few gaseous signaling molecules known and is additionally exceptional due to the fact that it is a radical gas. It is a key vertebrate biological messenger, playing a role in a variety of biological processes. Nitric oxide, known as the 'endothelium-derived relaxing factor', or 'EDRF', is biosynthesized endogenously from L-arginine, oxygen and NADPH by various nitric oxide synthase (NOS) enzymes (Fig. 4.1).

Reduction of inorganic nitrate may also serve to make nitric oxide. The endothelium of blood vessels uses nitric oxide to signal the surrounding smooth muscle to relax, thus resulting in vasodilatation and increasing blood flow. Nitric oxide is highly reactive, having a lifetime of a few seconds, yet diffuses freely across membranes. These attributes make nitric oxide ideal for a transient paracrine (between adjacent cells) and autocrine (within a single cell) signaling molecule [7]. The production of nitric oxide is elevated in populations living at high altitudes, which helps these people avoid hypoxia by aiding in pulmonary vasculature vasodilatation. Effects include vasodilatation, neurotransmission, modulation of the hair cycle, production of reactive nitrogen intermediates and penile erections (through its ability to vasodilate). Nitroglycerin and amyl nitrite serve as vasodilators because they are converted to nitric oxide in the body. Sildenafil citrate, popularly known by the trade name *Viagra*, stimulates erections primarily by enhancing signaling through the nitric oxide pathway in the penis. NO is made by different synthases all of which act in different parts of the body in order to react when needed. For instance,

4.1 Biological Functions of Nitric Oxide

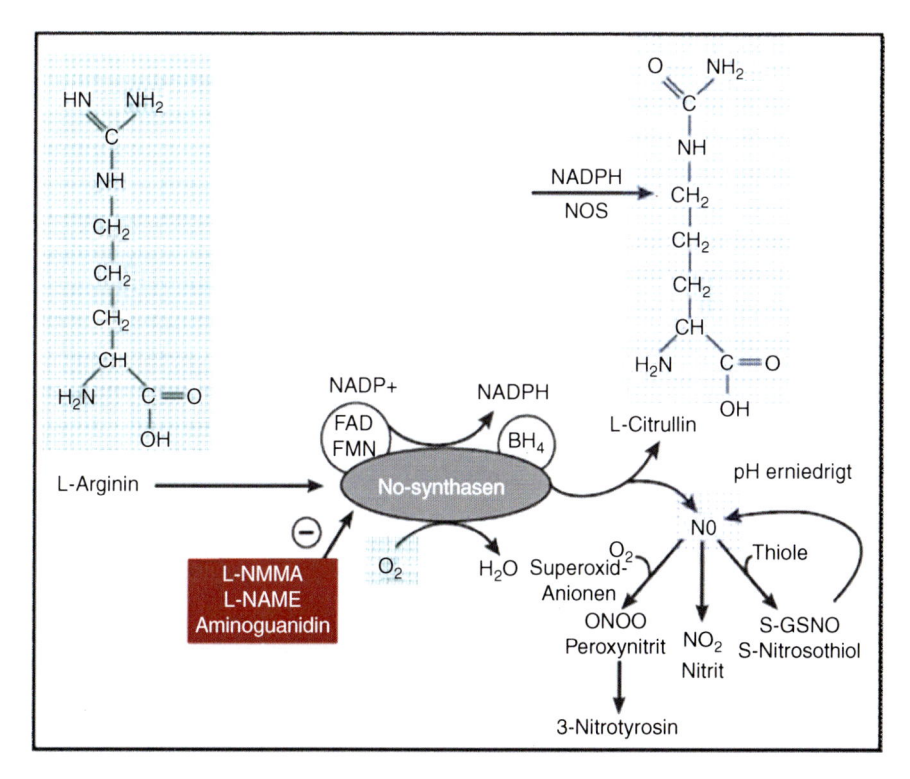

Fig. 4.1 Nitric oxide being made from arginine via the NO-synthetase and tetrahydrobioterin (BH$_4$), a necessary coenzyme for its formation, which when deficient results in increased levels of citrulline and peroxynitrite (ONOO$^-$), nitrite (NO$_2$), and S-nitrosoglutathione (S-GSHG)

endothelial NOS (eNOS) is found within the inner lining of arteries where the transmitter induces relaxation of the smooth vascular muscles with vascular dilatation followed by a drop in blood pressure. This principle is used by the agent nitroglycerin, used medically as a vasodilator to treat heart conditions, such as angina and chronic heart failure. It is one of the oldest and most useful drugs for treating heart disease by shortening or even preventing attacks of angina pectoris. Because of venous dilatation (reduction of preload) and to a smaller degree of the arteries, there is a reduction in energy requirement with a decline in myocardial oxygen consumption followed by subsiding in chest pain. **Neuronal NOS (nNOS)** is found within nerve cells where it acts as a neurotransmitter, being activated by hypoxia & cerebral malperfusion. The **induced NOS (iNOS)** serves within the cells of the immune system. It is activated as a defense system fighting off intracellular pathogens. iNOS is also activated during inflammation, infection and exposure to chemicals (Table 4.1). Lastly, **mitochondrial NOS (mtNOS)** is found in mitochondria where it serves as a metabolic regulator for the synthesis, the proliferation, and apoptosis (programmed cell death). It plays an important role in embryonic development,

Table 4.1 Nitric oxide (NO) and nitric oxide synthase (NOS), serving multiple specific tasks, i.e. transmitter and a metabolic regulator in different organ systems

Type	Regulation	Duration of action	Place of action
eNO	Ca^{++}-inflow dependent	1–5 s	Endothelial cell
nNO	Ca^{++}-inflow dependent	1–5 s	Glia cells
iNO	antigens, bacteria, viruses, parasites	days	Systemic
mtNO	Ca^{++}-inflow	days-weeks	Mitochondria

Table 4.2 Summary of physiological functions of nitric oxide (NO)

- Natural neurotransmitter -> activates the glutamate receptor
- Prevents lipid peroxidation by scavenging oxygen radicals
- Blocks thrombocytes aggregation
- Increases insulin receptor sensibility
- Dilates vessels via endothelial nitric oxide synthase (eNOS)
- Fights intracellular viruses, parasites (borellia, trichomonads, plasmodia, tuberculosis bacteria) via upregulation of iNOS
- Destroys cancer cells
- Lowers pulmonal vascular resistance
- Regulates mitochondrial metabolism (via mtNOS)
- Radical scavenger for all hydroxyl-, peroxyl-, tryoxyl groups and superoxide
- Induces programmed cell death (apoptosis)

disposal of old cells, as well as the regression of cancer cells, regulating oxygen consumption and ATP-formation within mitochondria (Table 4.2).

Nitric oxide (NO) contributes to vessel homeostasis by inhibiting vascular smooth muscle contraction and growth, platelet aggregation, and leukocyte adhesion to the endothelium. Humans with atherosclerosis, diabetes, or hypertension often show impaired NO pathways [8] and a high salt intake was demonstrated to attenuate NO production, although bioavailability remains unregulated [9].

Nitric oxide is also generated by phagocytes (monocytes, macrophages, and neutrophils) as part of the human immune response (Fig. 4.2). Phagocytes are armed with the inducible nitric oxide synthase (iNOS), which is activated by interferon-gamma (IFN-γ) as a single signal or by tumor necrosis factor (TNF-α) along with a second signal [10]. On the other hand, transforming growth factor-beta (TGF-β) provides a strong inhibitory signal to iNOS, whereas interleukin-4 (IL-4) and IL-10 provide weak inhibitory signals. In this way the immune system may regulate the armamentarium of phagocytes that play a role in inflammation and immune responses. Nitric oxide is secreted as an immune response since free radicals are toxic to bacteria (Fig. 4.2); the mechanism for this includes DNA damage [11–13] and degradation of iron sulfur centers into iron ions and iron-nitrosyl compounds [14].

The four important NO-isomers, which regulate cellular function and protection of the body (Table 4.1):

- Neuronal NO (nNO) serving as a vital neurotransmitter,
- Endothelial NO (eNO) the important transmitter inducing vascular dilatation with reduction in blood pressure and heart load,

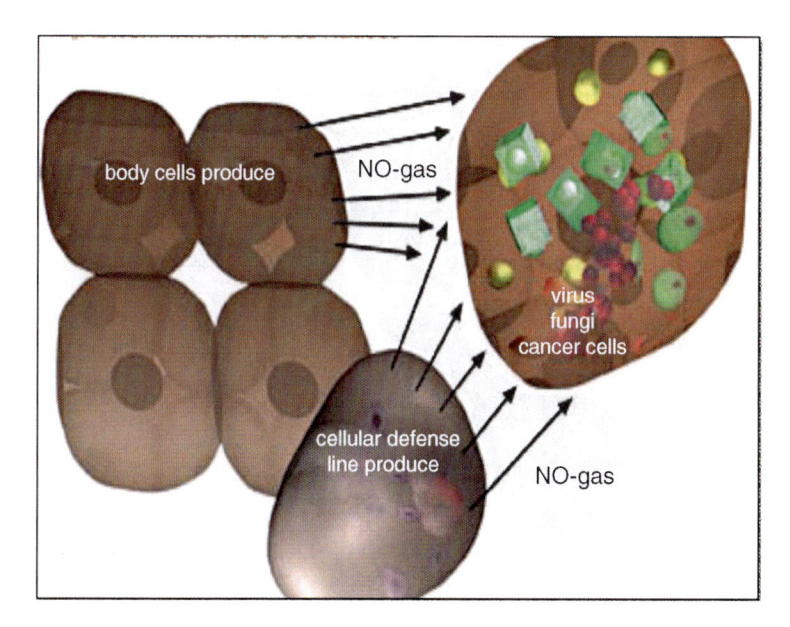

Fig. 4.2 Schematic representation of phagocytes, which produce NO for combat pathogens after having invaded the body

- Induced NO (iNO), where its release is activated through cytokines of immune cells following inflammation in order to exterminate viruses and fight all invaders such as bacteria, parasites and viruses via the poisonous gas NO released from immune cells (Fig. 4.2),
- Mitochondrial NO (mtNO), a metabolic modulator for synthesis, proliferation and apoptosis (programmed cell death) with the cellular organelles, which also increase NO synthesis in case of hypoxia, regulate the inflammatory response via a balanced formation of Th1- and Th2-cells, which formed from naive CD4-cells (Fig. 3.4). And while many bacterial pathogens have evolved mechanisms for nitric oxide resistance [15], proinflammatory responses with either a promotion or a prevention of autoimmune disease is regulated within this delicate balance.

4.2 Nitric Oxide-Peroxynitrite, Causes for Multisystem Ailments

Before getting to the pathology of reactive nitrosative oxygen species (NOS) and that of reactive oxidative species (ROS) one should look at the physiological aspect of NO and why and where it is used by the body. This is because the NO-system in the human has many physiological functions being synthesized within cells by the enzyme NOsynthase (NOS).

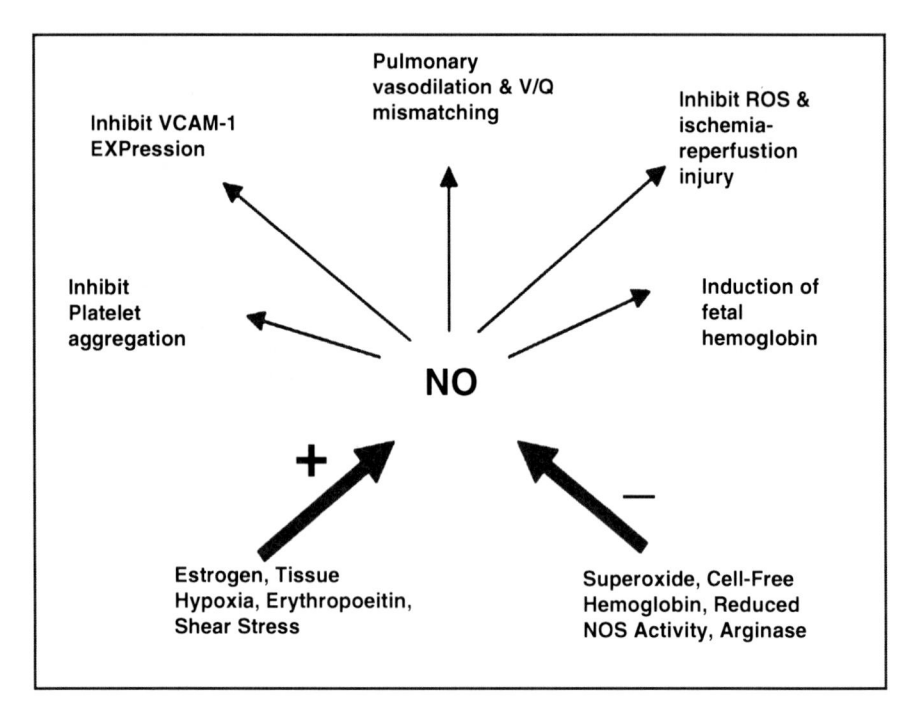

Fig. 4.3 Factors in the inhibition and release of nitric oxide radicals resulting in different clinical symptoms (Adapted from [16])

There are several common routes where nitric oxide is being used for bodily functions (Fig. 4.3):

- All types of NOS produce NO from arginine with the aid of molecular oxygen and NADPH.
- NO diffuses freely across cell membranes.
- There are so many other molecules with which it can interact, that it is quickly consumed close to where it is synthesized.
- Thus NO acts in a paracrine or even autocrine fashion, affecting only cells near its point of synthesis.
- NO relaxes the smooth muscle in the walls of the arterioles. At each systole, the endothelial cells that line the blood vessels release a draft of NO. This diffuses into the underlying smooth muscle cells causing them to relax and thus permit the flow of blood to pass through easily. This is a principal mode if action of antihypertensive agents in where the NO synthase in endothelial cells (eNOS) of patients does not work properly as they suffer from hypertension (Fig. 4.3).
- NO in penile erection. The erection of the penis during sexual excitation is mediated by NO released from nerve endings close to the blood vessels of the penis. Relaxation of these vessels causes blood to pool in the blood sinuses producing an erection (Fig. 4.4).

Fig. 4.4 Erection problems as a cause for taking Viagra® a word being derived from the sanskrit word *vyaghra* which relates to the tiger

- The NO produced by eNOS inhibits inflammation in blood vessels. It does this by blocking the exocytosis of mediators of inflammation from the endothelial cells.
- NO and peristalsis. The wavelike motions of the gastrointestinal tract are aided by the relaxing effect of NO on the smooth muscle in its walls.
- NO inhibits the contractility of the smooth muscle wall of the uterus. As the moment of birth approaches, the production of NO decreases.
- NO around the glomeruli of the kidneys increases blood flow thus increasing the rate of filtration and urine formation.
- NO affects secretion from several endocrine glands. For examples, it stimulates
  - The release of the gonadotropin-releasing hormone from the hypothalamus;
  - The release of pancreatic amylase from the exocrine portion of the pancreas;
  - The release of adrenaline from the adrenal medulla.
- NO and the autonomic nervous system. Some motor neurons of the parasympathetic branch of the autonomic nervous system release NO as their neurotransmitter. These nerves probably mediate the actions of NO on penile erection and peristalsis.
- NO and the medulla oblongata. Hemoglobin transports NO at the same time it carries oxygen. When it unloads oxygen in the tissues, it also unloads NO. In severe deoxygenation, NO-sensitive cells in the medulla oblongata respond to this release by increasing the rate and depth of breathing.

- NO and learning. In laboratory animals (mice and rats), NO is released by neurons in the CA1 region of the hippocampus and stimulates the NMDA receptors, that are responsible for long-term potentiation (LTP), activated in memory and learning.
- NO and fertilization. The acrosome at the tip of sperm heads activates its NO synthase when it enters the egg. The resulting release of NO in the egg is essential (at least in sea urchins) for triggering the next steps in the process:
  - blocking the entry of additional sperm and
  - orienting the pronuclei for fusion.
- NO aids in the killing of engulfed pathogens (e.g. bacteria) within the lysosomes of macrophages.
- Th1 cells, the ones responsible for an inflammatory response against invaders, secrete NO.
- NO and longevity. Mice whose genes for eNos have been knocked out
  - Show signs of premature aging;
  - Have a shortened life span;
  - Fail to benefit from the life-extending effect of a calorie-restricted (CR) diet.
- Harmless bacteria, living as commensals at the rear of our throat, convert nitrates in our food into nitrites. When these reach the stomach, the acidic gastric juice (pH ~1.4) generates NO from them. This NO kills almost all the bacteria that have been swallowed in our food.
- NO also inhibits the aggregation of platelets and thus keeps inappropriate clotting from interfering with blood flow.
- NO prevents inflammatory cells from invading into the vascular wall.
- NO increases insulin sensitivity.
- NO reduces pulmonary vascular resistance with a therapeutic effect in pulmonary hypertension.
- NO acts as a neurotransmitter between nerve cells.
- NO is used as a potent poison gas by immune cells via iNOS activation in order to fight invasion of bacteria, viruses and eliminate cancerous cells (Fig. 4.2).
- NO has also been implicated in many plant activities:
  - It is a weapon against invading pathogens. Infection of the plant triggers the formation of a NO that like the animal versions makes NO from arginine. Release of NO by the infected cell induces a number of defense responses.
  - A gradient of NO may also guide the pollen tube to its destination in the ovule. The popular prescription drug sildenafil citrate (Viagra®) enhances its effect on pollination (just as it does on penile erection).
  - NO inhibits flowering.
  - NO promotes recovery from etiolation.

Examples for NO-concentration Dependant Reactions

1. *NO in physiological concentrations*: NO-binds with the beta-subunits of hemoglobin arising to S-nitroso-hemoglobin, which favors the oxygenated status. Effect: Dilatation of vasculature.
2. *NO in supra-physiological concentrations*: NO-connects with the alpha-subunit of hemoglobin, which favors the deoxygenated status, i.e. oxygen is released from hem being available for further use. Effect: higher tissue-oxygen-partial pressure.
3. *Release of excess NO and superoxide (O_2^-) through chronic (silent) inflammation*: Indicator is a sedimentation rate higher than normal (3–13 mm/h). Also, found as an inflammatory response to acute head injury with excess NO-release.

References

1. Hou YC, Janczuk A, Wang PG (1999) Current trends in the development of nitric oxide donors. Curr Pharm Des 5:417–441
2. Giles TD (2006) Aspects of nitric oxide in health and disease: a focus on hypertension and cardiovascular disease. J Clin Hypertens 8:2–16
3. Furchgott RF, Zawadzki JV (1980) The obligatory role of endothelial cells in the relaxation of arterial smooth muscle by acetylcholine. Nature 288:373–376
4. Katsuki S et al (1977) Stimulation of guanylate cyclase by sodium nitroprusside, nitroglycerin and nitric oxide in various tissue preparations and comparison to the effects of sodium azide and hydroxylamine. J Cyclic Nucleotide Res 3:23–35
5. Ignarro LJ et al (1987) Endothelium-derived relaxing factor produced and released from artery and vein is nitric oxide. Proc Natl Acad Sci USA 84:9265–9269
6. Palmer RM, Ferrige AG, Moncada S (1987) Nitric oxide release accounts for the biological activity of endothelium-derived relaxing factor. Nature 327:524–526
7. Berg JM, Tymoczko JJ, Stryer L (2007) Biochemistry, 6th edn. W.H. Freeman and Company, New York
8. Dessy C, Ferron O (2004) Pathophysiological roles of nitric oxide: in the heart and the coronary vasculature. Curr Medi Chem Anti-Inflamm Anti-Allergy Agents Med Chem 3:207–216
9. Osanai T et al (2002) Relationship between salt intake, nitric oxide and asymmetric dimethylarginine and its relevance to patients with end-stage renal disease. Blood Purif 20:466–468
10. Gorczyniski RM, Stanely J (1991) Clinical immunology. Landes Bioscience, Austin/TX
11. Wink DA et al (1991) DNA deaminating ability and genotoxicity of nitric oxide and its progenitors. Science 254:1001–1003
12. Nguyen T et al (1992) DNA damage and mutation in human cells exposed to nitric oxide in vitro. Proc Natl Acad Sci USA 89:3030–3034
13. Li CQ et al (2006) Threshold effects of nitric oxide-induced toxicity and cellular responses in wild-type and p53-null human lymphoblastoid cells. Chem Res Toxicol 19:399–406
14. Hibbs JB et al (1988) Nitric oxide: a cytotoxic activated macrophage effector molecule. Biochem Biophys Res Commun 157:87–89
15. Janeway CA et al (2005) Immunobiology: the immune system in health and disease. Garland Science, New York
16. Schechter AN, Gladwin MT (2003) Hemoglobin and the paracrine and endocrine functions of nitric oxide. N Engl J Med 348:1483–1485

Chapter 5
Representative Ailments with Excess Nitric Oxide Formation

The notion that chronic fatigue syndrome (CFS) multiple chemical sensitivity (MCS), posttraumatic Stress syndrome (PTSD) and fibromyalgia (FMS) as well as several other diseases may share a common etiology has originally been proposed by Miller [1] and later by Buchwald and Garrity concluded in a study of CFS, MCS and FM patients that: "despite their different diagnostic labels, existing data, though limited, suggests that these illnesses may be similar if not identical conditions...." [2] and Donnay and Ziem proposed that CFS, FMS and MCS "may simply reflect different aspects of a common underlying medical condition" [3]. As today there are several causes that are accepted as the origin of major disease entities in western medicine:

1. Infectious diseases related to a virus or bacterial growth.
2. Genetically disorders related ailments.
3. Nutritional deficiency related diseases.
4. Diseases resulting from hormonal dysfunction.
5. Allergies resulting from the addition of preservatives, pesticides, artificial coloring and flavor enhancers in food
6. Autoimmune diseases ensuing from an overactive immune system.
7. Aberrant metabolism in cells resulting in cancer.
8. Cardiovascular diseases due to chronic inflammation.
9. Neurodegenerative diseases subsequent to oxygen radical/nitrosative stress and finally
10. Mitochondropathy related diseases due to excess activity of the NO/ONOO$^-$ cycle.

Especially in the latter, nitrosative stress, presents a pathological condition when highly reactive nitrogen-containing chemicals, such as nitrous oxide (NO) or superoxide radicals (O_2^-) with the formation of peroxynitrite (ONOO$^-$) are produced in excess, exceeding the ability of the human body to neutralize and eliminate them (Fig. 5.1). This excess can lead to reactions that alter protein structure of mitochondria thus interfering with normal body functions. In this context, the prevalence of

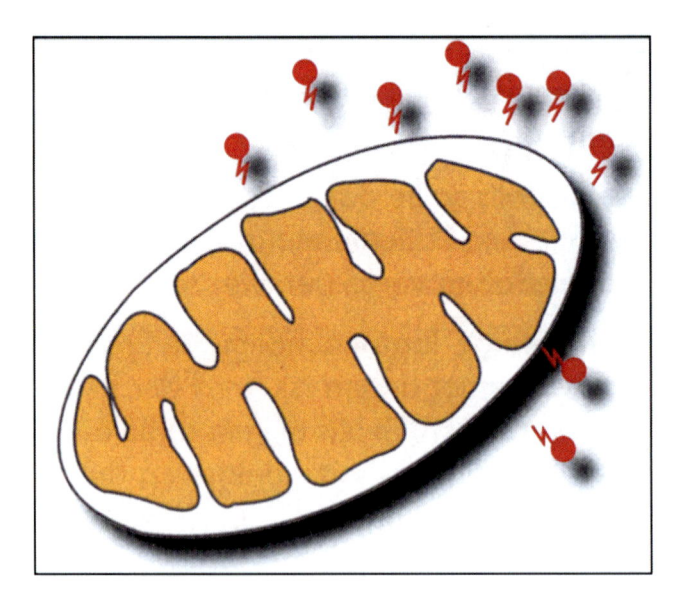

Fig. 5.1 Oxidative and nitrosative stress attack function of mitochondria

multiple chemical sensitivity (MCS), fibromyalgia (FMS), posttraumatic stress syndrome (PTSD) and chronic fatigue syndrome (CFS) with an important impact on the lives of sufferers, suggests that the NO/ONOO⁻ cycle paradigm may be among the most important findings in modern medicine, which is able to explain many current chronic ailments. Many other additional diseases may be found under the umbrella of this paradigm which expands our understanding of such ailments [4], especially when it comes to the treatment of the underlying cause of such a chronic disease. This assumption is underlined by the observation, that many of the following co-morbidities share similar aspects of the NO/ONOO⁻ cycle:

1. Tinnitus
2. Post-radiation syndrome (PRS)
3. Multiple sclerosis (MS)
4. Asthma
5. Irritable bowel syndrome (IBS)
6. Neuropathic pain (NP)
7. Arteriosclerosis
8. Diabetes type 2
9. Complex regional pain syndrome (CRPS)
10. Postherpetic neuralgia (PHN)
11. Chronic whiplash associated disorder
12. Amyotrophic lateral sclerosis (ALS)
13. Parkinson's disease (PD)
14. Alzheimer's disease (AD)
15. Broken Heart Syndrome, or Tako-Tsubo-cardiomyopathy.

The latter is characterized by an apical ballooning and functional impairment due to excessive emotional stress [5], where a common nominator is readily found by the effects of adrenaline inducing an excessive up-regulation in the formation of NO⁻ (NOS) and O_2^- (ROS) radicals.

5.1 Chronic Fatigue Syndrome, Symptom of Mitochondrial Failure

We are made up of lots of different cells – heart, blood, muscle nerve cells etc. All these cells are different because they all have a different job of work to do. To do this job of work requires energy. But the way in which energy is supplied is the same for every cell in the body. Indeed all animals share this same system. The mitochondria in a dog, a cat or a horse are exactly the same as in a human being. Mitochondria are a common biological unit across the animal kingdom. Energy is supplied to cells by mitochondria, which can be regarded as little engines giving power every cell in the body. The human body typically contains less than 100 g of ATP at any instant, but can consume up to 100 kg per day. Thus the recycling OxPhos process is extremely important since it produces more than 90% of the cellular energy. The main features and processes are illustrated in a simplified form in the following figure. The ETC culminates with the protein complex ATP synthase which is effectively a reversible stepping motor in which three ATP molecules are produced from ADP and inorganic phosphate (P_i) by every revolution [6]. Because of evolutionary history ATP is made inside the mitochondrial inner membrane but used outside in the cytosol where it releases energy by converting it to ADP and P_i. The Pi as a negative ion is co-transported back inwards together with H⁺, while ADP³⁻ is transported inwards through the translocator protein adenosine nucleotide translocase (TL or ANT) in exchange for ATP⁴⁻ moving out into the cytosol. Here, potential problems may arise because it is known that some specific molecules (e.g. atractyloside) block the transfer inwards and certain others can block the outward transfer [7]. Thus, there is the possibility that many chemicals stemming from environmental contaminants, can block these transfers in the mitochondrial wall (Fig. 5.2).

Two indicators (or downstream products) of excessive (phosphor) lipid peroxidation (and mitochondrial membrane damage), as discussed above, are the aldehyde derivates Malondialdehyde (MDA) and Crotonaldehyde. MDA is reactive and potentially mutagenic (changing one's DNA). Crotonaldehyde is a known irritant. According to John McLaren Howard of Acumen Laboratory, aldehydes from lipid peroxidation (when it occurs) tend to accumulate in the mitochondrial membranes. Such aldehydes, usually in the form of malondialdehyde or crotonaldehyde, can block some translocator (TL) protein sites in the inner mitochondrial membranes (Fig. 5.4). It is rare for them to accumulate to very high levels on the inner mitochondrial membranes but moderate or trace accumulation on TL sites does occur. The accumulation of these aldehydes on the mitochondrial membrane can impact

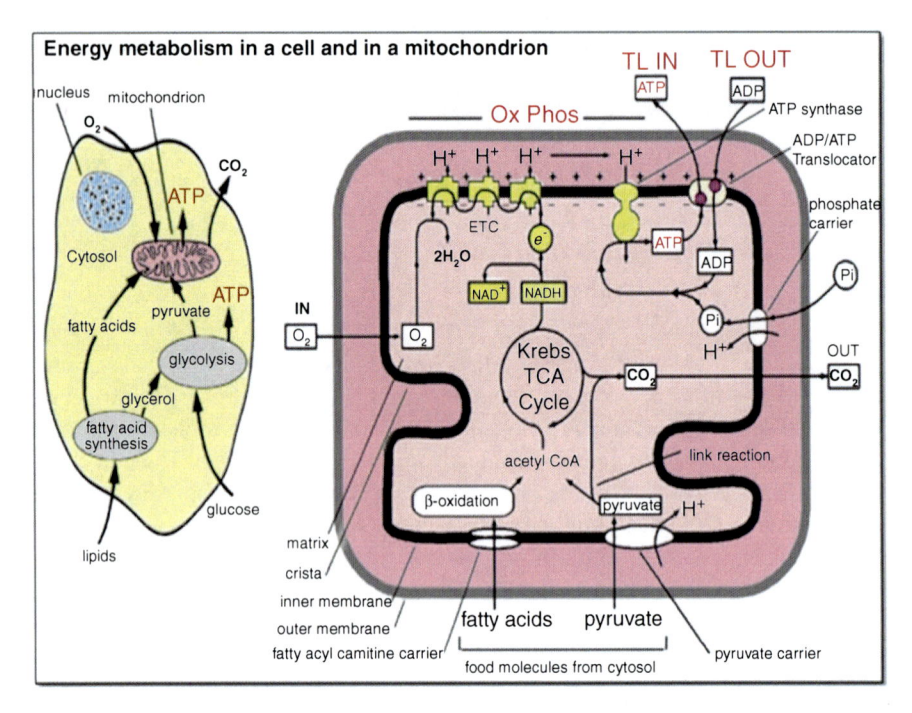

Fig. 5.2 Main stages and location of energy metabolism in a human cell (*left*), and simplified details of a mitochondrion showing the main metabolic cycles and the oxidative phosphorylation respiratory chain (*right*). The outer mitochondrial membrane is highly permeable whereas the inner membrane is permeable only to water and gases. Special carrier and translocator proteins (TL) pass reactants through it. At the top are the proteins involved in the respiratory electron transfer chain (ETC) and in the transfer of ATP and ADP between the cytosol and mitochondrion. ADP and P_i are combined by ATP synthase to make ATP. The ADP/ATP Translocator opens out (TL OUT) to transfer ADP into the matrix and opens in (TL IN) to transfer ATP to the cytosol. Nicotinamide adenine dinucleotide plays a key role in its oxidised form NAD^+ and its reduced form $NADH + H^+$ in carrying and transferring protons (H^+) and electrons (e^-) (Adapted from [8])

oxidative phosphorylation (i.e. ATP production) and energy production significantly. In other words, their presence not only tells us that the mitochondrial membranes themselves have been oxidized/damaged but their presence on the mitochondrial membrane itself may also may be actively preventing proper mitochondrial function. And although the body tends to clear MDA from the mitochondrial membranes, once they are being produced as quickly as they are being removed, then they will remain on the mitochondrial membranes. The aim therefore is to stop or reduces the source of oxidative stress which resulted in the oxidation of the lipids and hence their production, after which their levels will go down. The only exception to this is where MDA is adducted to the DNA (i.e. DNA Adducts) where it cannot be cleared very quickly.

Since it is the task of mitochondria to supply energy in the form of ATP (adenosine triphosphate) for cellular function any defect will result in insufficient

ATP production and if large enough by organ malfunction. In this respect ATP is the universal currency of energy. It can be used for all sorts of biochemical activity, ranging from muscle contraction to hormone production. When mitochondria fail, this results in a poor supply of ATP, and cells go slow because they do not have the energy to function at a normal speed. This means that all bodily functions go slow.

> **Every cell in the body can be affected by insufficient energy production of mitochondria**

The following explains what happens inside each cell: ATP (3 phosphates) is converted to ADP (2 phosphates) with the release of energy for work. ADP passes into the mitochondria where ATP is remade by oxidative phosphorylation (i.e. a phosphate group is stuck on). ATP recycles approximately every 10 s in a normal person – if this goes slow, then the cell goes slow and so the person goes slow and clinically has poor stamina or a chronique fatigue. Problems arise when such a system is stressed. If the CFS sufferer asks for energy faster than he can supply it, (and actually most CFS sufferers are doing this most of the time!) ATP is converted to ADP faster than it can be recycled. This means there is a build up of ADP. Some ADP is inevitably shunted into adenosine monophosphate (AMP −1 phosphate). However, this creates a real problem, resulting in a metabolic disaster, because AMP cannot be recycled and is lost in the urine. Once however, the ATP levels drops as a result of leakage of AMP, the body then has to make brand new ATP. ATP can be made very quickly from the sugar D-ribose, but D-ribose is only slowly made from glucose (via the pentose phosphate shunt). This takes anything from 1 to 4 days and is the biological basis for the delayed fatigue following strenuous exercise.

Such a lack in energy production has been nicely demonstrated in a study of CFS-sufferers by Myhill and coworkers [9], where the depleted mitochondrial energy supply correlated closely with a low ability in performance (Fig. 5.3).

In this study the Mitochondrial Energy Score, was calculated from the biochemical measurements of the lab "ATP profile" which separates the energy generation and recycling processes into five steps. As in any multistep process, for example electrical power production or an assembly line, the efficiency of the overall process is the product of the efficiencies of the individual steps. Any suggestion of relative weighting is irrelevant; it only results in an overall normalization factor. The product of ATP and ATP ratio is the cellular concentration of ATP complexes with magnesium and this is the available energy supply of ATP. OxPhos is the efficiency of the ETC, which converts ADP into ATP. However, for the recycling of ADP to make more energy available the translocator protein must efficiently have its binding site facing out to collect ADP (TL OUT) and alternately facing in (TL IN) to efficiently transmit ATP from the mitochondria into the cytosol where this energy can be used. These data corroborate others [11, 12] who propose that an elevated nitric

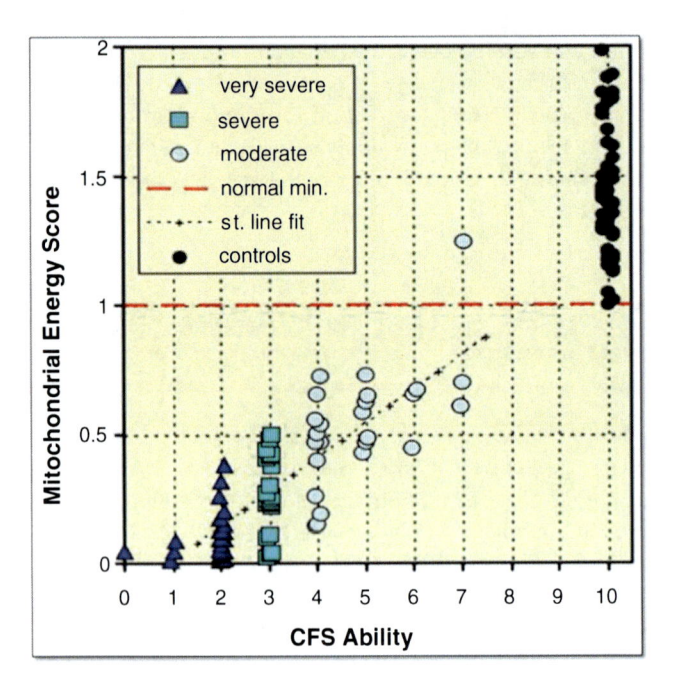

Fig. 5.3 The mitochondrial energy score, plotted against the ability with a point for each CSF patient when compared with a control where the ability to perform was regularly high reaching a major ten. This convincingly demonstrates that mitochondrial dysfunction is a major risk factor for chronic fatigue syndrome (CFS) (Adapted from [10])

oxide/peroxynitrite mechanism is the common etiology of multiple chemical sensitivity, chronic fatigue syndrome, and posttraumatic stress disorder [12]. They suggested that the elevated nitric oxide/peroxynitrite level is the actual cause for FMS, MCS, and PTSD and a depressed "ATP profile" in neutrophils of patients just indicates mitochondrial dysfunction in patients with such ailments. Moreover, the degree of such dysfunction strongly correlates with the severity of the illness. Since neutrophils are the major effector cells of the immune system, the observed mitochondrial dysfunction is bound to have a deleterious effect on this system. Indeed this is the biological basis of poor stamina, and a patient can only go at the rate at which mitochondria can produce ATP. If mitochondria go slow, stamina and performance (mental or physical) is poor.

However another problem may arise. If the body is very short of ATP, it can make a very small amount of ATP directly from glucose by converting it into lactic acid. This is exactly what many CFS sufferers do and indeed we know that CFS sufferers readily switch into anaerobic metabolism. This, however, results in two serious problems. Firstly, lactic acid quickly builds up especially in muscles causing pain, heaviness, aching and soreness ("lactic acid burn"). Secondly, no glucose is available in order to make D-ribose ! So new ATP cannot be easily made when the patient is really run down, and where recovery will take days! When mitochondria function well, as the person rests following exertion, lactic acid is quickly converted back to glucose (via pyruvate) and the lactic burn disappears. But this is an energy requiring process! The conversion of lactic acid to glucose needs six molecules of ATP of the body to work, but the reverse process requires only two molecules of ATP. If there is no ATP available, and this is of course what happens as mitochondria fail, then the lactic acid may persist for many minutes, or even hours causing pain before the reverse process takes place in the liver being called the Cori cycle.

5.2 Excess Formation of Nitric Oxide in Inflammation

If however, the gas "nitric oxide (NO)" is produced in excessive amounts than necessary for protozoan extinction or is taken up from the environment (e.g. inhalation of nitrosative gases, uptake of xenobiotics, medication resulting in NO release, ingestion of food fertilized with pesticides herbicides), than it will become a poison for body cells. This is because nitric oxide is one of the most volatile gases, which similar to heat, rapidly diffuses from the inside of the body (the source of production) to the exterior and the blood system, where it is inactivated by sulfur containing amino acids (produced by exobionts within the intestine). Within the blood system the body continuously measures the concentration of nitric oxide in order to protect itself from excessive loads, and when extremely high concentrations of NO are reached this results in a shut down of the cellular production. By shutting down its own NO production, the body prevents the activation of a natural apoptotic program, which otherwise would have been initiated by excess levels of nitric oxide. Once, however, this first defense line is inactivated approx. 60 trillions of cells suddenly are unprotected and open to attack by opportunistic pathogens. For example people with HIV often in their past have been exposed to excessive concentrations of NO by abusing amylnitrite, uninhibited use of antibiotics, or the long-term use of analgesics such acetaminophen (not however because of the exposure to the alleged HIV virus!). As a result, the body now is left defenseless which however, is compensated by a switch of the immune system of the lymphatic Th1-cells (responsible for intracellular NO production) to Th2-cells (responsible for antibody production outside the cells) in the blood and the lymph (Th1/Th2 switch; Fig. 5.4). As a result of this dominance in Th2-cells (with simultaneous Th1-depression) there is a weakening against all intracellular pathogens such as lyme, trichomonades, and plasmodiums, whereby attenuation of this intracellular immunologic defense line may also favor the development of carcinogenic cells (Fig. 5.4).

The main sites of ROS production within mitochondria are the complexes I and III of the electrical transport chain (ETC), where unpaired electrons "leak" and interact with oxygen to form highly reactive radical species such as superoxide ions (Fig. 5.5) and hydroxyl radicals [13].

Since damaged mitochondria have been shown to replicate faster and are only able to burn up glucose, their energy productivity drops to a mere of 16%. This is because

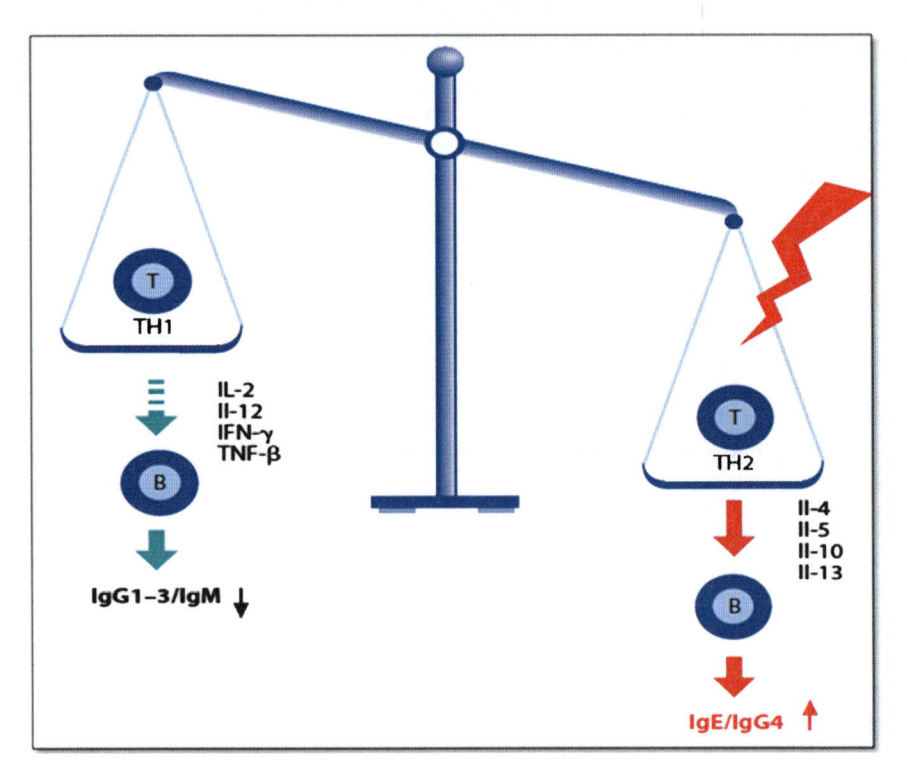

Fig. 5.4 Shift in the Th1/Th2 balance in favor of TH2 resulting in allergic reaction with symptoms affecting mucous membranes of the respiratory tract with an increase in allergic reaction

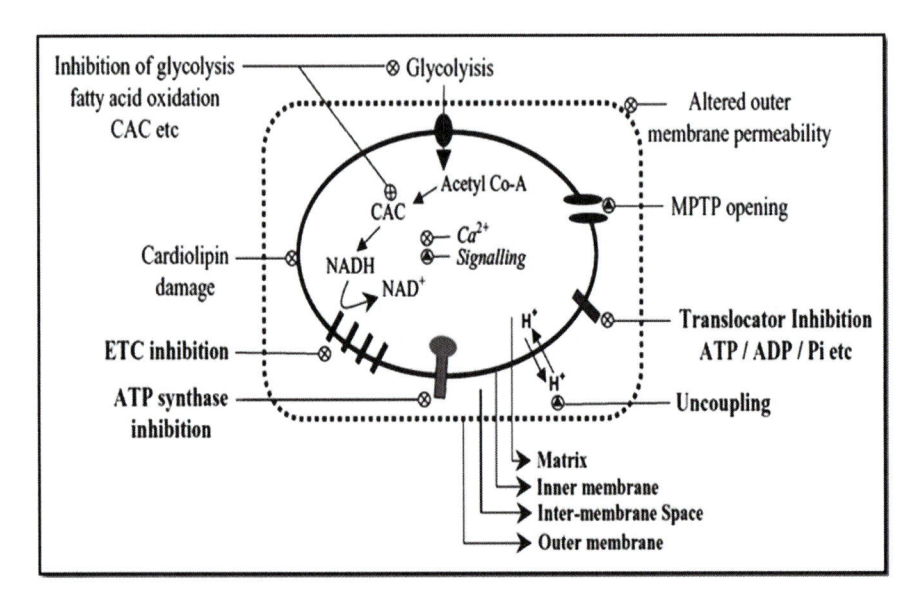

Fig. 5.5 Possible sites of xenobiotic interference with mitochondrial function leading to mitochondrial toxicity (Adapted from [14]). ETC-electron transport chain MPTP opening- Mitochondrial permeability transition pore opening

5.2 Excess Formation of Nitric Oxide in Inflammation

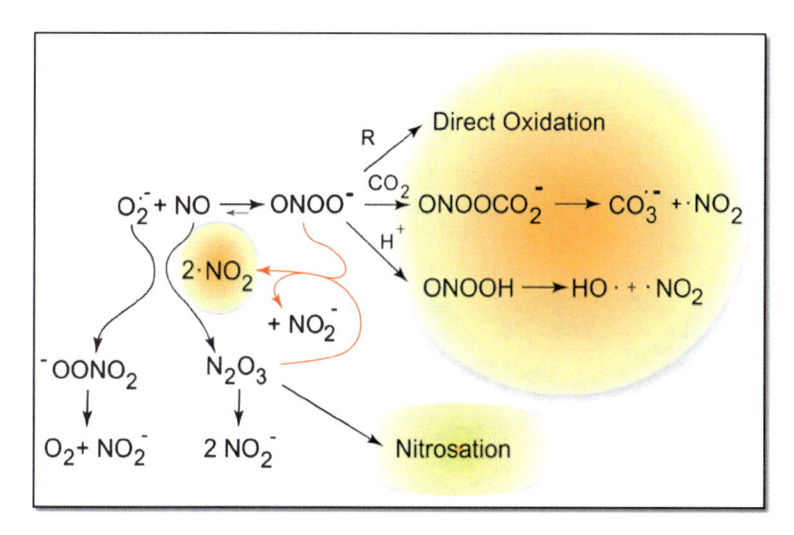

Fig. 5.6 The interaction of nitric oxide, superoxide, peroxynitrite, and nitrogen dioxide. When nitric oxide and superoxide are both present, they may also react with nitrogen dioxide to form N_2O_3 and peroxynitrate. Peroxynitrate decomposes to nitrite and oxygen, while N_2O_3 can react with thiols to give nitrosothiols or with hydroxide anion to give nitrite (Adapted from [15])

the sole use of glucose by mitochondria for energy production also results in the excess formation of free radicals atoms, molecules, or ions with unpaired electrons on an open shell configuration, i.e. nitric oxide which reacts with singlet oxygen to form the highly toxic peroxynitrite $ONOO^-$, hydrogen peroxide H_2O_2, hydroxyl radicals OH^-, and/or singlet oxygen O_2^- etc. Fig. 5.6).

Free radicals are oxygen-containing chemicals that have an impaired electron, and Goldstein and coworkers have demonstrated [16], that it also reacts at a diffusion-limited rate with peroxynitrite to yield two molecules of nitrogen dioxide and one of nitrite (Fig. 5.6). This creates a cycle to generate more nitrogen dioxide when bolus additions of peroxynitrite are added at neutral pH and substantially increases the number of potential reactions occurring. These same reactions will also occur in vivo, particularly when nitric oxide is produced faster than superoxide.

The impaired electron makes free radicals highly reactive to DNA, proteins, membranes, and other cell machineries, resulting in oxidative damages including DNA mutations, protein dysfunction, and destruction of membrane and other cell structures (Fig. 5.7). Due to the mutated mtDNA and the replication in inefficient mitochondria with an impaired electronic chain reaction, those mitochondria now act as ROS producing cannons heating up the deleterious $NO/ONOO^-$ cycle (Fig. 5.8).

Once NO-levels exceed normal values a feedback–mechanism sets in as a protective measurement with a block of NO-production; lasting for too long this block is genetically fixed

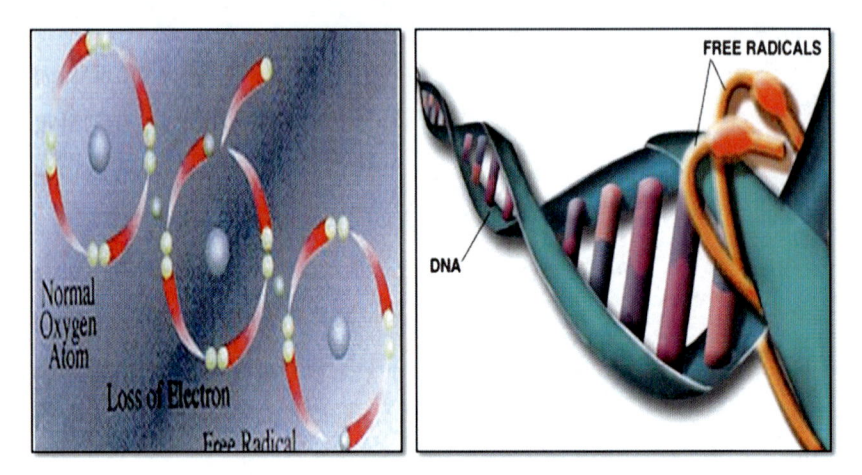

Fig. 5.7 Free radicals are atoms or molecules missing an electron. They must find another electron to make itself complete. A once stable oxygen atom loses its electron to a free radical, and now is damaged and open to disease. Free radical damage has been associated with over 60 known diseases and disorders. Free radicals are the major cause in nearly every known disease, from heart disease, arthritis, cancer and cataracts, and even the aging process itself

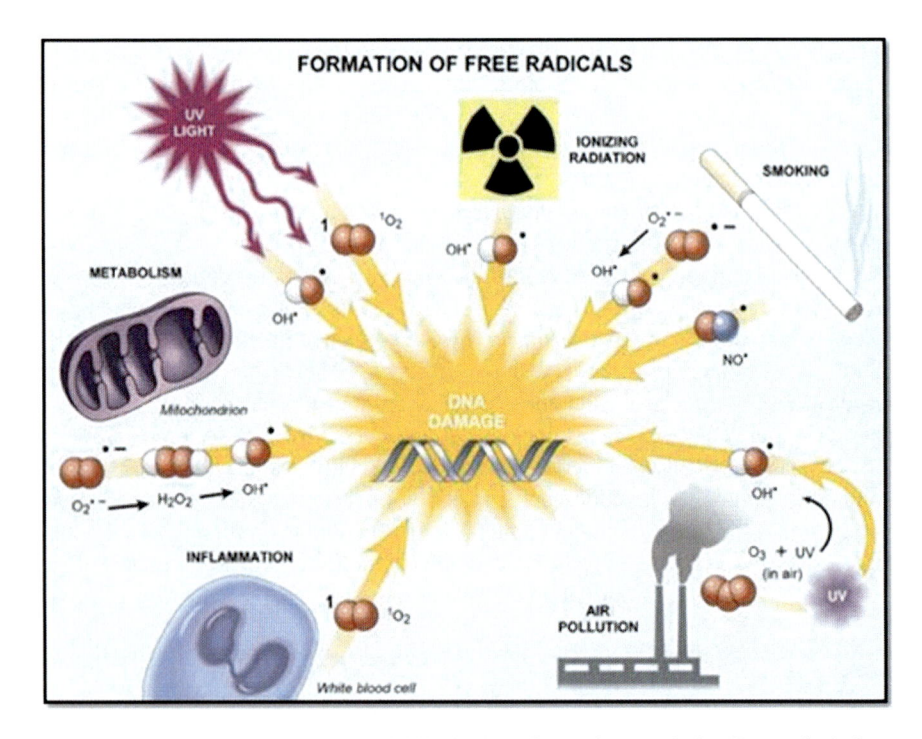

Fig. 5.8 Each human cell undergoes 10,000 hits from free radicals each day. Free radicals form in the body when it is exposed to pollution, chemicals, sunlight, ozone, cigarette smoke, food additives, and oxygen

5.2 Excess Formation of Nitric Oxide in Inflammation

A typical example of such blockade in the regulation of NO-synthesis is the inability of eNOS to locally produce sufficient NO in the inside wall of blood arteries for smooth muscle relaxation resulting in hypertonia. A feedback regulator is the existing preload with reactive nitrogen oxide species (RNS) and reactive oxidative species (ROS).

NO is formed in four different synthetic ways in the organism:

- Endothelial NO synthesis (eNOS). In this pathway, NO Neuronal NO synthesis (nNOS), the nerve cell is the place, where formation takes place.

Both NO forms are calcium-dependent and serve as neural chemical messengers during the transmission of neuronal signals.

- Induced NO synthesis (iNOS) a factor in the immune system. The induction iNOS is activated by bacterial, viral or parasite infections, but also by chemicals.
- Mitochondrial NO synthesis (mtNOS). Mitochondria form NO as a metabolic regulator for the oxidative phosphorylation and thus for the cellular energy-(ATP)-synthesis. Excess NO has the property, to bind itself on iron or especially on FeS-bearing enzymes and block them. This bondage is reversible and can be broken up by partial pressure of the oxygen [17]. This would explain the hunger for oxygen of affected patients.

In this context, biochemical effects of an increased NO synthesis are as follows:

1. Inhibition of the FeS-bearing enzymes in the mitochondrial respiratory chain, and particularly in the complexes II and I. As a result of this inhibition, less ATP is formed, but this inhibition means that at the same time an increased formation of superoxide from the mitochondria takes place, which is further augmented by physical strain, i.e. increased metabolic activity.
2. Inhibition of the FeS-bearing aconitase in the citric acid cycle. Through this inhibition the conversion of citric acid to isocitric acid is inhibited. The interruption of the citric acid cycle means a deficit in the availability of NADH, so that in this metabolic level the electron flow in the mitochondria is disturbed as well.
3. A further FeS-bearing enzyme is the liver-7α-hydroxylase, which achieves the conversion of cholesterol to bile acids. An inhibition of this activity leads therefore to an early increase in the cholesterol levels, an increase, which is completely resistant to any diet and has nothing to do with nutrition rich in cholesterol.

Other iron-bearing enzymes, which react with NO are:

- The cytochrome-c-peroxidase in the mitochondria
- The lactoperoxidase
- The myeloperoxidase
- The thyroid peroxidase
- The uterine peroxidase
- Hemoglobin and myoglobin
- The catalase
- The ferrochelatase
- and other enzymes

Due to the fact that in mitochondria play an important part in cell respiration, ATP-synthesis, the Krebs-citric acid cycle, in fatty acid ß-oxidation, glutamine synthesis, partially in steroid hormone synthesis and in the beginning of a new formation of glucose (gluconeogenesis), massive metabolic deficits can develop, which in turn are expressed clinically in form of disturbances in blood synthesis (e.g. porphyria), in lactose intolerance, and above all in a chronic energy deficit of affected cells. This effect in particular is noticeable in those organs with high-energy demand, such as the central nervous system and the musculature. Affected patients suffer from a chronic energy deficit as they are easier fatigued, need longer relaxation breaks, have a reduced ability to concentrate and can concentrate only for short periods of time. Stamina performances in particular, where the need of energy is derived from the fatty acid ß-oxidation, are almost impossible. The continuous feeling of hunger forces the affected persons to always eat; carbohydrates in this context, however, often block the energy production further (pyruvate dehydrogenase deficit in the mitochondria), so that this continuous urge to eat does not replace the energy deficit, but, on the contrary, augments it. Energy carriers of carbohydrates cannot be utilized, instead they lead to a compulsory fat transformation, which is deposited in the periphery. In contrast to refined carbohydrates, ingested proteins and fatty acids can be better utilized energetically, as they can enter the citric acid cycle through via acetyl coenzyme A. A clinical result of these metabolic changes is the tendency to increased fat deposition, chronic hunger, chronic hypoglycemia and a tendency to high cholesterol levels. Inhibition of the citric acid cycle further increases the energy deficit, since the acetyl coenzyme A cannot be expelled from mitochondria into the cellular matrix, being only possible via the citrate-shuttle. Further consequences of an excess of NO formation are disturbances in the formation of hemoglobin (hemopoiesis) so that porphyria may arise, a chronic immune insufficiency with a higher incidence for infections, a malfunction of the thyroid gland or a reduced availability of γ-amino butyric acid (GABA) resulting in an increase activity of the excitatory glutaminergic system. The chronic energy deficit in neuronal cells consequently is expressed in a chronic tiredness and a fatigue syndrome.

The amino acid citrulline, being a breakdown product from the chemical reaction of arginine with NO^- has the nature to be stored in certain proteins such as e.g. fibrinogen. Citrullinated fibrinogen, however, acts as an antigen and triggers an autoimmune reaction, i.e. an aseptic (not bacterial) inflammatory reaction, especially in the large joints and the small joints of spinal vertebra. As a consequence of such inflammation, this results in chronic pain of joints or in chronic pain of the lumbar spinal column, not being the result of excessive physical strain. If this condition lasts for a longer period of time, rheumatism of the joints is diagnosed.

Blockade of mitochondrial energy metabolism has a resulting effect, in so far that mitochondria literally become free radical cannons, which trigger oxidative stress. If additional stimulation of the immune systems by means of chemical contamination of food occurs, Th1-lymphocytes release increasingly interferon γ, which in turn is a strong stimulant for further NO synthesis. Aseptic inflammatory reactions trigger the release of the tumor necrosis factor alpha (TNF-α), which in the course of the inflammatory process via iNOS leads to a further increase of NO-formation all of which result in a chronic self-perpetuating $NO/ONOO^-$ cycle (Fig. 5.9).

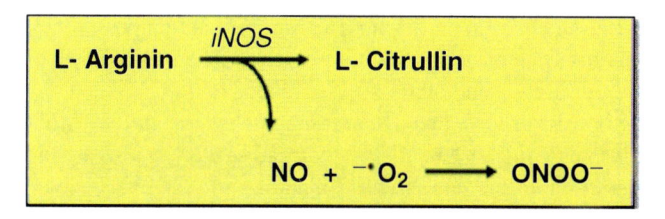

5.2 Excess Formation of Nitric Oxide in Inflammation

Fig. 5.9 Formation of the deadly cocktail NO/ONOO⁻ following decoupling of iNOS (induced NO-synthase), which now operates independently, acting like a peroxynitrate-producing cannon

The chronic energy deficit of the cells, in order to maintain their survival, forces them to turn on their secondary "emergency power supply aggregates" such as the membrane bound NADH-oxidoreductase or the aerobic glycolysis, i.e. the conversion of glucose to lactate in the presence of oxygen. At the same time protooncogenes are activated. In addition NO in the body leads to an increase in the formation of nitrosamine, resulting in a higher risk for genetic mutations. During such a chronic energy deficit brain cells have to face a particular dilemma, since the sedative GABA-effect no longer prevails, with an increase in glutamate receptors activity. The resulting calcium influx induced by glutamate induces a continuous calcium overload within the cells leading to hyperactivity, and a high irritability of brain neuronal cells. In addition, sensitization of the glutamate-NMDA-receptors generate reactive superoxide (O_2^+), a free radical with one unpaired electron, which interacts with nitric oxide generating the highly toxic product peroxynitrite. The increased formation of superoxide via the glutamate receptors plus the mitochondrial formation of superoxide within a chronic inflammatory processes, as well as the membrane bound NADH-oxidoreductase lead to a further worsening of the clinical situation, because NO together with the superoxide free radicals form the toxic peroxynitrite (ONOO⁻). Such reaction inevitably will take place, because in mitochondria NO has the property to bind with higher affinity to superoxide than superoxide to the detoxifying superoxide dismutase (manganese-dependent) and with superoxide dismutase in cell plasma (copper-zinc dependant superoxide dismutase). With the formation of peroxynitrite mitochondrial function is irreversibly impaired.

NO and O2⁺ create the lethal cocktail peroxynitrite (ONOO⁻)!

Peroxynitrite is highly toxic, it has an oxidative function and is stored on aromatic amino acids such as tryptophan or its by-products such as serotonine. The binding to phenylalanine leads to storages in tyrosine or for example with catecholamines, so that autoimmune reactions of the thyroid gland and of other neurotransmitter systems is triggered. A typical example is Hashimoto thyreoiditis where peroxynitrite binds to thyroxine, being identified by the immune system as an antigen. The products of peroxynitrite with amino acids, e.g. nitrotyrosine, can be detectable in blood plasma, similar to other citrullinated peptides. Peroxynitrite and nitrated amino acids can be traced at an early stage during a chronic inflammatory process, especially within the nervous system such as in multiple sclerosis, amyotrophic lateral sclerosis, but also in arteriosclerosis and other diseases related to mitochondopathy. As a result, there is a low hormonal level and a decrease in neurotransmitter activity within the brain. With chronical increase in ONOO⁻ formation there is a significant higher risk of development of autoimmune disease. In addition to the irreversible inhibition of mitochondrial function and the permanent production of free oxygen radicals within the cellular structure, the DNA genome of the cell nucleus is also injured.

Mitochondria possess a round-shaped genome, consisting of 37 DNA molecules, which due to their low histone protein content, cannot be repaired. Damage of the mitochondrial genome by peroxynitrite is not an immediate process, but becomes evident after a certain period of time. The reason is, that each mitochondrial gene possesses between 500 and 1,000 copies. If only 5–10% of the copies are damaged this is not a relevant deficit because of a sufficient compensation, and clinically there is not deficit in the performance of cellular function if the degree in heteroplasmy amounts to only from 5% to 10%. If this quantity, however, increases up to 40% or 50%, than an obvious deficit in functioning of mitochondrial energy production becomes evident. As a result, pyruvate kinase activity is inhibited accompanied by an inability in carbohydrate breakdown (Fig. 5.10). Consequently carbohydrates from glucose and other nutrients cannot be broken down by mitochondria for energy production, resulting in the end product H_2O. In contrast, carbohydrates are broken down via an anaerobic cycle resulting in the formation of lactic acid with lactacidosis, which can be detected in the plasma where the lactate to pyruvate ratio amounts to values higher than 10:1. Because of the increase in lactic acid formation, lactacidosis and metabolic disturbances with fatigue are worsened.

5.3 Excessive Nitric Oxide Formation Damage Mitochondria

This small and simple structured molecule nitric oxide (NO) has a Janus two-faced mode of action. While it mediates protective attributes on one hand, on the other side it demonstrates destructive properties [18, 19] (Table 4.2). Sustained high levels of NO production result in direct tissue toxicity and contribute to the vascular collapse associated with septic shock, whereas chronic expression of NO is associated with various carcinomas and inflammatory conditions including juvenile diabetes, multiple sclerosis, arthritis and ulcerative colitis [20]. There are several mechanisms by which NO has been demonstrated to affect the biology of living cells. These include oxidation of iron-containing proteins such as ribonucleotide reductase and aconitase, activation of the soluble guanylate cyclase, ADP ribosylation of proteins, protein sulfhydryl group nitrosylation, and iron regulatory factor activation [21]. NO has also been demonstrated to activate NF-κB in peripheral blood mononuclear cells, an important transcription factor in iNOS gene expression in response to inflammation [22]. It was found that NO acts through the stimulation of the soluble guanylate cyclase, which is a heterodimeric enzyme with subsequent formation of cyclic GMP. Cyclic GMP activates protein kinase G, which causes phosphorylation

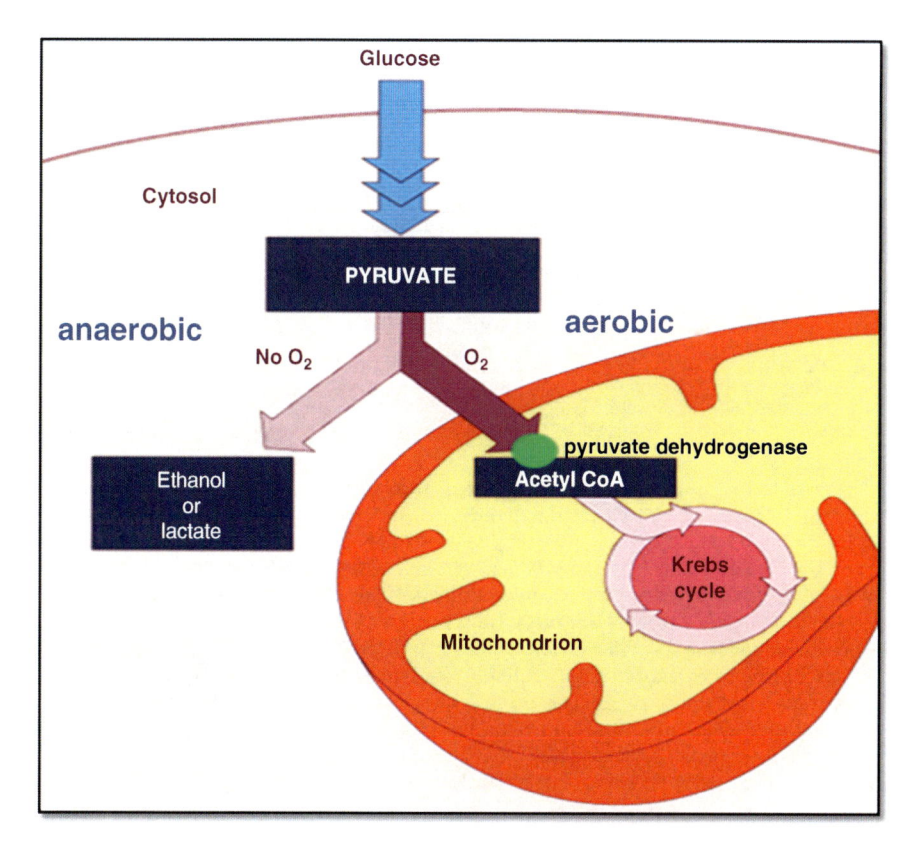

Fig. 5.10 The block in pyruvate dehydrogenase resulting in the inability of carbohydrate to enter the Krebs cycle as Acetyl CoA and be used for energy production in mitochondria. Instead the anaerobic pathway is used resulting in excess buildup of lactate with lactic acidosis, which is determined by change of the lactate/pyruvate ratio

of myosin light chain phosphatase, with inactivation of myosin light-chain kinase, ultimately leading into the dephosphorylation of the myosin light chain with smooth muscle relaxation [23]. Since NO binds to Fe and FeS-containing enzymes, it:

- Blocks aconitase in the citric acid-cycle resulting in insufficient production of NADH necessary for energy making the respiratory electron chain within mitochondria (Fig. 7.13).
- Blocks liver 7-α-hydroxylase, which converts cholesterine into bile acids resulting in a diet-resistant hypercholesterolemia.
- Blocks pyruvate-dehydrogenase activity with insufficient transformation of pyruvate into acetyl-CoA resulting in pyruvate congestion, which is succeeded in deprivation of the glycolysis metabolic pathway to the citric acid cycle and the release of energy via NADH (Fig. 5.10).

- Blocks ferrochelatase, with insufficient haem synthesis.
- Blocks myeloperoxidase with increased susceptibility to infections.
- Blocks thyroxin peroxidase necessary for coupling iodine with dysfunction of the thyroid gland.
- Blocks CYP 450 enzyme with reduced detoxification capacity of the liver.

References

1. Miller G (1999) Are we on the threshold of a new theory of disease? Toxicol Ind Health 15:284–294
2. Buchwald D, Garrity D (1994) Comparison of patients with chronic fatigue syndrome, fibromyalgia, and multiple chemical sensitivities. Arch Intern Med 154:2049–2053
3. Donnay A, Ziem G (1999) Prevalence and overlap of chronic fatigue syndrome and fibromyalgia syndrome among 100 new patients. J Chron Fatigue Syndr 5:71–80
4. Pall ML (2009) Explaining 'unexplained illnesses': disease paradigm for chronic fatigue syndrome, multiple chemical sensitivity, fibromyalgia, posttraumatic stress disorder, Gulf War syndrome and others. Harrington (Hayworth) Press, New York
5. Donohue D, Movahed MR (2005) Clinical characteristics, demographics and prognosis of transient left ventricular apical ballooning syndrome. Heart Fail Rev 10:311–316
6. Schultz BE, Chan SI (2001) Structures and proton-pumping strategies of mitochondrial respiratory enzymes. Annu Rev Biophys Biomol Struct 30:23–65
7. Fiore C et al (1998) The mitochondrial ADP/ATP carrier: structural, physiological and pathological aspects. Biochimie 80:137–150
8. Bradfield P et al (2002) A2 level biology. Pearson Education, Upper Saddle River
9. Myhill S, Booth NE, McLaren-Howard J (2009) Chronic fatigue syndrome and mitochondrial dysfunction. Int J Clin Exp Med 2:1–16
10. See reference [9]
11. Pall ML (2007) Nitric oxide synthase partial uncoupling as a key switching mechanism for the NO/ONOO⁻ cycle. Med Hypotheses 69:821–825
12. Pall ML, Satterlee JD (2001) Elevated nitric oxide/peroxynitrite mechanism for the common etiology of multiple chemical sensitivity, chronic fatigue syndrome andposttraumatic stress disorder. Ann N Y Acad Sci 933:323–329
13. Brand MD et al (2004) Mitochondrial superoxide and aging: uncoupling-protein activity and superoxide production. Biochem Soc Symp 71:203–213
14. Boelsterli U (2004) Disruption of cellular energy production by xenobiotics. In: Boelsterli U (ed) Mechanistic toxicology. Taylor & Francis, London, pp 282–300
15. Pacher P, Beckman JS, Liaudet L (2007) Nitric oxide and peroxynitrite in health and disease. Physiol Rev 87:315–424
16. Goldstein S et al (1999) Effect of *NO on the decomposition of peroxynitrite: reaction of N2O3 with ONOO. Chem Res Toxicol 12:132–136
17. Radi R et al (1991) Peroxynitrite oxidation of sulfhydryls: the cytotoxic potential of superoxide and nitric oxide. J Biol Chem 266:4244–4250
18. Suschek CV, Schnorr O, Kolb-Bachofen V (2004) The role of iNOS in chronic inflammatory processes in vivo: is it damage-promoting, protective, or active at all? Curr Mol Med 4:763–775
19. Suschek CV et al (2002) Even after UVA-exposure will nitric oxide protect cells from reactive oxygen intermediate-mediated apoptosis and necrosis. Cell Death Diff 8:515–527

References

20. Taylor BS et al (1997) Nitric oxide down-regulates hepatocyte-inducible nitric oxide synthase gene expression. Arch Surg 132:1177–1183
21. Shami PJ et al (1995) Nitric oxide modulation of the growth and differentiation of freshly isolated acute non-lymphocytic leukemia cells. Leuk Res 19:527–533
22. Kaibori M et al (1999) Immunosuppressant FK506 inhibits inducible nitric oxide synthase gene expression at a step of NF-κB activation in rat hepatocytes. J Hepatol 30:1138–1145
23. Surks HK (2007) cGMP-dependent protein kinase I and smooth muscle relaxation: a tale of two isoforms. Circ Res 101:1078–1080

Chapter 6
Principles in Nitrosative Stress

In clinical medicine, nitric oxide is considered an antianginal drug, since it causes vasodilation, which can help with ischemic pain known as angina by decreasing the cardiac workload, and by dilating the veins there is less blood returned to the heart per cycle [1]. This decreases the amount of volume that the heart has to pump. Nitroglycerin pills, taken sublingually are used to prevent or treat acute chest pain. The nitroglycerin reacts with a sulfhydryl group (–SH) to produce nitric oxide, which eases the pain by causing vasodilation. Recent evidence suggests that nitrates may be beneficial for treatment of angina due to reduced myocardial oxygen consumption both by decreasing preload and afterload and by some direct vasodilation of coronary vessels [1]. However, aside from its beneficial effects, nitric oxide can also contribute to reperfusion injury when an excessive amount produced during reperfusion (following a period of ischemia) reacts with superoxide to produce the damaging oxidant peroxynitrite. In contrast, inhaled nitric oxide has been shown to help survival and recovery from paraquat poisoning, which produces lung tissue–damaging superoxide and hinders NOS metabolism. Another important biological reaction of nitric oxide is S-nitrosylation, the conversion of thiol groups, including cysteine residues in proteins, to form S-nitrosothiols (RSNOs). S-Nitrosylation is a mechanism for dynamic, post-translational regulation of most or all major classes of protein. In addition, it has been demonstrated that the role of this intercellular gaseous signaling agent is involved in sleep homeostasis through nNO synthase initiating non-REM sleep (NREM). For instance, infusion of a NO donor produced an increase in NREM with dominant delta- and theta-EEG waves followed by a decline in cerebral ATP synthesis that closely resembled NREM recovery after prolonged wakefulness, while inhibition of NO synthesis completely abolished non-rapid eye movement (NREM) recovery sleep. NO production via nNO synthase in the basal forebrain therefore seems a prerequisite for the induction of recovery sleep [2].

6.1 Factors Triggering NO/ONOO Cycle – Inflammation, Radical Nitrosative and Oxygen Substrates

Many of the symptoms of mitochondropathy are caused by a poor antioxidant status. This is because during normal cell metabolism there is a constant production of free radicals, even within the mitochondria. Therefore one cannot live without producing free radicals. These are highly reactive and potentially dangerous unstable molecules, which have an unpaired electron. However, the body has evolved many systems for clearing these free radicals before they cause too much damage. Inside mitochondria the most important free radical scavenger systems are: coenzyme Q10 and the manganese-dependent superoxide dismutase (SOD). Outside the mitochondria radical scavenging is made possible via the zinc/copper dependent superoxide dismutase, the glutathione peroxidase (GSH), acetyl L-carnitine, as well as vitamins A, C and E and lots of other natural antioxidants found in nuts, seeds and vegetables. If, however, there is a poor antioxidant status, high dose vitamin B_{12} will take over many of these antioxidative functions. This is why the effect of B_{12} injections is often obvious and running out of B_{12} equally becomes obvious, because the residues of NO and hydrogen peroxide form the aggressive ONOO⁻ via nitric oxide synthase (NOS) leading into a further augmentation of inflammation, which by itself already results in an increased formation of NO.

$$NO + H_2O_2 \rightarrow ONOO^-$$
$$\text{nitric oxide} \quad \text{superoxide} \quad \text{peroxynitrite}$$

The common etiology therefore focuses on nitric oxide and its oxidant product peroxynitrite, a potent oxidant, and where upregulation of the NO/ONOO⁻ cycle results in a multisystem disease [3]. There are several lines of evidence for the existence of such a NO/ONOO⁻ cycle leading to a number of ailments in modern life. First, experiments done with drugs that act by releasing nitric oxide (i.e. nitroglycerine and nitroprusside) have been shown to increase nitric oxide synthesis via all three synthases. Second, all of the elements of the cycle are implicated in generating pain or hyperalgesia, a pattern which readily can be explained by N-Methyl-D-Aspartate (NMDA) receptor activation, and which regularly is involved in the upregulation of the NO/ONOO⁻ cycle. In addition, the action of NMDA within the cycle elements can be explained by implementing a mechanism as outlined in the NO/ONOO⁻ cycle (Fig. 6.1):

1. Citrullin (formed from Arginine + O_2^- (ROS) = NO + citrullin) combines with proteins, which are identified by the immune systems as foreign compounds and result in the systemic release of antibodies inducing inflammatory reactions in various parts of the body resulting in a true rheumatic reaction.
2. NO also activates the cyclooxygenase (COX)-system starting off an inflammatory cascade with release of interleucines followed by inflammatory signs at joints, tendons and muscles and the potential development of an autoimmune disease [5]. At the same time NO activates macrophages and other inflammatory cells.

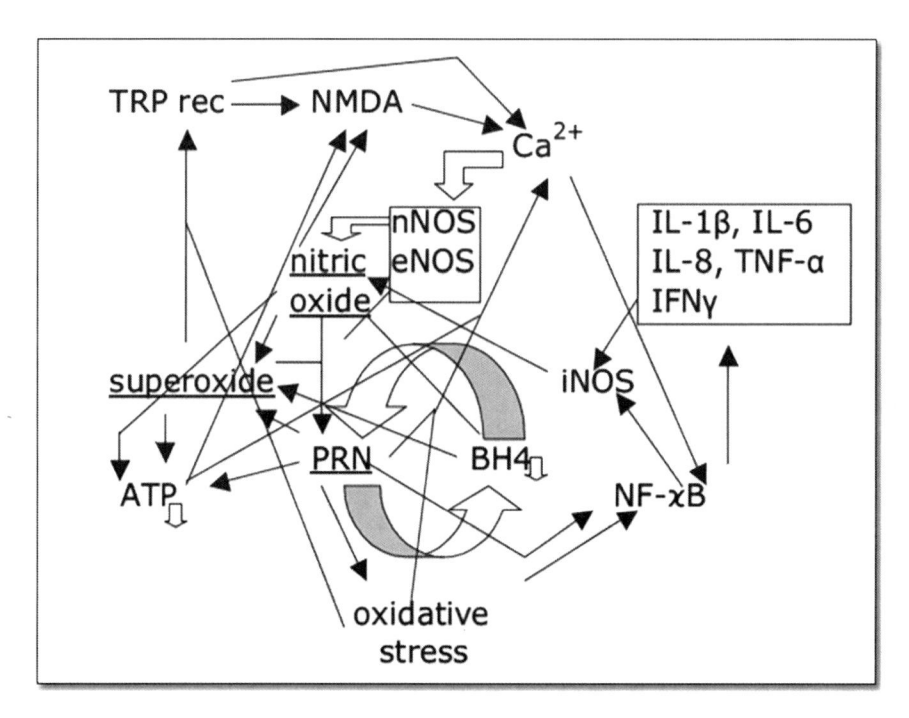

6.1 Factors Triggering NO/ONOO Cycle – Inflammation... 57

Fig. 6.1 The vicious cycle in the self-perpetuation of increased formation of NO and peroxynitrite (PRN = OONO⁻) (Adapted from [4])

3. Due to the NO-induced block of glycolysis where conversion of pyruvate into acetyl CoA takes place in order to enter the Krebs-cycle and the obstruction of beta-oxidation of fatty acids, there is a deficiency in NADH and NASDH formation, necessary to spark off regular ATP-formation in mitochondria [6].
4. Upregulation of the NMDA- and the vanilloid-production with extensive receptor binding in the pain propagating pathways of the CNS, there is an increase in transmission of afferents with an increase in susceptibility and sensitivity to otherwise non-noxious stimuli [7].
5. Due to NMDA-receptor activation there is also an increase inflow of Ca^{2+} ions into cells of the hippocampal area and the striatum within the CNS with ensuring apoptotic reactions with decay in neuronal function followed by the development of Parkinson and Alzheimer´s disease [8]. And only recently a study demonstrated that mitochondrial dysfunction and oxidative stress are pathophysiologic mechanisms implicated in experimental models and genetic forms of Parkinson's disease (PD). Certain pesticides may affect these mechanisms, and the pesticides rotenine and paraquat have has been definitively associated with PD in humans [9, 10].
6. Nitrosation of aromatic amino acids such as tyrosin leads into the formation of nitrotyrosin and nitrosylation of hormones such as thyroxin leads into autoimmune reaction with Hashimoto thyreoditis, an autoimmune disease. The autoimmune attack on the thyroid makes the gland slowly less able to function, and eventually, the thyroid becomes underactive. When hypothyroidism itself can be measured

by blood tests, many practitioners will finally diagnose the hypothyroidism, and treat the patient with thyroid hormone replacement drugs.
7. Due to blockade of an important coenzyme in hormone synthesis, i.e. tetrahydroxybioterin (BH_4; Fig. 6.3), there is a reduced formation of melatonin, tryptophan, melanine, and phenylalanine since this coenzyme presents a necessary component in the synthesis of these neurotransmitters (Fig. 6.4).

Several basic principles are suggested that underly the formation and activation of the vicious NO/ONOO⁻ cycle as an explanatory model, which has been demonstrated as being a valid explanation for a number of ailments:

1. Short-term stressors that initiate cases of multisystem illnesses act by raising nitric oxide synthesis and via NF-κB activation, a rise in inflammatory mediators (i.e. interleukins) affecting joints, tendons and musculature.
2. If nitric oxide reactive oxygen species (NOS) and other inflammatory mediators are produced in excess, they can directly inhibit mitochondrial respiration [11]. This is because NO⁻ competes with oxygen in binding to cytochrome oxidase c at complex IV, thereby decreasing the activity of the enzyme and blocks the electron transport chain leading to an overproduction of superoxide (O_2^-).
3. Superoxide reacts with NO to generate peroxynitrite (ONOO⁻) and other nitrogen species that are able to alter the structure and function of several other mitochondrial proteins, notably complex I [11].
4. Nitrosation targets transcription factors, where over 200 different proteins have been found to be targeted by S-nitrosation, including metabolic enzymes, phosphatases, transcription factors, and others [12]. Nitric oxide (NO), reactive oxygen species and other inflammatory mediators are produced in excess and can directly inhibit mitochondrial respiration. NO competes with oxygen in binding to cytochrome oxidase (complex IV), thereby decreasing the activity of the enzyme. This will block the electron transport chain and lead to overproduction of superoxide [11].
5. NO contributes to protein misfolding in mitochondria, since recent studies have suggested that nitrosative stress due to generation of excessive nitric oxide (NO) can mediate excitotoxicity by triggering protein misfolding and aggregation. S-Nitrosylation, or a covalent reaction of NO with specific protein thiol groups, represents a convergent signal pathway contributing to NO-induced protein misfolding and aggregation, compromised dynamics of mitochondrial fission-fusion process, thus leading to neurotoxicity [13].
6. Downregulation of the glutathione (GSH) system by NO. In the presence of oxygen, endogenously produced NO can rapidly lead to S-nitrosation of protein thiols (SNO) and glutathione (GSNO), presumably via the formation of other reactive nitrogen oxide species, such as N_2O_3, NO2, and thiyl radicals [14]. All these biochemical changes are converted into a chronic illness through the activation of the vicious cycle, with chronic elevation of nitric oxide and peroxynitrite and other cycle elements being produced permanently and maintaining the ailment(s). Because various elements of the NO/ONOO⁻ cycle are elevated in the chronic phase of each of these ailments, they can be measured in blood plasma and/or the urine (see Appendix 1).

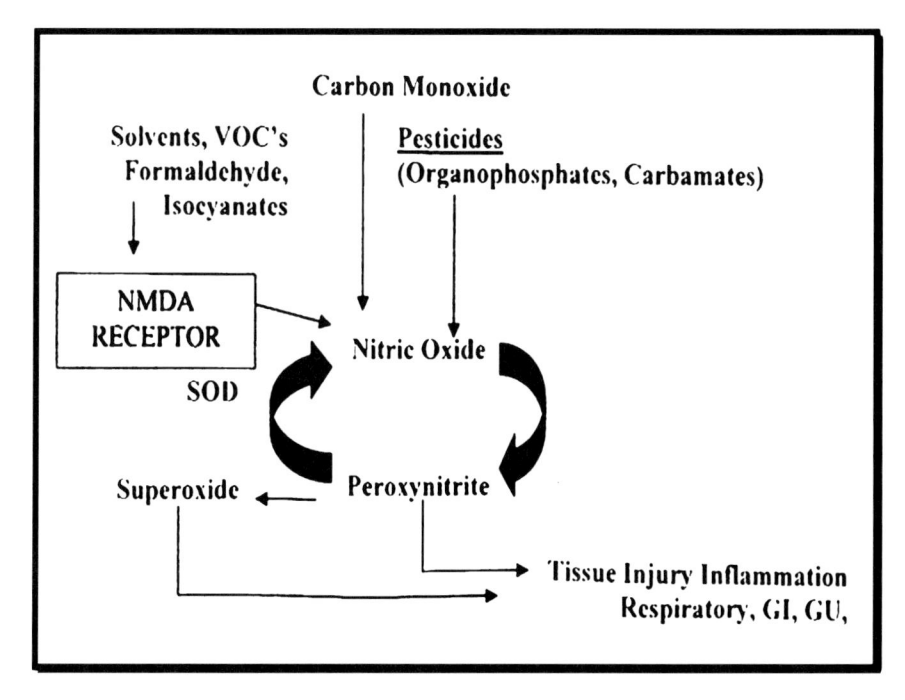

Fig. 6.2 Principle of neural sensitization in toxic volatile organic compounds (VOCs) and solvents with activation of the NMDA-receptor and the NO/ONOO⁻ cycle. *Gi* gastrointestinal, *Gu* gastric ulcer

7. Downregulation of melatonine synthesis, another important radical oxygen/nitrosative substrate scavenger system of the organisms. Any lack in this homonal-like agent results in less NO inactivation, loss of sleep-inducing effects, loss of protection from the destructive effects of nitric oxide/peroxynitrite radicals on the ß-cells of the pancreas and lastly a higher incidence of autoimmune diseases [15].

Clinical symptoms and signs of these ailments are generated by the elevated nitric oxide and/or an elevated peroxynitrite level, accompanied by inflammatory cytokines, an elevated NMDA and vanilloid receptor activity, all of which result in central nervous system effects:

1. Since these receptor sites are found in abundance in pain processing pathways and the thalamus, their sensitization with longterm potentiation clarifies why in patients with CFS, FMS and MCS chronic painful sensations are a leading symptom [16].
2. Neuronal sensitization occurs by activation and an increase in production of brain and nerve cell N-methyl-D-aspartate (NMDA) transmitter, which in return increases brain nitric oxide (NO) synthesis [17–19].
3. Because several biochemical cycles are set in motion, nitric oxide then forms a tissue damaging free radical known as peroxynitrite [20–22], resulting in neurodegeneration [23], which deplete ATP energy [24, 25], further increasing sensitization of the NMDA-receptor [26, 27] (Fig. 6.2).

Among the several factors that are able to induce NO/ONOO⁻production and sensitization of the NMDA-receptor, the exposure to chemicals such as pesticides (i.e. organophosphates) is a major contribution as they inhibit acetylcholine, thus activating the muscarinic receptors followed by an increase in nitric oxide or formaldehyde which by itself activates the formation of NMDA [17, 28]. For instance, it has been demonstrated that various petrochemicals (volatile organic compounds, solvents) disrupt energy production in the mitochondria with an increase in superoxide, which in return increases peroxynitrite production [29], followed by an increase in tissue-damaging free radicals within the central nervous system [30]. Mitochondrial disruption is the underlying cause in such chemically injured patients [31] and exposure of such petrochemicals and many other chemicals irritants cause inflammation [32], which when being exposed for a sufficient time can lead to chronic neurogenic inflammation with increased cytokine and free radical release and an elevated nitric oxide level [33, 34]. Among others, thyroid and sex hormones, insulin, glucocorticoids and leptin positively modulate mitochondrial energy production, protein synthesis and biogenesis [35–37].

6.2 Leptin, Link for Inflammation in NO-Related Diseases

Leptin, a novel link between obesity, diabetes, cardiovascular risk, and ventricular hypertrophy is produced predominantly by adipocytes in overweight patients [38]. Increases in NO-production by intravenous injection of leptin increased nitrite and nitrate concentrations in blood serum of normotensive Wistar rats but not in obese Zucker rats, which have a mutation in their leptin receptor gene[39]. In obesity, leptin increases eNOS expression and decreases intracellular L-arginine, resulting in eNOS-uncoupling and depletion of endothelial NO and an increase of cytotoxic ONOO⁻[40].

Leptin aside from its angiogenic properties and proinflammatory immune responses, induces oxidative stress with ROS formation in human endothelial cells leading to artherogenic processes [41]. Also NF-κB another transcription factor is activated by leptin and there are specific agonists increasing leptin levels. These include TNF alpha and other pro-inflammatory cytokines, insulin, glucose, angiotensin II and estrogens [42]. Because of its dual nature as a hormone and a cytokine, leptin links the neuroendocrine and the immune system. The role of leptin in the modulation of immune response and inflammation has recently become increasingly evident. The increase in leptin production that occurs during infection and inflammation strongly suggests that leptin is a part of the cytokine network, which governs the inflammatory-immune response and the host defense mechanisms. Higher leptin levels affect the immune status and inflammation since leptin induces cytokine producing capacity switch towards Th1 producing cells, particularly by increasing interferon gamma, TNF-α and IL-2 producing capacity [43–44]. Therefore leptin can be considered a pro-inflammatory stimulator as well as other proinflammatory mediators such as the tumor necrosis factor alpha (TNF-α) and IL-1, all of which increase the expression of leptin mRNA in adipose tissue [45]. Free leptin surrogates are associated with masked hypertension in non-obese normoglycemic subjects and free leptin is almost equally increased in masked and sustained hypertension, suggesting a similar leptin-related vascular impairment [46].

Leptin has also been found to be overexpressed in human breast tumors and is produced by breast cancer cells in response to obesity-related stimuli [47]. Colonic leptin may also be a source of a novel proinflammatory cytokine involved in inflammatory bowel disease (IBD), since leptin induced epithelial wall damage and neutrophil infiltration that represent characteristic histological findings in acute intestinal inflammation. These observations provide evidence for an intra-luminal biological signaling of leptin and a new pathophysiological role for intraluminal leptin during states of intestinal inflammation such as inflammatory bowel disease [48]. And lastly, leptin may be involved in neuropathic pain development. This is because there is accumulating evidence suggesting the existence of a molecular substrate for neuropathic pain produced by neurons, glia, and immune cells showing that adipocytes (leptin) associated with primary afferent neurons may be involved in the development of neuropathic pain through adipokine secretion [49].

6.3 Tetrahydrobiopterin Impaired by Nitrosative Stress

Tetrahydrobiopterin (BH$_4$, THB, trade name Kuvan®; Fig. 6.3) or sapropterin (INN) is a naturally occurring essential cofactor of the three aromatic amino acid hydroxylase enzymes, used in the degradation of amino acid phenylalanine and in the biosynthesis of the neurotransmitters serotonin (5-hydroxytryptamine, 5-HT), melatonin, dopamine, norepinephrine (noradrenaline), epinephrine (adrenaline), and nitric oxide (NO) [50].

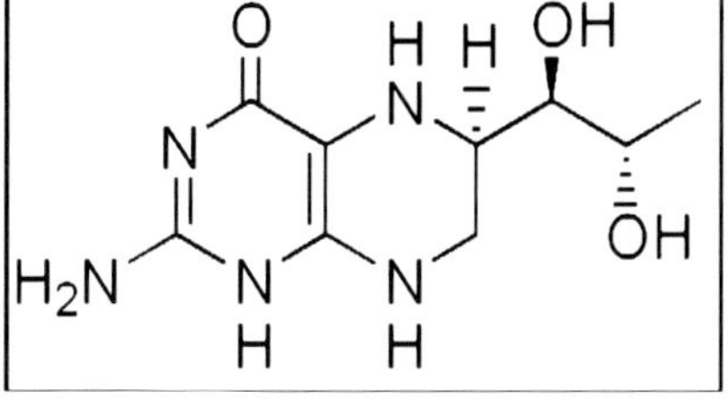

Fig. 6.3 The molecular structure of tetrahydrobiopterin (BH$_4$) or (6R)-2-Amino-6-[(1R,2S)-1,2-dihydroxypropyl]-5,6,7,8-tetrahydro-4(1H)-pteridinone an essential cofactor in the synthesis of hormones and also of nitric oxide

Tetrahydrobiopterin has the following responsibilities acting as a cofactor with:

- Tryptophan hydroxylase (TPH) for the conversion of L-tryptophan (TRP) to 5-hydroxytryptophan (5-HTP)
- Phenylalanine hydroxylase (PAH) for conversion of L-phenylalanine (PHE) to L-tyrosine (TYR)
- Tyrosine hydroxylase (TH) for the conversion of L-tyrosine to L-DOPA (DOPA)
- Nitric oxide synthase (NOS) for conversion of guanidino nitrogen of L-arginine (L-Arg) to nitric oxide (NO)
- Glyceryl ether monooxygenase (GEMO) for the conversion of 1-alkyl-sn-glycerol to 1-hydroxyalkyl-sn-glycerol

Clinically inborn errors of BH_4 synthesis are found with deficit in tetrahydrobiopterin biosynthesis and/or regeneration which results in phenylketonuria (PKU) from excess L-phenylalanine concentrations or hyperphenylalaninemia (HPA), as well as monoamine and nitric oxide neurotransmitter deficiency and chemical imbalance. The chronic presence of PKU can result in severe brain damage, including symptoms of mental retardation, microcephaly, speech impediments such as stuttering, slurring, and lisps, seizures or convulsions, and behavioral abnormalities, among other effects. In the present case acquired BH_4 deficiency because of oxidation by means of perroxynitrite results in a myriad of consequences which presently are a hot topic in research and in treatment of various chronic ailments especially in regard to cardiovascular disease [51].

The purpose of tetrahydrobiopterin is not only to reduce levels of L-phenylalanine, but more directly to convert amino acids to neurotransmitter precursors, and the production of NO. Due to its role in the conversion of L-tyrosine in to L-dopa, which is the precursor for dopamine, any deficiency in tetrahydrobiopterin can also cause other neurological issues unrelated to a toxic buildup of L-phenylalanine, such as dopamine which is a vital neurotransmitter the precursor of epinephrine and norepinephrine. One of the primary conditions resultant from GTPCH-caused tetrahydrobiopterin deficiency is dopamine-responsive dystonia. Currently, this condition is typically treated with carbidopa/levodopa, which directly restores dopamine levels within the brain (Fig. 6.4).

This new but important finding is that the coenzyme tetrahydrobiopterin (BH_4) in the synthesis of neurotransmitters (Table 6.1) is oxidized by peroxynitrite resulting in a depletion [52–54] in other NO/OONO$^-$ related diseases (Fig. 6.5).

Short-term stressors, capable of increasing nitric oxide levels, act to initiate cases of illnesses including chronic fatigue syndrome, multiple chemical sensitivity, fibromyalgia and posttraumatic stress disorder. These stressors, acting primarily through the nitric oxide product, peroxynitrite, are thought to initiate a complex vicious cycle mechanism, known as the NO/ONOO$^-$ cycle that is responsible for chronic illness. The complexity of the NO/ONOO$^-$ cycle raises the question as to whether the mechanism that switches on this cycle is this complex cycle itself or whether a

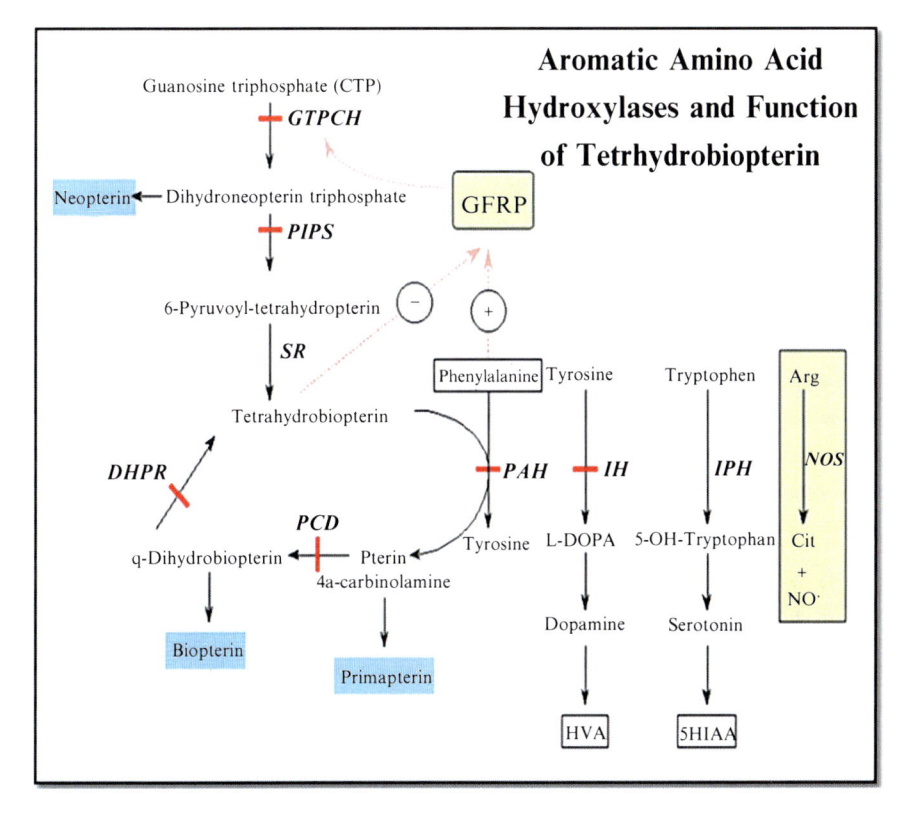

6.3 Tetrahydrobiopterin Impaired by Nitrosative Stress

Fig. 6.4 Tetrahydrobiopterin (BH$_4$) the necessary coenzyme for the production of essential neurotransmitters dopamine, melatonin, and serotonin in the CNS

Table 6.1 Summary of biochemical reactions where tetrahydrobiopterin is a necessary cofactor

BH4 a necessary cofactor (electron transmission) in several enzymatic redox-reactions
- Synthesis of adrenalin from tyrosins via BH4-dependant tyrosin-hyroxylase
- Synthesis of noradrenaline from tyrosine via the BH4-dependant tyrosine-hydroxylase
- Synthesis of L-DOPA from tyrosine via the BH4-dependant tyrosine-hydroxylase
- Synthesis of serotonin from tryptophan via the tryptophan-hydroxylase
- Formation of nitric oxide (NO) from arginine via the nitric-Oxide-synthase (iNOS) resulting in NO and citrulline
- In contrast to pteridin-derivatives folic acid and riboflavine, the human body can synthesize BH4 and therefore it is not essential
- Only the tetrahydroform of biopterin is biologically active.

simpler mechanism is the primary switch. It is proposed that the switch involves a combination of two variable switches, the increase of nitric oxide synthase (NOS) activity and the partial uncoupling of the NOS activity, with uncoupling caused by a tetrahydrobiopterin (BH$_4$) deficiency [56]. NOS uncoupling causes the NOS enzymes to produce superoxide, the other precursor of peroxynitrite, in place of nitric oxide. Thus partial uncoupling will cause NOS proteins to act like peroxynitrite

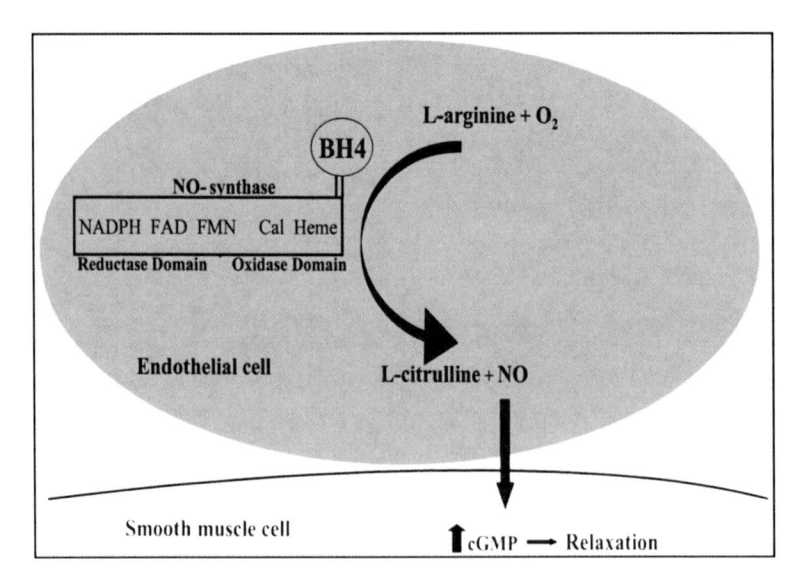

Fig. 6.5 Under physiological conditions, following binding of tetrahydrobiopterin (BH$_4$) to the oxidase domain of nitric oxide (NO) synthase (NOS), the enzyme is activated and generates NO and L-citrulline from L-arginine and O$_2$. NO diffuses rapidly to the underlying smooth muscle cells leading to relaxation through an increase of cGMP levels (Adapted from [55])

synthases, leading, in turn to increased NF-κB activity. Peroxynitrite is known to oxidize BH$_4$, and consequently partial uncoupling may initiate a vicious cycle, propagating the partial uncoupling over time. The combination of high NOS activity and BH4 depletion will lead to a potential vicious cycle that may be expected to switch on the larger NO/ONOO$^-$ cycle, thus producing the symptoms and signs of chronic illness. The role of peroxynitrite in the NO/ONOO$^-$ cycle also implies that such uncoupling is part of the chronic phase cycle mechanism such that agents that lower uncoupling will be useful in treatment (Ref. [11] of Chap. 5) [57].

It is important to point out:

Tetrahydrobiopterin is a necessary cofactor in the activity of nitric oxide synthases (NOSs), and when depleted there is partial uncoupling of all NOSs with uncontrolled toxic superoxide and perroxynitrite production in place of nitric oxide

The consequence of this is that in cells and tissues that have high NOS activity and partial uncoupling, one has many adjacent enzyme molecules, some producing nitric oxide and others producing superoxide and these will react rapidly with each other to form excess peroxynitrite [58]. Thus BH$_4$ deficiency has been found to play

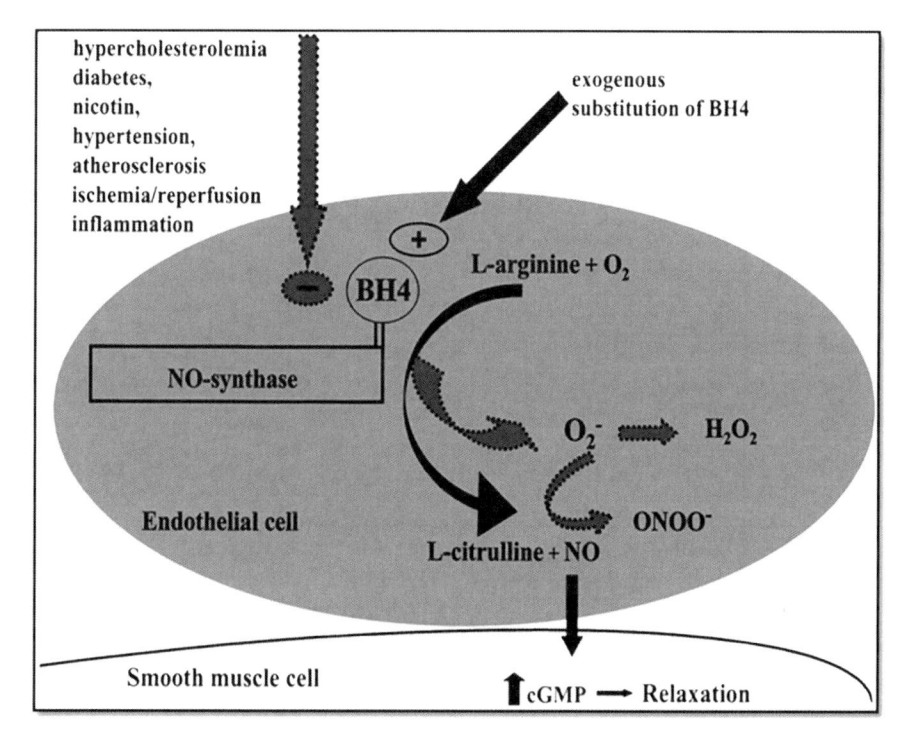

6.3 Tetrahydrobiopterin Impaired by Nitrosative Stress

Fig. 6.6 Under the influence of different risk factors of atherosclerosis as well as ischemia-reperfusion injury and inflammation, the bioactivity of BH_4 is reduced. In the presence of suboptimal levels of BH_4, NOS generates both NO and superoxide anions (O_2^-). This leads to the formation of hydrogen peroxide (H_2O_2) from O_2 and peroxynitrite (OONO$^-$) from O_2^+ and NO. Under these conditions, substitution of BH_4 would restore the original activity of NOS and lead to increased production of NO (Adapted from [55])

an important factor in diabetes [59], coronary heart disease associated with diabetes and arteriosclerosis [60].

Although, the exact role of tetrahydrobiopterin in the control of NOS catalytic activity is not completely understood, existing evidence suggests that it can act as alosteric and redox cofactors. Thus, suboptimal concentration of tetrahydrobiopterin reduces formation of nitric oxide and favors "uncoupling" of NOS leading to NOS-mediated reduction of oxygen and formation of superoxide anions and hydrogen peroxide. Recent findings suggest that accelerated catabolism of tetrahydrobiopterin in arteries exposed to oxidative stress may contribute to pathogenesis of endothelial dysfunction present in arteries exposed to hypertension, hypercholesterolemia, diabetes, smoking, and ischemia-reperfusion (Fig. 6.6). Beneficial effects of acute and chronic tetrahydrobiopterin supplementation on endothelial function have been reported in experimental animals and humans.

It appears that beneficial effects of some antioxidants (e.g. vitamin C or 5-methyltetrahydrofolate) on vascular function could be mediated via increased intracellular concentration of tetrahydrobiopterin [61, 62].

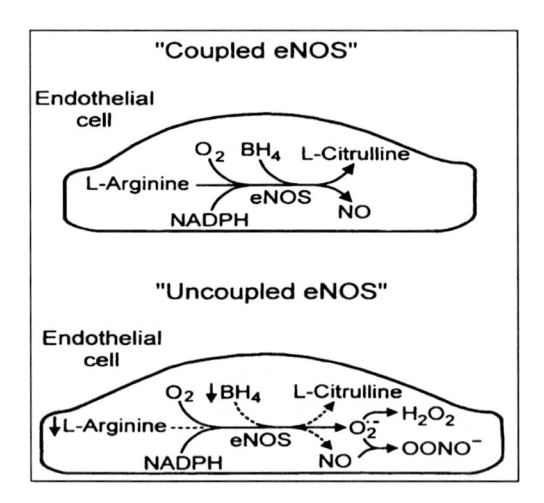

Fig. 6.7 Schematic representation of nitric oxide synthesis via endothelial NO synthesis (eNOS). "Uncoupling" of synthesis by suboptimal concentrations of L-arginine and/or tertahydrobiopterin (BH$_4$) results in excess formation of aggressive superoxide (O$_2^+$), hydrogen peroxide (H$_2$O$_2$) and peroxynitrite (OONO$^-$) (Adapted from [63])

Imbalances within the biopterin-metabolism with low BH$_4$ levels are now being been associated with a number of ailments, where instead of a regular NO-synthesis (necessary for vasodilatation) nitric oxide synthases are uncoupled yielding excess formation of superoxide anions und peroxynitrite (Fig. 6.7).

If due to proceeding morphological changes the mitochondrial apoptosis-induced channel (MAC), located in the outer mitochondrial membrane [64] is activated, cytochrome c is also released from mitochondria. Associated with such release, apoptosis by peroxynitrite caspases is turned on resulting in programmed cell death. Apoptotic reaction differs from necrosis, in which the cellular debris can damage the organism. Therefore, induced NOS (iNOS) is thought of being activated by the following processes

- Inflammation in cells of the immune system
- As a regulator to fight pathogens
- Ingestion of chemical toxins (xenobiotics)
- Proinflammatory cytokines and lipopolysaccharides* (LPS)
- An insufficient BH$_4$ formation

*Lipopolysaccharides also known as lipoglycans, are large molecules consisting of a lipid and a polysaccharide joined by a covalent bond; they are found in the outer membrane of Gram-negative bacteria, act as endotoxins and elicit strong immune responses.

References

1. Abrams J (1996) Beneficial actions of nitrates in cardiovascular disease. Am J Cardiol 77:C31–C37
2. Kalinchuk AV et al (2006) Nitric oxide production in the basal forebrain is required for recovery sleep. J Neurochem 99:483–498
3. Preyor WA, Squadritro GL (1997) The chemistry of peroxynitrite: a product from the reaction of nitric oxide with superoxide. Am J Physiol 268:699–722
4. Pall ML (2009) Explaining "unexplained illnesses": disease paradigm for chronic fatigue syndrome, multiple chemical sensitivity, fibromylagia, post-traumatic stress disorder, Gulf War syndrome and others. Harrington Park (Haworth) Press, New York
5. Abramso SB (2008) Nitric oxide in inflammation and pain associated with osteoarthritis. Arthritis Res Ther 10(Suppl 2):S2
6. Colussi C et al (2000) H2O2-induced block of glycolysis as an active ADPribosylation reaction protecting cells from apoptosis. FASEB J 14:2266–2276
7. Chang HM et al (2003) Upregulation of NMDA receptor and neuronal NADPH-d/NOS expression in the nodose ganglion of acute hypoxic rats. J Chem Neuroanat 25:137–147
8. Greenamyre JT, Young AB (1989) Excitatory amino acids and Alzheimer's disease. Review. Neurobiol Aging 10:593–602
9. Kamel F et al (2007) Pesticide exposure and self-reported Parkinson's disease in the agricultural health study. Am J Epidemiol 165:364–374
10. Tanner CM et al (2011) Rotenone, paraquat and Parkinson's disease. Environ Health Perspect. doi:10.1289/ehp. 1002839
11. Liaudet L, Soriano FG, Szabo C (2000) Biology of nitric oxide signaling. Crit Care Med 28:N37–N52
12. Hess DT et al (2005) Protein S-nitrosylation: purview and parameters. Nat Rev Mol Cell Biol 6:150–166
13. Gu Z, Nakamura T, Lipton SA (2010) Redox reactions induced by nitrosative stress mediate protein misfolding and mitochondrial dysfunction in neurodegenerative diseases. Mol Neurobiol 41:55–72
14. Schrammel A et al (2004) S-nitrosation of glutathione by nitric oxide, peroxynitrite, and NO*/O2*. Free Radic Biol Med 34:1078–1088
15. Kolb H, Kolb-Bachofen V, Roep BO (1995) Autoimmune versus inflammatory type I diabetes: a controversy. Immunol Today 16:170–172
16. Leea JH et al (1993) Nitric oxide synthase is found in some spinothalamic neurons and in neuronal processes that oppose spinal neurons that express Fos induced by noxious stimulation. Brain Res 608:324–333
17. Haley JE et al (1990) Evidence for spinal N-methyl-D-aspartate receptor involvement in prolonged chemical nociception in the rat. Brain Res 518:218–226
18. Lafon-Ca M (1999) Nitric oxide, superoxide and peroxynitrite: putative mediation of NMDA-induced cell death in cerebellar cells. Neuropharmacology 32:1259–1266
19. Reynolds IJ, Hastings TG (1995) Glutamate produces production of reactive oxygen species in cultured forebrain neurons following NMDA receptor activation. J Neurosci 15:3318–3327
20. Lafon-Cazal M et al (1999) Nitric oxide, superoxide and peroxynitrite: putative mediation of NMDA-induced cell death in cerebellar cells. Neuropharmacology 32:1259–1266
21. Beckman JS (1991) The double edged role of nitric oxide in brain function and superoxide-mediated injury. J Dev Physiol 15:53–59
22. Lafon-Cazal M et al (1993) NMDA-dependent superoxide production and neurotoxicity. Nature 364:535–537
23. Coyle JT, Puttfarken P (1993) Oxidative stress, glutamate and neuro generative disorders. Science 262:689–699
24. Beckman JS, Crow JP (1993) Pathologic implications of nitric oxide, superoxide and peroxynitrite formation. Biochem Soc Trans 21:330–333

25. Pryor WA, Squadrito GJ (1995) The chemistry of peroxynitrite: a product of the reaction of nitric oxide and superoxide. Am J Physiol 268:L699–L722
26. Novelli A et al (1988) Glutamate becomes neurotoxic via the NMDA receptor when intercellular energy levels become reduced. Brain Res 451:205–212
27. Schultz JB et al (1997) The role of mitochondrial dysfunction and neuronal nitric oxide in animal models of neurodegenerative diseases. Mol Cell Biochem 174:171–184
28. MaMahon SB et al (1993) Central excitability triggered by noxious inputs. Curr Opin Neurobiol 3:602–610
29. Garbe TR, Yukama H (2001) Common solvent toxicity: auto oxidation of respiratory redox-cyclers enforced by membrane derangement. Z Naturforsch 56:483–491
30. Mattia CJ et al (1993) Toluene-induced oxidative stress in several brain regions and other organs. Mol Chem Neurophysiol 18:313–328
31. Ziem GE (1997) Profile of patients with chemical injury and sensitivity. Environ Health Perspect 105:417–436
32. Lenga RE (1988) In: S-A. Corporation (ed) Library of chemical safety data. Sigma-Aldrich Corporation, Milwaukee
33. Ziem G, Mc Tamney J (1997) Profile of patients with chemical injury and sensitivity. Environ Health Perspect 105:417–436
34. Kehrl HR (1997) Laboratory testing of the patient with multiple chemical sensitivity. Environ Health Perspect 105:443–444
35. Weitzel JM, Iwen KA, Seitz HJ (2003) Regulation of mitochondrial biogenesis by thyroid hormone. Exp Physiol 88:121–128
36. Stirone C et al (2005) Estrogen increases mitochondrial efficiency and reduces oxidative stress in cerebral blood vessels. Mol Pharmacol 68:959–965
37. Orci L et al (2004) Rapid transformation of white adipocytes into fat-oxidizing machines. Proc Natl Acad Sci USA 101:2058–2063
38. Sader S, Nian M, Liu P (2003) Leptin: a novel link between obesity, diabetes, cardiovascular risk, and ventricular hypertrophy. Circ Res 108:644–646
39. Frühbeck G (1999) Pivotal role of nitric oxide in the control of blood pressure after leptin administration. Diabetes 48:903–908
40. Korda M et al (2008) Leptin-induced endothelial dysfunction in obesity. Am J Physiol Heart Circ Physiol 295:1514–1521
41. Bouloumie A et al (1999) Leptin induces oxidative stress in human endothelial cells. FASEB J 13:1231–1238
42. Takamura T et al (2007) Gene expression profiles in peripheral blood mononuclear cells reflect the pathophysiology of type 2 diabetes. Biochem Biophys Res Commun 361:379–384
43. Otero M et al (2005) Leptin, from fat to inflammation: old questions and new insights. FEBS Lett 579:295–301
44. Matarese G (2000) Leptin and the immune system: how nutritional status influences the immune response. Eur Cytokine Netw 11:7–14
45. Grunfeld C et al (1996) Endotoxin and cytokines induce expression of leptin, the ob gene product, in hamsters. J Clin Invest 97:2152–2157
46. Thomopoulos C et al (2009) Free leptin is associated with masked hypertension in nonobese subjects: a cross-sectional study. Hypertension 53:965–972
47. Terrasi M et al (2009) The leptin promoter polymorphism Lep-2548G/A can be associated with increased leptin secretion by adipocytes and elevated cancer risk. Int J Cancer 125:1038–1044
48. Sitaraman S (2004) Colonic leptin: source of a novel proinflammatory cytokine involved in IBD. FASEB J 18:696–698
49. Maeda T et al (2009) Leptin derived from adipocytes in injured peripheral nerves facilitates development of neuropathic pain via macrophage stimulation. Proc Natl Acad Sci USA 106:13076–13081
50. Kaufman S (1958) A new cofactor required for the enzymatic conversion of phenylalanine to tyrosine. J Biol Chem 230:931–993

51. Silberman GA et al (2010) Uncoupled cardiac nitric oxide synthase mediates diastolic dysfunction. Circulation Res 121:519–528
52. Fox RA et al (2007) The impact of a multidisciplinary, holistic approach to management of patients diagnosed with multiple chemical sensitivity on health care utilization costs: an observational study. J Altern Complement Med 13:223–229
53. Nichol CA, Smith GK, Duch DS (2003) Biosynthesis and metabolism of tetrahydrobiopterin and molybdopterin. Exp Biol Med 228:1291–1302
54. Werner ER et al (2003) Tetrahydrobiopterin and nitric oxide: mechanistic and pharmacological aspects. Exp Biol Med 228:1291–1302
55. Tiefenbacher CP (2001) Tetrahydrobiopterin: a critical cofactor for eNOS and a strategy in the treatment of endothelial dysfunction? Am J Physiol Heart Circ Physiol 280:H2484–H2488
56. Sun H, Patel KP, Mayhan WG (2001) Tetrahydrobiopterin, a cofactor for NOS, improves endothelial dysfunction during chronic alcohol consumption. Am J Physiol Heart Circ Physiol 281:H1863–H1869
57. Touyz RH, Briones AM (2011) Reactive oxygen species and vascular biology: implications in human hypertension. Hypertens Res 34:5–14
58. Kinoshita H et al (1997) Inhibition of tetrahydrobiopterin biosynthesis impairs endothelium dependent relaxations in canine basilar artery. Am J Physiol Heart Circ Physiol 273:H718–H724
59. Shinozaki K et al (2000) Oral administration of tetrahydrobiopterin prevents endothelial dysfunction and vascular oxidative stress in the aortas of insulin-resistant rats. Circ Res 87:566–573
60. Nyström T, Nygren A, Sjöholm A (2004) Tetrahydrobiopterin increases insulin sensitivity in patients with type 2 diabetes and coronary heart disease. Am J Physiol Endocrinol Metab 287:E919–E925
61. Hyndman ME et al (1991) Interaction of 5-methyltetrahydrofolate and tetrahydrobiopterin on endothelial function. Am J Physiol Heart Circ Physiol 282:H2167–H2172
62. Katusic ZS (2001) Vascular endothelial dysfunction: does tetrahydrobiopterin play a role? Am J Physiol Heart Circ Physiol 281:H981–H986
63. See reference [63]
64. Dejean LM et al (2006) Regulation of the mitochondrial apoptosis-induced channel, MAC, by BCL-2 family proteins. Biochim Biophys Acta 1762:191–201

Chapter 7
Nitrosative Stress in Diverse Multisystem Diseases

As a result of such decoupling, NO-synthases (NOSs) now become "radical cannons" producing reactive oxygen (ROS) and nitrosative oxygen substrates (NOS) with all the sequelae for the whole organism (Fig. 7.1). Only recently multiple sclerosis is being discussed as not being caused by a loss of the myelin sheaths of nerve cells with axon damage however, by macrophage-derived excess release of reactive oxygen and nitrogen species (ROS and RNS) that trigger mitochondrial pathology and initiate focal axonal degeneration [1]. Indeed, neutralization of ROS and RNS rescued axons that have already entered the degenerative process.

Another major disease entity where nitric oxide and/or peroxynitrite play a pivotal role is the seemingly increase in the number of patients with diabetes type 2 with insulin resistance where glucose can be regarded as an inflammatory mediator [2] and asymmetric dimethylarginine (ADMA) and other endogenous nitric oxide synthase (NOS) inhibitors acting as an important cause of vascular insulin resistance [3]. Thus, overflow of glucose contributes to nitrosative stress in CNS and in obesity, the brain becomes insulin resistant and can have too much glucose, which is associated with accelerated brain aging involving NO-induced oxidative damage to neuronal mitochondria [4]. In addition, BH_4 deficiency is correlated with insufficient production of nitric oxide and an increase of superoxide production in hypertensive patients with diabetes mellitus [5].

In **diabetics**, high blood sugar effects mitochondrial function and induces cytokine release from tiny collar perivascular fat which by itself downregulates BH_4 synthesis [6]. In addition, high blood glucose levels lead to NO-induced neuronal mitochondrial damage switching them off (Fig. 7.2) accompanied with accelerated brain aging [7]. It also has to be considered that hyperglycemia activates the inflammatory system via advanced glycation end-products (AGE), which are known to increase NF-κ B activity [8].

Phenylketonuria, Hyperphenylalaninemia [9]

Cardiovascular diseases with endothelial dysfunction [10, 11] and **hypertonus** [12]. It should be noted, that under the influence of different risk factors of atherosclerosis

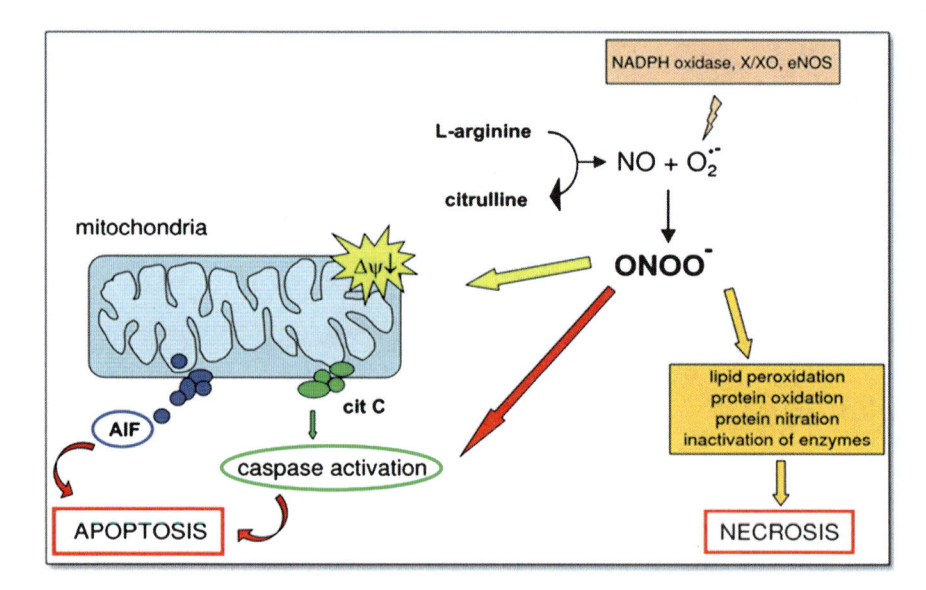

Fig. 7.1 Formation of toxic peroxynitrite (ONOO⁻) attacking mitochondria and lipid membranes of the body with protein nitration (resulting in an antigen), as well as an activation of enzymes necessary for ATP synthesis

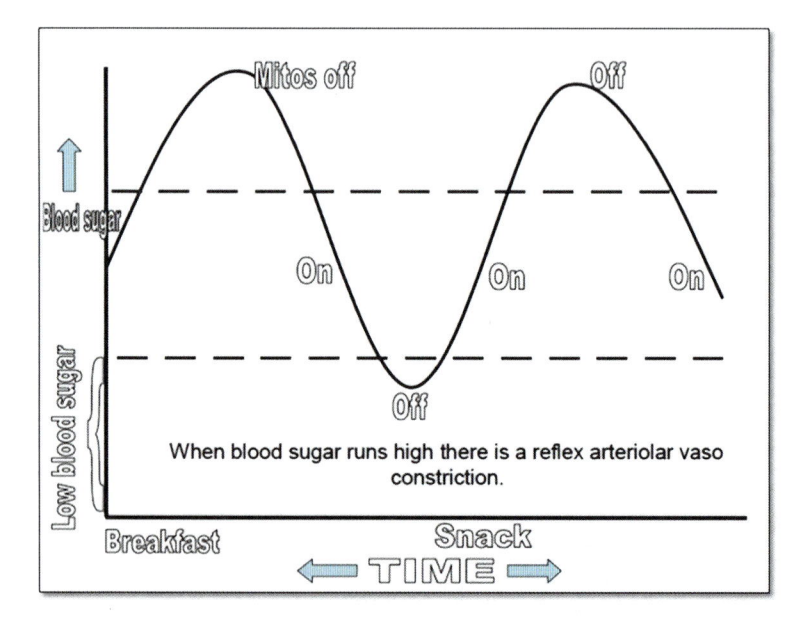

Fig. 7.2 The deleterious effects of either a high and a low blood sugar level affecting arterial vascular reflex and mitochondrial function, which is being shut off in both instances

as well as ischemia-reperfusion injury and inflammation, the bioactivity of BH_4 is reduced. In the presence of suboptimal levels of BH_4, NOS generates both NO and superoxide anions (O_2^+). This leads to the formation of hydrogen peroxide (H_2O_2) from O_2^+ and peroxynitrite ($OONO^-$) from O_2^+ and NO [13]. Under these conditions, substitution of BH_4 would restore the original activity of NOS and lead to increased production of NO. The significance of BH_4 in patients with cardiovascular artery disease was demonstrated by intracoronary infusion of BH_4, which restored coronary endothelial function by improving the bioavailability of endothelium derived nitric oxide through bioavailability of eNOS in hypercholesteremic patients [14].

> Depletion of the cellular NAD^+ leads to inhibition of cellular ATP-generating pathways leading to cellular dysfunction. The PARP-triggered depletion of cellular NADPH directly impairs the endothelium-dependent relaxations of the blood vessels. The effects of elevated glucose are also exacerbated by increased aldose reductase activity leading to depletion of NADPH and generation of reactive oxidants.

And since BH_4 depletion is observed in **vascular dysfunction** associated with artery narrowing in diabetics, substitution of this coenzyme resulted in improved endothelial function [14–18].

The obvious link of BH_4 to hypertension is readily explained by the loss of tetrahydrobiopterin (BH_4) resulting in an uncoupling of endothelial NO synthase (eNOS) with diminished NO production, increased production of ROSs & hypertension [19]. In such situation only tetrahydrobiopterin, and not L-arginine, decreased NO synthase incoupling in cells expressing high levels of endothelial NO synthase [20, 21]. Aside, BH_4 seems to improve immune therapy in cancer, which may be a rational target for antiangiogenesis in tumor [22, 23]. And lastly, BH_4 also seems to play a significant role in chronic neuropathic pain. Novel approaches for treating pain are required to address a widely recognized, yet largely underserved and unmet, clinical need. The recently discovered link between tetrahydrobiopterin (BH_4) synthesis and pain in preclinical models and humans provides a promising new approach for treating neuropathic and other forms of chronic pain. The rate-limiting enzyme in BH_4 synthesis, guanosine triphosphate cyclohydrolase 1 (GCH1), and sepiapterin reductase (SPR) are both promising drug targets based on initial active-site characterization of the SARs of these two enzymes. Reducing the elevated BH_4 levels associated with pain to baseline, while maintaining sufficient BH_4 levels to limit side effects is the goal of discovery programs for novel therapeutics targeting GCH1 or SPR [24]. Although all preclinical data clearly established a link between BH_4 and eNOS, clinical relevant therapeutic so far have not evolved. This seems irrational since all data clearly point for a substitution of tetrahydrobiopterin (BH_4) in order to recouple NOS and

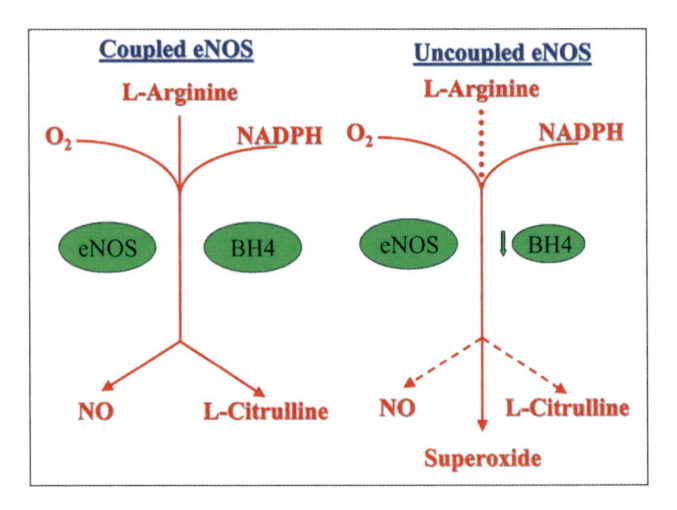

Fig. 7.3 Coupled and uncoupled eNOS. Coupled eNOS utilizes O_2, L-Arginine and NADPH to produce NO and L-Citrulline eNOS can be uncoupled by BH_4 deficiency to produce superoxide rather than NO, which may further reduce available NO by being used to form peroxynitrite

reverse the underlying pathology (Fig. 7.3). For instance, it has been demonstrated in preestablished advanced hypertrophy, fibrosis, and dysfunction of the heart that BH_4 was able to reverse such pathology [25].

7.1 Factors Leading to a Reduction of BH_4

Aside from substitution with BH_4, it is important to reduce the underlying cause of uncoupling of eNOS caused by BH_4 deficiency in preclinical studies conclusively have demonstrated that hypertension produces a cascade involving production of ROSs from the NADPH oxidase leading to oxidation of tetrahydrobiopterin and uncoupling of endothelial NO synthase (eNOS) and where substitution with BH_4 or its protection from oxidation results in a relief from hypertension [19]. The following pathological states should be considered for a potential cause of BH_4 deficit:

1. **A pro-inflammatory metabolic state** resulting in cytokine-induced upregulation of GTP-cyclohyrolase, however not of PTPS (protein tyrosine phosphatases), a group of enzymes that remove phosphate groups from phosphorylated tyrosine residues on proteins, with relative BH_4-deficiency (Fig. 7.4).
2. **An increase in oxidative stress,** where the proinflammatory metabolic state coincides with an increased formation of ROS and peroxynitrite resulting in oxidation of BH_4 and a reduction in biological activity.

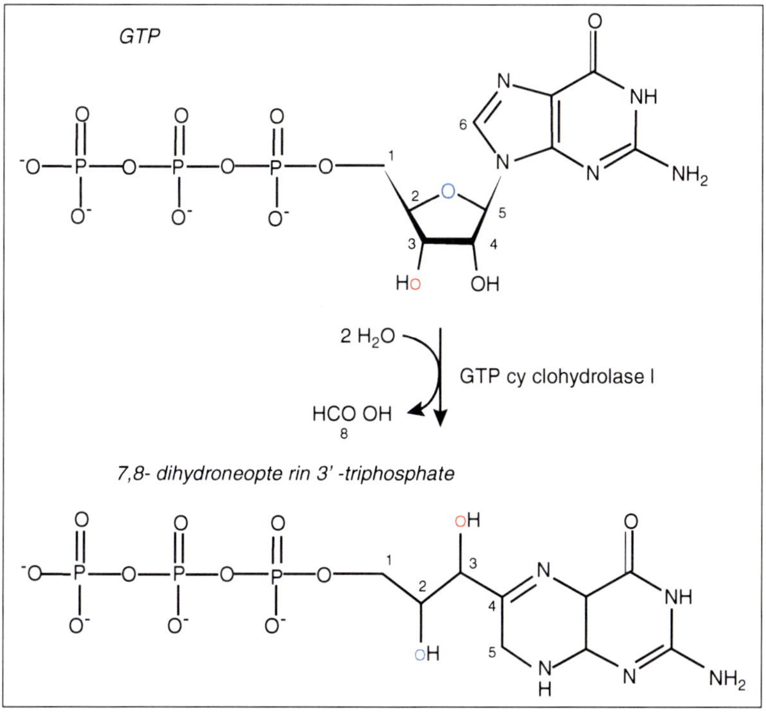

7.1 Factors Leading to a Reduction of BH$_4$ 75

Fig. 7.4 The chemical reaction performed by GTP-cyclohydrolase. Structural drawing of the GTP cyclohydrolase reaction to transform GTP into 7,8-dihydroneopterin 3'-triphosphate, the precursor of tetrahydrobiopterin (BH$_4$)

3. **Genetic mutations of enzyme involved in BH$_4$ synthesis and/or regeneration,** when the enzymes GTP-cyclohydrolase and PTPS involved in synthesis and the enzymes PCD und DHPR involved in regeneration result in a deficient BH$_4$ formation.
4. **Hyperglycemia,** which contributes to nitrosative stress in CNS, activating the inflammatory system.

Such effect is underlined in obesity, where the brain becomes insulin resistant since too much glucose is associated with accelerated brain aging and may involve NO-induced oxidative damage to neuronal mitochondria [26].

Since peroxynitrite in turn will oxidize more BH$_4$, this will result in an increase in partial uncoupling of NOS. Such reciprocal relationship between peroxynitrite and BH$_4$-depletion is a potential vicious cycle within the larger NO/ONOO⁻ cycle and constitutes the essential core of the derangement leading into mitochondropathy [27]. Lowering of this central unit will be expected to produce a clinical improvement in any NO-related disease, but will produce an increase in nitric oxide. While the net effect of any turned on excess formation of nitric oxide is a negative one, agents that

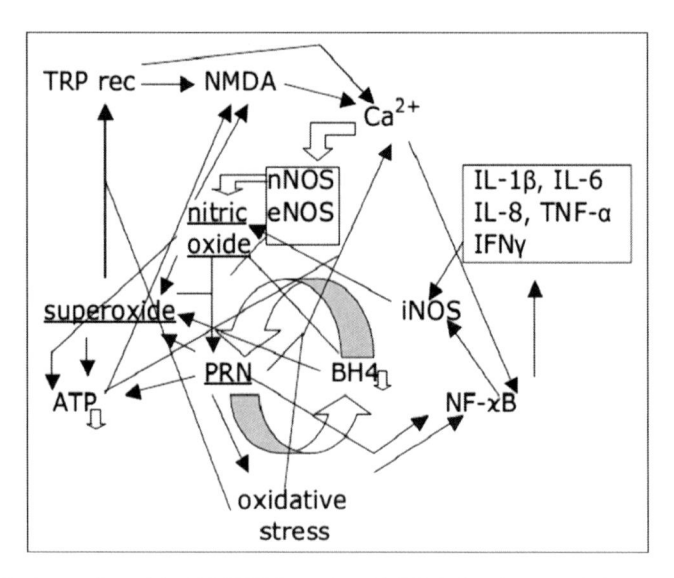

Fig. 7.5 The interaction of the NO/ONOO⁻ cycle with BH₄ being the central players and responsible for the generation of different symptoms and signs within multisystem diseases (Adapted from (Ref. [10] of Chap. 5))

increase nitric oxide by lowering this central uncoupling of NOS with toxic superoxide production is helpful, since formation of the higher aggressive superoxide and especially formation of peroxynitrite is downregulated. An extended overview of this peroxynitrite (ONOO⁻) tetrahydrobiopterin (BH₄) interaction as being the central core is outlined in the following figure. Once peroxynitrite (PRN) induces a depletion of BH₄, this further results in a malfunction of ATP synthesis in mitochondria. In addition, in the upper left hand corner, the TRP channels are a large family of cation channels. They influence a multitude of physiologic processes in the body from pain to thermosensation. There are six subfamilies, one of which is the Transient Receptor Potential Vanilloid (TRPV 1–6). When activated, these receptors conduct mostly calcium, with some sodium and potassium conductance. The TRPV1 receptor, found in the central nervous system (CNS), in sensory ganglia (dorsal root ganglia), and peripheral C and A delta sensory fibers, is of particular interest in transformation of pain. Under physiological conditions, this receptor is activated by temperatures > 43 °C (noxious), low pH ≤ 5.2, low voltage, and low lipid levels. Of note, certain venomous insects and arachnids also activate this receptor, although the exact mechanism for this is unknown. One agent that specifically interacts with the TRPV-receptor is capsaicin, the chemical that gives jalapeño peppers their "heat," has been found to be an exogenous agonist. The benefits of capsaicin include the fact that it is lipophilic, readily crosses cell membranes, and is relatively easily synthesized being an agent that is used in neuropathic pain. But aside, this receptor family is linked to mitochondrial disease because two other members of the TRP family of receptors are actively involved in symptoms of mitochondrial malfunctions, and not just the vanilloid (TRPV1) receptor with eNOS and nNOS activation (Fig. 7.5).

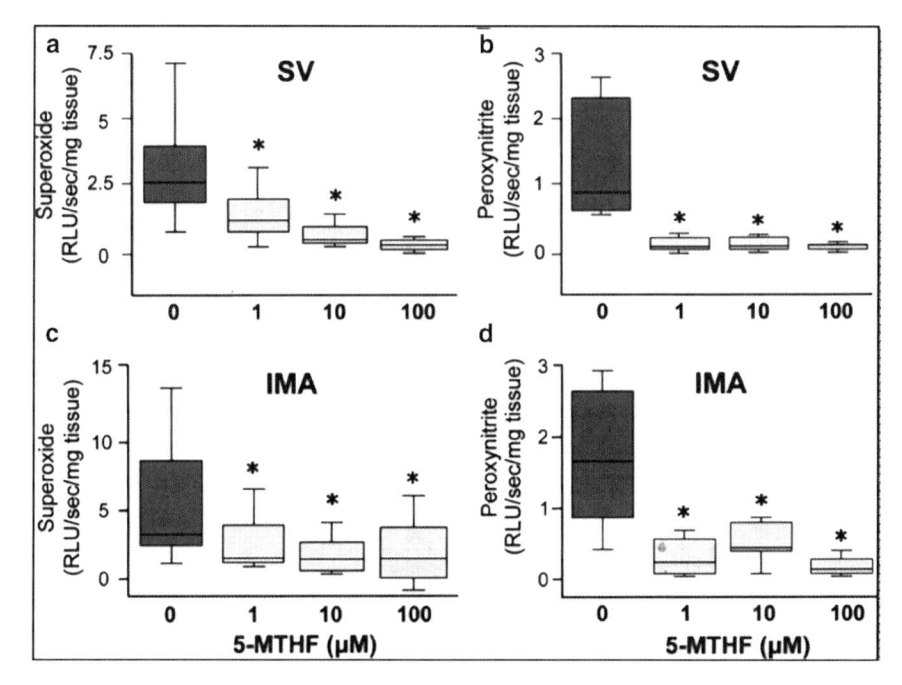

Fig. 7.6 5-MTHF reduces vascular superoxide production by prevention of ONOO⁻-mediated BH_4 oxidation improving eNOS coupling. Note, the reduction of peroxynitrite in saphenous vein (SV), and internal mammary artery (IMA) following 5-MTHF administration (Adapted from [28])

7.1.1 BH_4 Level Restoration Results in Cardiovascular Improvement

There are several ways how a deficient tetrahydrobiopterin level can be augmented resulting in recovery of the present ailment:

- Substitution of tetrahydrobiopterin (generic name sapropterin dihydrochloride or Kuvan®, BioMarin Pharmaceutical, Merk Serino) in a dose of 10–20 mg/kg per day. Such high dosages are necessary since the agent BH_4 demonstrates low bioavailability.
- Supplement with precursors necessary for BH_4 synthesis such as tyrosine, or 5-hydroxytryptophane.
- Substitute cofactors necessary for BH_4-synthesis such as magnesia, folic acid or 5-MTHF (=5-methyl-tetrahydro-folate), the circulating form of folic acid, which also acts like a potent peroxynitrite scavenger (Fig. 7.6).
- Use 5-MTHF = 5-Methyltetrahydrofolate as a scavenger of excessive peroxynitrite formation as demonstrated in saphenous veins (SV) and internal mammary artery (IMA; Fig. 7.7).

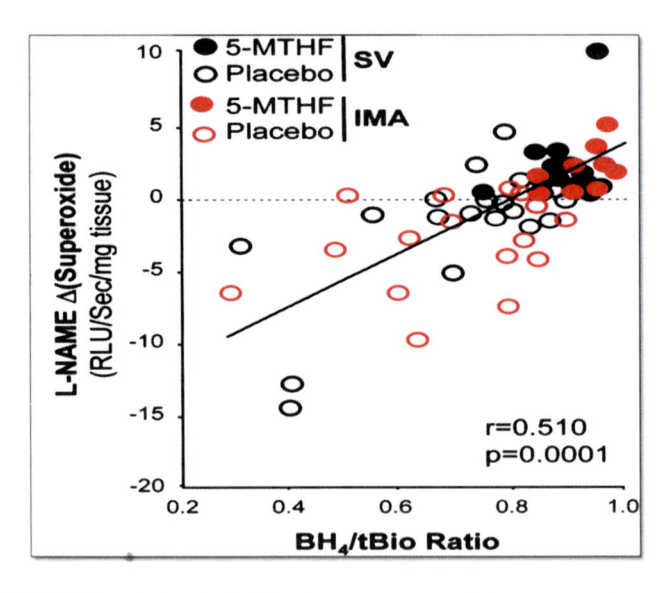

Fig. 7.7 5-MTHF (5-methyltetrahydrofolate), a strong peroxynitrite scavenger increases vascular BH_4 and the BH_4/total biopterin ratio (Adapted from [28])

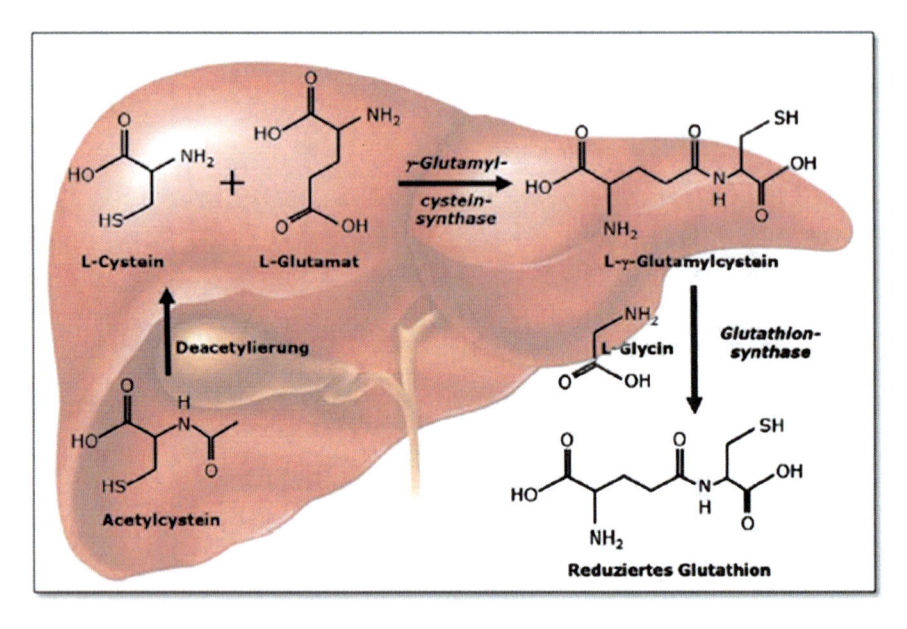

Fig. 7.8 Increased synthesis of the oxygen radical-scavenger glutathion by means of high dose N-acetyl-cysteine (NAC 2 g/day)

7.1 Factors Leading to a Reduction of BH$_4$

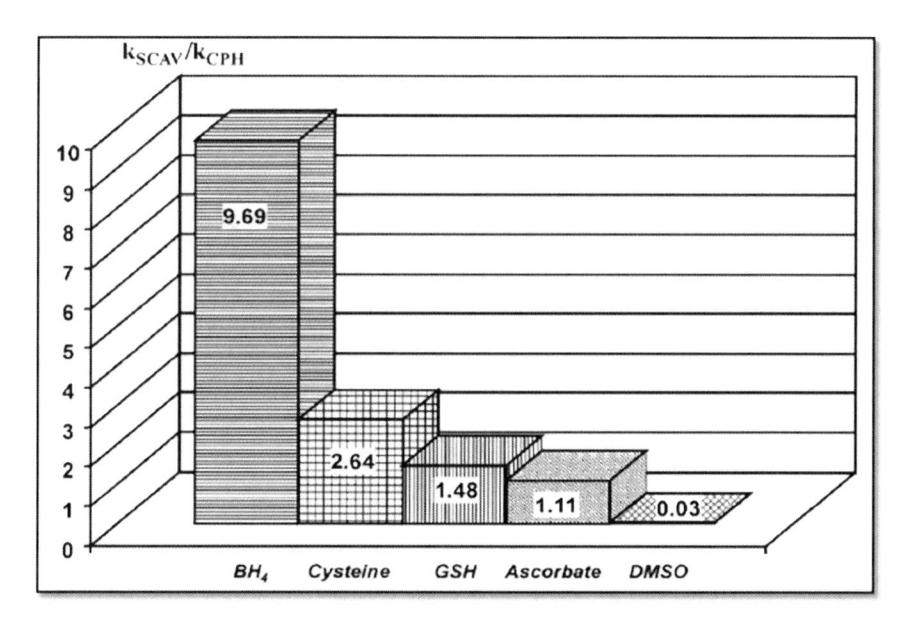

Fig. 7.9 Reactivity of different peroxynitrite scavengers on electron spin resonance (ESR) spectroscopy demonstrating highest efficacy for BH$_4$ when compared with N-acetyl-cysteine, glutathione (GSH), vitamin C (ascorbate) or DMSO (Adapted from [29])

- Avoid the digestion of toxic chemicals, which inhibit BH$_4$ metabolism and lead into mitochondropathy.
- Increase antioxidative protective capacity by giving neutraceuticals such as vitamin E, C, N-acetyl-cysteine (Fig. 7.8), add minerals such as Zn, Cu, Mn, and give polyphenoles to reduce oxidative stress.
- Treat the proinflammatory process, since any additional viral or bacterial infection acts like a trigger releasing proinflammatory cytokines, which by themselves activate nitrite oxygen synthase (iNOS) with excessive NO formation, resulting in further accumulation and activation of the nitrite/peroxynitrite (NO/ONOO⁻) cycle.
- Give scavengers for peroxynitrite since endothelial BH$_4$ is a crucial target for oxidation by ONOO⁻, especially since it had been demonstrated, that BH$_4$ is a necessary part and the reaction rate constant for vasodilatation exceeds those of other thiols or even ascorbate (Fig. 7.9).

7.1.2 Local Malfunction of Mitochondria Related to Multisystem Diseases

According to present understanding (although not generally accepted), the following ailments most likely are connected to nitration, cytokine release and activation of the NO/ONOO⁻ cycle (Fig. 7.5) followed by malfunction of mitochondria and/or a defect in the electron transport chain with a deficit in energy production:

1. **Attention deficit hyperactivity disorder (ADHD),** where cutting down on sugar and food dyes as well as refined products can help to calm down the kids [30].
2. **AIDS related therapy** with HAART (Highly Active Anti-Retroviral Therapy), where in contrast to conventional theory, fundamental mitochondrial defects are considered as the major cause of immune deficiency and all so called HIV-related diseases [31, 32].
3. **Mitochondrial toxicity induced by HIV nucleoside reverse transcriptase inhibitors.** Since the introduction of highly active antiretroviral therapy (HAART), the life expectancy of HIV-infected patients has improved enormously. However, the chronic administration of antiretroviral medications has led to the recognition of long-term complications of these therapies [33–37]. Mitochondrial toxicity is now recognized as a major adverse effect of nucleoside analogue treatment and can lead to myopathy, peripheral neuropathy, and hepatic steatosis with lactic acidosis, which can be life-threatening.
4. **Arteriosclerosis,** which is due to a NO/ONOO⁻ induced inflammatory reactions within the endothelium [38–40].
5. **Autism,** caused by exposure to toxic metals, xenobiotics, drug side effects/ nutrient interactions, food toxicants (allergens, stimulants, etc.), excess nutrient imbalance, insufficiency and energy imbalance with hyperglycemia, hypoglycemia as a functional approach to such brain disorder resulting in abnormal methylation, neuroinflammation, and mitochondropathy [41].
6. **Gastroesophageal reflux disease (GERD),** a condition in which the stomach contents leak backwards from the stomach into the esophagus. This action can irritate the esophagus, causing heartburn and other symptoms. GERD is usually caused by changes in the barrier between the stomach and the esophagus, including abnormal relaxation of the lower oesophageal sphincter, which normally holds the top of the stomach closed. Since excess nitric oxide levels result in relaxation of the smooth circular muscle there is leakage of gastric content into the esophagus.
7. **Circulatory shock** where immunohistochemical and biochemical evidences demonstrate the production of peroxynitrite in various experimental models of endotoxic and hemorrhagic shock both in rodents and in large animals. In addition, biological markers of peroxynitrite have been identified in human tissues after circulatory shock. Peroxynitrite can initiate toxic oxidative reactions in vitro and in vivo. Because of lipid peroxidation, there is a direct inhibition of mitochondrial respiratory chain enzymes [42].
8. **Reduced immune response** with evidence that innate immune receptors are dysregulated in patients with chronic granulomatous disease (CGD) resulting in NADPH oxidase inhibition on receptor expression [43].
9. **Depression,** where the asymmetric dimethylarginine ADMA, a metabolic by-product of continual protein modification processes in the cytoplasma of all

human cells and closely related to L-arginine, interferes with L-arginine in the production of nitric oxide. It is considered a major cause of sufficient neurotransmitter (e.g. serotonin) action, which are methylated by asymmetric dimethylarginine [44].

10. **Diabetes mellitus** with quantitative decrease in mtDNA copy numbers linked to the pathogenesis of diabetes and followed by influence of the mitochondria on nuclear-encoded glucose transporters and influence of nuclear encoded uncoupling proteins on mitochondria [45–47].
11. **Friedreich ataxia**, where peroxynitrite and NMDA activation induces neurotoxicity [48–50].
12. **Epilepsia** as a result of an increased release of excitatory transmitters at the NMDA receptor [51, 52].
13. **Alzheimer** with overexpression of the NMDA-, the TRP- (transient receptor potential ion channel) or vanilloid-receptor followed by an increased Ca^{++} influx resulting in neurodegeneration [53–57].
14. **Hereditary fructose-, gluten-, lactose intolerance** with a NO/ONOO-related defect in gluconeogenesis [58].
15. **Glaucoma,** caused by oxidative/nitrosative stress [59–61].
16. **Heart insufficiency with** energetic abnormalities in cardiac muscle fibers leading into heart failure, which correlates closely with clinical symptoms and mortality. Most likely the cellular mechanism leading to energetic failure is related to mitochondrial dysfunction and a defect in oxidative phosphorylation within the electron transport chain (ETC) [62–64]. Such assumption was corroborated in a pilot study where the coenzyme Q10 was given to patients with latent cardiac insufficiency (n=12) in daily doses of 10 mg/kg. With increasing plasma concentrations of Q10 there was an inverse correlation in brain natriuretic peptide (BNP; a sensitive marker for myocardial impairment) while mitochondrial membrane potential, a sensitive marker of an energy deficit the driving force of contractility, demonstrated an increase (Fig. 7.10).
17. **Amyotrophic lateral sclerosis** (ALS) patients demonstrated immunoreactivity for nitrotyrosine densely detected in motor neurons of ALS which was not or only minimally detected in controls [65–68].
18. **Multiple chemical sensitivity** (MCS), as demonstrated in numerous studies, very likely is due to an ATP deficit in mitochondria induced by xenobiotics (Ref. [30] of Chap. 6) [69–74].
19. **Cancer,** where it is postulated that due to a blockade within the electronic transport chain (block of aconitase in Krebs-cycle) mitochondria (with mtDNA alterations) switch to anaerobic glycolysis resulting in an archaic anaerobic fermentation for ATP synthesis resulting in uncontrolled mitosis with lack of apoptosis [75–77]. Thus primary or secondary OXPHOS failure shifts cell metabolism towards ATP generation by glycolysis (Warburg effect) with mitochondrial dysfunction in cancer morphogenesis and faulty signaling. Also, inflammatory mediators that regulate OXPHOS are linked to cancer-induced morphogenesis [78].

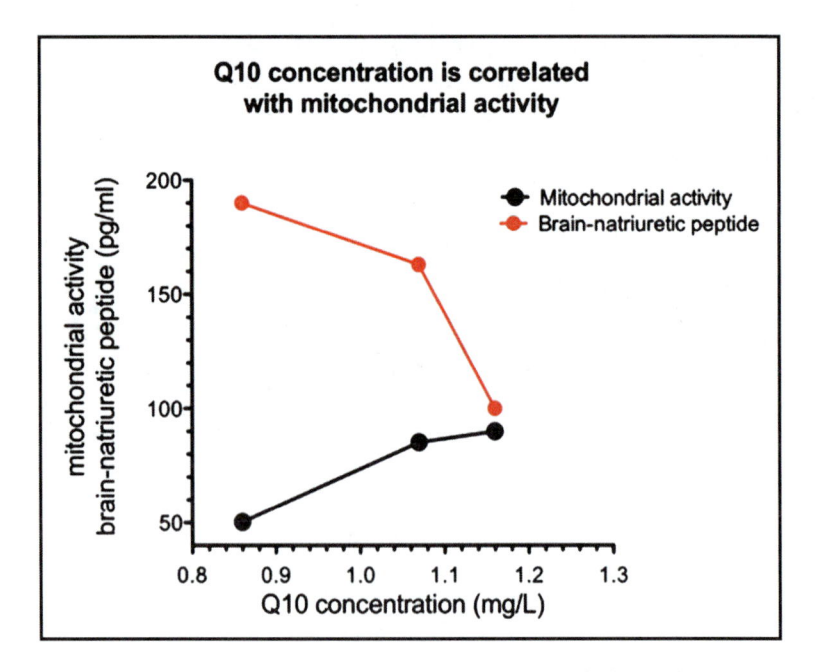

82 7 Nitrosative Stress in Diverse Multisystem Diseases

Fig. 7.10 Close correlation of increasing concentrations of plasma Q10 and mitochondrial membrane energy (in%), where cardiac impairment is reflected in brain natriuretic peptide (BNP), a sensitive marker released from functionally impaired myocardial cells. With higher Q10 levels there is an increase in mitochondrial activity, which is mirrored in a reduction of BNP, suggesting increase in contractility (own observations)

20. **Ventricular Arrhythmia** with **Hypertrophic Cardiomyopathy**. In hypertensive cardiac hypertrophy, the heart energy metabolism is reverted from the normal adult type that obtains the majority of its requirement for adenosine triphosphate (ATP) from metabolism of fatty acids and oxidative phosphorylation (OXPHOS), to mitochondrial-derived reactive oxygen species (ROS) that contribute to cardiomyopathy, the fetal form, which metabolizes glucose to lactate [79]. Also, mitochondrial dysfunction enhances the generation of radical oxygen species (ROS), which damage mtDNA, nDNA, proteins, and lipid membranes, since mice lacking the mitochondrial antioxidant enzyme manganese-superoxide dismutase (SOD) developed dilated cardiomyopathy and where palliative mitochondrial therapy with L-acetyl-carnitine and coenzyme Q10 improved cardiac function in patients with cardiomyopathy [80]. Also, it was demonstrated that cardiomyocyte function impairment is caused by ROS formation from mitochondria [81, 82].
21. **Acute pain** with a negative correlation, however in chronic, inflammatory pain with a direct correlation of NO production via induced nitric oxide synthase activation (iNOS; Fig. 7.11) [83], findings that are similar to opioid-induced hyperalgesia [84].
22. **Opioid-induced hyperalgesia (OIH)**, with a causal role for the ONOO⁻ pathways culminating in antinociceptive tolerance to opiates associated with the

7.1 Factors Leading to a Reduction of BH$_4$ 83

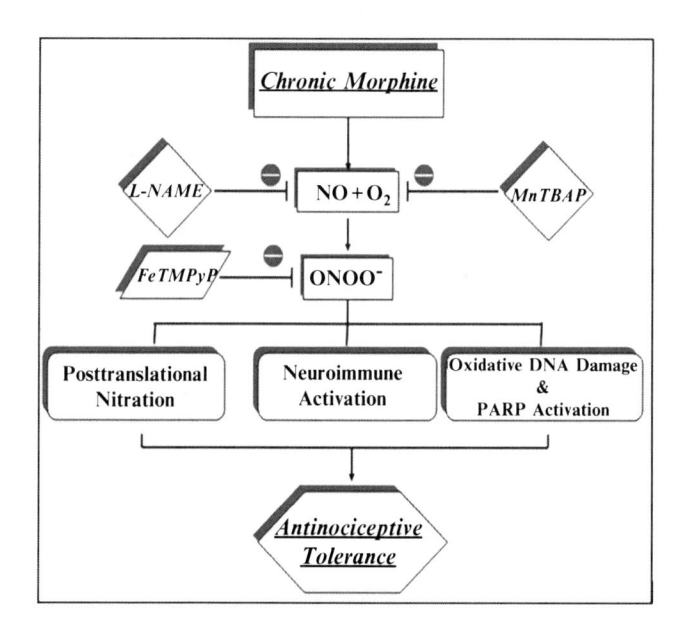

Fig. 7.11 Formation of ONOO⁻ in the spinal cord during repeated administration of morphine plays a critical role in the development of morphine-induced antinociceptive tolerance through at least three biochemical pathways: posttranslational nitration, neuroimmune activation, and release of proinflammatory cytokines, and oxidative DNA damage and PARP activation. Inhibition of its formation by removal of NO and O$_2^-$ or by catalytically decomposing it by ONOO⁻ decomposition catalysts such as FeTM-4-PyP^{5+} blocks these pathways, leading to inhibition of antinociceptive tolerance (Adapted from [171])

appearance of several tyrosine-nitrated proteins in the dorsal horn of the spinal cord (Fig. 7.12). Peroxynitrite (ONOO⁻) decomposition catalysts may have therapeutic potential as adjuncts to opiates in relieving suffering from chronic pain [85]. In contrast, L-arginine, the precursor of nitric oxide (NO), when co-administered with morphine accelerates opioid tolerance and when given prior to the opioid decreases morphine's potency [86].

23. **Macula degeneration** with pathology-related increase in ROS levels leads to detrimental effects that are involved in impairments in synaptic activity, plasticity and function of the retina [88, 89].
24. **Tinnitus,** initiated by at least 11 short-term stressors increase nitric oxide or other mechanisms such as N-methyl-D-aspartate activity, oxidative stress, nitric oxide, peroxynitrite level, vanilloid receptor activity, NF-κB activity, and intracellular calcium levels are all reported to be elevated in tinnitus [90].
25. **Metabolic Syndrome** with impaired functional capacity of skeletal muscle mitochondria in type 2 diabetes, with some impairment also present in obesity[91, 92].
26. **Multiple Sclerosis,** with various mitochondrial mechanisms being involved in the pathogenesis of the ailment leading to therapeutic considerations,

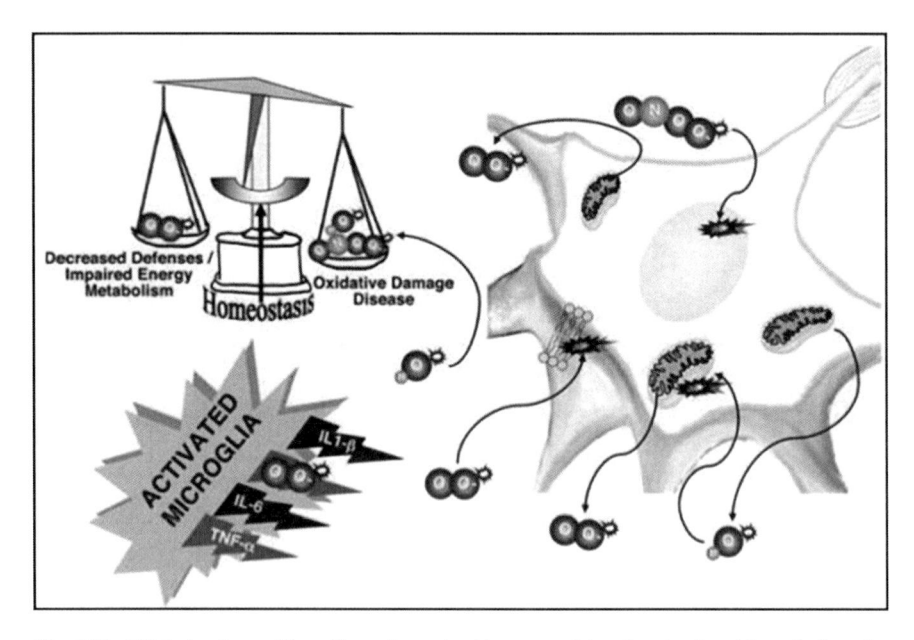

Fig. 7.12 NO being formed from lipopolysaccharides or cytokines in mitochondria and glia cells, either increases/decreases the effects of opioids resulting in the development of tolerance and/or withdrawal (Adapted from [87])

>   reemphasizing the importance of early neuroprotection with antioxidants in combination with approved means to combat inflammatory demyelination [93, 94].
> 27. **Neurodermatitis,** where the COX activity and NO free radical production at the site of inflammation might account for this activity [95].
> 28. **Polyarthralgia with muscle stiffness,** related to propanolol use, is the likely cause of a resultant mitochondrial disorder [96].
> 29. **Osteoarthrosis,** where the chondrocyte matrix synthesis and mineralization are modulated by the balance between ATP generation and consumption, a mechanism by which chondrocytes generate energy [97, 98]. Analysis of mitochondrial respiratory chain (MRC) activity in OA chondrocytes showed a significant decrease in complexes II and III compared to normal chondrocytes. On the other hand, mitochondrial mass is increased in OA, as demonstrated by a significant rise in chondroitin sulfate activity, and OA cells demonstrated a reduction in the mitochondrial membrane potential (delta ψ) difference (normal 200 mV) using the fluorescent probe JC-1 [99]. From this it was concluded that the pathogenesis of OA includes elaboration of increased amounts of NO as a consequence of up-regulation of chondrocyte-inducable NO synthase via IL-1, TNF-alpha and other factors induction. In addition, NO reduces chondrocyte survival and induces cell death with morphologic changes characteristic of chondrocyte apoptosis. NO also reduces the activity of complex IV and decreases the delta psi as measured as the ratio of red/green fluorescence. Furthermore, NO induces the mRNA expression

7.1 Factors Leading to a Reduction of BH$_4$

of caspase-3 and -7, and it reduces the expression of mRNA bcl-2 and the bcl-2 protein synthesis.

30. **Polyneuropathia**, where previous use of statins is contributed as a casual factor in insufficient neuronal mitochondrial function coinciding with a lack in Q10 formation [100].
31. **Asthma**, where aggressive formation of peroxynitrite results in dysfunction of pulmonary cellular layers with insufficient formation of the cells own antioxidative GSH-system [101, 102].
32. **Rheumatic disease,** with an excess formation of citrulline and nitrotyrosine directing towards an increased NO and ONOO$^-$ synthesis as demonstrated by specific antibody evidence [103, 104].
33. **Preeclampsia,** where the formation of asymmetric dimethyl arginine is an indicator for excessive oxygen radical formation [105].
34. **Hypoxia and solvent-related encephalopathy**, with upregulation of nitric oxide synthase [106] or a direct inhibition of mitochondrial ATP formation through intoxication through organic solvents [107–112].
35. **Hypercholesterinemia** being due to excess NO, which binds to the Fe-containing enzyme within the liver resulting in an inhibition of cholesterol 7-α-hydroxylase, a diet resistant excess formation of cholesterol [113].
36. **Septic shock, endotoxemia,** where the excess formation of NO/ONOO$^-$ results in the symptoms of shock [114–116].
37. **Parkinson,** with NO/ONNO$^-$ related inhibition of sufficient ATP- and NADH-formation, the spark for the activation of the mitochondrial electrical chain in basal ganglia [117–120].
38. **Chronic (neuropathic) pain.** The management of neuropathic pain continues to be a challenge for clinicians and despite taking prescribed medication for pain, patients with neuropathic pain continue to have pain of moderate severity and new strategies are being screened involving the glutamate excitatory transmitter system [121, 122], which has a role in the mitochondrial electron transport chain in neuropathic and other forms of inflammatory pain [123].
39. **Porphyria,** where results suggest a possible defect in mitochondrial NADH oxidation in acute intermittent porphyria [124]. Data from lymphocytes of variegate porphyria women have shown that they are more susceptible to producing mitochondrial ROS and suffering from oxidative damage when submitted to stressful situations [125].
40. **Fibromyalgia Syndrome** (FMS) with abnormal *mitochondrial* function, possibly because membranes of the *mitochondria* are disrupted [126], with elevated concentrations of ROS in patients and where higher levels tend to correspond with more pain, while muscles of *fibromyalgia* patients had a substantial drop in the number of *mitochondria* showing signs of mitophagy [127]. In addition, fatigue as demonstrated by others is closely related with a lack in energy production of mitochondria [128].
41. **Chronic Fatigue Syndrome** (CFS) with a remarkable correlation between the degree of mitochondrial dysfunction and the severity of the illness. The individual factors indicate which remedial actions, in the form of dietary supplements,

drugs and detoxification, are most likely to be of benefit, and what further tests should be carried out (Ref. [9] of Chap. 5) [129]. Also, persons with CFS were twofold as likely to have a metabolic syndrome (odds ratio = 2.12, confidence interval = 1.06, 4.23) compared with the controls. In addition, there was a significant graded relationship between the number of metabolic syndrome factors and CFS with each additional factor being associated with a 37% increase in the likelihood of having CFS [130]. This study supports the hypothesis that different ailments altogether have the same underlining pathology.

42. **Irritable bowel syndrome** (IBS) with mucosal barrier dysfunction, a promoting bacterial translocation has been observed. Finally, an altered mucosal immune system with dysfunction of enteric nerves has been associated with the disease [131].
43. **Hashimoto thyreoiditis,** an autoimmune disease being induced by nitrosation of the thyreoid hormone which now acts like an antigen [132] as corroborated by detection of antibodies [133].
44. **AIDS and AIDS-treatment neuropathies** with mitochondrial toxicity secondary to gamma-DNA polymerase inhibition and subsequent abnormal mitochondrial DNA synthesis [134, 135] involving mitochondrial impairment, which leads to pyruvate and NADH accumulation, enhancing the conversion of pyruvate into lactate, ultimately leading to lactic acidosis [136].
45. **Stroke and other CNS-related neurological disorders** (i.e. Parkinson, Alzheimer) with a growing body of evidence to implicate excessive or inappropriate generation of nitric oxide (NO). It is now well documented that NO and its toxic metabolite, peroxynitrite (ONOO$^-$), can inhibit components of the mitochondrial respiratory chain leading, if damage is severe enough, to a cellular energy deficiency state in neurons [137, 138].
46. **Bladder-interstitial cystitis** via Ca^{++} buffering from endoplasmic reticulum within microdomains between both organelles [139, 140].
47. **Interchangeable chronic fatigue syndrome and metabolic syndrome X (MSX)** in Chernobyl accident survivors, which clearly demonstrated that CFS and MSX are considered to be different stages of another neuropsychiatric and physical pathological development, and where CFS can transform towards MSX, suggesting mitochondrial genome disorders together with changes of transmembrane ionic transport are suggested to be the basis for CFS and MSX [141].

7.1.3 Factors Driving Nitrosative Stress

Due to the excessive formation of NO and peroxynitrite (ONOO$^-$) the radical nitrosative molecules bind to Fe- or FeS-containing enzymes, which are necessary in regular mitochondrial oxidative phoshorylation (cellular respiration). Especially mitochondrial enzymes such as aconitase (an enzyme containing the SH-group in

7.1 Factors Leading to a Reduction of BH$_4$

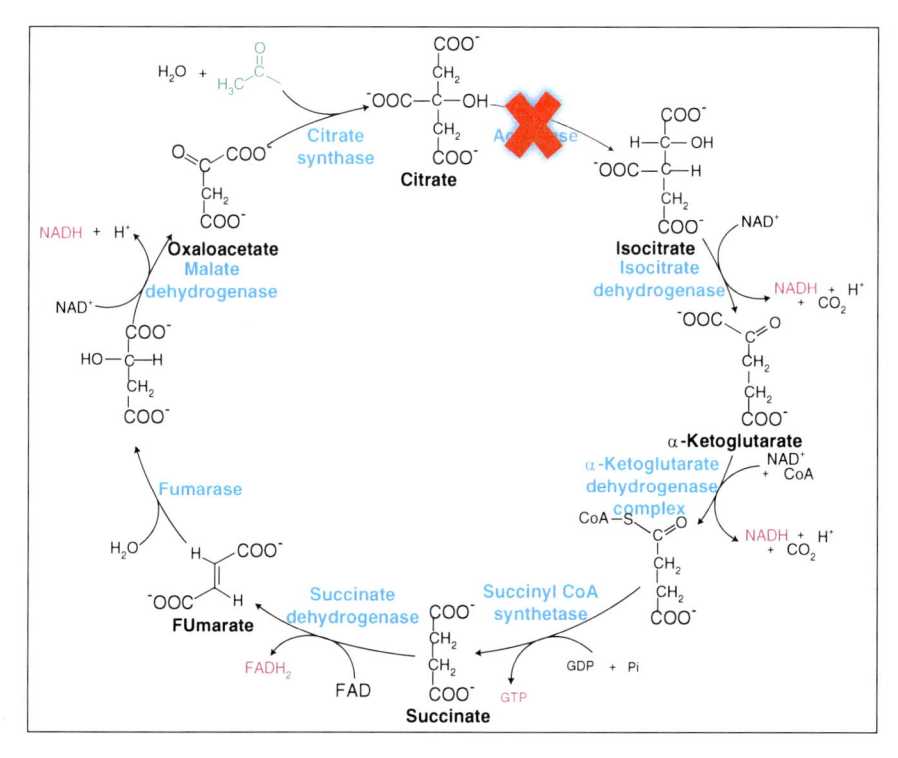

Fig. 7.13 Site of the SH-containing enzyme aconitase blocked within the Krebs-cycle due to excessive nitrosative stress

the Krebs Cycle; Fig. 7.13) is readily inactivated by peroxynitrite, but not by its precursor, nitric oxide and there is an insufficient formation of high energy substrate ATP, all of which results into a blockade of cytochrome c [142].

However, aside from aconitase blockade, peroxynitrite also inhibits function of other organs or organelles:

1. Cytochrome P-450 enzyme within the liver
2. Synthesis of hem- and myoglobin
3. Cytochromhydroxylase (Tph2) a rate-limiting enzyme in synthesis of the central neurotransmitter serotonin
4. Thyreoperoxidase
5. Inhibition of mitochondrial cytochrome c-oxidase a rate limiting step in the formation of ATP, resulting in
   a. Insufficient ATP-production
   b. Intracellular loss of energy
   c. Loss of stamina, memory impairment, and fatigue

6. Activation of cyclooxygenase (COX), an enzyme necessary for biosynthesis of prostaglandins such as thromboxane and prostacyclines involved in chronic inflammatory process of joints with immune activation, inducing:

   a. Increased formation of inflammatory factors i.e. TNF-α, IL-1, and *lipopolysaccharides* (LPS).
   b. Chronic (silent) inflammation of joints seen in osteoarthrosis, polyarthrosis, which already by themselves result in
   c. Increased release of NO/OONO⁻ resulting in a windup and a vicious cycle.

In addition, regular intake of certain medication has been demonstrated to be potentially toxic, as it can block the transformation of ATP within the electron transport chain of mitochondria (for further information see following chapter).

7.2 Exogenic Toxicity Leading to Mitochondrial Dysfunction

Anything can be toxic if it inhibits the electron transport chain (ETC). Oxidative-phosphorylation disease is affected by any disruption in the ETC. What appears to be a necessity, i.e. oxygen, in itself can be damaging. There are many ways that free radicals can be formed and these free oxygen radicals can be toxins if not handled appropriately. Free radical damage can cause increased energy needs with a cascading effect of further damage. When medication is needed, the key is to attempt a balance between the treatment needs and the side effects of such treatment. While the pathobiology of mitochondrial toxicity is not well understood in detail, toxicity can be exacerbated by additional underlying problems and treatments. The types of toxic agents include: pharmaceutical products (medications), anesthesia, surgery, environmental agents, diet, stress related endogenous agents, and mitochondrial cofactors (Table 7.1).

7.2.1 Pharmaceutical Products

Establishing mitochondrial toxicity is not an FDA requirement for drug approval, so there is no real way of knowing which agents are truly toxic. Nor is there an absolute contraindication against any particular agent, but there are a number that we know should be avoided. Some agents, either through research studies or anecdotal evidence, have been shown to induce direct toxicity to mitochondria. These agents inhibit or disrupt either the electron transport chain, protein production, mitochondrial DNA transcription, or enzyme activity. Agents that cause indirect toxicity are those that increase free radicals, decrease the production of endogenous antioxidants, or deplete nutrients that are needed to scavenge radical oxygen species (ROS).

Anticonvulsants, specifically, most anticonvulsants are well tolerated except the anticonvulsant and mood stabilizing agent valproic acid (Depakote™). This drug can inhibit several mitochondrial functions, since it is known to play an important role in

7.2 Exogenic Toxicity Leading to Mitochondrial Dysfunction

Table 7.1 Summary of reported drugs used in medicine with potential mitochondrial toxicity

Barbiturates	Amytal. Nembutal, Seconal, Phenobarbital
Neuroleptics	Chlorpromazine, Fluphenazine, Haloperidol, Rispiridone
Chemotherapeutics	Dixorubicin, Mitomycin C
Local anesthetics	Lidocaine, Bupivcaine
Antidiabetics	Metformin, Phenformin, Trogiltazine, Rosigiltazine, Pioglitazone
NSAIDs, Analgesics	Diclofenac, Indomethacin, Naproxen, Finipofen, Salicylic acid, Ibuprofen, Acetaminophen
Anesthetics	Halothane, Propofol
Insecticides, Herbicides, Fungicides	Dinitrophenol, Pentachlorphenol, Rotenoid
Ca^{2+}-Antagonists	Flunirazine, Cinnarizine
Antibiotics	Tetracyclines, Antimycin
Tricyclic antidepressants	Fluoxetine, Amitriptyline, Amoxapan, Citalopram
Fibrates	Ciprofibrate,
Antianginals, Cardiotonics	Amiadarone, Perhexeline, Diethylaminoeethoxyhexestrol, ß-Blocker
Antivirals	Zidovudine (AZT), Stavudine, Didanosine, Zalcitabine, Lamivudine, Abacavir, Interferon
Antituberculostatics	Isoniazid
Steroids	Estrogen, Cortisol, Progesterone, and Testosterone
Antibiotics	Aminoglycosides (i.e. Neomcin, Streptomycin, Kanamycin et aliter)
Antiepileptics	Phenytoin, Valproate
Statins or HMG-CoA reductase inhibitors	Clofibrate, Atorvastin, Fluvastin, Lovastin, Pitavastin, Pravastin, Rosuvastin, Simvastatin
Antiolytic agents	Diazepam, Aprazolam

ETC electron transport chain

carnitine utilization by the mitochondria and it has been shown to particularly inhibit complex IV. In addition, it can cause liver dysfunction. This does not mean it should never be used, but caution needs to be taken regarding liver function. If used, plasma carnitine levels need to be monitored and maintained carefully.

Psychotropics: Certain psychotrophic drugs have been shown to be potentially toxic. For instance, antidepressants such as Prozac™, Elavil™, and Cipramil™ can cause autonomic dysfunction. Other psychotropic drugs such as antipsychotics, barbituates, and antianxiety medications (i.e. antidepressants) also inhibit various mitochondrial functions.

Any *cholesterol-lowering* medications, especially statins, have been shown to reduce the endogenous coenzyme Q10 production, since it is synthesized in the same metabolic pathway as cholesterol. Statins therefore can be considered as contraproductive, since Q10 levels are necessary for sufficient cardiac function in patients with chronic heart failure (CHF) [143]. Therefore, such agents potentially can be considered as toxic. Also, it had been demonstrated in a placebo-controlled study, that regular statin intake for 2 months results in memory impairment in 56% of patients, an effect, which resided, when statins intake was stopped [144]. In addition, a word of caution regarding other cholesterol medications such as cholestyramine that binds to bile acids as they can disrupt the electron transport chain within mitochondria.

Analgesics and Anti-inflammatory agents. Pain relievers such as acetaminophen (Indocin®, Naproxen®), acetylsalicylic acid (Aspirin®) and all NSAIDS (non steroidal anti-inflammatory drugs) [145], increase oxidative stress, and therefore are potentially toxic. Aspirin is contraindicated for children, but it can be harmful for patients with an already existing mitochondropathy while the risk for a potential Reye Syndrome with acute liver failure should be considered. However, patients should keep in mind that it is important to avoid fevers in patients demonstrating mitochondrial deficiency. Therefore, the benefits of some of these medications as fever-reducers may outweigh their potential side effects.

Antibiotics, specifically tetracycline, minocycline, chloramphenicol, and aminoglycosides, can be harmful to the mitochondria because they inhibit the mtDNA translation and protein synthesis. In addition, they can cause hearing loss as well as cardiac and renal toxicity.

Steroids Steroids may reduce the transmembrane mitochondrial potential. In contrast, steroids when used in local delivery such as inhaled steroids that only target the lungs or injected steroids that target specific locations (e.g. the knee in osteoarthritis) generally can be recognized as safe.

Cardiotonic agents, which are used by a wide range of population with cardiac problems such as *amiodarone*, have potential toxicity as it is used as an anti-arrhythmic over a long period of time.

Also, *antivirals* like interferon, or *antiretrovirals* used by HIV/AIDS patients [146], and all *chemotherapeutics* used in cancer patients, demonstrate significant toxicity for mitochondria [147].

Metformin, and also thiazoline derivatives used in diabetes [148], are also considered toxic to mitochondria.

Beta-blockers used in heart insufficiency and antihypertonics have toxicity due to increased oxidative stress, and may also contribute to feelings of fatigue since they directly reduce the formation of ATP [149–154]

Diuretics used for heart insufficiency and beta-blockers are usually not harmful to the mitochondria themselves, but they may cause fluid imbalances, which can be potentially difficult for patients having a pre-existing mitochondrial impairment. In all cases of the drugs mentioned, the objective is to balance the need for use of these drugs with the damage they may cause.

Cholesterol-lowering agents like statins indirectly impair mitochondrial function by inhibiting endogenous Q10 synthesis [155].

Once statins are prescribed the addition of Q10 is mandatory in order to avoid mitochondrial malfunction.

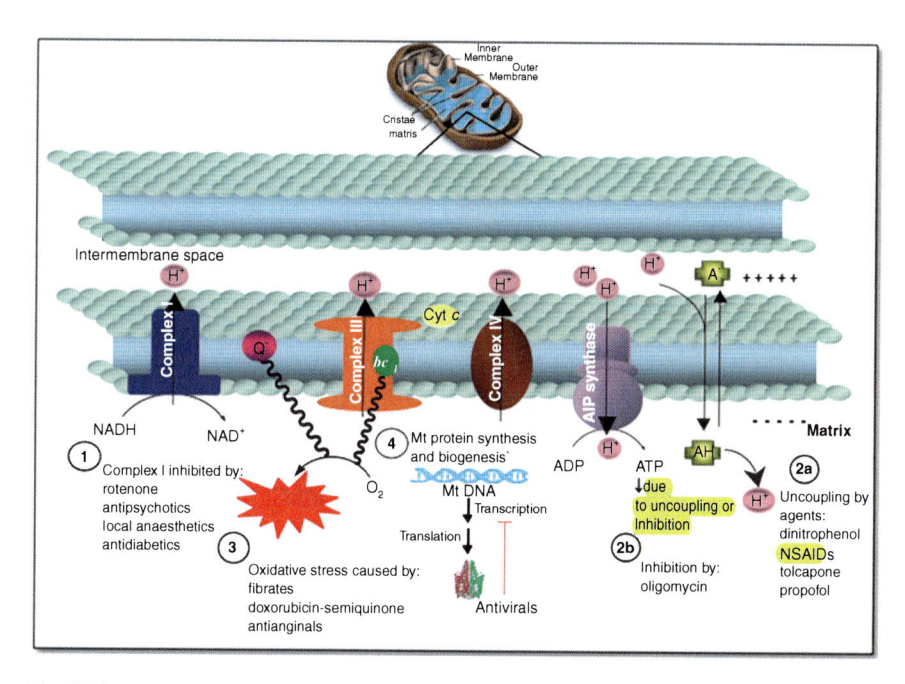

7.2 Exogenic Toxicity Leading to Mitochondrial Dysfunction

Fig. 7.14 Mechanism and site of drug-induced mitochondrial toxicity. Cyt *c*: Cytochrome *c*; Mt: Mitochondria (**1**) Inhibition of complexes I and III can result in the generation of reactive oxygen species due to auto-oxidation of reduced complexes. (**2a**) Uncoupling of mitochondrial oxidative phosphorylation occurs when protonophoric drugs mediate the transport of protons through membrane causing the membrane potential to dissipate, thus disconnecting the electron transport chain from ATP formation. (**2b**) Mitochondrial oxidative phosphorylation inhibition occurs when drugs bind to ATP synthase. (**3**) Mitochondrial oxidative stress is caused by auto-oxidation of doxorubicin semiquinone radicals formed by complex I or by inhibition of complex I. (**4**) Antiviral drugs impair mitochondria by targeting mitochondrial DNA polymerases, inhibiting mitochondrial DNA replication and protein synthesis (Adapted from [160])

In summary the following typical ailments and the agents used to treat just the symptoms and not the causative factor in one way or the other all are related to western life style (Fig. 7.14):

- Diabetes type 2, since ß-cells of the pancreas are very sensitive to NO while high blood sugar levels results in stress inducing the release of NO⁻ through decoupling of iNOS. The close correlation of diabetes and ROS/NOS has been demonstrated conclusively in offspring's of patients with diabetes type 2. Compared to a control group these subjects had a 60% lesser uptake of glucose by muscle cells, a 80% higher accumulation of lipids within the musculature, and a 30% reduction of oxidative phosphorylation in mitochondria [156], corroborating clinical observations that glucose when not transposed into energy is converted into fat.
- Insufficient methylation resulting not only in reduced formation of melatonin, serotonin and adrenalin but also increase in asymmetric dimethylarginine (ADMA) which is considered a sensible marker for cardiovascular disease (CD) using the ADMA®Elisa test [157, 158].

- Regular intake of cholesterine-lowering agents, which aside from an increase formation of NO also induce a decline in the formation of Q10.
- Regular use of the selective ß-blocker nibivitol (Nebelid®) results in an increase of NO-synthesis.
- Long-time coronary dilators used as nitrates or niroglycerine for angina, both of which induce NO-synthesis.
- Acetylsalicylic acid and all proton pump inhibitors (PPI) reduce the absorption of sufficient amounts of Vit B_{12} impairing mitochondrial function.
- Triglyceride lowering agents such as fibrates are potentially toxic for mitochondria.
- Ágents used in rectile dysfunction such as sildenafil (Viagra®) increase NO-synthesis.
- The anti-arrhythmogenic agent amiodaron (Cordarex®) induces lactate acidosis resulting in mitochondrial deficiency
- The antihypertonic ACE-inhibitor Enalapril®, which increases NO-formation.
- Long-time use of the oral antidiabetic agent metformin results in lactate acidosis and mitochondrial damage.
- Arginine, an amino acid often added to supplemental medication results in the formation of NO and citrulline (see page 49). Therefore, the amino acid arginine is **contraindicated** in all patients with coronary heart disease [159].
- Non-steroidal anti-inflammatory drugs (NASIDs) result in a destruction of the mucous membranes of the gut with a reduced absorption of micronutrients and especially of the vitamins B_{12} and folic acid. A similar reduction in absorption is also observed with the intake of any histamine- antagonist.
- Diabetic type 2 patients with metformin therapy demonstrate a Ca^{++}-dependent reduced endocytosis related to a reduced absorption of the vitamin B complex (B_1, B_6, B_{12}, folic acid).

In summary, chronic intake of any medication in selective risk populations often goes in hand with a deficit in essential vitamins and micronutrients. The groups include epileptics, patients with rheumatoid arthritis, with hyper tonus, cardiovascular disease, cancer patients, diabetics, chronic pain patients as well a pregnant or breastfeeding women. This is because the daily intake of the above medication competes with the sufficient reabsorption, with increased metabolisation and elimination of micronutrients, reducing their efficacy and result in a deficit of necessary vitamins. Such deficit instantaneously does not become visible especially in pregnant women, elderly patients, or the very young and is the reason why supplementation with vitamins in general should be recommended in such risk population.

Although mankind stands to obtain great benefit from *nanotechnology*, it is important to consider the potential health impacts of nanomaterials (NMs). This consideration has launched the field of nanotoxicology, which is charged with assessing toxicological potential as well as promoting safe design and use of NMs. Although no human ailments have been ascribed to NMs thus far, early experimental studies indicate that NMs could initiate adverse biological responses that can lead to toxicological outcomes with one principal mechanisms resulting in the generation of reactive oxygen species and oxidant injury. Because oxidant injury is also a major mechanism by which ambient ultrafine particles can induce adverse health effects, it is useful to consider the lessons learned from studying ambient particles [161]. For instance, being increasingly used in various fields, including biomedicine and electronics, one application utilizes the pacifying effect of nano-TiO$_2$ (titandioxide), which is frequently used as pigment in cosmetics. Although TiO$_2$ is believed to be biologically inert, an emerging literature reports increased incidence of respiratory diseases in people exposed to TiO$_2$ In this respect nanoparticles activate the NLR pyrin domain containing three (Nlrp3) inflammasome and cause pulmonary inflammation through the release of IL-1α and IL-1β. Inhalation of nano-TiO$_2$ provoked lung inflammation, which was strongly suppressed in IL-1R– and IL-1α–deficient mice. Thus, the inflammation caused by nano-TiO$_2$ in vivo is largely caused by the biological effect of IL-1α. The current use of nano-TiO$_2$ may present a health hazard due to its capacity to induce IL-1R signaling, a situation reminiscent of inflammation provoked by asbestos exposure [162].

Anesthesia. In mitochondrial disease, there seems to be an increased sensitivity to anesthesia, especially the volatile agents. For this reason, there should be very close management of any anesthesia used even when anesthetic agents are given i.v. For example, the hypnotic propofol in long-term administration has been shown to induce the so-called propofol infusion syndrome with acute cardiac insufficiency related to mitochondrial toxicity [163]. Often a decreased dosage is adequate, and the smallest dose over the shortest period of time should be the goal of all anesthetics for patients already demonstrating mitochondrial impairment. Patients with mitochondrial injury should make sure that their anesthesiologist is informed and knowledgeable about their condition so that they use the upmost caution and safety while using anesthetics.

Environmental Agents. Tobacco smoke (primary or secondary inhalation) and alcohol are both potentially toxic for mitochondria. Other environmental factors may not be as controllable, but patients should be aware of their toxicity. These include rotenone, a chemical used in insecticides and pesticides and fat-soluble chemicals with benzene rings such as used in hair dye and paint fumes.

Diet. The ketogenic diet (used sometimes for those with seizure disorders) can be a stressor for those with an already existing (silent) mitochondrial impairment. Such diet increases cellular utilization of the beta-oxidation pathway, and the diet can be helpful for those with a pyruvate dehydrogenase disorder (Fig. 7.15). It is important for patients to have a balanced diet with supplements including pyridoxine (B$_6$), ferrodoxin, iron, copper, riboflavin (B$_2$), zinc, and selenium, as well as other vitamins and/or minerals. In addition, fasting (!) should be avoided, since a state of *hypoglycemia* can also be toxic and lead to problems similar to those of diabetics. Also, *hyperglycemia* by itself can cause increased superoxide production with the formation of aggressive nitric oxide and superoxide in the endothelium, which can lead to a vascular endotheliopathy with vessel wall dysfunction and metabolic derangement resulting in local inflammatory reactions, a prerequisite for later arteriosclerosis [164]. Certain foods like peas, beans, legumes, and almonds have substances in them that can be toxic to the mitochondria.

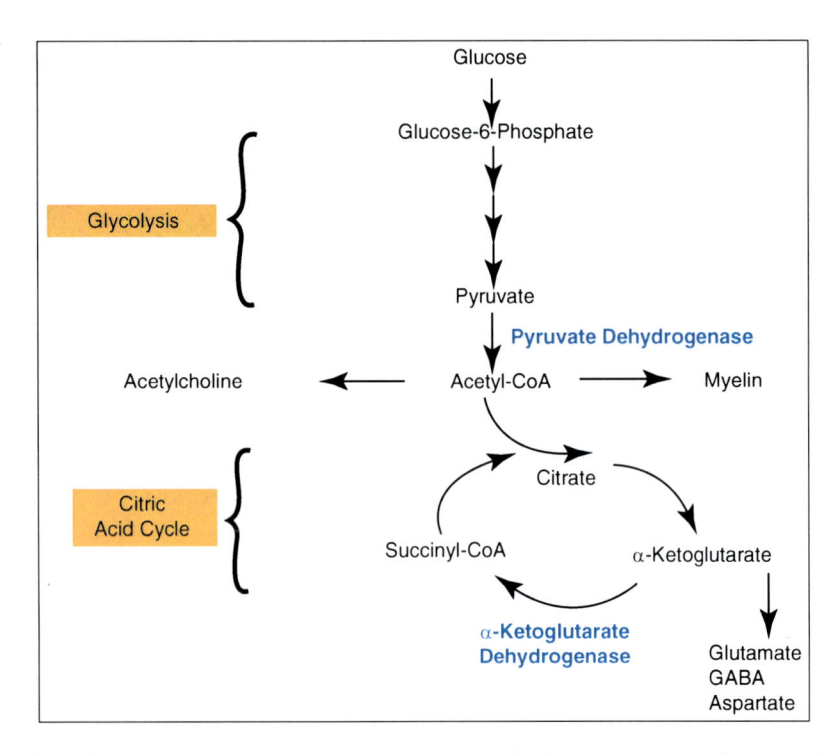

```
                              Glucose
                                 ↓
                        Glucose-6-Phosphate
                                 ↓
         ⎧                       ↓
Glycolysis⎨                      ↓
         ⎩                    Pyruvate
                                 │  Pyruvate Dehydrogenase
                                 ↓
  Acetylcholine  ←──────────  Acetyl-CoA  ──────────→  Myelin
                                  ↘
                                   Citrate
         ⎧
Citric   ⎨
Acid Cycle⎩   Succinyl-CoA            α-Ketoglutarate
                                            ↓
                 α-Ketoglutarate
                  Dehydrogenase          Glutamate
                                         GABA
                                         Aspartate
```

Fig. 7.15 Pyruvate dehydrogenase a necessary constituent for channeling pyruvate from carbohydrate glycolysis into the Krebs-citric acid cycle

7.2.2 Stress Hormones on Mitochondrial Function

Stress hormones such as adrenalin, catecholamines, and testosterone can be harmful for patients with any kind of impaired mitochondrial function. Stress, and especially psychological stress, should be avoided as much as possible so as not to further compromise mitochondrial function. Even the coenzyme Q10 by itself can become an oxygen radical and cause additional stress on mitochondrial function if the dosage is too high. The most common dosage is 10–20 mg/kg/day. There is a small study that used 60 mg/kg/day for a short term and it was tolerated well. Usually, however, dosages above 20 mg/kg of ubiquinone are NOT tolerated well. For ubiquinol (the reduced form with higher bioavailability) the dosage is usually below 10 mg/kg/day (range of 6–8 mg/kg/day). Also, high doses of riboflavin, L-carnitine and L-arginine can be harmful for mitochondrial function, so there is the need for careful monitoring regarding signs of fatigue and memory impairment, as mitochondria in the CNS are the first to be affected.

Since the elderly population (50 plus) in general is getting long-term prescription medication, the following agents should be taken with caution, as they all induce a further decline in mitochondrial function, or when given in high dosages, even by themselves induce malfunction in the electron transport chain (Table 7.1).

In conclusion, while there are some substances, which are clearly toxic to the mitochondria and there is evidence how they exhibit the toxic effect; these should be avoided by any means. However, there are many more that are only suspected of being toxic (Fig. 7.14). Of these potentially toxic agents one should take care in their use, monitor the effects carefully, recording any observed deterioration in mental capacity, and vigilance. Thus, when using a new treatment schedule, one should add only one new medication at a time in order to monitor both the positive and the negative effects. All of this information should be made known to the general practitioner who is in charge of ones health.

7.2.3 Additives in Food Chain with Mitochondrial Toxicity

Many drugs or their electrophilic metabolites can inhibit the mitochondrial electron transport chain and cause reactive oxygen species (ROS) formation, which can trigger MMP and cytochrome *c* release. Alternatively, ROS can be formed when pro-oxidant drug radicals are formed by peroxidase-catalysed drug metabolism. Inhibition of any of the four respiratory complexes in the respiratory steps results in the generation of ROS, as only cytochrome oxidase (complex IV) can reduce oxygen with four electrons to form water and inhibit the mitochondrial electron transport chain. In addition, additives which are found in daily food chain are suspected to induce NO/OONO-activation, ranging from colorants (E100-E180) like tartrazine (E102), chinoline yellow (E104), canthaxanthine (E161) can also result in a high potential for allergy (Table 7.2).

Especially aspartame has been a subject of big debate as it was shown not to be a diet product, it was originally banned by the American FDA, it actually makes you gain weight, it is not safe for anyone, it cannot be considered a food additive, and is poisoning when taken in cumulative doses. Controversy arose because of scientific work related to animal and epidemiological studies, which conclusively have shown that aspartame induces brain tumors [165]. All such deleterious effects on cells and their mitochondria not only holds true for food additives but also for so called functional foods, which is claimed to be of health benefit and especially for so called *trans fats*, a common name for unsaturated fats with trans-isomer (ε-isomer) fatty acid(s). Because the term refers to the configuration of a double carbon-carbon bond, trans fats may be monounsaturated or polyunsaturated but never saturated. In the process of hydrogenation a hydrogen atoms to *cis*-unsaturated fats is added, eliminating double bonds and making them into partially or completely saturated fats. However, partial hydrogenation, if it is chemical rather than enzymatic, converts a part of *cis*-isomers into *trans*-unsaturated fats instead of hydrogenating them completely. Trans fats also occur naturally to a limited extent: Vaccenyl and conjugated linoleyl (CLA)

Table 7.2 Summary of food additives, which are also claimed of being toxic to mitochondrial function

- Preservatives (E200-E297) like sulfides (E220-E228) induce asthma attacks and benzoic acid (E210-E219) leads to allergic reactions
- Flavor enhancer (E620-E625) like monosodium glutamate affects satiety feeling. It induces the "China-restaurant-Syndrome" with nausea, headache, feeling of heat, lasting for 2 h
- Sugar substitutes (E950-E1518) like saccharine (E954), aspartame (E951; known as nutra-sweet, candarel, sanecta or spoonfull is degraded to formaldehyde) or cyclamate (E952) are found in all sorts of "light" drinks, are suspected to induce cancer and cerebral tumor, respectively. The former is also being used for hog masking!
- Emulsifying and stabilizing agents, thickeners and gelling agents (E400-E440, E460-495) like alginate (E400-405), agar-agar (E406), and locust bean gum, a thickening agent (E410), result in reduced mineral absorption. Especially the degraded carageene (E407) has been shown to induce allergies and ulcerative colitis
- Acidifiers (E338-E341) used as preservatives like organic phosphorus compounds, found in all Cola drinks inhibit Ca^{++}-absorption and have been shown to reduce bone density by 3.8%.
- Cake glaze used in baking of tarts and cakes (E173) contains aluminum and citric acid, crosses the blood-brain barrier resulting in aluminum deposits in Alzheimer patients
- Methomyl (Lannat® or Nudrin®), a pesticide used for cultivation of all sorts of fruits, being used extensively outside the EU, however tested in > 1,000% above the permitted limit, induces cancer, neurotoxicity and DNA damage
- Azo dye stuff yellow-orange (E 110), used in puddings, potato chips, fruit gum, or yellow is manufactured from crude oil and results in attention deficit syndrome (ADH)

containing trans fats occur naturally in trace amounts in meat and dairy products from ruminants, although the latter also constitutes a cis fat. Only two essential fatty acids are known for humans: alpha-linolenic acid (an Ω-3 fatty acid) and linolenic acid (an Ω-6 fatty acid). Other fatty acids that are only "conditionally essential" include gamma-linolenic acid (an Ω-6 fatty acid), lauric acid (a saturated fatty acid), and palmitoleic acid (a monounsaturated fatty acid). Trans fats are not essential fatty acids; they are found in margarine, and snacks such as biscuits, crumpets and potato chips.

> **A word of warning – consumption of trans fats increases the risk of coronary heart disease!**

Health authorities worldwide recommend that consumption of trans fat be reduced to trace amounts (Fig. 7.16). Trans fats from partially hydrogenated oils are more harmful than naturally occurring oils [166] by raising levels of "bad" LDL cholesterol and lowering levels of "good" HDL cholesterol [167]. Even the cholesterol-lowering agents do not seem to hold the promise as originally tooted since a large size study demonstrated that within 5 years of intake of statins there was a higher incidence of melanoma when compared to placebo [168].

Fig. 7.16 Look for this sign as it indicates that there is a food conscious population, which knows and respects the importance of a healthy diet

7.2.4 *Chemical Substances – Originators of Mitochondropathy*

References and further reading may be available for this article. To view references and further reading you must purchase this article.

In daily life, from the environment human mankind is exposed to approximately 60 000 different chemical substances, of which 4,000–6,000 demonstrate cancer-inducing properties:

(A) Contamination via the skin

- Several preservatives (lindan, pentachlorphenol, halogenated fungicides),
- 8,000 dyers, of which 2,000 nitrosative azo dyes and 6,000 textile dyes (halogenated hydrocarbons, phosphoric acid ester, formaldehyde, ammoniac, etc.).

(B) Inhalation via the lungs

- Nitrogen oxides,
- Nitrosamine,
- Ozone,
- Aromatic hydrocarbons like benzpyrene,
- Benzanthrazen and others,
- Metal dust,
- Organic solvents,
- Plutonium, radon, tritium, and others

(C) Contamination of food products from conventional agriculture and foodstuff industry

- Heavy metals,
- Pesticides, insecticides etc.
- Nitrates, nitrosamine,
- Aliphatic and aromatic hydrocarbons
- Softeners from plastics,
- Colorings and preservatives

In general, there are some agents, which are known to be toxic both for children and adults, and one can use this knowledge to guide recommendations. However, there is a lot more that is not known, because many of the recommendations are based on anecdotal evidence or very small studies. For example, though there is a long list of psychotropic drugs which "*may be toxic*", many of them probably are not contraindicated but rather should be used with frequent observations and records about how they affect the patient, and a switch should be done one drug at a time. Again, a key point to keep in mind is that the life-saving benefit of some of these drugs outweighs their possible side effects in an acute situation. It however, is imperative to isolate any potentially toxic agent being used over a long period of time, outweighing its potential benefit in medical treatments, to its obvious toxicity to mitochondrial function. Although it appears that some effects from a mitochondrial toxin are reversible, any deficiencies need to be identified and dealt with because regularly the problems of toxicity to mitochondria will only come as clinical signs of malfunction of a certain organ, when at least 60% of all mitochondria have deteriorated. The problem often is finding out just what the particular vulnerabilities are for each particular patient. Certain medications may trigger toxic reactions, but the challenge is to figure out whether these symptoms are from ongoing mitochondrial damage, natural history of the disease, or the use of new drugs. When toxic agents can't be avoided, improving overall mitochondrial function is possible by including key elements such as coenzyme Q10, L-carnitine, vitamin B complex (especially vitamin B_{12}, since it acts as a NO scavenger), antioxidants (vitamin C, vitamin E, α-lipoic acid, selenium), and NADH, while it is also imperative to minimize stress and illness! Patients, who have a high level of mercury stored in their bodies, apparently react negatively because the α-lipoic acid mobilized stored mercury, resulting in an increased impact on areas in the body including the brain. Those suffering from high levels of mercury in their bodies will have to undergo thorough mercury detoxification before they can follow any micronutrient therapy. This is because mercury can up-regulate the $NO/ONOO^-$ cycle and its methyl mercury metabolite can increase NMDA activity resulting in pain. Also, agents that are designed to lower the $NO/ONOO^-$ cycle biochemistry will only be effective if people taking them avoid all toxins that otherwise rise the cycle and thus exacerbate their symptoms. Potential agents that raise the cycle will include all chemicals in the MCS group, excessive exercise leading to post-exertional malaise in the CFS group, chronic infections particularly in the CFS and fibromyalgia group, food allergens in those with food allergies, and psychological stress in many but especially in those suffering from post-traumatic stress disorder and FMS.

7.2.4.1 Complex Regional Pain Syndrome, a Mitochondrial Disease?

Characterized by allodynia that occurs in response to external stimuli, hyperalgesia, an affected edematous limb, cold and purple in appearance, after an acute damage, being unable to operate, this entity of a painful disease even for a pain specialist presents an intriguing task for treatment. Neuropathic pain may result from disorders of the peripheral nervous system or the central nervous system (brain and spinal cord). Thus, neuropathic pain may be divided into peripheral neuropathic pain, central neuropathic pain, or mixed (peripheral and central) neuropathic pain. Central neuropathic pain is found in spinal cord injury, multiple sclerosis, and some strokes. In this aspect fibromyalgia is a disorder of chronic widespread pain, is potentially a central pain disorder and is responsive to medications that are also effective for neuropathic pain [169].

The starting point for neuropathic pain is a lesion or dysfunction within the somatosensory cortex. Current knowledge regarding the mechanisms of neuropathic pain is incomplete and is biased by a focus on animal models of peripheral nerve injury. Pain management specialists advocates the use of antidepressant, anticonvulsants together with potent opioids or selective, noradrenaline-serotonin-reuptake inhibitors (SSRI), tricyclic antidepressants as well as topical lidocaine or capsaicin with or without cannabinoids have demonstrated variable results, however, with regular side effects. Contrary, others consider this ailment as a so-called mitochondrial-related functional disorder where micronutrients (also labeled as neutraceuticals) such as Q10, (200 mg), vitamin B1 (100 mg) plus L-carnitine (3×330 mg) together with vitamin C (2×500 mg/day) have been shown to be effective without inducing any side effect. As to the mode of action of micronutrients in such a debilitating ailment upregulation of excitatory receptors involved in the mitigation of painful stimulation, together with an increase in nociceptive afferents is considered as mutual cause for such a disorder. And since the NMDA- as well as the increase of sensitivity of the TRP/vanilloid-receptor seem to be involved in neuropathic pain as outlined in the Fig. 7.17, presents an alternative therapeutic approach, which downregulates the hyperactive NO/ONOO- cycle. This is because four TRPVs (the transient receptor potential channel vanilloid 1–4; TRPV1, TRPV2, TRPV3, and TRPV4; Fig. 7.18) are expressed in afferent nociceptors pain sensing neurons, where they act as transducers of thermal and chemical stimuli [170]. Hence antagonists or blockers of these channels may find application for the prevention and treatment of painful diseases.

A different point of view in neuropathic pain is that nitric oxide (NO) can increase peroxynitrite (PRN) levels, which by itself stimulates oxidative stress from where NF-κB is upregulated resulting in the activation of iNOS, which in turn increases nitric oxide levels. The loop by itself comprises a potential vicious cycle and, as outlined in the figure there are a number of other loops that can collectively result in a much larger vicious cycle. There is now ample evidence that in a number of cases a preexisting mitochondropathy, which was followed by a mechanical insult results in neuropathic pain. For instance, excitotoxicity is a pathological process by which nerve cells are damaged and killed by excessive stimulation by means of neurotransmitters such as glutamate, glycine and/or kainic acid. In such instant receptors for the excitatory neurotransmitter glutamate (glutamate receptors) such as the NMDA receptor and AMPA receptor are overactivated. Excitotoxins like NMDA and kainic acid which bind to these receptors, as well as a pathological high level of glutamate, can cause excitotoxicity by an increased inward flux of calcium ions (Ca^{2+}) into the cell (Fig. 7.19).

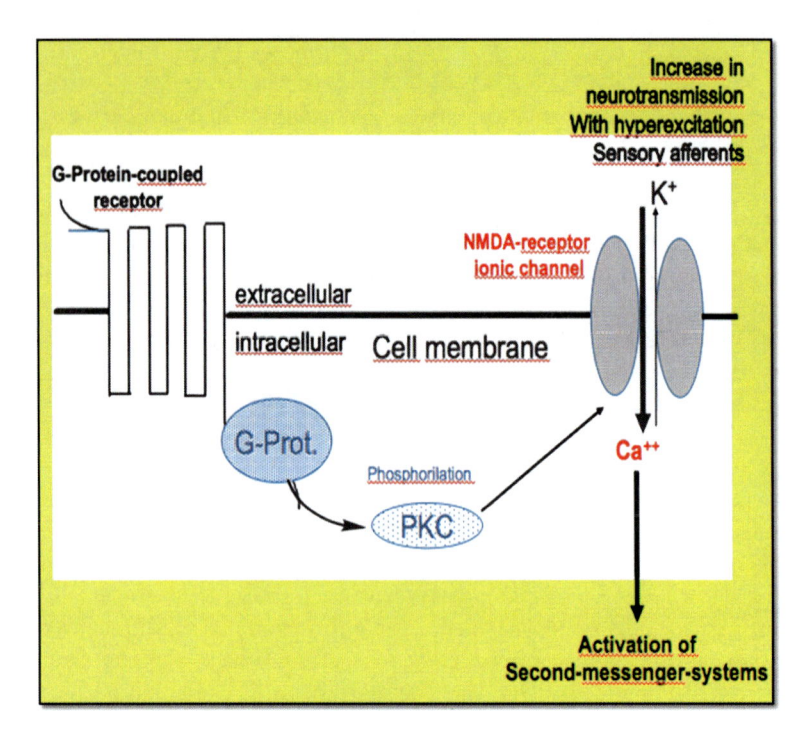

100   7 Nitrosative Stress in Diverse Multisystem Diseases

Fig. 7.17 Principle of the excitatory N-methyl-D-asparate (NMDA) receptor system resulting in an increased inward shift of Ca^{++} ions into the neuronal cell

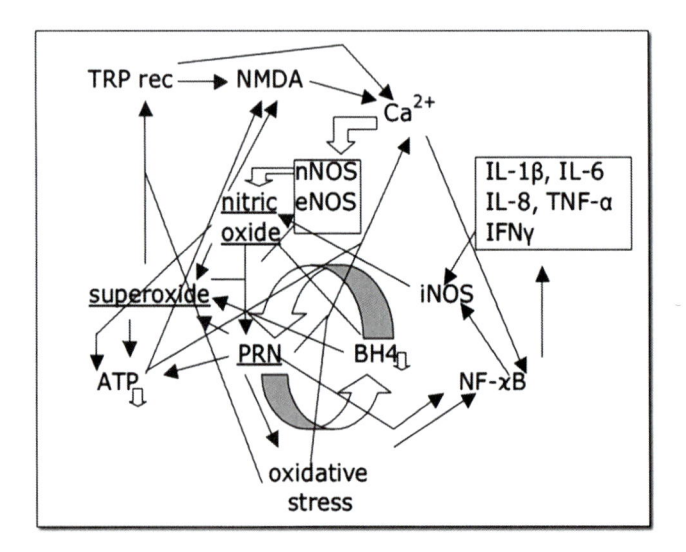

Fig. 7.18 Diagram of the vicious NO/ONOO$^-$ cycle resulting in chronic pain via NOS upregulation followed by activation of the NMDA- and the vanilloid receptor sites (TRP rec). Each arrow represents one or more mechanisms by which the variable at the foot of the arrow can stimulate the level of the variable at the head of the arrow. It can be seen that these arrows form a series of loops that can potentially continue to stimulate each other (Adapted from Ref. [10] of Chap. 5)

Fig. 7.19 The excitatory N-Methyl-D-aspartate (NMDA) receptor, a glutamate-gated ion channel, ubiquitously distributed throughout the brain, fundamental to excitatory neurotransmission with its different binding sites and where magnesium acts as an inhibitory substance being involved in neuropathic pain but also in neurodegeneration

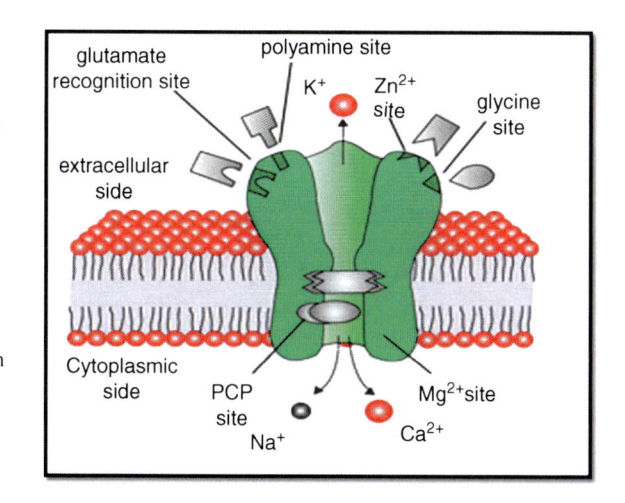

Since excess Ca^{2+} influx into cells activates a number of enzymes, including phospholipases, endonucleases, and proteases such as calpain, they all damage cell structures such as components of the cytoskeleton, membrane, and the DNA of mitochondria resulting in chronic pain with an increase in sensitivity of the excitatory system. Thus, a causal role for $ONOO^-$ in pathways culminating in antinociceptive tolerance to opiates has been demonstrated [171]. Peroxynitrite ($ONOO^-$) decomposition catalysts may have therapeutic potential as adjuncts to opiates in relieving suffering from chronic pain.

7.3 Neurodegeneration Resulting from Blockade in the Energy Pathway

In neurological diseases, where BH_4 is limited in some cellular sources of NOS it may generate superoxide whilst other BH_4 saturated NOS enzymes may be generating NO. Such a scenario favors peroxynitrite generation, and if peroxynitrite is not scavenged, e.g. by antioxidants such as reduced glutathione, the natural scavenger system for radical oxidative/nitrosative radical substrates, irreversible damage to critical cellular enzymes could ensue, which as demonstrated in various studies can result in

- **Alzheimer** [172]
- **Parkinson** [173]
- **Autism** [174]
- **Depression** [175]
- **DOPA- responsive dystonia** [176, 177]

Any neuronal sensitization is therefore associated with self-perpetuating neuroexcitation and excessive response especially when the patient is exposed to chemical exposure such as organic solvents, pesticides and herbicides found in fertilizers as well as heavy metal contamination [178, 179]. Such continuous exposure of

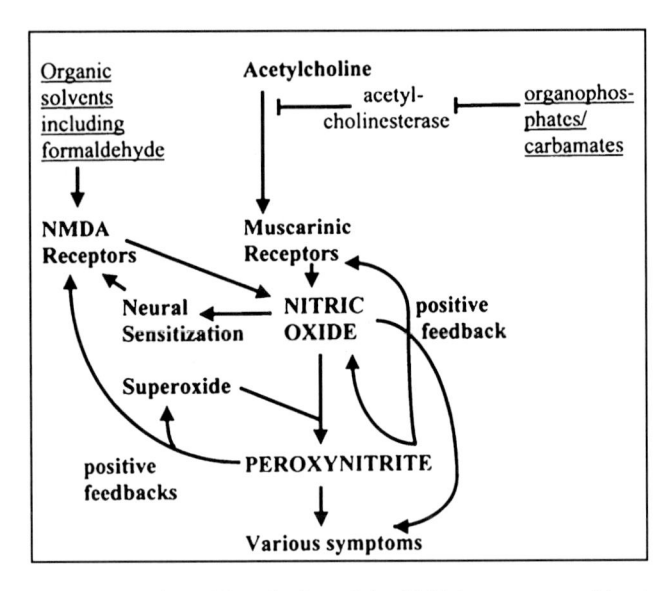

```
                    Acetylcholine
Organic                 │   acetyl-        organophos-
solvents                ├──cholinesterase──phates/
including               │                  carbamates
formaldehyde            │
     │                  ▼
     ▼              Muscarinic
  NMDA              Receptors
  Receptors              │
        ╲                ▼
         Neural ←── NITRIC     positive
         Sensitization OXIDE   feedback
          ╲             │
           Superoxide   │
                        ▼
  positive ──── PEROXYNITRITE
  feedbacks          │
                     ▼
                Various symptoms
```

Fig. 7.20 Neural sensitization with activation of the NMDA receptor resulting in an increase influx of Ca^{2+} resulting in an increased risk of inflammation, neurodegeneration, lung damage, and other cellular injury, and where magnesium is a potent blocker (Adapted from [73, 185])

mitochondria to toxins of various origin is of importance, since the resulting long-term NMDA activation with increased nitric oxide and peroxynitrite production induces neuronal cell death in brain (Fig. 7.20), which clinically impress as neurodegenerative disease (Refs. [20, 23, 28] of Chap. 8) [180–183]. Peroxynitrite being an aggressive radical, also weakens the blood-brain barrier, allowing more chemicals to enter the brain in excess [184].

In addition, nitric oxide also damages the first step of detoxification within the liver involving the cytochrome p450 system, allowing chemicals, and many other agents to accumulate to a higher amount within the body [186].

This vicious cycle must under all circumstances be interrupted to the maximum possible degree. The symptoms of sensitization through the effects of an overexpressed NO/ONOO⁻ cycle are just initial warnings of other temporarily still silent, toxically-induced mitochondropathies can erupt in different organs when defects in mitochondrial function exceeds that of 60%. It is only than, when clinical symptoms become obvious in for instance the liver, the pancreas, the immune system, the adrenals, and/or the CNS [187–192]. Any covering and/or blockade of the symptoms of the NO/OONO⁻ cycle by means of medication is obsolete without a simultaneous approach to correct the underlying disturbed biochemical mechanism within the mitochondria. Otherwise this would be like turning off a fire serene without extinguishing the underlying problem.

Aside from symptoms of advanced neurodegeneration, which is reflected in Parkinson or Alzheimer, due to the extensive impaired activity of mitochondria within the myocardium, any significant enhanced peroxynitrate production results

7.3 Neurodegeneration Resulting from Blockade in the Energy Pathway

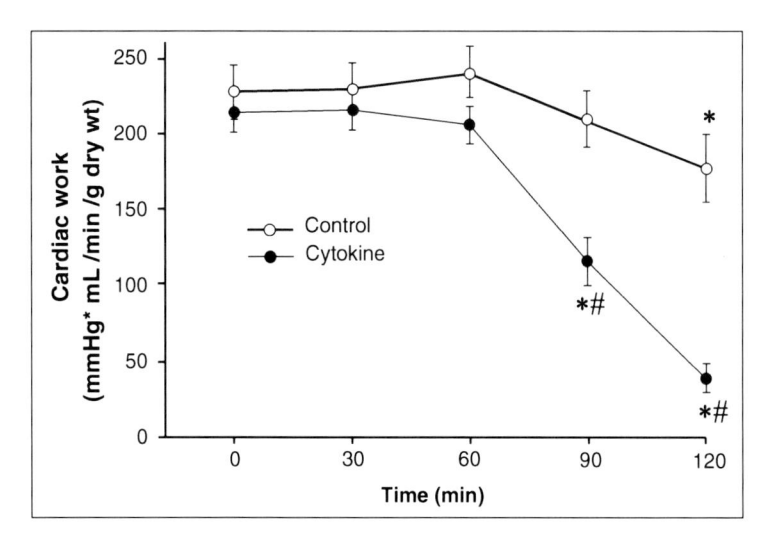

Fig. 7.21 Time course of changes in cardiac work in control and cytokine-treated isolated working rat hearts. *$P<0.05$ vs 0-minute value within the same group (1-way repeated measures ANOVA); #$P<0.05$ vs control (unpaired t test, n = 57–8 in each group)

in deleterious effect on myocardial contractility. Such assumption is underlined by studies on isolated heart preparations, which conclusively have demonstrated that proinflammatory cytokines stimulate the concerted enhancement in superoxide and NO-generating activities in the heart, thereby enhancing peroxynitrite generation, which causes myocardial contractile failure [193]. Such data were further underlined by administration of the NO synthase inhibitor *N*G-nitro-L-arginine, and the superoxide scavenger CuZn-SOD, Mn-SOD and L-propionyl-carnitine inhibited the decline in myocardial function and decreased perfusate nitrotyrosine levels [194, 195].

Therapeutic use of superoxide scavengers in medicine has become a great topic and numerous patents and agents are being tested for efficacy, among them tiron and templon [196] as well as purified recombinant human NAD(P)H: quinone oxidoreductase one, which in preliminary experiments have demonstrated oxygen radical scavenger activity [197, 198].

Because nitric oxide, superoxide and peroxynitrite involved in mitochrondropathy have quite limited diffusion distances in biological tissues and because the mechanisms involved in the cycle is at the level of individual cells, the pathological changes of mitochondropathy first are restricted to a local area (Fig. 7.21). The consequences of this primarily local mechanism, however, later show up in multisystem ailments through which variations are sees in regard to symptoms and signs from one patient to another. This difference in tissue impact of the NO/ONOO⁻ cycle and its subsequent effect on mitochondrial function leads to exactly such dissimilarities in symptoms and signs as they are observed in different sets of patients. There is evidence from brain scan studies where variable tissue distribution in the brains of patients suffering from one of these illnesses could be directly visualized (Fig. 7.22).

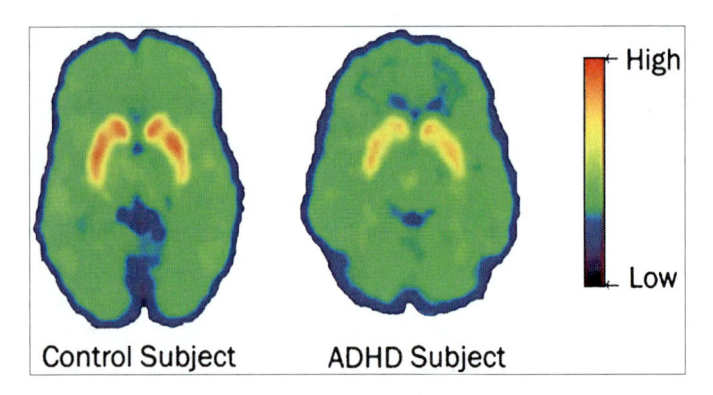

Fig. 7.22 Example of a brain scan of a patient with ADHD (attention deficit hyperactivity disorder) which is also suspected to be induced by the NO/ONOO⁻ cycle resulting in areas with a lowered perfusion when compared to control

Because of such similar cause, therapy primarily should focus on the downregulation of the activated NO/ONOO⁻ cycle aiming to treat the cause, and not just the symptoms. Nitric oxide, superoxide and peroxynitrite are the three compounds centrally involved in the NO/ONOO⁻ cycle, which because of their relatively short half-lives in biological tissues, and their tendency not to diffuse very far from their site of origin in the body, there is a good chance for treatment. And although nitric oxide has the longest half-life, it only diffuses about one millimeter from its site of origin. Most of the mechanisms of action, as outlined by arrows in the following figure, act at the cellular levels, with the consequence that the NO/ONOO⁻ cycle may be elevated in one tissue of the body but may show little or no elevation in adjacent tissues, having little or no impact on mitochondrial function in a neighboring organ. This local nature in biochemistry illustrates, that one may have various effects on organ malfunction from one patient to another, leading to different symptoms and signs from one individual to another with Parkinson, Alzheimer, CFS, FMS, ALS or even diabetes. The striking variation in symptoms from one subject to another can even be observed in patients having the same kind of ailment and presents a great puzzle to the physician. Such variations, however, can easily be explained by the local nature of the NO/ONOO⁻ cycle and its ensuing detrimental effects on mitochondrial function (Fig. 7.23).

Yet, the primarily local nature does not imply that there are no systemic effects. The antioxidant depletion of superoxide dismutase (SOD) and/or glutathione (GSH) produced by local oxidative stress, which to a substantial extent is due to the release of inflammatory cytokines, results in systemic effects. Such systemic effects may, in turn, produce an inefficient neuroendocrine and immune function, resulting in systemic symptoms. However, the primary local nature helps us to understand the profound variations in symptoms and signs seen from one patient to another. And although the different ailments may differ from one patient to another and may also

7.3 Neurodegeneration Resulting from Blockade in the Energy Pathway

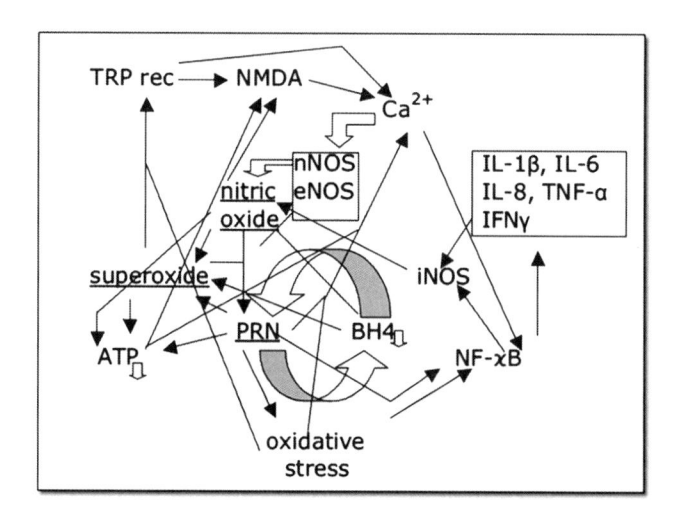

Fig. 7.23 Activation of the vicious nitric oxide (NO) and peroxynitrite (ONOO⁻) cycle via the endothelial and the neuronal nitric oxide synthase (eNOS, nNOS) resulting in mitochondropathy and an insufficient ATP formation. Simultaneous nuclear factor kappa B (NF-κB) activation in the immune system results in an increased release in proinflammatory cytokines (IL-1β, IL-6, TNF-α, etc.), and an overexpression of the NMDA-, the TRP- (transient receptor potential ion channel) and the vanilloid-receptor followed by an increased Ca^{++}-influx, being the cause for neurodegeneration

differ in regard to the complaints expressed by the patient, although they all share the same etiology of an overexpressed NO/ONOO⁻ cycle, resulting in additional ailments all of which are caused by insufficient mitochondrial function. For instance, after the heart, the brain is the most voracious consumer of energy in the body. A study found that deficiencies in the ability to fuel brain neurons might lead to some of the cognitive impairments associated with autism. Mitochondria are the primary source of energy production in cells and carry their own set of genetic instructions, mitochondrial DNA (mtDNA), to carry out aerobic respiration. Thus, any dysfunction in mitochondria of the neuronal network is associated with a number of other neurological conditions, including Parkinson's or Alzheimer's disease (Fig. 7.24), schizophrenia and bipolar disorder. Researchers found that mitochondria from children with autism consumed far less oxygen than mitochondria from the group of control children, a significant sign of lowered mitochondrial activity.

Another example for insufficient energy production is the oxygen consumption in one critical mitochondrial enzyme complex, the NADH oxidase. In autistic children was only a third of that found in control children. Reduced mitochondrial enzyme function proved widespread among the autistic children. Eighty percent had lowered activity in NADH oxidase than did controls, while 60%, 40% and 30% had low activity in succinate oxidase, ATPase and cytochrome c oxidase, respectively. The researchers went on to isolate the origins of these defects by assessing the activity of each of the five enzyme complexes involved in mitochondrial respiration [41]. Complex I was the site of the most common deficiency, found in 60% of autistic subjects, and occurred five out of six times in combination with complex V, and

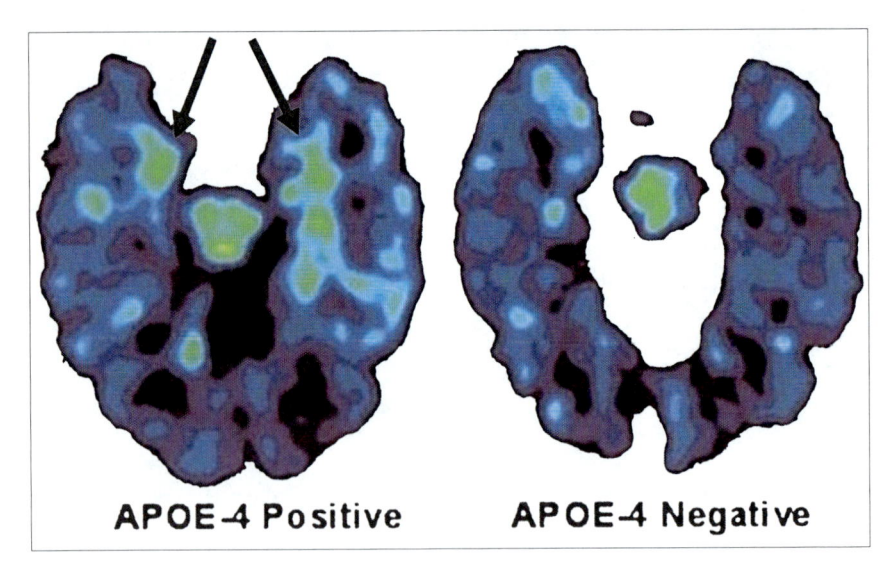

Fig. 7.24 PET brain scan of a patient revealing plaque and tangle accumulation, abnormal protein deposits and hallmarks of neurodegeneration, demonstrating advanced Alzheimer's

other children had problems in complexes III and IV. Levels of pyruvate, the raw material mitochondria transform into cellular energy, also were elevated in the blood plasma of autistic children. This suggests the mitochondria of children with autism are unable to process pyruvate fast enough to keep up with the demand for energy, pointing to a novel deficiency at the level of an enzyme named pyruvate dehydrogenase. Mitochondria also are the main intracellular source of oxygen free radicals. Free radicals are very reactive species that can harm cellular structures, including DNA. Although, cells in general are able to repair typical levels of such oxidative damage, Giulivi and her colleagues found that hydrogen peroxide levels in autistic children were twice as high as in normal children. As a result, these cells of children with autism were exposed to higher oxidative stress [41].

7.3.1 Neurodegenerative Diseases Have a Common Denominator

It has been demonstrated that Parkinson, Alzheimer and amyotropohic lateral sclerosis (ALS) all seem to share many properties in common, and much of the research on the three ailments has focused on how they differ, rather than how they are similar. Yet, similarities are more striking when one focus on the elevation of NO/ONOO$^-$ cycle elements (Fig. 7.25). They all have in common the oxidative stress, an elevated nitric oxide and superoxide concentration, an elevated peroxynitrite level, the increase in intracellular calcium, the raised NF-κB activity, the increase in inflammatory cytokines, plus excitotoxicity with an excess in NMDA-activity

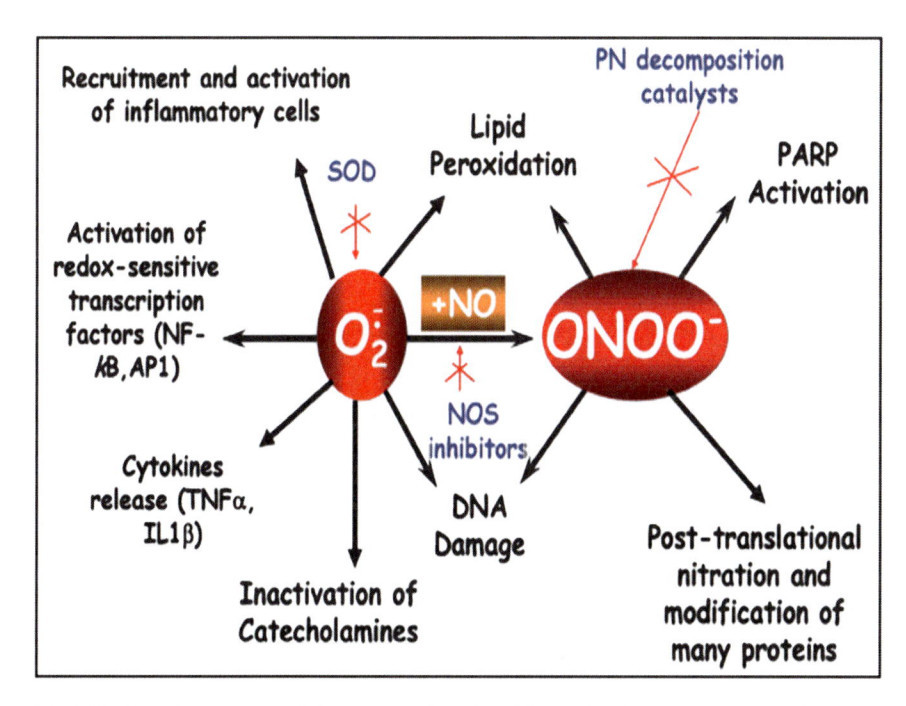

7.3 Neurodegeneration Resulting from Blockade in the Energy Pathway

Fig. 7.25 Reactive oxygen and nitrogen species (nitroxide species) in particular superoxide (O_2^-), nitric oxide (NO) and/or peroxynitrite (ONOO⁻) and the products of their interaction, are potent pro-inflammatory mediators involved in a variety of diseases. *PARP* poly(ADP-ribose) polymerases induce cellular necrosis

all of which result in mitochondrial dysfunction. One aspect of the cycle that has only recently been discovered is the depletion in the coenzyme tetrahydroxybioterin (BH_4), which has been shown to occur in Alzheimer's disease and in Parkinson, but so far has not been studied in ALS. All these three neurodegenerative diseases produce neurodegeneration via apoptotic cell death of neurons. It is well known that peroxynitrite, if sufficiently elevated, can induce apoptosis, a fact that is consistent with the overexpressed NO/ONOO⁻cycle and mitochondropathy(Fig. 7.25). It has also been demonstrated, that elevation of nitric oxide synthesis by mitochondrial nitric oxide synthase (mtNOS) not only leads to elevated nitric oxide but also to an elevated level of superoxide and peroxynitrite plus various types of mitochondrial damage induced by the activated NO/ONOO⁻ cycle. If this elevation of mitochondrial nitric oxide is sufficiently high, it leads to apoptosis (programmed cell death), another way in which the cycle can lead to apoptotic cell death with consequent neurodegeneration [199].

In some cases the apoptotic stimuli comprise extrinsic signals such as the binding of death inducing ligands to cell surface receptors called death receptors (Fig. 7.26). These ligands can either be soluble factors or can be expressed on the surface of cells such as cytotoxic T lymphocytes. The latter occurs when T-cells recognise damaged or virus infected cells and initiate apoptosis in order to prevent

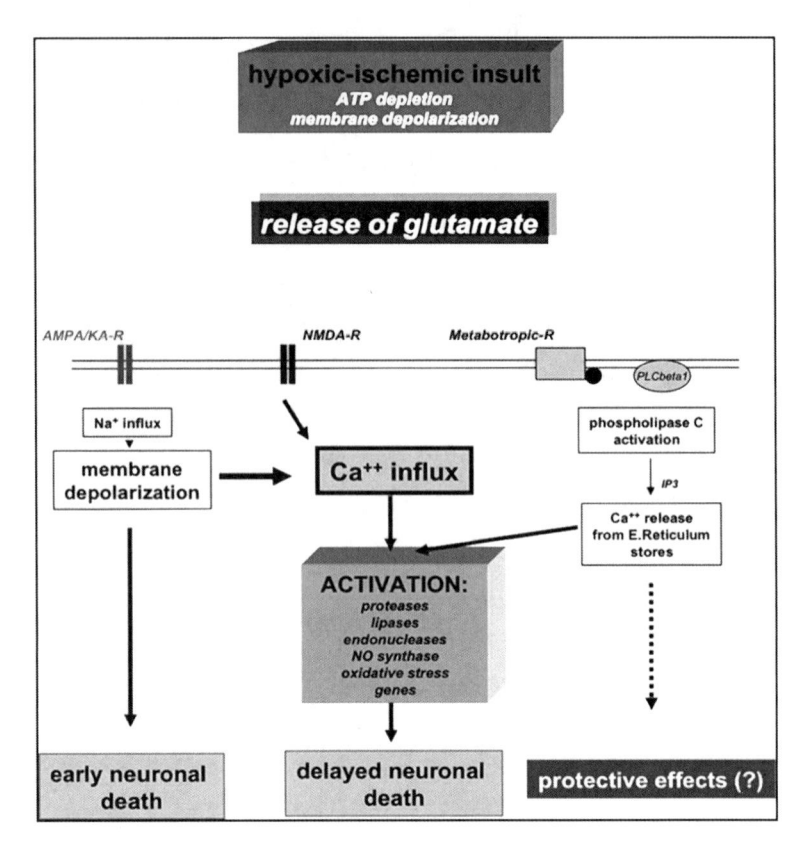

Fig. 7.26 Diagram of the molecular cascade leading to neuronal cell death after hypoxic–ischemic insult in neuronal cells. *R* receptors

damaged cells from becoming neoplastic (cancerous) or virus-infected cells from spreading the infection. Apoptosis can also be induced by cytotoxic T-lymphocytes using the enzyme granzyme. In other cases apoptosis can be initiated following intrinsic signals that are produced following cellular stress. Cellular stress may occur from exposure to radiation or chemicals or to viral infection. It might also be a consequence of growth factor deprivation or oxidative stress caused by free radicals. In general intrinsic signals initiate apoptosis via the involvement of mitochondria. The relative ratios of the various bcl-2 proteins can often determine how much cellular stress is necessary to induce apoptosis (Fig. 7.26).

Induced by oxygen and glucose deprivation is followed by a severe decrease in adenosine triphosphate content, causing failure of the sodium–potassium–adenosine triphosphate pump that maintains the polarity of the neuronal membrane. Membrane depolarization induces excessive glutamate release, leading to a massive influx of sodium and calcium via the NMDA receptor. The increase in intracellular calcium activates several enzymes including phospholipases, proteases, and endonucleases, and neuronal nitric oxide synthase (nNOS). Further deleterious effects occur at the

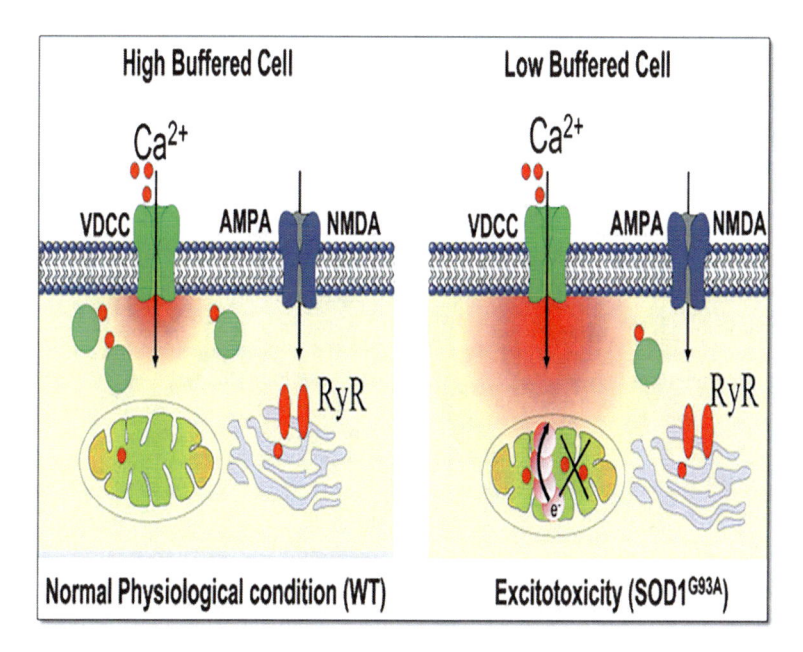

7.3 Neurodegeneration Resulting from Blockade in the Energy Pathway

Fig. 7.27 Casual role of excess ROS and NOS with low Ca^{2+} buffering followed by excitotoxicity under physiological stress and pathophysiological conditions in motor neurons (MNs). *VDCC* Voltage-dependent calcium channels. *RyR* Ryanodine receptors form a class of intracellular calcium channels in various forms of excitable animal tissue like muscles and neurons. It is the major cellular mediator of calcium-induced calcium release (CICR) in animal cell. AMPA = (α-amino-3-hydroxy-5-methyl-4-isoxazolepropionic acid receptor) a subgroup of the glutamate -, the NMDA-, and the kainate-receptors. They affection channels, which mediate the fast component of postsynaptic transmission. Their activation increases a change in conductivity of the postsynaptic membrane

reperfusion phase as a result of excessive production of superoxide, which damages the mitochondria and produces peroxynitrate by forming a complex with NO and the generation of free radicals. Reactive oxygen species cause oxidation of lipids and proteins and damage to DNA. The most important change in neurodegenerative diseases is that neuronal apoptosis (programmed cell death) is induced via these two elevated levels of nitric oxide and peroxynitrate. And while it has been shown that the amyloid Aß has an important role in the causal relationship of Alzheimer, the rate-limiting step in the production of Aß is the proteolytic clipping of its precursor protein APP by a protease known as beta-secretase or BACE1. The production of BACE1 is known to be stimulated by both oxidative stress and by NF-κB activity, thus underlining the importance of the two elements NO and OONO$^-$. It follows that beta-secretase (BACE1) activity will be elevated by this cycle, in turn leading to a tissue-specific production of the Aß protein (Fig. 7.27; Table 7.3). Furthermore, a paper by Vassar and coworkers has shown that BACE cleaves membrane bound amyloidal precursor protein (APP), leading to the generation of amyloidal beta protein fragments (Aβ) [200]. Although a normal process in healthy neurons, this pathway becomes pathogenic in Alzheimer's Disease due to an increase in BACE

Table 7.3 Correlates in neurodegenerative diseases that can be viewed as tissue-specific consequences of the NO/ONOO- and mitochondrial disease

Histological changes observed in neurodegenerative disease	Results supporting the involvement of the NO/OOHNO- mitochondrial disease
Parkinson with Lewy body formation	Peroxynitrite-mediated nitration and oxidation of α-synuclein [204, 205]. Nitric oxide-mediated nitrosylation of the protein parkin, converting it from a protective protein to a component of the protein aggregate that makes up Lewy bodies, also containing the modified α-synuclein and other proteins
Alzheimer with hyperphosphorylated tau protein aggregates leading to neurofibrillary tangles	Peroxynitrite-mediated nitration and oxidation of tau protein, followed by aggregation [206–208]. Phosphorylation is stimulated by NO, excessive NMDA activity and other elements of ROS and NOS. Hyperphosphorylated tau protein aggregates leading to formation of neurofibrillary tangles
Amyotrophic lateral sclerosis with neurofilament aggregates	Nitric oxide-Peroxynitrite-mediated apoptosis [209] with loss of down-regulation of NMDA receptor activity [210] results in neurofilament aggregate formation, being part of the motor neuron cell death in ALS

Activity [201, 202]. The excessive amounts of Aβ produced accumulate and form extracellular amyloidal plaques. The plaques and free Aβ fragments induce oxidative stress, resulting in membrane damage and an influx of calcium, which activates proteases involved in the generation of neurofibrillary tangles (NFTs). The lowered energy metabolism induced by the mitochondrial breakdown also induces an increase in BACE1 synthesis which consequently is followed by Aβ synthesis, acting by increasing eIF2a phosphorylation [203]. Therefore, additional mechanisms seem to play a role in the elements of the NO/OONO- cycle resulting in an elevation of BACE 1 activity. The Aβ protein in turn is known to activate nitric oxide, iNOS induction, superoxide, peroxynitrite, oxidative stress, excitotoxicity, energy depletion and inflammatory cytokine levels on neuronal cells in culture. From such data it seems clear that Aβ by itself can stimulate the cycle via protein aggregates that insert themselves into the plasma membrane and function as calcium channels, resulting in an increase of calcium ions into the cell. Such excessive intracellular calcium initiate and stimulate the NO/OONO- cycle. It follows that Aβ levels increased by the cycle in return stimulate this cycle, presenting a deleterious effect on mitochondria in neuronal cells where it is synthesized. It there does not present any conflict between the notion that Aβ has a (one) causal role in Alzheimer's and the concept that Alzheimer's disease is a radical stress disease resulting in mitochondropathy and a programmed neuronal cell death (Table 7.3). The role of Aβ as part in Alzheimer's disease may explain some of the differences between Alzheimer and other mitochondrion related diseases. In Alzheimer there is a trend to spread over time, impacting not only the entire cerebrum but also some adjacent tissues, so it is not highly localized. The localized nature of the NO/ONOO- cycle and the corresponding mitochondrial diseases is caused by the short half-life of superoxide, peroxynitrite and nitric oxide, allowing them to diffuse only a short distance from

7.3 Neurodegeneration Resulting from Blockade in the Energy Pathway

their site of origin. In contrast, because of the stability of Aß, there not only is an accumulation over time but also an expansion into tissues that are of substantial distance from the site of start, with a steady progression not only in severity but also in expanding the area of tissue apoptosis (Table 7.3). Because of this twofold origin, Alzheimer should be treated not only with the aim to lower the concentration or the activity of Aß, but also by using agents that act to lower the NO/ONOO⁻ toxicity and the mitochondrial impairment.

It can be seen from the above Table 7.3 that correlates of these four neurodegenerative diseases do exist: Aß accumulation and aggregation in Alzheimer, Lewy body formation in Parkinson, hyperphosphorylated tau protein leading to neurofibrillary tangles in Alzheimer as well as formation of neurofilament aggregates in ALS appear to be tissue consequences of the NO/ONOO⁻ induced specific mitochondrial disease. Three of them, Aß accumulation, Lewy body formation and formation of neurofilament aggregates appear to be tissue-specific elements of the NO/ONOO⁻ specific mitochondrial disease, and contribute, in a tissue specific manner, to the etiology of Alzheimer's and ALS. Therefore, there is strong evidence supporting the view that Alzheimer's, Parkinson's and ALS are of specific NO/ONOO⁻ radial mitochondrial nature.

First of all there was strong evidence coming from studies related to CFS, MCS and FMS being due to a NO/ONOO⁻ mitochondrial cause. This was followed by substantial evidence for mitochondrial impairment within the three neurodegenerative diseases Parkinson, Alzheimer and ALS. By putting together some the individual mechanisms demonstrated in biochemistry and physiology, it all results in a complex and vicious cycle, which predicts a variety of properties, based simply on the combination of interactive mechanisms within the cycle, leading into excitotoxicity (Fig. 7.27). Such excitoxicicity may stem from previous exposure to organophosphates, pesticides, pyrethroid, mercury, carbon monoxide, and/or hydrogen sulfide. In addition, intake of so called preservatives in diet, artificial coloring and flavoring in the various products of daily food, all of which after intake over several years are able to spark silent mitochondropathy into a vicious cycle with impairment and chronic disability.

For instance, low Ca^{2+} buffering in amyotrophic lateral sclerosis (ALS) with vulnerable hypoglossal motor neurons exposes mitochondria to higher Ca^{2+} loads compared to highly buffered cells (Fig. 7.27). Under normal physiological conditions, the neurotransmitter opens glutamate, NMDA and AMPA receptor channels, and voltage dependent Ca^{2+} channels (VDCC) with high glutamate release, which is taken up again by EAAT1 and EAAT2. This results in a small rise in intracellular calcium that can be buffered in the cell. In ALS, a disorder in the glutamate receptor channels leads to high calcium conductivity, resulting in high Ca^{2+} loads and increased risk for mitochondrial damage. This triggers the mitochondrial production of reactive oxygen species (ROS), which then inhibit glial EAAT2 function, leading to further increase in glutamate concentration at the synapse and further rises in postsynaptic calcium levels, contributing to the selective vulnerability of motor neurons in ALS [211]. Excitotoxicity may also be involved in the retina [212], in spinal cord injury, stroke, traumatic brain injury and neurodegenerative diseases of the central nervous system (CNS) such as multiple sclerosis, Alzheimer's disease, Parkinson's disease, and alcoholism [213] or alcohol withdrawal [214] and Huntington's disease [215, 216]. Other common conditions that cause excessive glutamate concentrations around neurons are hypoglycemia [217] as well as in status epilepticus [218]. Excitotoxicity, however, can also occur from substances produced within the body (i.e. endogenous excitotoxins). Glutamate is a prime example of an excitotoxin in the brain, and it is also the major excitatory neurotransmitter in the mammalian CNS [219]. During normal conditions, glutamate concentration can be increased up to 1 mM in the synaptic cleft, which is rapidly decreased in the slip of milliseconds [220]. When the glutamate concentration around the synaptic cleft cannot be decreased or reaches higher levels, the neuron kills itself by the process called apoptosis [221]. Such a pathologic cascade can also occur after brain injury. Brain trauma or stroke causes ischemia, in which blood flow is reduced to inadequate levels followed by accumulation of glutamate and aspartate in the extracellular fluid, resulting in cell death, which is aggravated by lack of oxygen and glucose. The biochemical cascade caused by ischemia and involving excitotoxicity is the ischemic cascade, where a series of biochemical reactions are initiated in the brain and other aerobic tissues after seconds to minutes of ischemia. This is typically secondary to stroke, injury, or cardiac arrest due to a heart attack. Most ischemic neurons that die do so due to the activation of chemicals produced during and after ischemia. The ischemic cascade usually goes on for 2–3 h but can last for days, even after normal blood flow returns [222]. Normally, a cascade is a series of events in which one event triggers the next, in a linear fashion. Thus "ischemic cascade" is actually a misnomer, since in it events are not always linear: in some cases they are circular, and sometimes one event can cause or be caused by multiple events one of which also comprises mitochondrial breakdown and/or glutamate overexcitation (Table 7.4).

The fact that the ischemic cascade involves a number of steps has led scientists to suspect that neuroprotectants such as calcium channel blockers or glutamate antagonists could be used to interrupt the cascade at a single one of the steps, blocking the downstream effects. Though initial trials for such neuroprotective drugs were hopeful, until recently, human clinical trials with neuroprotectants such as NMDA receptor antagonists were unsuccessful. Because of the events resulting from ischemia and glutamate receptor activation, a deep chemical coma may be induced in patients with brain injury to reduce the metabolic rate of the brain and save energy to be used to remove glutamate actively.

Another approach is based on the pathogenesis that one of the damaging results of excess calcium in the cytosol is the opening of the mitochondrial permeability transition pore, a pore in the membranes of mitochondria that opens when the organelles absorb too much calcium. Opening of the pore may cause mitochondria to swell and release proteins that can lead to apoptosis. The pore can also cause mitochondria to release more calcium. In addition, production of adenosine triphosphate (ATP) may be stopped, and ATP synthase may in fact begin hydrolysing ATP instead of producing it [223]. And since inadequate ATP production resulting from brain trauma it can eliminate electrochemical gradients of certain ions. Glutamate transporters, however require the maintenance of these ion gradients to remove glutamate from

7.3 Neurodegeneration Resulting from Blockade in the Energy Pathway   113

Table 7.4 The cascade with mitochondrial breakdown, glutamate overexpression and excitotoxicity following neuronal ischemia

1. Lack of oxygen causes the neuron's normal process for making ATP for energy to fail.
2. The cell switches to anaerobic metabolism, producing lactic acid.
3. ATP-reliant ion transport pumps fail, causing the cell to become depolarized, allowing ions, including calcium (Ca^{++}), to flow into the cell.
4. The ion pumps can no longer transport calcium out of the cell, and intracellular calcium levels get too high.
5. The presence of calcium triggers the release of the excitatory amino acid neurotransmitter glutamate (Fig. 7.28).
6. Glutamate stimulates AMPA receptors and Ca^{++}-permeable NMDA receptors, which open to allow more calcium into cells.
7. Excess calcium entry overexcites cells and causes the generation of harmful chemicals like free radicals, reactive oxygen species and calcium-dependent enzymes such as calpain, endonucleases, ATPases, and phospholipases in a process called excitotoxicity. Calcium can also cause the release of more glutamate.
8. As the cell's membrane is broken down by phospholipases, it becomes more permeable, and more ions and harmful chemicals flow into the cell.
9. Mitochondria break down, releasing toxins and apoptotic factors into the cell.
10. The caspase-dependent apoptosis cascade is initiated, causing cells to "commit suicide."
11. If the cell dies through necrosis, it releases glutamate and toxic chemicals into the environment around it. Toxins poison nearby neurons, and glutamate can overexcite them.
12. If and when the brain is reperfused, a number of factors lead to reperfusion injury.
13. An inflammatory response is mounted, and phagocytic cells engulf damaged but still viable tissue.
14. Harmful chemicals damage the blood brain barrier.
15. Cerebral edema (swelling of the brain) occurs due to leakage of large molecules like albumins from blood vessels through the damaged blood brain barrier. These large molecules pull water into the brain tissue after them by osmosis. This "vasogenic edema" causes compression of and damage to brain tissue.

the extracellular space. The loss of ion gradients results not only in the stopping of glutamate uptake, but also in the reversal of the transporters, causing them to release glutamate and aspartate into the extracellular space. This results in a buildup of glutamate and further damaging activation of glutamate receptors [224].

On the molecular level, calcium influx is not the only factor responsible for apoptosis induced by excitotoxicity. Recently [226] it has been noted that extrasynaptic NMDA-receptor activation, triggered by both glutamate exposure or hypoxic/ischemic conditions, activate a CREB (cAMP response element binding) protein shutoff, which in turn causes a loss of mitochondrial membrane potential with apoptosis. On the other hand, activation of synaptic NMDA receptors only being triggered by the CREB pathway, stimulate the BDNF (brain-derived neurotrophic factor), not activating apoptosis. BDNF acts on certain neurons of the central nervous system and the peripheral nervous system, helping to support the survival of existing neurons, and encourage the growth and differentiation of new neurons and synapses [227, 228]. In the brain, it is active in the hippocampus, cortex, and basal forebrain, areas vital to learning, memory, and higher thinking, being important for long-term

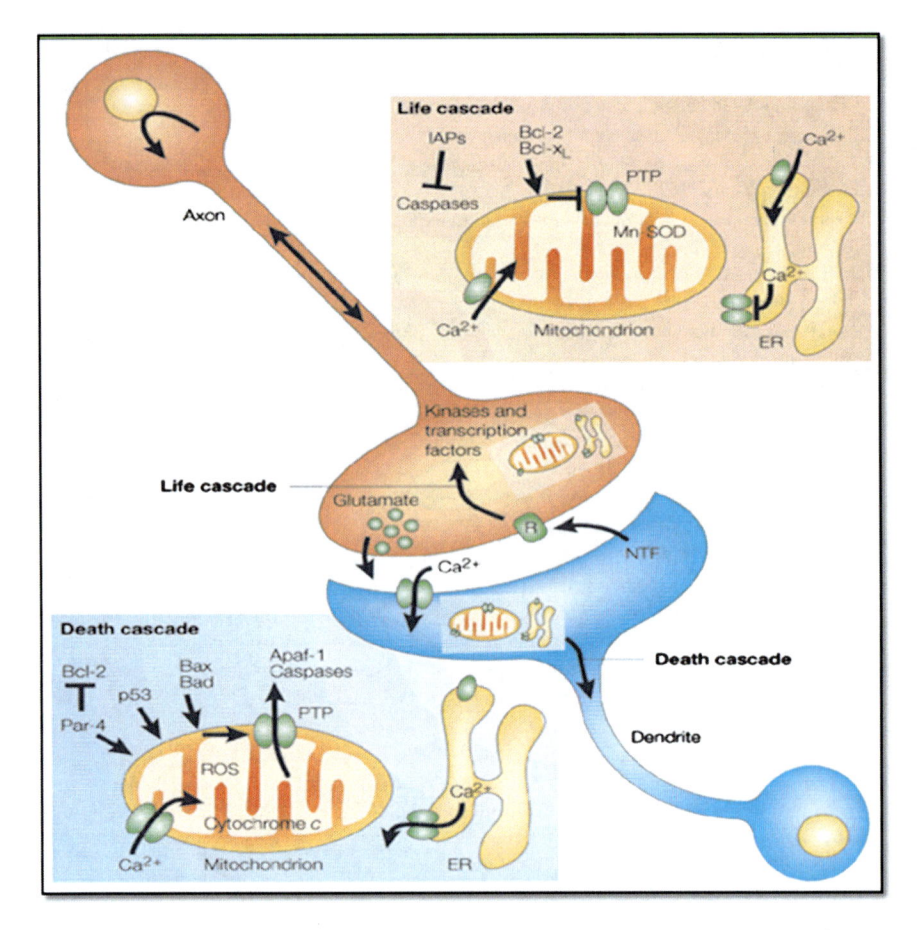

Fig. 7.28 Principle of cascade in traumatic CNS injury with glutamate-induced excess Ca^{2+} influx into mitochondria resulting in programmed cell death (Modified from [225])

memory [229]. BDNF was the second neurotrophic factor to be characterized after nerve growth factor (NGF).

Such fall in mitochondrial inner membrane potential can be measured by flow cytometry (Fig. 7.29). Normal mitochondrial function has a requirement for a negative charge on the inner side of the mitochondrial inner membrane. This is maintained by the asymmetrical distribution of H$^+$ ions across the membrane, giving rise to both a potential difference and a chemical pH gradient. Loss of mitochondrial membrane potential is detected in cells by measuring a loss of dye from the inner mitochondrial matrix or a change in fluorescence of the dye as the transmembrane potential drops (Fig. 7.29). For instance the toxin staurosporine induces apoptosis loosing mitochondrial function over time. Labeling such cells with annexin V-FITC, DAPI and DiIC(5) shows that not only do apoptotic cells lose mitochondrial function, a high proportion of live cells progressively lose mitochondrial function with increasing exposure to staurosporine.

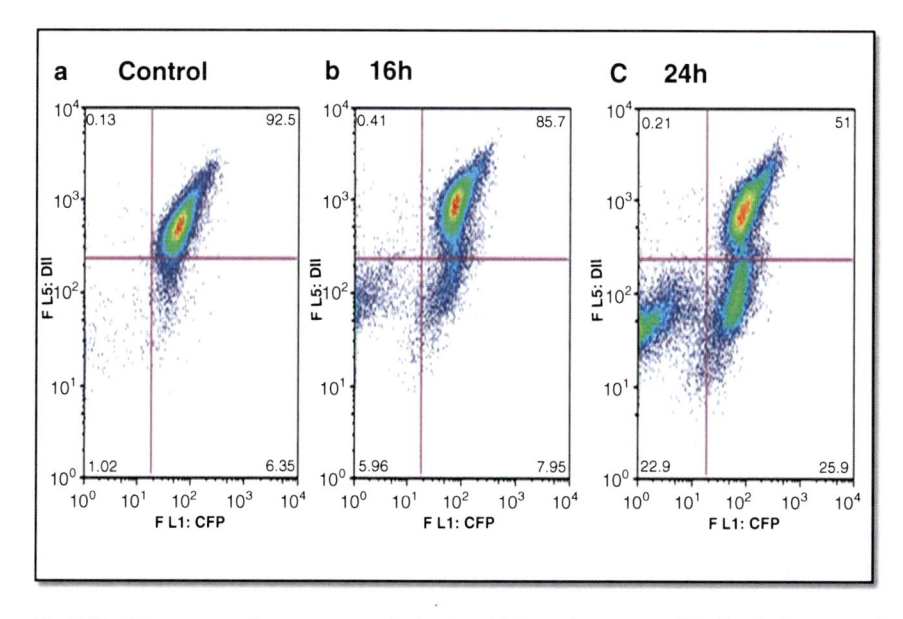

7.3 Neurodegeneration Resulting from Blockade in the Energy Pathway

Fig. 7.29 Time course of measurement of mitochondrial membrane potential when being exposed to a toxin using flow cytometry and luminescence

7.3.2 Mitochondropathy and Neurodegeneration in Different Disease States

As demonstrated, short-term stressors initiate various cases of diseases by increasing nitric oxide levels and/or other cycle elements like peroxynitrite and NF-κB. The increase in NO and peroxynitrite activates the NO/ONOO⁻ cycle which then causes the chronic mutiple chemical sensitivity, during which all elements of the cycle will all be elevated. All symptoms and signs of this ailment are caused by the elevated NO/ONOO⁻ cycle, involving nitric oxide, superoxide, peroxynitrite, NF-κB, oxidative stress, and an increase in vanilloid- as well as NMDA-receptor activity. But most important, all these changes finally result in a decreased synthesis of ATP at the cellular level of mitochondria. The biochemical changes of the cycle can be considered as local, because nitric oxide, superoxide and peroxynitrite diffuse only a few cm within the surrounding tissue, they have limited biological half lives and because of a positive feedback the cycle is maintained solely at the cellular level. Therapy should focus on a down-regulation of parts of the NO/ONOO⁻ cycle, rather than just on the relief of symptoms in this group of neurodegenerative diseases. In addition, there are five principles in MCS that also are observed in other acquired (neurodegenerative) mitochondrial diseases:

- Short-term stressors initiate various ailments by increasing nitric oxide levels and/or other cycle elements like peroxynitrite.
- This increase in NO and peroxynitrite initiates a NO/ONOO⁻ cycle, which eventually results in a chronic ailment where all elements of that cycle are elevated.

- The symptoms and signs of these ailments are primarily caused by elevated elements of the NO/ONOO⁻ cycle, specifically nitric oxide, superoxide, peroxynitrite, NF-κB, oxidative stress, and an increase in vanilloid- and NMDA-receptor activity.
- The biochemical changes within the cycle are of local nature, because nitric oxide, superoxide and peroxynitrite have limited half-lives in biological tissues and because there is a positive feedback that maintains the cycle strictly at a cellular level.
- Therapy should focus on the down-regulation of parts of the NO/ONOO⁻ cycle, rather than on the relief of symptoms as done in modern medicine.

Multiple chemical sensitivity (MCS) has been historically viewed as the most challenging of this group of neurodegenerative ailments and its properties have in the past been a great challenge in medicine. Commonly suspected substances include smoke, pesticides, plastics, synthetic fabrics, scented products, petroleum products and paints. Symptoms may be vague and non-specific, such as nausea, fatigue, and headaches. Now there is a detailed a explanation for its diverse properties, one that is supported by a wide array of evidence, coming up with a mechanism that explains the many puzzling properties of this disease. But most of all, the convincing arguments for this mechanism is the explanation for the cause of so many unknown characteristics.

7.3.2.1 Evidence for Multiple Chemical Sensitivity

There are seven classes of chemicals associated in the initiation of a MCS, while the same classes of chemicals are also involved in triggering an increase in sensitivity response in those patients already presenting subtle signs of the ailment:

1. The large class of organic solvents and related compounds
2. Organophosphorous/carbamate pesticides
3. Organochlorine pesticides including chlordane & lindane
4. Pyrethroid pesticides
5. Mercury (when working with its derivative methylmercury)
6. Hydrogen sulfide
7. Carbon monoxide

Members of these seven classes of chemicals are known to produce an increase in NMDA activity and the degree of toxicity within the body is known to be greatly decreased by NMDA antagonists. Such observations show that the seven classes of chemicals all produce a common response in the body, i.e. excess in NMDA-activity. Besides, there are other observations suggesting an excess in NMDA activity in MCS patients (Fig. 7.30):

- MCS patients are hypersensitive to monosodium glutamate, being the main physiological agonist of the NMDA receptor site.
- The NMDA antagonists dextromethorphane and ketamine are reported to lower reactions to chemicals in MCS patients
- People carrying an allele of the CCK-B receptor producing increased NMDA activity are more susceptible to getting MCS.

7.3 Neurodegeneration Resulting from Blockade in the Energy Pathway

Fig. 7.30 Interaction with NMDA receptor activation and the development of mitochondropathy through excess formation of NO

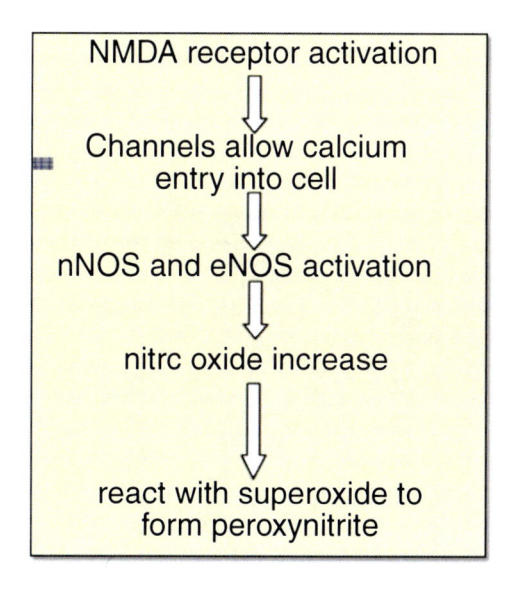

```
NMDA receptor activation
          ⇓
Channels allow calcium
     entry into cell
          ⇓
nNOS and eNOS activation
          ⇓
   nitrc oxide increase
          ⇓
  react with superoxide to
    form peroxynitrite
```

These observations show, that not only these seven classes of chemicals all produce an excessive NMDA activity; they also show that this response is essential for their action as toxicants in the body.

There are clinical observations that connect excessive NMDA activity with MCS:

- Bell and others have proposed that neural sensitization has a key role in MCS and the probable mechanism for such neural sensitization, called long-term potentiation (LTP), is known to involve increased NMDA activity [230, 231].
- Elevated NMDA activity has been shown to play an essential role in several animal models of MCS [73].
- Elevated NMDA activity also appears to play a part in related illnesses such as fibromyalgia, chronic fatigue syndrome and post-traumatic stress disorder, with the most widespread evidence for such a role in fibromyalgia syndrome (Ref. [4] of Chap. 5) [232].

All these observations argue in favor of excessive NMDA activity being the key role in MCS, and that chemicals involved act as toxicants by producing excessive NMDA activity. This also provides an explanation for the question in patients with MCS, how so many chemicals are able produce a common response in the body? The finding that all seven classes of these chemicals all act as toxicants largely through a common denominator, i.e. excessive NMDA activity, has important implications in toxicology and in environmental medicine that go far beyond MCS. These chemicals, after all, implicate that they may be a cause in other diseases. Excessive NMDA activity should be viewed as a common end point in toxicology, following chemicals (mainly carcinogens) acting as "genotoxins" and other chemicals acting as endocrine disruptors (Fig. 7.31).

One of the breakthroughs in understanding the nature of MCS came from a comparison of the NO/ONOO$^-$ cycle model of these illnesses with the neural

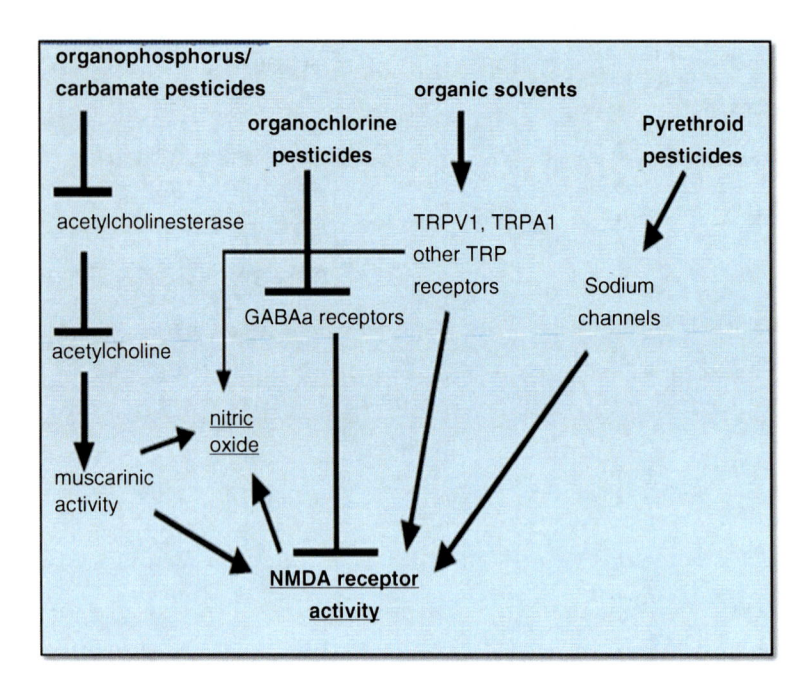

Fig. 7.31 Action of pesticides and organic solvents resulting in multiple chemical sensitivity (MCS) (Adapted from (Ref. [11] of Chap. 5)

sensitization model of MCS developed by Bell and coworkers [230]. They argued that the most important mechanism of MCS was neural sensitization in the hippocampus region of the brain. This is the same region that has key functions in learning and memory. The idea Bell and coworkers developed was that the synapses in the brain, the contacts between neurons by which one stimulates another, may become both hypersensitive and hyperactive in response to chemical exposure. The basic idea here is that this process of neural sensitization, which is involved on a very selective basis in learning and memory, appears to be activated especially in MCS. The main mechanism of neural sensitization is known as long term potentiation (LTP). LTP is known to involve increased NMDA receptor activity, increased cellular calcium influx, nitric oxide and also superoxide accumulation. From this one can derive the major connections between the NO/ONOO⁻ cycle mechanism and the neural sensitization mechanism. Having chemicals, which produce an increased NMDA activity, they in turn stimulate the long term potentiation mechanism in neuronal activity. Several elements of the NO/ONOO⁻ cycle have an impact on LTP, including NMDA activity, intracellular calcium accumulation, as well as nitric oxide and superoxide production. The following diagram may depict how some of these elements interact among each other, all resulting in mitochondropathy with insufficient ATP formation (Fig. 7.32).

7.3 Neurodegeneration Resulting from Blockade in the Energy Pathway

Fig. 7.32 The neural sensitization cycle where NO is a physiological messenger system reactivating NMDA stimulation, the development of nociception and a depletion in ATP

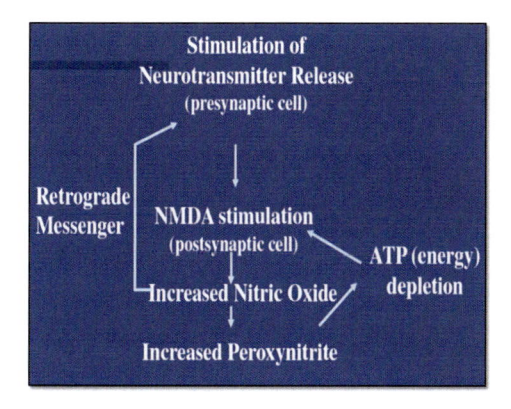

In addition to toxic material, there are also different short-term stressors that are reported to putatively initiate CFS:

1. Viral infections involving multiple viruses
2. Bacterial infections
3. Physical trauma
4. Severe psychological stress
5. Toxoplasmosis (protozoan infection)
6. Ciguatoxin poisoning
7. Exposure to solvents
8. Carbon monoxide exposure
9. Ionizing radiation exposure

Any infection and the exposure to ionizing radiation all act via an uncoupling of iNOS and NF-κB elevation, whereas the others act, at least in part via excessive NMDA activity with the capacity to activate the NO/ONOO⁻ cycle (Ref. [10] of Chap. 5). There is no other explanation how all of these stressors can produce a common response in the body. All these nine mechanisms have important roles in generating the chemical sensitivity reported in MCS: Nitric oxide (NO) acting as a retrograde messenger, increasing NMDA stimulation. Peroxynitrite acting to decrease energy metabolism in mitochondria, while producing an increase in NMDA sensitivity to stimulation. And peroxynitrite acting to decrease energy metabolism, with resulting in lesser production of the inhibitory transmitter GABA, leading to increased decoupling of excitatory NMDA activity. In addition, peroxynitrite can nitrate a residue on the NMDA receptor, producing a permanently open channel, which clinically is related to an increase in pain sensations. Peroxynitrite can also nitrate the glutamine synthase protein leading to glutamate accumulation in the cell and in the extracellular fluid, leading to increased NMDA stimulation with pain as seen in FMS. Chemical actions increase NMDA activity in regions of brain where the NO/ONOO⁻ cycle due to previous chemical exposure is already upregulated. Nitric oxide also inhibits cytochrome P450 metabolism resulting in a reduced detoxification with the likelihood of an increased sensitivity to some chemicals metabolized by the liver. Above all, oxidants lead to increased TRPV1 and TRPA1 (vanilloid receptor) activity, with an upregulation of sensitivity to chemicals acting via these receptors. Since peroxynitrite produces a breakdown of the blood brain barrier (BBB), there is an increase in chemical access to neuronal cells of the brain. In addition, chemical sensitivity in other parts of the body have been demonstrated by others with peripheral sensitivity occurring in the lower lungs, in the upper respiratory tract, in the gastrointestinal (GI) tract and neurogenic inflammation in the skin with edema, vasodilatation, and infiltrates of leukocytes [233]. These sensitivity responses are initiated by previous chemical exposure and the chemicals involved are similar to those involved in central (brain) sensitization. This suggests that similar mechanisms are involved. And while some MCS patients demonstrate each of these peripheral regions, others lack sensitivity in these regions. Other researchers [190, 234] have suggested two additional mechanisms involved in the process of peripheral sensitization, termed neurogenic inflammation and mast cell activation, both of which are compatible with the underlying activation of the NO/ONOO⁻ mechanism and is followed by mitochondropathy. There are different types of evidence supporting the formation of NO/ONOO⁻ excess and the upregulated NMDA activity having a significant role in the outbreak and specifically are part in transducing chemical actions in MCS. In this respect, nitric oxide and peroxynitrite as well as oxidative stress all play a role in the outbreak of the ailment. Several authors conclusively have demonstrated that the cause of MCS are organic solvents and other toxic compounds that act via the TRPV1 (vanilloid) receptor in MCS patients [235, 236]. Here the transient receptor action potential of the channel V subfamily member 1 (TRPV1), also known as the capsaicin receptor or the vanilloid receptor 1, is activated. It is a nonselective cation channel that may be activated by a wide variety of exogenous and endogenous physical and chemical stimuli. In this respect, Ashford and Miller described the striking similarities between neural sensitization and MCS, each of which can be taken as an evidence for a role of neuronal sensitization [237]. There are also two additional types of evidence indicating the role of inflammation in this disease, where Bell and coworkers demonstrated changes in brain EEG patterns in response to low level chemical exposure [238] and Kimata [239] reported on changes in both nerve growth factor (NGF) levels and histamine levels being elevated in MCS. In addition, Millqvist [240] reported on increased cough sensitivity to capsaicin in MCS patients, and Shinohara [241] reported on hypersensitivity reactions in individuals being exposed to chemicals while Joffres [242] described changes in skin conductivity with low level chemical exposure. In addition, there are a number of studies with nasal lavage measurements, showing that chemically sensitive individuals respond with increased inflammatory markers on the exposure to a chemical. Each of these may be specific for MCS patients and each is compatible with the NO/ONOO⁻ mitochondropathy model.

7.3 Neurodegeneration Resulting from Blockade in the Energy Pathway 121

All clinically reported observations also occur in one or more animal model for MCS:

1. Neural sensitization and cross sensitization, where
2. Sensitization to one chemical also produces sensitization to a second chemical.
3. Progressive sensitization, where sensitivity progresses with increasing numbers of chemical exposures.
4. Chemical agents acting via decrease in acetylcholinesterase, GABA activity or via
5. increased TRPV1 or sodium channel activity (see Fig. 7.31).
6. Increase in oxidative stress and/or increase in NMDA activity related to MCS.
7. Increase in nitric oxide.
8. Increase in peroxynitrite.
9. Elevated inflammatory cytokine levels or increased levels of other inflammatory markers.
10. Elevated levels of intracellular calcium.
11. Breakdown of the blood brain barrier.
12. Neurogenic inflammation.
13. Airway sensitivity (i.e. increased reactive airways).
14. Chemical linkage to the sensory irritation response, which is thought to be involve a number of TRP receptors including the vanilloid TRPV1.

While all the described changes have been reported for MCS a large number of these symptoms can also be observed in FMS, which suggests that they are caused by the very same NO/ONOO⁻ cycle.

7.3.3 *Chronic Fatigue Syndrome, Fibromyalgia Caused by Mitochondrial Failure*

The work of mitochondria is to supply energy in the form of ATP (adenosine triphosphate), which is the universal currency of energy. ATP is used for all sorts of biochemical work ranging from muscle contraction to hormone production. When mitochondria fail, the net result is an insufficient production of ATP. As a consequence the cells go slow because they do not have sufficient energy supply to function at normal speed. The consequence of a poor mitochondrial function is that all bodily functions now go slow with all typical signs and symptoms since every cell in the body can be affected. Such lowered energy production is readily explained by looking at the synthesis of adenosine triphosphate (ATP, the molecular unit of currency). When ATP is converted to ADP the energy being released is being used up

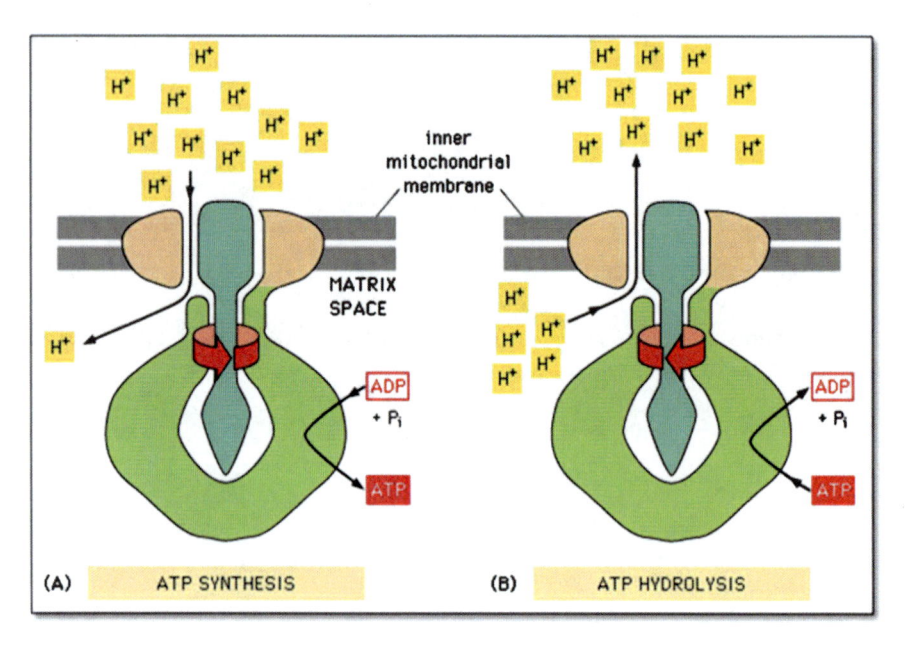

Fig. 7.33 A necessary constituent of mitochondria – the translocator protein moves ADP-ATP to and fro via the ATP synthase, a spherical protein extending into the matrix

for work. ADP passes back into mitochondria via the translocase in complex V where a new phosphorus group is attached to ADP resulting into the formation of ATP (i.e. a phosphate group is attached; Fig. 7.33).

The energy for this conversion comes from oxidative phosphorylation within mitochondria. There, ATP recycles approximately every 10 s in a normal person – if this goes slow, then the cell goes slow and so the person goes slow and clinically has poor stamina which clinically is diagnosed as CFS. In order to use ATP, it must exit the mitochondrial inner membrane, while ADP must enter the inner mitochondrial membrane to supply the cell with newly synthesized ATP. ATP/ADP translocase couples the movement of ATP out while moving ADP into the inner. Once ATP is exchanged for one ADP molecule simultaneously. The charge on ATP is −4, while ADP's charge is only −3. Due to the large positive charge generated by electron transports build up of hydrogen ions in the inner membrane space, the more negatively charged ATP is preferentially moved across the inner membrane, towards the cytosol (Fig. 7.34). The cost of moving ATP/ADP is approximately one hydrogen ion.

The problem arises when the system is stressed either by an increased need or by an insufficient production due to mitochondrial insufficiency. If the CFS sufferer asks for energy faster than he can supply it, (and actually most CFS sufferers are doing this most of the time!) ATP is converted to ADP faster than it can be recycled. This means there is a build up of ADP. Some ADP is inevitably shunted into adenosine monophosphate (AMP −1 phosphate). But this creates a real problem, indeed a

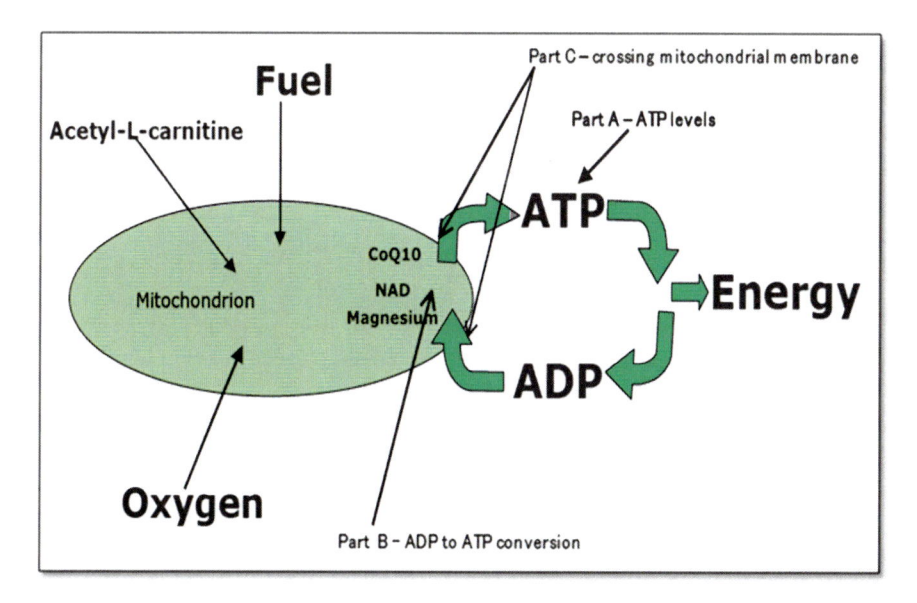

7.3 Neurodegeneration Resulting from Blockade in the Energy Pathway

Fig. 7.34 The importance of recycling ADP back into mitochondria for conversion to ATP, with the necessary micronutrients (Acetyl-L-carnitine, Q10, Mg, NAD) for the conversion process

metabolic disaster, because AMP, largely speaking, can only by recycled within 4 days and are lost in the urine. This loss of energy is the biological basis of poor stamina in CFS sufferers. One can only go at the rate at which mitochondria can produce ATP. If mitochondria go slow, stamina is poor. If ATP levels drop as a result of leakage of AMP, the body then has to make new ATP, which can be made very quickly from a sugar D-ribose (Fig. 7.35). However, D-ribose is only made slowly from glucose via the pentose phosphate shunt, taking anything from 1 to 4 days. So this is the biological basis for delayed fatigue.

Yet there is another problem, since when being very short of ATP the body can only make up a very small amount of ATP directly from glucose via conversion to lactic acid. This is exactly why CFS sufferers readily switch into anaerobic metabolism. This conversion however results in two serious problems – lactic acid quickly builds up especially in muscles to cause pain, heaviness, aching and soreness ("lactic acid burn"), secondly no glucose is available in order to make D-ribose! Therefore new ATP cannot readily by made when those patients are really in need of energy, and recovery will take days!

When mitochondria function well, as the person rests following exertion, lactic acid is quickly converted back to glucose (via-pyruvate) and the lactic burn disappears. But this is an energy requiring process! Glucose to lactic acid produces two molecules of ATP for the body to use, but the reverse process requires six molecules of ATP. If there is no ATP available, and this is of course what happens as mitochondria fail, then the lactic acid may persist for many minutes, or indeed hours causing pain. This reverse process takes place in the liver and is called the Cori cycle,

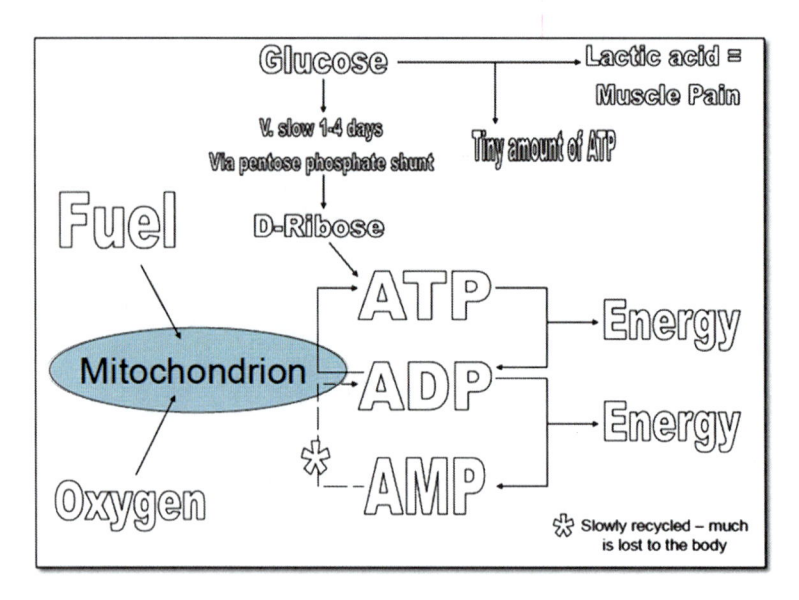

Fig. 7.35 If mitochondria go slow there is less turnover of ADP into ATP resulting in less stamina the action of which depends on the translocator protein in complex V of mitochondria which scoops ADP from cytoplasma

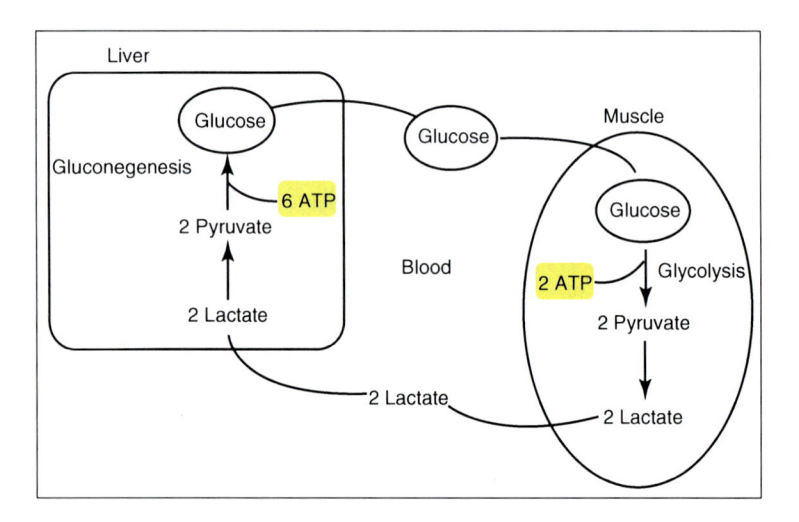

Fig. 7.36 The classical Cori cycle which converts lactic acid back into glucose within the liver

The Cori cycle, named after its discoverers, Carl and Gerty Cori, refers to the metabolic pathway in which lactate produced by anaerobic glycolysis in the muscles moves to the liver and is converted to glucose, which then returns to the muscles and is converted back to lactate (Fig. 7.36).

7.3.3.1 Brain Fog, Poor Memory, Difficulty Thinking Clearly – Similarities in CFS and FMS

What allows the brain to work quickly and efficiently is its energy supply. If this is impaired in any way, then the brain will go slow. Initially, the symptoms would be of a foggy brain; but if symptoms progress, one will end up with dementia. We all see this in our everyday life, with the effect of alcohol being the best example. Short-term exposure gives us a deliciously foggy brain, we stop caring, we stop worrying, and it alleviates anxiety. However, it also removes one's drive to do things, one's ability to remember; it impairs judgment and our ability to think clearly. Medium-term exposure results in mood-swings and anxiety, only to be alleviated by more alcohol. Longer-term use could result in severe depression and then dementia. Incidentally, this example also illustrates how most drug side-effects result from energy and nutritional deficiencies!

7.3.3.2 Low Cardiac Output Secondary to Mitochondrial Malfunction in CFS

Two papers have underlined the assumption that CFS primarily is a disease of mitochondrial failure [243] (Ref. [9] of Chap. 5) with published results demonstrating, that the mitochondrial dysfunction is linked to the production of energy through adenosine triphosphate or ATP. The researchers state that depending severity of the patient's symptoms there is a higher degree of mitochondrial dysfunction. In the study it is also pointed out that now it is possible to determine whether fatigue and lack of energy is from CFS or from other conditions like sleep disorders and hormonal deficiencies. Since mitochondria supply the cell with energy in the form of ATP, a universal currency of energy, it is used for all sorts of biochemical action from muscle contraction to hormone production. When mitochondria fail, it results in poor supply of ATP, and the cells go slow because they do not have the energy supply to function at normal speed. As a result all bodily functions go slow. There are however, two main symptoms in CFS that make the diagnosis; first of all there is very poor stamina and second there is a delayed fatigue. Now one can explain what is going on inside cells and the related effects on major organs of the body, primarily the heart. If mitochondria, the small power organelles found inside every cell in the body, do not operate properly, then the energy supply to every cell in the body is impaired. Such effects will also take place in the heart and the central nervous system, organs that require a lot of energy to function properly. Many of the symptoms of CFS can be explained by insufficient heart function because the myocardium does not contract properly. And while cardiologists are used to deal with heart failure due to poor blood supply to the myocardium, in CFS heart malfunction is caused by insufficient muscle contraction and therefore is of cardiomyopathy origin. And although traditional tests of heart failure, such as ECG,

ECHOs, angiograms etc., will be normal the function of the heart is insufficient. Peckerman and coworkers have demonstrated this in a study, where patients with severe CFS had significantly lower stroke volume and cardiac output than controls and less ill patients. In the study postexertional fatigue and flu-like symptoms of infection differentiated the patients with severe CFS from those with less severe CFS (88.5% concordance) and were predictive (R2 = 0.46, $P<0.0002$) for lower cardiac output This indicates, that cardiac output in CFS patients is impaired, and furthermore the level of impairment correlates very closely with the level of disability in patients [243]. Consequently, CFS patients feel much better when lying down, because there is an acceptable cardiac output when being in a supine position, however, when standing up they are at borderline heart failure. CFS is therefore the symptom, which prevents the patient from developing severe heart insufficiency.

There are three stages of chronic fatigue syndrome, which are characterized by different immunological and biochemical changes. Chronologically these stages are not completely distinct, and one runs into the other. However, if one can identify the stage in, this will have implications in the type of treatment.

1. **The acute alarm stage** – this follows the original trigger for chronic fatigue, which may be of sudden onset following viral exposure or chemical exposure, or maybe of gradual onset over some months or years. This stage is characterized by number of symptoms, which can be anything from acute fever, lymphadenopathy and malaise to headache, irritable bowel, muscle aching, anxiety, sleeplessness or whatsoever. This stage probably reflects immune overactivity, which might be appropriate in the case of the virus or bacterial infection, inappropriate in the case of allergy and autoimmunity and is accompanied by a heightened stress response with high levels of stress hormones to allow the body to engage and cope with these increased energetic demands. Chronic work or emotional stress also results in such an alarm reaction.
2. **The second exhausted phase** is characterized by biochemical and hormonal failures. Nearly always there is foggy brain, the person cannot think clearly, there is an inability to perform multitask work, and an inability to learn. In addition, there is exhaustion with poor stamina and delayed fatigue, while the list of symptoms starts to lessen as they are be replaced by chronic background pain and malaise. This is the stage where there are obvious biochemical failures due to exhaustion of micronutrients, toxic stress with poor healing and repair. Typically one see's poor mitochondrial function, poor antioxidant status, poor quality of sleep, poor digestion of foods with bouts of hypoglycemia, and episodes of low activity of the hypothalamic pituitary adrenal axis, which correlates with depression.
3. **The third stage of maladaptation**. This stage arises after several years of the above second stage and is characterized by relative freedom from the above symptoms so long as one stays strictly within limits. However, this stage is characterized by marked symptoms of a push being followed by a crash. It indicates that the

patients can only go very slightly above their permitted limits (which may even be a stage of bedridden!), and results in a potential to make them ill for days.

The researcher Dr Cheney hypothesizes, that the normal stress response which involves the release of cortisol, thyroid hormones, growth hormones, insulin, is deficient. Normally if the body is stressed, this hormonal response allows us to adapt to the situation, cope with that stressful situation and then drop back to baseline functioning. In the third maladapted stage the stress hormones do result in a complete opposite effect. Instead of allowing the sufferer to cope and respond via a normal stress response, they make him collapse instead. This is where Cheney and coworker are focusing their efforts in therapy. But in order to understand this third stage one need's to look a little more closely at the second stage of biochemical failures. Cheney et al. believe that all these stages result from how the body deals with free radical stressors, i.e. nitric oxide and peroxynitrite. While in the first stage the patient coping with increased demand greatly increases the output of energy. This, however, results in additional formation of free radical stressors and it is the ability to deal with such stress, which determines between those patients that recover and those who goes into a classical fatigue. This is underlined by considerable success when treating this lack in energy and by addressing all the issues that can result from poor energy supply to mitochondrial function, all of which are closely linked with the formation of free radical stressors. Since mitochondria are also a major source of free radical formation, and if not scavenged efficiently (for instance by the endogenous SOD), are easily damaged by them. This is why determination of the antioxidant status is important to understand the cause of the underlying disease. The balance between producing free radicals and the ability to deal with them is called the "redox" state. It is important because the redox state determines how well the body deals with acute and chronic infections. Being made up of different cells, all have to fulfill a special function, which requires energy. The way in which energy is generated is the same for every cell in the body and the mitochondria are the very organelles, which supply the cells with energy in form of ATP.

7.3.4 Chronic Fatigue Caused by Mitochondrial Failure

Since the mitochondria supply energy in form of the universal currency of energy called ATP (adenosine triphosphate), it is used for all sorts of biochemical reactions, from muscle contraction to hormone production. When these mitochondria fail, it results in a poor supply of ATP, so cells go slow because they do not have the energy supply to function at normal speed. This means that all bodily functions go slow and every cell in the body can be affected. During this process of energy discharge, ATP (3 phosphates) is converted to ADP (2 phosphates) with the release of energy for work. ADP via the translocase protein complex passes back into mitochondria where ATP is remade by oxidative phosphorylation (i.e. a phosphate group is stuck on). ATP recycles approximately every 10 s in a normal person, and the portion of ATP being synthesised approximates the body weight of a person (ca 70 kg!!!). Once the synthesis goes slow, then the cell function goes slow, the person goes slow and clinically there is poor stamina. Problems arise when such system is stressed, i.e. the CFS sufferer asks for energy faster than the cellular organelles can supply them. Actually most CFS sufferers are doing this most of the time with ATP being converted to ADP faster than it can be recycled. This results in a build up of ADP and some ADP is inevitably shunted into adenosine monophosphate (AMP −1 phosphate). But this creates a further problem, resulting in a metabolic disaster, because AMP, cannot be recycled back into ATP and is lost in the urine. All these changes are the biological basis for poor stamina, since the patient cannot go at the rate at which mitochondria can produce ATP. If mitochondria go slow, stamina is poor. If however ATP synthesis further drops as a result of leakage of AMP, the body then has to make new ATP, which can be made quickly from the sugar D-ribose. However, D-ribose is only slowly made from glucose via the pentose phosphate shunt, which takes anything from 1 to 4 days, being the biological basis for delayed fatigue. Once the body is very short of ATP, it can make up a very small amount of ATP directly from glucose by converting it into lactic acid. This is exactly what is done within the metabolism of CFS sufferers and indeed is has been demonstrated that CFS sufferers readily switch into an anaerobic metabolism. This conversely results in two additional serious problems – firstly, there is a build up of lactic acid especially in muscles causing pain, heaviness, aching and soreness ("lactic acid burn"), and secondly no glucose is available in order to make D-ribose! As a result new ATP cannot be easily produced when there is a real need for it and recovery can take several days! Once there is sufficient mitochondrial function, and the patient rests following exertion, lactic acid via pyruvate is quickly converted back to glucose via the Cori cycle within the liver and the lactic acid burn disappears. But this is an energy requiring process! Converting glucose to lactic acid produces two molecules of ATP for the body to use, but the reverse process requires six molecules of ATP. If there is no ATP available, and this is exactly what happens as mitochondria fail, then the lactic acid may persist for many minutes, or hours causing pain. To sum up, in CFS patients, one of the main causative factors is inefficiency in recycling ADP back to ATP again. This pathway is often the bottleneck in energy production in such individuals. If the cell is not efficient at recycling ADP to ATP, then the cell runs out of energy very quickly, which causes the symptoms of weakness and poor stamina. The cell must then go into a 'rest' period until more ATP can be manufactured/recycled from ADP. At any one time, the cells in the heart muscle only have enough ATP in reserve for around 10 contractions. If a cell is pushed to produce energy when no ATP is available, then it will use the ADP instead, and converts this into AMP (adenosine monophosphate). AMP consists of a phosphate group, the sugar ribose, and the nucleobase adenine. AMP however, cannot be recycled, which is why the body does not normally use AMP to produce energy. Any ATP, which is converted to AMP, is considered as being "lost". So in order to regain ATP it must be recycled from any ADP that remains, and a resting period must be used to create ADP from scratch using fresh raw ingredients. To create ATP from scratch, the body must first break down the various proteins, triglycerides, fatty acids and sugars into their constituent parts, after which mitochondria must build up ATP from these components using the enzyme ATP synthase (for further information see the Krebs Cycle on page 5).

Dr. Cheney has developed a tool for looking at the energy available to cells in the heart. This cellular free energy can be measured using echocardiography, where the interval in milliseconds between aortic closure and mitral opening, known as the IVRT, the *isovolumetric relaxation time* is being measured. Physiologically, it is the time it takes to pump free calcium out of the myocardium to produce relaxation of the myofibrils to allow ventricular filling, and it takes all the available free energy in the heart to do this work. IVRT is inversely related to the cellular free energy. The higher the IVRT, the lower the cellular free energy. The heart is the most energetic organ in the body, which is reflected by the fact than 50% of the heart is made up by mitochondria. Mitochondria make energy in the form of ATP and this ATP is used to pump calcium into the sarcoplasmatic reticulum where the muscle fibers are. If this is done slowly, then the concentration becomes critical, as calcium rushes back and the energy generated by this process allows the muscle fibers to contract. If mitochondria go slow it takes longer for this recharging to take place. So the time between the start of atria charging and mitral valve closing (called the isometric volume relaxation time) is an indirect reflection of mitochondrial function. The more efficient the mitochondria, the more cellular free energy is available and the shorter the IVRT. A normal value is around 75 ms while a value of 150 ms is indicative for heart failure. This test is highly accurate and reproducible within a 2% range, and allows assessing within minutes the effect of various stressors on the availability of energy to cells in the heart, being an instant measure of mitochondrial function. It is now known that people with chronic fatigue syndrome are in a low cardiac output state, which is secondary to poor mitochondrial function as demonstrated in the echocardiogram and being clearly visible in the maladapted stage.

The **third and maladapted stage** of chronic fatigue syndrome arises because of abnormal control mechanisms. Normally in a stress situation there is a release of stress hormones from the thyroid gland, the adrenal gland, and the pituitary gland in the brain. By looking at the heart response when these hormones are applied transdermally to controls and to patients with fatigue syndromes there is a response within 30 s. What one finds is that after application of these extracts to normal people, the heart improves its function. However, when one applies these extracts to patients with chronic fatigue syndrome the heart response always gets worse. There is a further interesting peculiarity with respect to the energy supply to another organ, the brain which is different from that used by the rest of the body. Although the brain weighs just 2% of total body weight, when being in use it absorbs 20% of the body's oxygen requirement and 25% of its energy needs! This cannot be explained by the number of mitochondria in the brain (there are not enough), which

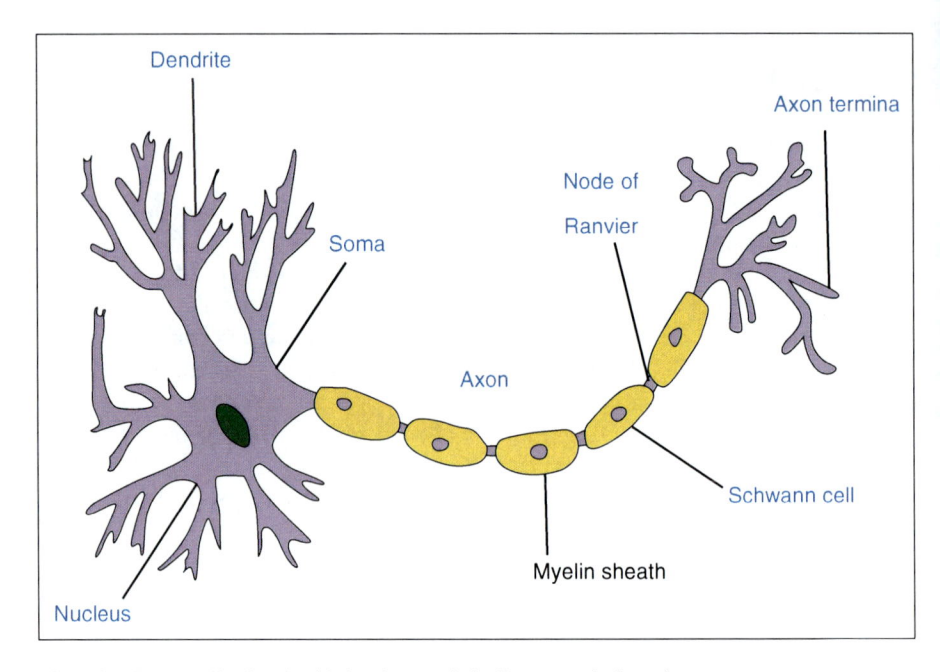

Fig. 7.37 The myelin sheath with its characteristically anatomical portions

means there must be another energy-generating source. Brain cells are also very different from normal cells. They have a cell body, and very long tails – or dendrites -, which communicate with other cells. Indeed, if a nerve-cell body from the spinal cord that supplied one's toes was sitting on my desk and was the size of a football, the tail would be in New York! These tails (dendrites) are too small to contain mitochondria, but it has been suggested that the energy supply comes directly from the myelin sheaths themselves. They too can produce ATP and it is this that supplies the energy for neurotransmission. Myelin sheaths are made up almost entirely of fats, so one needs to look at oils and fats in order to improve energy supply to brain cells (Fig. 7.37).

This assumption is corroborated by the fascinating work of Dr. Mary Newport who has shown that coconut oils can cure dementia with the key ingredients medium-chain triglycerides (MCT) [244]. It suggests that they are an essential part within the process of energy generation and healing. Thus, MCTs are an excellent energy supply to the brain, which aside from glucose can also use short-chain fatty acids or ketones for energy supply. As to they mode of action it is suggested that the myelin sheath membranes act as an energy source which allows them to synthesize ATP. In such case, it would explain how general anesthetics work, since presently no one really knows the exact mechanism of action, which presently is thought to be through a nonspecific perturbation of lipid membrane of CNS neurons. If however, general anesthetics deranges the consistency of myelin-sheath cell-membranes, then the energy supply to nerve cells is switched off and nerve transmission altogether would cease leading not only to just a foggy-brain but leading into unconsciousness. Thus, one should not be concerned about dietary fat and cholesterol causing arterial disease! The results of some studies of Polynesian peoples for whom coconut is the chief source of energy should put the mind at rest. Also, it has been suggested that statins, by reducing the cholesterol that the brain loves, are contributing to the current epidemic of Alzheimer's disease [144]. And although it is rare for CFS patients to tolerate statins, nearly always they make them ill. So, adding coconut oil to the regime of anyone with symptoms of foggy brain or dementia is likely to be helpful, partly by improving fuel supply and partly by improving membrane function.

It is important to realize that CFS is a protective adaptive state. If a person would not switch into a CFS, then the uncontrolled free radical stress would kill him! For instance, compensation for this lack by sufficient sleep for 2 weeks is sufficient to regain normal mitochondrial function. However, if one forces the system against its will there is the risk of creating more free radicals making symptoms much worse. By applying the idea of maladaptation and insufficient energy production of mitochondria, all therapeutical implications are based on sound logical basis:

- Reduce generation of free radicals
- Balance rest and pacing
- Use a low carbohydrate diet
- Get rid of toxins such as pesticides, heavy metals, volatile organic compound[s] (VOCs) and drug medication all of which generate free radicals
- Improve ability to scavenge free radicals
- Improve micronutrient status
- Improve length and quality of sleep

Since the Redox state is greatly affected by the pH, there is a enormous change in the oxidation/reduction balance resulting from a small change in the pH. This is particularly true when major components of the redox system are heavily protein-buffered. In CFS patients, the degree to which they switch to anaerobic metabolism is one of the major factors, since in their struggle to cope with the physical demands, they soon begin to depend on anaerobic metabolism and build up a tolerance to increased lactate acidosis. Such acidosis, also when being build up only locally, has to be considered a major stress factor in general and specifically in the cellular redox balance. This connotation is underlined in leukocyte respiration studies [245], where one sees another major derangement in the uncoupled electron transport/ oxidative phosphorylation chain.

7.3.4.1 Additional Symptoms in Chronique Fatigue and Fibromyalgia

Chest Pain

This is a common symptom in CFS. Chest pain results when energy delivery to the muscles is impaired. There is a switch to anaerobic metabolism, lactic acid is produced and this results in the symptom of angina. Doctors recognize this as a major cause of poor blood supply, i.e. when the supply of fuel and oxygen is impeded. However this fuel and oxygen has to be converted to ATP by mitochondria, so if this is slow, the same symptom of angina will result. One molecule of sugar, when burnt aerobically by mitochondria, will produce 36 molecules of ATP. In anaerobic metabolism, only two molecules of ATP are produced. This is very inefficient and lactic acid builds up quickly. The problem is that to convert lactic acid back to sugar (pyruvate) six molecules of ATP are needed (via the Cori cycle). So in CFS the chest pain is longer lasting because this conversion back is so slow. Clinically this does not look like typical angina. Many patients are told they have non-typical chest pain with the implication that nothing is wrong! Actually they have mitochondrial failure of their heart.

Effects on the Skin

If one shuts down the blood supply to the skin, this has two main effects. The first is that the skin is responsible for controlling the temperature of the body. This means that CFS patients become intolerant to heat. If the body gets too hot then it cannot loose heat through the skin (because there is a reduction blood perfusion) and the core temperature increases. The only way the body can compensate for this is by switching off the thyroid gland, which is responsible for the level of metabolic activity in the body and hence heat generation. As a result one gets a compensatory underactive thyroid, which by itself worsens the problems of fatigue.

The second problem is that if the microcirculation in the skin is shut down, then the body cannot sweat. This is a major way through which toxins, particularly heavy metals, pesticides and volatile organic compounds are excreted. Therefore the CFS sufferer is much better at accumulating toxins, which of course further damage mitochondria.

Symptoms in Muscles

If the blood supply to muscles is impaired, then muscles quickly run out of oxygen when one starts to exercise. With no oxygen in the muscles the cells switch to anaerobic metabolism, which produces lactic acid, which is why the muscles ache. Similar to the above problem of myocardial activity of the heart, muscles in CFS patients have very poor stamina because the mitochondria, which supply them with energy, are malfunctioning.

Symptoms in the Liver and Gut

Poor blood supply to the gut results in inefficient digestion, poor production of digestive juices and a leaky gut syndrome. Leaky gut syndrome causes many other problems such as allergies, autoimmunity, malabsorption, etc., which further increase the problems of CFS. If liver circulation is inadequate, this will result in poor detoxification, not just of heavy metals, pesticides and volatile organic compounds, but also of toxins produced as a result of fermentation in the gut, again further poisoning mitochondria.

Effects on the Brain

Functional scans of the brains of CFS patients look as if they are diagnosed as patients after a stroke. This is because the blood supply to some areas of the brain is impaired. The default is temporary and with rest, blood supply recovers. However, this explains the multiplicity of brain symptoms in patients with CSF, such as poor short-term memory, difficulty in multi-tasking, slow mental processing. Furthermore, brain cells are not particularly well stocked with mitochondria and therefore they run out of energy very quickly.

Effects on the Heart

There are two effects related to mitochondrial dysfunction of the heart. The first effect is a poor energy production within cells of the conductive system to the heart resulting in malfunction with disturbance of the electrical conductivity causing dysrhythmias. Many patients with chronic fatigue syndrome complain of palpitations, missed heart beats or bigeminus. This is particularly the case in patients having a toxic load and a poisoning by chemicals since those chemicals are directly toxic to nerve cells resulting in arrhythmias. The second obvious effect is a marked reduction of tolerance to exercise. Heart muscle fatigues in just the same way that other muscles fatigue. Symptomatically this causes chest pain and fatigue. In the long term it can cause heart valve defects because the muscles, which normally hold the mitral valve, open also fatigues. The difference between this type of heart failure and medically recognized congestive cardiac failure is that patients with CFS protect themselves from organ failure because of their fatigue symptoms. Patients with congestive cardiac failure initially do not get fatigue and often present themselves with organ failures such as kidney or overt heart failure with pulmonary congestion.

The approach in treating CSF is exactly the same as in congestive heart failure, regardless of the diagnosis.

This is because patients with angina, high blood pressure, congestive heart failure, cardiomyopathy, valve defects as well as patients with cardiac dysrhythmias also have mitochondrial problems and will respond in the same way to a nutritional therapy and a detoxification therapy as patients with chronic fatigue syndrome.

Effects on Lung and Kidneys

The lung and the kidneys are relatively protected against poor microcirculation because they have the renin-angiotensin system, which keeps the blood pressure up in these vital organs. Therefore clinically one does not see patients with kidney failure or pulmonary hypoperfusion in CFS.

7.3.4.2 Explaining the Fatigue Problems to CFS Patients

Clinically there are two different stages of fatigue:

1. *The mild chronic fatigue syndrome* – in mild fatigue there is mild failure of mitochondria. If mitochondria go slow then cells go slow. If cells go slow then organs go slow. The body will become generally less efficient. For example if somebody were mildly affected, they would not be able to increase their fitness; if they try to exercise they would quickly switch into lactic acid metabolism and would be forced to stop. Indeed it is now know that mitochondria are responsible for controlling the normal ageing process. Therefore many of the symptoms and diseases associated with ageing are actually the result of a declining mitochondrial function. Many of the ageing diseases are now being attributed to mitochondrial failure such as loss of tissues (loss of muscle bulk called sarcopenia), organ failures, neurodegenerative conditions, heart disease and even cancer. And many symptoms, which are attributed to ageing, are due to mitochondrial dysfunction. The question is not that we can stop the mitochondria from ageing, but we certainly can slow this process down by using a natural nutrition, an unprocessed, contaminated diet, and freedom from toxic stress, healthy lifestyles, and a reduction of toxic load.
2. *Severe chronic fatigue syndrome* – in severe chronic fatigue all the above factors apply. However, there is an additional problem. The most metabolically demanding organ in the body is the heart and if mitochondria cannot supply the heart with sufficient energy then the heart will go into a low output state. This compounds the problem of all mitochondria. If the heart is in a low output state then blood supply is poor and therefore the uptake and distribution of fuel and oxygen necessary for the pump to work are also impaired. So this explains all the above problems and makes them proceed even faster with people ending up in a greater disability.

The underlying poor mitochondrial function precipitates to a much more severe illness which suddenly becoming critical, when there is a further decline in cardiac output in someone who is already compromised. This is because all energy to the body is supplied by mitochondria, which firstly produce NAD^+ (nicotinamide adenosine diphosphate) from the Kreb's-citric acid cycle and which is used to power oxidative phosphorylation to generate ATP (adenosine triphosphate). These molecules are the "currency" of energy in the body. Almost all energy requiring processes in the body has to be "paid for" with NAD^+ and ATP, but largely by ATP.

Fig. 7.38 The principle of recycling ADP through mitochondria to generate ATP, which is used as the energy supply of the cell

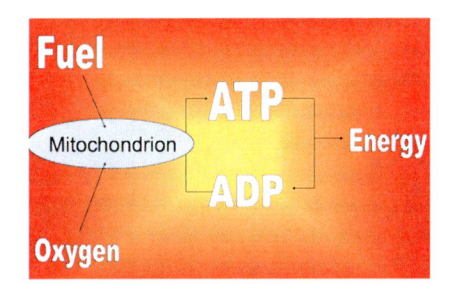

These reserves of ATP in cells are very small. At any moment in the myocardial cells there is only enough ATP to last for about ten contractions. Thus the mitochondria have to be extremely good at recycling ATP to keep the cell constantly supplied with energy. If the cell is not very efficient at recycling ATP, then the cell runs out of energy very quickly, which causes the symptoms of weakness and poor stamina. The cell literally "cuts down on function" and waits until more ATP has been manufactured. In producing energy, ADP (adenosine-two-phosphates) is converted into ATP (adenosine-three-phosphates) by recycling ADP back through mitochondria to produce ATP (Fig. 7.38). However, if the cell is pushed (i.e. stressed) when there is no ATP available, then it will start to use ADP instead. The body can create energy from ADP resulting into the formation of AMP (adenosine-mono-phosphate). But the problem is that AMP cannot be recycled. The only way that ADP can be regenerated is by making it from fresh ingredients, but this will take days and explains the delayed fatigue seen in patients with chronic fatigue syndrome.

To summarize, the basic pathology in CFS is the slow recycling of ATP to ADP and back to ATP again. If patients push themselves too much and make more energy demands, then ADP is converted to AMP, which cannot be recycled resulting in a delayed fatigue. The reason for such delay is the fact that it takes the organism several days to make new ATP from the nutritional ingredients. When the CSF patient overdoes things he "hits a brick wall" or comes to an end because of insurmountable problems, because there is no ATP or ADP to function at all.

7.3.5 Hormonal Problem in CFS – Hypothyroidism, Adrenal Insufficiency

It is now quite clear that there is a distinct hormonal disturbance in CFS patients with a general suppression of the hypothalamic – pituitary – adrenal (HVA) axis. Since it is the pituitary which is the "conductor of the endocrine orchestra", and if the pituitary is malfunctioning (Fig. 7.39), then this will have a major effect on the thyroid gland, the adrenal gland (with reduced formation of cortisol), on sex hormones, possibly the pineal gland (which produces melatonin for normal sleep), as well as on hormones for growth and urine production. In practice one invariably should measure thyroid hormones (TSH, T4 and T3), prescribe melatonin, check

Fig. 7.39 The pituitary gland, the conductor in the release of other hormones of the body (TSH, FSH, LH, STH, ACTH, MSH Prolactin). It however is governed by the hypothalamus, which through interconnections either induce the release (oxytocin, ADH) or stores releasing hormones

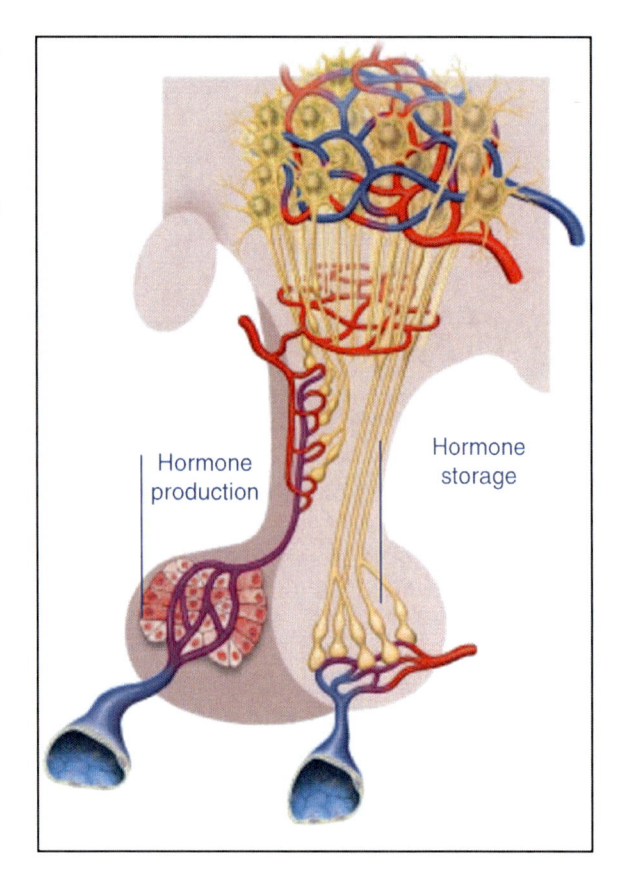

adrenal function by determining cortisol in saliva and very occasionally use sex hormones (mainly testosterone in men). Many CFS patients are substantially improved by correcting their thyroid or cortisol hormones, which is a reason why one should insist in all CFS patients to get a full hormonal check-up. With low levels of thyroid hormones in the blood cause the problem in CFS, it can be an underactivity for three reasons:

1. Either the gland itself has failed (primary thyroid failure). In **primary thyroid failure**, the blood tests show high levels of thyroid stimulating hormone TSH and low levels of T4 and T3.
2. The pituitary gland, which drives the thyroid gland into action, is underfunctioning. In **pituitary failure**, the blood tests show low levels of TSH, T4 and T3.
3. There is failure to convert inactive T4 to active T3. If the **conversion problem** is apparent, then TSH and T4 may be normal, but T3 is low.

The symptoms of these three problems are the same. Yet, there is another problem, which is when the so-called "normal range" of T4 is set too low. This is because many patients with low to normal T4 often improve substantially when they are

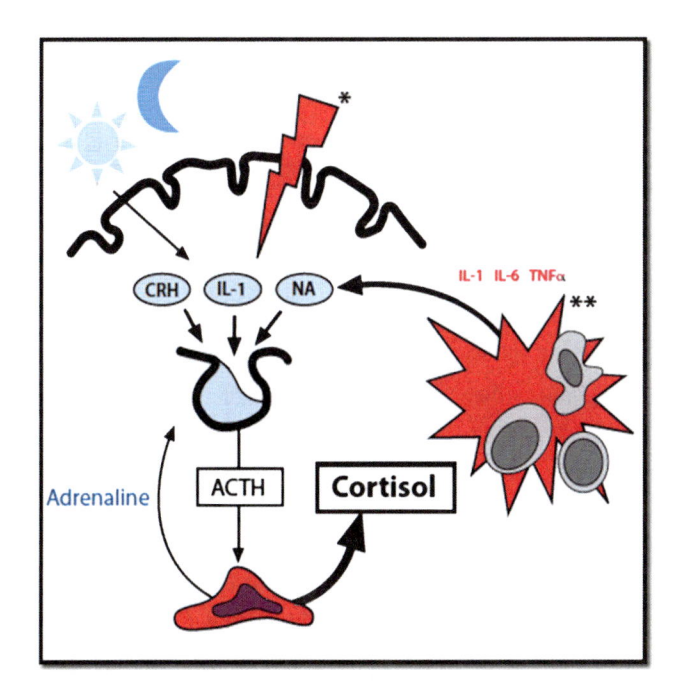

7.3 Neurodegeneration Resulting from Blockade in the Energy Pathway

Fig. 7.40 The stress hormone cortisol also activates inflammatory response as cytokines interacts with the pituitary gland where continuous activation results in hypocortisolism with burnout, which is a forerunner of chronique fatigue

started on thyroid supplements to bring levels up to the top end of the normal range. Many patients with CFS have low to normal levels of thyroxine (T4), do well when their levels are increased to the average levels. And although laboratory indicates a normal range of 12–22 pmol/l many CFS patients have levels around 12–15 pmol/l. In such patients there is an indication for trying a T4 medication, especially if symptoms suggest that there is a need for higher values. It should be emphasized, however, that this is a trial, and it does not commit patients taking thyroid hormones for the rest of their life.

- **Adrenal insufficiency and cortisol** – the task of the adrenal gland is to produce the stress hormones to allow us to move up and adapt when the stress comes on (Fig. 7.40). Cortisol raises blood sugar levels. It is largely excreted during mornings and declines as the day progresses – this is why we should feel at our best early in the day, and blood sugar problems get worse as the day progresses. Often people compensate for this by eating more as the day goes on and explains why many hypoglycemic do not need or eat breakfast with supper being the largest meal of the day. Changing all of the above will help. But it may be appropriate to do an adrenal stress profile and actually measure output of the stress hormones cortisol and DHEA since a small supplement may be very helpful.
- **Sex hormones**, such as the anticontraceptive pill and hormone replacement therapy (HRT). These hormones all have the effect of raising blood sugar levels.

Indeed this is the mechanism, which is responsible for gestational (pregnancy) diabetes. The problem is that stopping these hormones will cause hypoglycemia and one gets into withdrawal symptoms. Researchers are beginning to suspect that this part of the mechanism that makes these hormones so addictive.

7.3.5.1 Toxins from Inside Affecting Mitochondrial Function

- **Toxins and pollutants**. There was a breakthrough paper in the Lancet that showed that the biggest risk factor for diabetes (and this is the end product of years of hypoglycemia as insulin resistance results) is the level of pollutants in the body (pesticides, volatile organic compounds and heavy metals). The paper indicated that chemical pollutants were a greater risk factor than being overweight [246]. It was suggested that the overweight problem reflected a larger chemical burden as the body tried to "dump" chemicals where they would be out of the way. When people who have the highest levels of Persistent Organic Pollutants (POPs) in the blood were compared with people having the lowest levels of Persistent Organic Pollutants in the blood, they were found to be 38 times more likely to be a diabetic. Persistent Organic Pollutants (POPs) are chemical substances that persist in the environment, bioaccumulate through the food web, and pose a risk of causing adverse effects to human health and the environment. With the evidence of long-range transport of these substances to regions where they have never been used or produced and the consequent threats they pose to the environment of the whole globe, the international community has now, at several occasions called for urgent global actions to reduce and eliminate the release of such chemicals.
- **Heavy metals in wine**. Red and white wines from most European nations carry potentially dangerous doses of at least seven heavy metals, UK researchers found [247]. And while a single glass of even the most contaminated wine isn't poisonous, drinking just one glass of wine a day, a common habit in Europe and the Americas, might be very hazardous. By calculating the "target hazard quotients" (THQs) for wines from 15 countries in Europe, South America, and the Middle East using a measure designed by the US Environmental Protection Agency to determine safe levels of *frequent,* long- term exposure to various chemicals with a THQ over 1 indicates a health risk. Typical wines had a THQ ranging from 50 to 200 per glass. Some wines had THQs up to 300. By comparison, THQs that have raised concerns about heavy-metal contamination of seafood typically range between 1 and 5. The metal ions that accounted for most of the contamination were vanadium, copper, and manganese. But four other metals with THQs above one also were found: zinc, nickel, chromium, and lead. Some 30 other metal ions were measured in the wines, but THQs could not be calculated because safe daily levels for these metals are not known. All of these oxidizing metal ions pose potential problems. But the manganese contamination particularly worries, since manganese accumulation in the brain has been linked to Parkinson's disease. Hungary and Slovakia had maximum

7.3 Neurodegeneration Resulting from Blockade in the Energy Pathway

potential THQ values over 350. France, Austria, Spain, Germany, and Portugal, all nations that import large quantities of wine to the US, had maximum potential THQ values over 100.

- Specific facts about each of the five mostly used heavy metals

Aluminum – The specific gravity is 2.80. The best way to test for contamination is blood, urine or feces for recent exposure; hair or fingernail analysis will detail the last 3–6 months. Hair with coloring in it will not give a reliable reading of toxicity. Because of its low specific gravity, aluminum is not a heavy metal. Aluminum can leach calcium from the bones. It can also stop the body's ability to digest and make use of calcium, fluoride and phosphorous.

Sources of Contamination	Symptoms and Illnesses	Preventative and Detoxifying Elements	Body Organs Targeted
antacids, over-the-counter drugs, douches, cookware, aluminum foil (especially when storing acidic foods), underarm antiperspirants, baking powders, bad water, food additives, automobile exhaust, tobacco smoke, fireworks	headaches, cognitive problems, learning disabilities, poor bone density, ringing in the ears, gastrointestinal disorders, colic, hyperactivity in kids, imbalance when walking, poor memory, degenerative muscular conditions, cancer, Alzheimer's	calcium, magnesium, iron, manganese, vitamin B complex, vitamin C	central nervous system, kidneys, digestion system

Arsenic – The specific gravity is 5.7. The best way to test for contamination is urine for recent exposure; hair or fingernail analysis will detail the last 3–6 months. Hair with coloring in it will not give a reliable reading of toxicity. During the Victorian era arsenic was believed to prevent aging.

Sources of Contamination	Symptoms and Illnesses	Preventative and Detoxifying Elements	Body Organs Targeted
smelting process of copper, zinc and lead, manufacturing chemicals, pesticides and glass, bad water, fish, shellfish, paints, rat poisoning, fungicides, wood preservatives	sore throat, red skin at contact point, abdominal pain, vomiting, diarrhea, anorexia, fever, mucosal irritation, arrhythmia, cardiovascular collapse, numbness, tingling, darkening of the skin, birth defects, cancer, diabetes, Raynaud's syndrome	vitamin C, alpha-lipoic acid	blood, kidneys, central nervous system, digestion system, skin, liver, lung, bladder

Cadmium – The specific gravity is 8.65. The best way to test for contamination is blood or urine for recent exposure; hair or fingernail analysis will detail the last 3–6 months. Hair with coloring in it will not give a reliable reading of toxicity. In the liver it is bonded to protein forming complexes, which are then transported to the kidneys causing damage to the filtration process. The damage allows essential proteins and nutrients to be excreted from the body causing even further damage. Cadmium is very difficult to remove from the body.

Sources of Contamination	Symptoms and Illnesses	Preventative and Detoxifying Elements	Body Organs Targeted
tobacco smoke, instant coffee and tea, nickel-cadmium batteries, bad water, some soft drinks, refined grains, fungicides, pesticides, some plastics	fatigue, irritability, headaches, high blood pressure, enlarged prostate, increased risk of cancer, hair loss, learning disabilities, kidney disorders, liver disorders, skin disorders, painful joints, decreased immune functions, lung damage	zinc, iron, vitamin C, amino acids (L-methionine, L-cysteine, L-lysine)	gastrointestinal system, liver, placenta, kidneys, lungs, brain, bones, central nervous system, bones, reproductive organs

Lead – The specific gravity is 11.34. The best way to test for contamination is blood. Lead suppresses neuron clusters in the brain.

Sources of Contamination	Symptoms and Illnesses	Preventative and Detoxifying Elements	Body Organs Targeted
tobacco smoke, eating leaded paint, lead based ceramic glazed cookware, leaded gasoline, eating contaminated liver, inner city living, canned food (lead soldered), some bone meal supplements, some insecticides, batteries, plumbing, pencils, crystal glass production	poor bone growth, learning disabilities, fatigue, poor task performance, irritability, anxiety, high blood pressure, weight loss, susceptibility to infections, headaches, ringing in ears, lack of concentration, gastrointestinal problems, constipation, muscle and joint pain, tremors, decreased immune functions, insomnia, hallucinations, birth defects, autism, colic	calcium, iron, zinc, vitamin C, amino acids (L-lysine, L-cysteine, L-cystine)	bones, brain, kidneys, thyroid gland, liver, central nervous system

7.3 Neurodegeneration Resulting from Blockade in the Energy Pathway 141

Mercury – The specific gravity is 13.546. The best way to test for contamination is blood or urine for recent exposure; an x-ray will show if it is flowing through your system. Mercury prevents zinc from performing its normal function in the body, even where there is plenty of zinc. Contaminated fish and seafood have become major issues. Never eat Mackerel King, Shark, Swordfish, Tilefish or Grouper. In general, the older and larger the fish the more mercury they will contain.

Sources of Contamination	Symptoms and Illnesses	Preventative and Detoxifying Elements	Body Organs Targeted
dental amalgam fillings, laxatives containing calomel, hemorrhoidal suppositories, some printer inks, tattoo inks, some paints, some cosmetics, some fabric softeners, wood preservatives, some solvents, some drugs, some plastics, contaminated fish and seafood, volcanic emissions, mining operations, paper mills, contaminated rainfall, thermometers, thermostats, some childhood vaccines	cognitive problems, memory problems, irritability, fatigue, insomnia, gastrointestinal problems, decreased immune functions, numbness, tingling, muscular weakness, impaired vision and hearing, allergic conditions, asthma, headaches, lung irritation, contributes to multiple sclerosis, autism	selenium, chlorella, vitamin C, amino acids (L-glutathione, L-methionine, L-cysteine, L-cystine)	gastrointestinal tract, brain, kidneys, liver, central nervous system

- **Helpful food, herbs and supplements for detoxification.** The following is a list of other detoxifying and strengthening edibles: kelp, spirulina, cilantro, garlic, green tea, alfalfa, chlorella, high fiber foods, dandelion root, yellow dock root, sarsaparilla root, echinacea, licorice root, vitamin E, vitamin A, alpha-lipoic acid, glutathione, lactoferrin, selenium, zinc, essential amino acids, essential minerals, MSM (methylsulfonylmenthane), rutin, SAMe (S-adenosylmethionine) and silibinin.

7.3.5.2 Toxins from Outside Affecting Mitochondrial Function

- **Heavy metals in the exhaust fumes** of motor vehicles with areal pollution. The most important emissions are those of Cu, Pb and Zn. Cu comes mainly from brake lining wear, Pb from exhaust fumes and Zn from tire wear, crash barriers corrosion and brake lining wear [248].
- **Household poisons** also contains toxic chemicals found in everyday household products that, when absorbed through the skin (as practically all chemicals are), lead directly to liver toxicity, nervous system disorders, and cancer. The bathroom is one of the most toxic rooms in the house for most American families. People use deodorants containing aluminum (leading to Alzheimer's disease), shampoos

containing harsh solvents (liver toxicity), toothpaste containing non-organic fluoride (osteoporosis), mouthwash with aspartame (brain tumors) or saccharin (cancer), and to top it off, most people slap on a dab of perfume or cologne containing highly toxic cancer-causing chemicals. In a laboratory analysis, one popular perfume was found to contain more than 40 chemicals classified as hazardous to the liver, and yet the perfume manufacturers give no warning to consumers about the toxic chemicals found in their products.

- **The bathroom** is only the beginning with the laundry room being highly toxic, containing the same chemical perfumes in both the laundry detergent and especially the dryer sheets. Dryer sheets coat all your clothes with a layer of toxic chemicals. When one wears those clothes, the body moisture causes those chemicals to come into contact with ones skin and are absorbed directly into your bloodstream. It's an easy way to poison the system with cancer-causing chemicals.
- **The kitchen** is also highly toxic: consumers purchase antibacterial soap products made with a potent nerve chemical similar to agent orange – that's what kills the bacteria. They also use automatic dish washing detergent containing yet more chemicals and toxic fragrance compounds that coat the plates, glasses and silverware with a thin layer of cancer-causing chemicals. Subsequently, families then eat off those dishes and ingest the chemicals.
- **In the yard**, people use horrific quantities of pesticides and herbicides with seemingly no care whatsoever about the health consequences of doing so. Of course, its obvious to the fact that when wiping out the all-important biodiversity of his lawn this would thereafter be dependent on a long list of chemicals to battle one lawn disease after another, arising from the fact that all the worms were dead. (Your lawn needs worms to be healthy.).
- The most dangerous poisons are not the ones labeled as such in the household as most people aren't even aware that their **perfumes and colognes** are poisons. They have no clue that most deodorants cause Alzheimer's disease. They're not even aware that dryer sheets coat their clothes in a thin layer of chemicals that promote liver cancer. So they keep buying and using all these products, day after day, oblivious to the reality. Product manufacturers, meanwhile, absolutely deny the health consequences of their products. They acknowledge that the chemicals are present, but they claim the skin doesn't absorb them. That's nonsense, of cause: the skin absorbs practically all chemicals. That's why the "patch" medicines work in the first place: the medicine is absorbed through the skin.
- **Fragrances** are the source of many toxic chemicals. One should avoid the fragrance at all costs. For perfumes and colognes, you'll have to buy natural products made exclusively with essential oils, not artificial chemicals. These can be very expensive, so why not consider just wearing no perfumes at all. Everyone around will greatly appreciate this anyway, since most people put on **far too much** fragrance as their senses are dulled to the smell of their selected fragrance. Fragrance by the way actually dulls the mind and the senses. That's a completely different topic, but that if one wears perfume and if people keep on using fragrance in the laundry, the mind gets dulled. By using only fragrance-free products, one will literally become more intelligent!

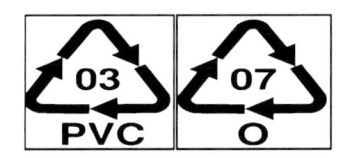

7.3 Neurodegeneration Resulting from Blockade in the Energy Pathway

Fig. 7.41 Plastics with the label type 3 and type 7 that may leak bisphenol A. There are seven classes of plastics used in packaging applications. Type 7 is the catch-all "other" class, and some type 7 plastics, such as polycarbonate (sometimes identified with the letters "PC" near the recycling symbol) and epoxy resins, are made from bisphenol A monomer

- **Bisphenol A**, commonly abbreviated as **BPA**, is an organic compound with two phenol functional groups. It is used to make polycarbonate plastic and epoxy resins, along with other applications. Bisphenol A is used primarily to make plastics, and products containing bisphenol A-based plastics have been in commerce use since 1953. Polycarbonate plastic, which is clear and nearly shatterproof, is used to make a variety of common products including baby and water bottles, sports equipment, medical and dental devices, dental fillings and sealants, eyeglass lenses, CDs and DVDs, and household electronics. BPA is also used in the synthesis of polysulfones and polyether ketones, as an antioxidant in some plasticizers, and as a polymerization inhibitor in PVC. Epoxy resins containing bisphenol A are used as coatings on the inside of almost all food and beverage cans, however, due to BPA health concerns, in Japan epoxy coating was mostly replaced by PET film. Bisphenol A is also a precursor to the flame retardant tetrabromobisphenol A, and was formerly used as a fungicide. Bisphenol A is a preferred color developer in carbonless copy paper and thermal paper, with the most common public exposure coming from some thermal point of sale receipt paper. BPA-based products are also used in foundry castings and for lining water pipes. In general, plastics that are marked with recycle codes 1, 2, 4, 5, and 6 are very unlikely to contain BPA. Some, but not all, plastics that are marked with recycle codes 3 or 7 may be made with BPA (Fig. 7.41). Type 3 (PVC) can also contain bisphenol A as an antioxidant in plasticizers. This is particularly true for "flexible PVC", but not true for PVC pipes.

Bisphenol A is an endocrine disruptor, which can mimic the body's own hormones and may lead to negative health effects [249–251]. Early development appears to be the period of greatest sensitivity to its effects. Regulatory bodies have determined safety levels for humans, but those safety levels are currently being questioned or under review as a result of new scientific studies.

- **Nickel toxicity**. Nickel toxicity is a very common problem and nickel is a substance often found to irreversibly bind to DNA. A test to measures chemicals that have stuck on to DNA should regularly been done in patients who have either been exposed to chemicals, or who have developed cancer. Almost invariably one finds toxic chemicals with the most common being lindane, nickel, or PBBs used as fire retardants (= Polybrominated biphenyls, also called brominated biphenyls or polybromobiphenyls), a group of manufactured chemicals of the polyhalogenated compounds with their chlorine analogs the PCB's and other heavy metals. It is possible to get rid of these toxins, either by using high doses of beneficial minerals, by using chelation therapy, or by a sweating detoxification regime, or a combination of these factors. Nickel biochemically looks very much like zinc and so enzymes, which normally incorporate zinc into them, in the presence of zinc deficiency, will take up nickel instead. This prevents the enzyme or the hormone from functioning normally. Clinically nickel toxicity often presents with hypoglycemia.

7.3.5.3 Getting Rid of Toxins to Improve Mitochondrial Function

All these toxic chemicals literally get in the way of many biochemical processes and prevent the body from normal function. So for some people applying detoxification regimes is very helpful using for instance infrared sweating/saunaing and improves liver detoxification with vitamins and minerals. We can easily test for pollutants in fat by doing a fat biopsy – this is a simple test, easier than a blood test. As part of normal metabolism, the body produces toxins, which have to be eliminated otherwise they poison the system. Therefore, the body has evolved a mechanism for getting rid of these toxins and the methods that it uses. The antioxidant system of the body for clearing up free radicals and the three most important **frontline antioxidants** are:

- **Co-enzyme Q10**. This is the most important antioxidant inside mitochondria and also a vital molecule in oxidative phosphorylation. Co-Q10 deficiency may also cause oxidative phosphorylation to go slow because it is the most important receiver and donator of electrons in oxidative phosphorylation. People with low levels of Co Q 10 have low levels of energy. According to experience, levels in CFS sufferers are almost always down and they can be corrected by taking the Co-enzyme Q10 300 mg daily for 3 months, after which continue with a maintenance dose of 100 mg.
- **Superoxide dismutase (SODase)** is the most important super oxide scavenger in muscles (zinc and copper SODase inside cells, manganese SODase inside mitochondria and zinc and copper extracellular SODase outside cells; Fig. 7.42). Deficiency can explain muscle pain and easy fatigability in some patients. SODase is dependent on copper, manganese and zinc, which is why people taking a physiological mix of minerals are on a safer side. However, when there is a deficiency, these minerals are taken separately. Experience shows that the best results are achieved by copper 1 mg in the morning, manganese 3 mg midday and zinc 30 mg at night. Low dose SODase may also result from gene blockages and these are also looked at when the SODase test is done. Blockages are most often caused by toxic stress, such as heavy metals and pesticides.
- **Glutathione peroxidase (GSH-Px)**. They are made up of glutathione, combined with selenium. There is a particular demand in the body for glutathione. Not only is it required for GSH-Px, which is an important frontline antioxidant, but also it is also required for the process of detoxification. Glutathione conjugation in the

7.3 Neurodegeneration Resulting from Blockade in the Energy Pathway

Fig. 7.42 The different superoxide dismutases for inactivation of toxic radicals and their inhibition by peroxynitrite

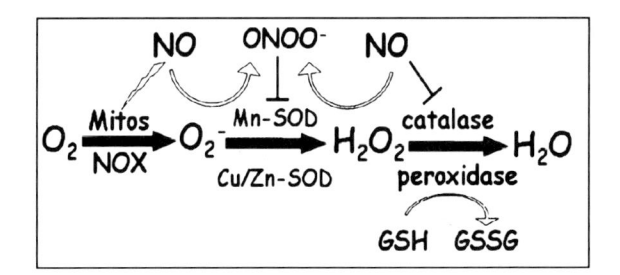

$$O_2 \xrightarrow[NOX]{Mitos} O_2^- \xrightarrow[Cu/Zn\text{-}SOD]{Mn\text{-}SOD} H_2O_2 \xrightarrow[peroxidase]{catalase} H_2O$$

with NO, ONOO⁻, NO inhibiting the SOD step, and GSH → GSSG coupled to the peroxidase step.

liver is a major route for excreting xenobiotics. This means that if there are demands in one department, then there may be depletions in another, so if there is excessive free radical stress, glutathione will be used up and therefore less will be available for detoxification and vice versa. Of course, in patients with chemical poisoning or other xenobiotic stress, there will be problems in both departments, so it is very common to find deficiencies in glutathione: If there is a deficiency of GSH-Px, then I recommend that patients eat a high protein diet (which contains amino acids for endogenous synthesis of glutathione), take a glutathione supplement of 250 mg daily, together with selenium 200 μg daily, or take the precursor in glutathione synthesis, i.e. N-acetyl-cysteine in a dose of 2 g/day.

- Taking a **"Caves Man Diet"** or taking the standard recommendations for nutritional supplements largely provides the second and third line antioxidants. These molecules are present in parts of a million and are in the frontline process of absorbing free radicals. When they absorb an electron from a free radical both the free radical and the antioxidant are effectively neutralized, but the antioxidants reactivate themselves by passing that electron back to **second line antioxidants** such as vitamins A and beta carotene, some of the B vitamins, vitamin D, vitamin E, vitamin K and probably many others. These are present in parts per thousand. Again, accepting an electron neutralizes these, but that is then passed back to the ultimate repository of electrons, namely vitamin C, which is present in higher concentrations. Most mammals can make their own vitamin C, but humans, fruit bats and guinea pigs are unable to do so. They have to get theirs from the diet and Linus Pauling, the world authority on vitamin C, reckoned that people need vitamin C in gram doses everyday. A minimum of 2 g of vitamin C daily is recommended and for some patients up to 6 g. Government recommend intake of 30 mgs a day is just sufficient to prevent scurvy, but insufficient for optimal biochemical function, especially when there is an excess formation of ROS and NOS together with peroxynitrite.

- **Paraoxonases** are a group of enzymes involved in the hydrolysis of organophosphates are antioxidants that sit on good cholesterol (HDL) and protects this and the bad cholesterol (LDL) from oxidation [252]. Levels of paraoxonase are determined genetically. These enzymes detoxify organophosphate pesticides so, if deficient, this makes the organophosphate very much more toxic. It is deficient in about one third of the population and this explains why about one third of those farmers exposed to organophosphates become ill.

- **Melatonin** also has been found to have a profound antioxidative property and where available, is being used as remedy in patients with an increase in ROS or NOS.
- **Vitamin B$_{12}$** is an excellent antioxidant and if there is a patient with particularly poor antioxidant status then B$_{12}$ injections are recommend. The injectable way of administration provides instant antioxidant cover and protects the patient from further damage while at the same time these necessary micronutrients helps to heal and repair its own antioxidant system. Vitamin B$_{12}$ is also an excellent treatment for foggy brain.

All the above antioxidants can be measured and almost routinely these frontline antioxidants (i.e. Co-enzyme Q10, superoxide dismutase, and glutathione peroxidase) should be determined. This antioxidative-detoxification system has worked perfectly well for thousands of years. Problems now arise because of toxins, which we are absorbing from the outside world. This is inevitable since we live in equilibrium with the outside world. The problem is that these toxins may overwhelm the system for detoxification (such as alcohol), or they may be impossible to break down (e.g. silicone, organochlorine pesticides), or they may get stuck in fatty organs and cell membranes and so not be accessible to the liver for detoxification (many volatile organic compounds). We all carry toxins as a result of living in our polluted world. However, much can be done to get rid of them or decrease our load, and the mechanisms that we can employ.

7.4 Symptoms of Mitochondropathy of the CNS

What allows the brain to work quickly and efficiently is its energy supply. If this is impaired in any way, then the brain will go slow. Initially, the symptoms would be of foggy brain, but if symptoms progress, one will end up with dementia. We all see this in our everyday life, with the effect of alcohol being the best example. Short-term exposure gives us a deliciously foggy brain – we stop caring, we stop worrying, it alleviates anxiety. However, it also removes one's drive to do things, one's ability to remember; it impairs judgment and our ability to think clearly. Medium-term exposure results in mood-swings and anxiety (only alleviated by more alcohol). Long-term use could result in severe depression ending up in dementia with examples including Korsakoff's psychosis and Wernike's encephalopathy. Incidentally, these two examples also illustrate how most drug side-effects result from nutritional deficiencies!

A normal nerve will pass a nerve impulse in 75 μs. The slower the time, the more there is a loss of information. For instance, reaction times are slowed with alcohol. If this interval extends to 140 μs, one has dementia; any time longer than that and the person gets unconscious as exemplified by the effects of a general anesthetic. The only cell organelle that is able to convert fuel and oxygen into ATP (Adenosine Triphosphate) the energy supply to the brain is the mitochondrion.

> **What allows energy supply to the brain is good mitochondrial function.**

Alcohol intolerance is almost universal in CFS patients. It can partly be explained by poor detoxification, but clinically there seems to be some direct effect on the brain, which can be explained by its physicochemical properties! Alcohol readily dissolves fats and is excellent in changing the consistency of the myelin sheath cell membranes. It works just like a general anesthetic and, indeed, many CFS patients are intolerant of volatile anesthetics. This is because the more fat-dissolving a chemical (e.g. an anesthetic), the greater its ability to produce foggy brain in CFS sufferers. This could also apply to a range of prescription medication. Such knowledge is very useful clinically, since people who are intolerant of alcohol need brain fats and oils as outlined above. And as we age, such intolerance increases, which just indicates the greater need for the correct fats and oils as we become metabolically less efficient with the passage of time! The following symptoms are characteristic for a brain fog all of which may be due to different causes, however, in the majority of cases this is due to an impaired energy production within mitochondria (Ref. [9] of Chap. 5):

1. Difficulty learning new things
2. Poor short term memory
3. Poor mental stamina and concentration – there may be difficulty reading a book or following a film story or following a line of arguments
4. Difficulty finding the right word
5. Thinking one word, but saying another

7.4.1 *General Approach for Diagnosing and Restoring Mental Health*

When diagnosing a possible mental deficit one should consider the following possible cause:

- **Arteriosclerosis** resulting in short breath and early exhaustion. The best test for generalized arteriosclerosis is one's ability to gain fitness. All normally healthy people should be able to gently jog one mile without distress.
- **Supply of energy** to vital organs, and can this supply of fuel and oxygen sufficient to be translated into ATP, the energy for brain cells to work ? (see CFS and its central cause mitochondrial failure on page 121 ff).
- **Low blood sugar level**. Is there a steady supply of this carbohydrate to the brain? (See hypoglycemia on page 170 ff, 222 ff).
- **Inability of neuronal cells to use glucose for energy generation** (neuronal diabetes). Coconut oil in this respect is likely to be very helpful, since the brain works well on short chain fatty acids and ketones, which are in abundance in

coconut oil. These must be pure cold-pressed organic virgin oils, which are semisolid at room temperature – not hydrogenated oils, which are hard at room temperature (see coconut oil and Alzheimer on page 230 ff).
- **Allergy** can certainly cause foggy brain. Wheat is a common cause of brain fog. Many religious groups will fast for several days in order to "clear the brain". Food allergy usually causes more than one symptom, and often several can be identified (see allergy on page 183 ff).
- **Getting deprived in Sleep.** Are you getting enough good quality sleep or is there sleep deprivation? During sleep, healing and repair takes place, and without these one gradually goes downhill. Sleep is vital for good health – especially in CFS patients (See maintaining sufficient sleep on page 166 ff)
- **Hormonal disturbances** (see hypothyroidism, underactive adrenal glands at page 153)
- **Physical and mental activity** is vital. If one does not use it, than you will lose it! There is a desirable balance between physical and mental activity. Often the best "treatment" for mental stress is physical activity, rather than being inactive and be a consumer just watching TV!.
- **Depression and anxiety** are often accompanied by foggy brain; but these are symptoms, which can be found in other psychiatric ailments and therefore should be further investigated to find the cause.

7.4.1.1 Foggy Brain from Getting Poisoned Inside/Outside the Body

- **Gut fermentation** – fermented foods result in the production of alcohol, D-lactate, hydrogen sulphide and other toxins. Any tendency to constipation will make this worse. This may be one explanation why colonic irrigation results in a "clearing" of the brain.
- **Caffeine** is a short-term mental stimulant. This can be helpful if one has to "perform", so long as one can rest and recover afterwards. If people are having more than three cups a day (tea, coffee, coca cola), then very likely this may have an overall deleterious effect.
- **As a result of a detoxification regime**, toxins are mobilized which have been previously stored in fatty tissue and after being redistributed into the circulation result in a prelim worsening of symptoms.

7.4.1.2 Foggy Brain from Getting Poisoned Outside the Body

- **As a result of alcohol,** which can be considered a potent neurotoxin. In addition, if one drinks more than one glass of wine daily, then a potential thiamine deficient may develop. Thiamine (vitamin B_1) is essential for normal brain function acting within the electronic transport chain of mitochondria to generate ATP.
- **Prescription medication** very often has profound effects on the brain, especially in older and younger people and when taken over a longer period of time (for further information see Sect. 7.2 on chemical poisons and toxins).

7.4 Symptoms of Mitochondropathy of the CNS   149

- **As a result of a long-term contaminated diet,** consisting of trans fats, growth promoting hormones or antibiotics in beef and chicken, or consuming herbicide contaminated food (e.g. Rebell®, ButisanStar®, Arelon®Top, or the plant growth regulator Cardinal™) in lettuce, fruits and cabbage. Look and check the list of possibilities (see Sect. 7.2.3, 7.2.4 on chemical poisons and toxins) the more one looks for it, the more one sees!

If, however, there are progressive brain symptoms this demands for additional diagnostic evaluation, such as using an MRI scan to exclude tumors or other anatomical lesions.

7.4.1.3 Forms of Detoxification of the Body

- **The liver** – detoxification by oxidation and conjugation (amino-acids, sulphur-compounds, glucuronide, glutathione, etc.) for excretion in urine.
- **Excretion via the bile** especially of fat-soluble toxins. The problem here is that many of these are recycled because they are reabsorbed in the intestine.
- **Sweating** – many toxins and heavy metals can be lost through the skin a major reason why sauna is a recommended method for detoxification.
- **Dumping chemicals** in hair, nails and skin, which is then shed.

7.4.2 Fibromyalgia Closely Related to Mitochondropathy

Fibromyalgia (from Latin, *fibro-*, fibrous tissues, geek *myo-*, muscle, greek *algos-*, pain) meaning muscle and connective tissue pain; also referred to as **FM** or **FMS** is a medical disorder characterized by chronic widespread pain and allodynia, a heightened and painful response to pressure [253]. Fibromyalgia symptoms are not restricted to pain (Fig. 7.43), leading to the use of the alternative term fibromyalgia syndrome for the condition. Other symptoms include debilitating fatigue, sleep disturbance, and joint stiffness. Some patients [254] may also report difficulty with swallowing, bowel and bladder abnormalities [255], numbness and tingling [256] and cognitive dysfunction [257]. Fibromyalgia is frequently comorbid with psychiatric conditions such as depression and anxiety and stress-related disorders such as posttraumatic stress disorder [258, 259]. Not all people with fibromyalgia experience all associated symptoms. Fibromyalgia is estimated to affect 2–4% of the population [258], with a female to male incidence ratio of approximately 9:1. As to the etiology of fibromyalgia, various short-term initiating stressors include infections, mostly viral but occasionally bacterial and physical trauma, particularly head and neck trauma are discussed. All of which has been shown to elevate nitric oxide in the body. The same is true of severe psychological stress that in some cases of fibromyalgia causes the outbreak of this disease. Some cases of fibromyalgia are secondary to autoimmune diseases, such as in lupus, and the autoimmune attack is known to produce substantial inflammation in the tissue attacked. Low thyroid hormone

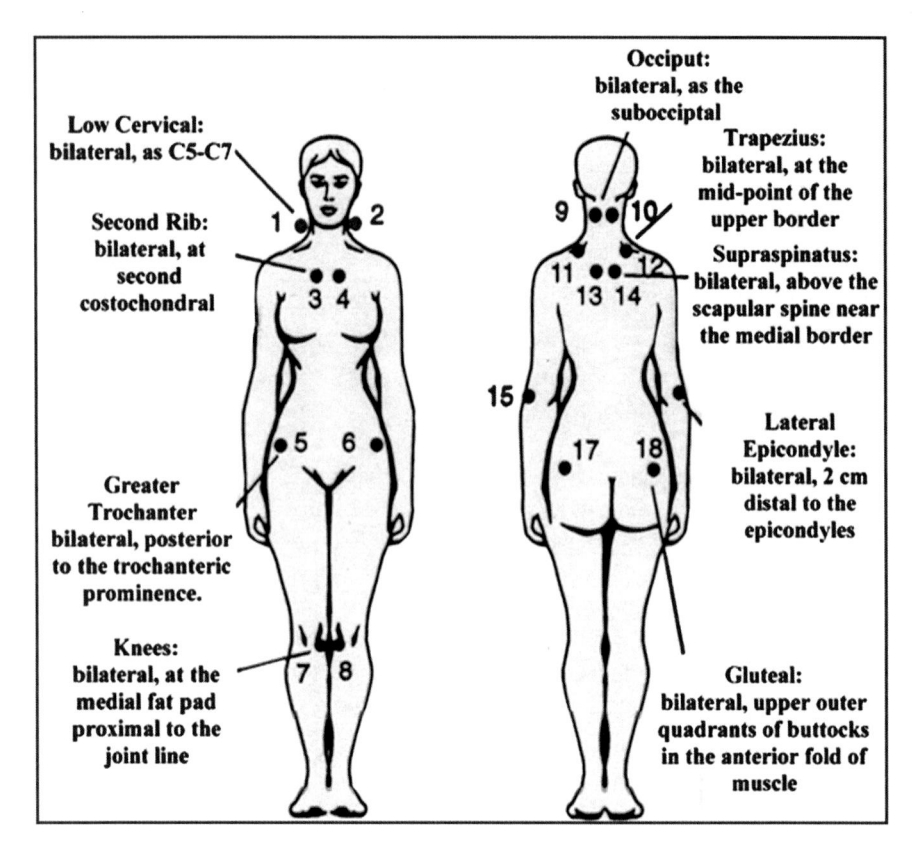

Fig. 7.43 Typical localization of painful tender points in patients with fibromyalgia where pain is induced by pressure, indicative for an sensitized pain transmission system

level has also been discussed as a cause in some cases of fibromyalgia, since low levels have been shown to greatly increase nitric oxide synthase activity in mitochondria. There are several studies reporting oxidative stress in fibromyalgia and the same number reporting that inflammatory cytokine levels are elevated [260].

There are other studies supporting the origin of mitochondria dysfunction with NMDA-activation being the common denominator and underlying mechanism that explains the symptoms of CFS/FMS [261–265], and where pain (Fig. 7.43) is reduced by means of non-selective NMDA-antagonists (i.e. ketamine, detropropoxyphene). Two other studies report citrate accumulation as a possible consequence of cycle mechanisms leading to lowered aconitase activity. There are a large number of studies showing that NMDA antagonists and other agents that lower NMDA activity produce substantial improvements in fibromyalgia patients, strongly suggesting that NMDA levels are elevated and that such elevated levels have causal roles [266]. The data on nitric oxide levels in fibromyalgia are mixed, with two studies reporting elevation but two others not showing elevation [267]. There is a single study reporting the NF-κB activity seems to be elevated in fibromyalgia and three studies report on the elevation

7.4 Symptoms of Mitochondropathy of the CNS

Fig. 7.44 Putative role of the thalamus in widespread pain perception of fibromyalgia syndrome (FMS) (Adapted from (Ref. [4] of Chap. 6)

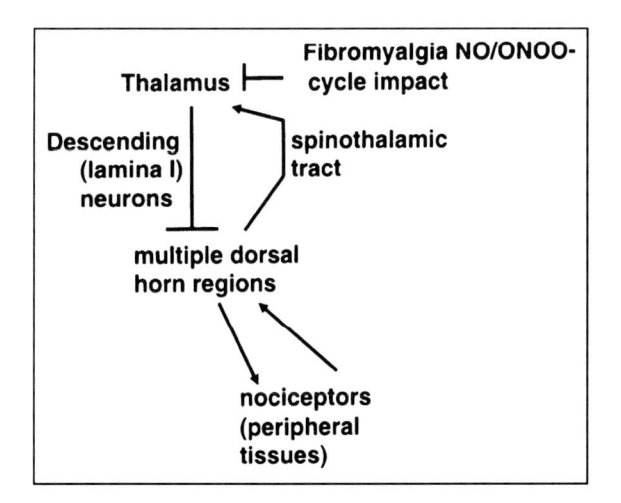

of one aspect of the cycle that so far has rarely been studied in these diseases, namely TRPV1 activity [268–270]. In terms of the primarily local nature of the cycle, this is supported in fibromyalgia by the same two types of observations that we have for CFS: In brain scan studies, where one can visualize the tissue impact, there is a large variation in the tissue being affected among different patients with fibromyalgia, suggesting a primarily local mechanism with variation in tissue localization. This may provide an explanation for the huge variation in symptoms reported among cases of fibromyalgia. One of the organs that appear to be impacted in all or almost all cases of fibromyalgia is the thalamus, a region of the brain having a special role in regulating pain processing. How can such widespread excessive pain be generated by the impact of the cycle (a primarily local mechanism) on a single region of the body? After all, unlike the excessive pain seen in other related diseases, the properties of fibromyalgia pain is that it seems to be turned on almost like a switch, over most of the body. The likely impact of the increased NO/ONOO$^-$ cycle seems to be the thalamus being the cause of this widespread, excessive pain and is depicted in the following Fig. 7.44. While the thalamus is not the only part of the brain, which is often impacted in fibromyalgia, it does seem to be impacted in most if not in all cases.

This interpretation of fibromyalgia pain is supported by a rat model of fibromyalgia [271] involving reduced serotonin release in the ventro-basal thalamus, which leads to increased NMDA activity and nitric oxide levels, producing widespread neuropathic pain. Consequently, both fibromyalgia and chronic fatigue syndrome are excellent examples for NO/ONOO$^-$ cycle diseases and the cycle provides a stringent explanatory mechanism, with the apparent common etiology, being proposed by many different research groups. The widespread excessive pain, that is the cardinal symptom of fibromyalgia (Fig. 7.43) provides a major challenge for pain specialists who are just in the beginning to understand the mechanism of the NO/ONOO$^-$ cycle in this disease.

The challenge in these illnesses therefore is to lower the whole pattern of elevated NO/ONOO$^-$ and turn them back into normal range. The cycle not only includes the elements nitric oxide, superoxide and peroxynitrite but a series of others, including the transcription factor NF-κB, accumulation of radical oxidative substrates (ROS), inflammatory cytokines (the box on the upper right of Fig. 6.1), three different forms of enzymes involved in the making of nitric oxide (i.e. nitric oxide synthases, iNOS, nNOS and eNOS), and two receptors within the central nervous system (CNS), the vanilloid (TRPV1)- and the N-methyl-D-Aspartate (NMDA)-receptor, which are also involved in the transition of pain.

7.4.2.1 Implications for Treatment in CFS/FMS

To support mitochondria, first and most important of all D-ribose should be administered, followed by CoQ10, acetyl-l-carnitine, NADH, magnesium and B_{12} injections. All these agents must be put in place to repair and prevent ongoing damage to mitochondria allowing them to recover. For mitochondria to recover in addition they need all the essential vitamins, minerals, essential fatty acids and amino acids to manufacture the cellular machinery to restore normal function. A mitochondrial function test (see chap. 9) allows to identify the exact lesion(s) which can be corrected by adding the nutritional supplements, improving antioxidant status, detoxification, and reset the function (Table 7.5). CFS as well as FMS sufferers have limited reserves of physical, mental and emotional energy and the mitochondrial function test allows the physician to direct and add the most needed supplements for recovery.

7.4.2.2 Treatment Package for Failing Mitochondria in CFS and FMS (Table 7.5)

- **Avoid physical overactivity** – do not use up energy faster than mitochondria can supply it. The individual range may vary but the patient can tell for himself when energy supply decreases
- **Feed the mitochondria** – supply the raw material necessary for the mitochondria to heal themselves and work efficiently. This means feeding the mitochondria correctly so they can heal and repair (for further information see section on therapy of mitochondropathy).
- **Address the underlying causes** as to why mitochondria have been damaged. This must also be put in place to prevent ongoing damage to mitochondria. In order of importance this involves:
- **Control any activity** to avoid undue stress to mitochondria
- **Get excellent sleep** so mitochondria can repair during the deep plane of slumber
- **Eat a non-contaminated** nutrition with respect to:
  - Taking a good range of micronutrient supplements consisting of Q10, magnesium, L-carnitine, curcuma, vitamin B-complex (especially B_{12}) plus D-ribose,
  - Stabilize blood sugar levels, avoiding the ups and downs
  - Identify any possible allergy to foods.

7.4 Symptoms of Mitochondropathy of the CNS

Table 7.5 Summary of symptoms and signs of evidence for an elevated NO/ONOO⁻ cycle in fibromyalgia

Symptoms-Signs	Explanation based on theory of elevated nitric oxide/peroxynitrite levels
Low energy metabolism/mitochondrial dysfunction	Inactivation and inhibition of several coenzymes in several complexes within mitochondria by peroxynitrite, nitric oxide and superoxide
Oxidative stress	Peroxynitrite, superoxide and other oxidants
PET scan changes	Energy metabolism dysfunction; changes in perfusion by nitric oxide, peroxynitrite and isoprostanes
SPECT scan changes	Depletion of reduced glutathione by oxidative stress; similar to perfusion changes seen under PET scan
Low NK cell function	Superoxide and other oxidants induce lowering of NK cell function
Elevated cytokines	NF-κB stimulation induces activity of inflammatory cytokine genes
Anxiety	Excessive NMDA activity in the amygdala
Depression	Elevated nitric oxide leading to depression; cytokines and NMDA increase, acting in part or in whole via nitric oxide.
Rage	Excessive NMDA activity in the periaqueductal gray region of the midbrain
Cognitive, learning and memory dysfunction	Low energy production by mitochondria in neurons of the brain, which is most susceptible to such changes.
Multisystemic pain	Activation of the TRP (vanilloid)- and NMDA-receptors involved in pain processing, induced in part through nitric oxide and cyclic GMP elevation
Chronique fatigue, low stamina	Energy metabolism dysfunction in mitochondria with reduced formation of ATP; ATPase dysfunction
Sleep disturbance	Sleep obstructed by inflammatory cytokines, NF-κB activity and nitric oxide
Orthostatic intolerance	Two mechanisms: Nitric oxide-mediated vasodilatation leading to blood pooling in the lower body; nitric oxide-mediated sympathetic nervous system dysfunction
Irritable bowel syndrome	Increase in sensitivity produced by excessive vanilloid (TRP)- and NMDA activity, and increase in nitric oxide formation
Intestinal permeability in gut leading to food allergies	Permeability increased by excessive local nitric oxide, inflammatory cytokines, NF-κB activity and peroxynitrite; peroxynitrite acts in part by stimulating polyADP-ribose polymerase activity

- Detoxify to unload heavy metals, pesticides, drugs, social poisons (alcohol, tobacco, etc.) and volatile organic compounds, all of which poison the mitochondria.
- Address the secondary damage caused by mitochondrial failure such as immune disturbances resulting in allergies and autoimmunity, poor digestive function, hormonal gland failure, low liver detoxification capacity.

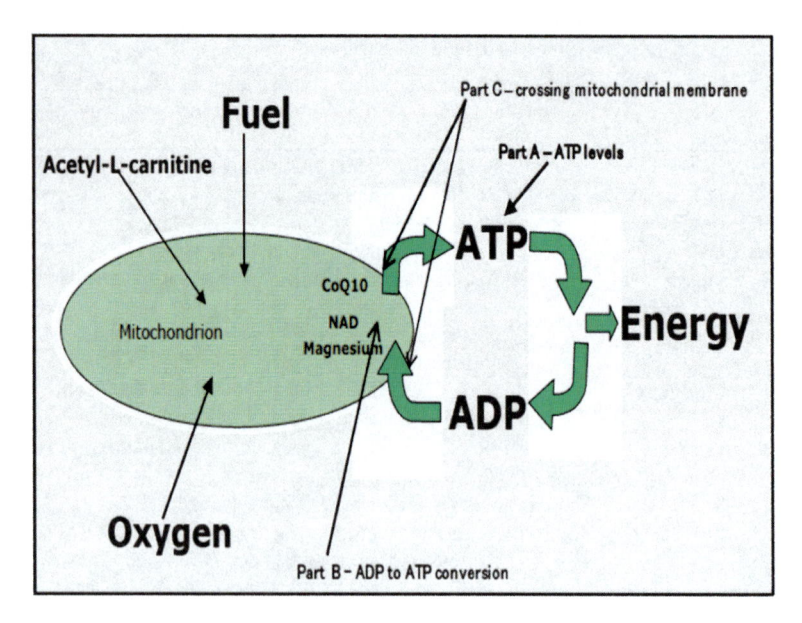

Fig. 7.45 Different steps in determining the efficacy of the translocator protein in mitochondrial function with the levels of Q10, NAD and magnesium necessary for sufficient function

- Restore ATP levels by recycling of AMP. And while AMP can be recycled, this process is slow and may take up to several days. Interestingly, the enzyme that does this (i.e. cyclic AMP) is activated by caffeine! So the perfect refreshment for CFS sufferers would be black organic coffee with a teaspoon of D-ribose!

7.4.2.3 How to Identify Problems of Mitochondria

The root causes of the mitochondria dysfunction in CFS/FMS is the glutathione depletion, coupled with the partial methylation cycle block. These two deficiencies allow the toxins and the oxidative stress to build up, and deplete the de novo production of L-carnitine, coenzyme Q10, phosphatidylcholine and creatine, all of which need the methylation step in their synthesis, later to be incorporated in the energy production of mitochondria. An *ATP function profile test* can be used to confirm the clinical picture of CFS, assess the level of disability objectively, identify where the biochemical lesion lies and give the physician an idea as to how to further clarify and correct that biochemical lesion(s). Even if the results do not show a marked impairment, mitochondrial function can always be improved further by taking the suggested supplements. The *ATP function profile* test comprises of three parts (Fig. 7.45):

A. The level of ATP
B. Oxidative phosphorylation within the Krebs's-Citric Acid Cycle where ADP is converted to ATP, and
C. Movement of ATP and ADP across the mitochondrial membranes.

7.4 Symptoms of Mitochondropathy of the CNS

Not only does this test measure the rate at which ATP is made, it also looks at where the problem lies. Part A of measures levels of ATP in the cell. Release of energy from ATP is a magnesium-dependent process and the first part of the test studies this aspect. The second aspect of the test (Part B) measures the efficiency with which ATP is made from ADP in the mitochondrion. If this is abnormal then this could be as a result of magnesium deficiency, of low levels of Co-enzyme Q10, low levels of vitamin B_3 (NADH) or of acetyl L-carnitine. It is also possible that ADP to ATP conversion is blocked and this is also seen on this part of the test.

The third possibility is that the protein, which transports ATP and ADP across mitochondrial membranes is impaired and this is also measured (Part C). Next look to evaluate is the rate at which ADP is converted to ATP. This is a magnesium dependent process; therefore if the rate of conversion is slow, it results from magnesium deficiency. Thus, low intracellular magnesium is a symptom of CFS and also a cause of it. This is because 40% of resting energy simply fuels the sodium/potassium (Na/K) and the calcium/magnesium (Ca/Mg) membrane ion pumps. So when the energy supply is diminished, there is insufficient energy to fire these pumps, and magnesium cannot be drawn into cells for oxidative phosphorylation to work, resulting in a further diminished supply of energy. This represents one of the many similar vicious cycles in CFS. The movement of ATP and ADP across mitochondrial membranes (Part C in Fig. 7.45) is dependent on the translocator protein (TL), which is located in the mitochondrial membrane and shunts ATP and ADP back and forth. Indeed 80% of mitochondrial membranes are made up of the translocator (TL) protein! If this is malfunctioning then this suggests a blockage by a toxin. At this date, there is no relevant test, which conclusively can demonstrate the toxins most likely involved. Many toxins can do this, there are endogenous toxins (from free radicals) and there are exogenous toxins from heavy metals, pesticides, volatile organic compounds (VOCs), etc. (for further information see Sect. 7.2). However, from these exogenous toxins one can all get rid off by sweating regimes. Exercise is obviously the most physiological method but this is impossible for CFS or FMS patients! Therefore, a sweating detoxification at least three times a week is a recommend technique using saunas, Turkish baths and spa therapies. However, the problem with these treatments is not that they warm up the skin and subcutaneous tissues, but the whole body is being warmed up. This means that chemicals are mobilized from the fat (which largely speaking lies underneath the skin), they get into the blood stream and may cause acute poisoning symptoms. Furthermore, many people with CFS do not tolerate heat at all for reasons of poor cardiac output (further information see pages on mitochondrial failure). As a result, many patients are unable to tolerate these sweating therapies. An alternative technique is far *infrared saunaing (FIRS)* where the main energy source that comes from the sun and is responsible for warming our skin when we are exposed to direct sunshine. The rays penetrate several centimeters through the skin and heat up the subcutaneous tissues. With enough sun on the skin the skin will warm up and sweat, and chemicals from subcutaneous tissues will be mobilized and pass out through the sweat. The sunshine does this without heating up the core temperature (although if you lie in the sun for long enough then the core temperature will eventually rise) therefore chemicals can be mobilized and excreted without causing the core temperature to rise.

For example, patients with severe heart disease who certainly do not tolerate a conventional sauna were able to tolerate FIRS very comfortably.

Far infrared saunaing does not damage the skin. The sun's ultraviolet rays at the other end of the spectrum, which do that, so have no fears on this count. Indeed primitive man evolved running naked under the African sun and so to have far infrared playing on their skin and warming it as an entirely natural process. The joy of sweating is that one gets rid of most toxins. This is because sweat is blood but without the cellular and protein content. So sweat contains everything – a range of all toxins including pesticides, volatile organic compounds and heavy metals. It is a very natural and physiological way of detoxification. Indeed it is very likely that many of the health benefits of sweating arise from the fact that one is undergoing detoxification. It is never too late to start sweating, and often one can find organophosphates in fat biopsies decades after the last exposure, with levels coming down after only a few weeks of sweating.

Another option is the intake of the algae Chlorella protothecoides being able to accelerate the detoxification process. This has been demonstrated in chlordecone poisoned rats, decreasing the half-life of the toxin from 40 to 19 days. The ingested algae passed through the gastrointestinal tract unharmed, interrupted the enteric recirculation of the persistent insecticide, and subsequently eliminated the bound chlordecone with the feces. The detoxification was similar to that obtained with cholestyramine. The active component in the process is sporopollenin, a carotenoid polymer of limited natural occurrence among microorganisms and plants [272]. Plant sporopollenin was not active, but algal cell walls and sporopollenin retained the therapeutic activity of the whole cells. Thus cells and cell walls of Chlorella protothecoides have potential as detoxifying drugs and other poisonous xenobiotic compounds and proves that *Sporopollenin* is a very potent ingredient in heavy metals *detoxification* process eliminating also heavy metals such as Arsenic (As), Cadmium (Cd), Plumb (Pb) and Mercury (Hg) (Fig. 7.46).

Key Points on Saunaing and Sweating

Firstly, not only are toxins excreted in the sweat when saunaing, but also so are beneficial minerals. Therefore after a sweat it is vital to rehydrate with a physiological mix of minerals containing all essential minerals in water and take a small supplement of salt (an eighth of a teaspoon salt on food). This phenomenon explains why even trained athletes sometimes drop dead inexplicably. They sweat so much that they deplete themselves of essential elements, in particular magnesium. They only need to rehydrate themselves with "electrolytes", namely sodium and potassium. Magnesium is necessary for muscles to relax and if levels are very low, the cardiac muscle (myocardium) cannot relax, and the heart just stops in systole. At post mortem study the heart appears normal. Secondly, it is important to shower immediately after a sweat in order to wash away toxins, which may otherwise be reabsorbed back through the skin.

Thirdly, the most excretion of toxins occurs in the first few minutes of sweating as they move onto the surface of the skin – thus, the best results come from many

7.4 Symptoms of Mitochondropathy of the CNS

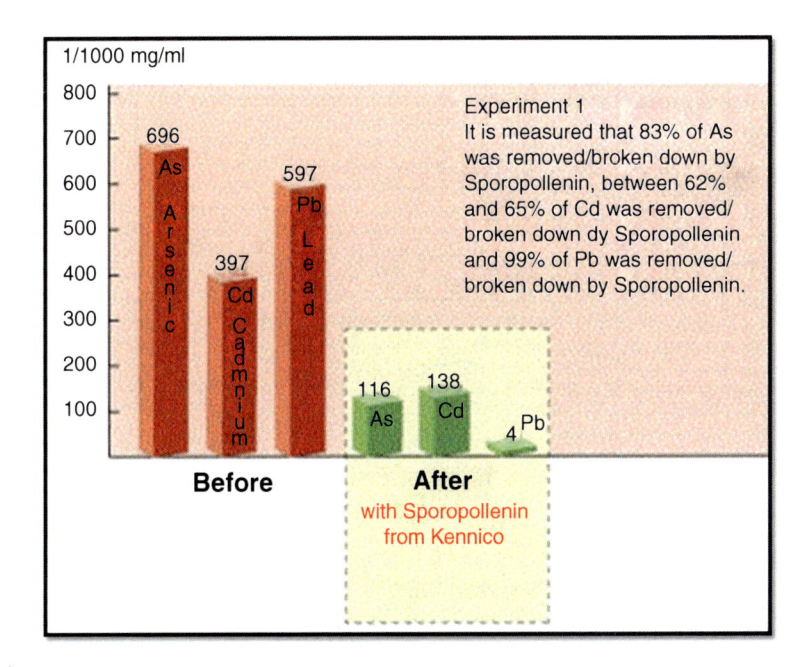

Fig. 7.46 Heavy metal removal utilizing sporopollenin

short sessions (e.g. one daily just to the point of sweating) rather than protracted sweating which may make the patient feel ill. Measuring levels of pesticides and VOCs in fat biopsies, or by measuring mercury excretion, have demonstrated that 50 sweating sessions reduce the toxic burden consistently well. One can expect levels to come down exponentially as more toxins are shifted in the earlier than in the later sessions. However, we are all inevitably exposed to toxins in the environment and there is an equilibrium between us and our environment so sweating regimes should be a regular part of keeping a healthy system.

Because the central problem of chronic fatigue syndrome is mitochondrial failure resulting in poor production of ATP, i.e. the currency of energy within the body and its impaired production, then all cellular processes will go slow. It is not good enough to measure absolute levels of ATP in cells since this will simply reflect how well relaxed the sufferer is. The perfect test is to measure the rate at which used ADP is recycled back into mitochondria (the translocator protein test), by using the so called "mitochondrial function profiles", which determines the exact site of mitochondrial failure:

1. The rate of ATP synthesis,
2. The efficacy of ADP recycling back into mitochondria,
3. The ATP transfer back into the cytosol,
4. The concentration of magnesium within mitochondria,
5. The efficacy of oxidative phosphorylation.

Table 7.6 The *ATP function profile test* necessary to identify the site of insufficient energy production within mitochondria

The first part is called "ATP profiles" and has been developed by Dr John McLaren-Howard at Biolab in London. It measures the rate at which ATP is recycled in cells and because production of ATP is highly dependent on magnesium status so the first part of the test studies this aspect.

The second part of the test measures the efficiency at which ATP is made from ADP. If this is abnormal, then this could be as a result of magnesium deficiency, of low levels of Co-enzyme Q10, low levels of vitamin B_3 (NAD^+) or of acetyl L-carnitine.

The third part being measured is the protein, which transports ATP and ADP across mitochondrial membrane (the translocator protein), is necessary for sufficient production of ATP.

To get a full picture of the mitochondrial function profile, also levels of Co-enzyme Q10, SODase, NAD^+, L-carnitine and cell free DNA are being measured.

Determining the Mitochondrial Function Profile

This blood test combines several tests, which together assess mitochondrial function and identifies the problem of energy production within the cell. It is exceptionally useful not only for chronic fatigue sufferers but also for any other patient where mitochondropathy is being suspected or is considered the underlying cause of the ailment. Since it gives clear indications for a treatment regime the test should be performed before starting with any kind of nutritional supplementation. This blood test combines several tests, which together assess mitochondrial function, and identifies where the problem areas within the energy production cycle are located. This is exceptionally useful for chronic fatigue sufferers, in patients with neurotoxicity and fibromyalgia as it gives clear indications for the treatment regime (Table 7.6).

The advantage of the *"ATP profiles test"* is that one now has an objective test of chronic fatigue syndrome (and also in FMS), which clearly shows that this ailment has a physical basis (Fig. 7.47). This test clearly shows that cognitive behavior therapy (CBT), graded exercise and the use of anti-depressants are inappropriate in addressing the origin of this illness. To get the full picture, its useful to combine this test with measuring levels of Co-enzyme Q10, SODase, Glutathione Peroxidase, L-carnitine, NAD and cell-free DNA. Cell free DNA is very useful because it reflects severity of the illness. When cells are damaged and die, they release their contents into the blood stream and cell free DNA measures the extent of this damage. The levels from CFS sufferers are similar to those from patients recovering from major infections, trauma, surgery or chemotherapy, putting CFS firmly in the area of a major organic pathology. SODase is an important antioxidant, which cleans up the free radicals (ROS or NOS) produced by unproductive chemical reactions in the cells. In addition, a serum L-carnitine test is available, which makes sense when determining mitochondrial function. All seven tests are now been combined as a "Mitochondrial Function Profile" and can be ordered online from

http://drmyhill.co.uk/wiki/Mitochondrial_Function_Profile

7.4 Symptoms of Mitochondropathy of the CNS

Blood test result:

Test	Result	Units	Reference rangs
Function test	34	%	Over 40 (mostly 41-47)
Zn/Cu-SOD	178	Enzyme activity (u)	240 - 410
Mn-SOD	242	Enzyme activity (u)	125 - 208
EC-SOD	21	Enzyme activity (u)	28 - 70

Function test → Very poor level

Gene studies

Sod form	Gene(s)	Comments
Zn/Cu-SOD chromosome 21	?Partly blocked	Low enzyme activity
Mn-SOD chromosome 6	Possibly polymorphic	High activity-?Polymorphism
EC-SOD chromosome 4	Normal	Low enzyme activity

Glutathione peroxidase(GSH-PX) → Low normal level

Red cell Glutathione peroxidase(GSH-PX)   67 U/gHb   67 - 90

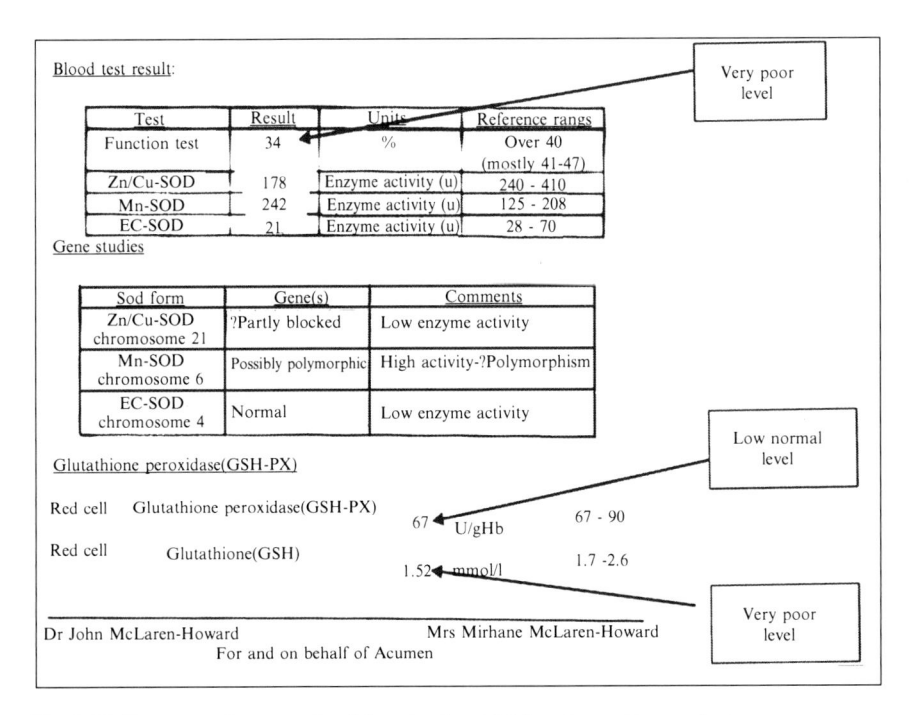

Red cell Glutathione(GSH)   1.52 mmol/l   1.7 -2.6 → Very poor level

Dr John McLaren-Howard   Mrs Mirhane McLaren-Howard
For and on behalf of Acumen

Fig. 7.47 Representative example of the mitochondrial function test in a patient

The test cost £225 plus £70 for a written analysis with a statement for therapy by a physician.

Interpretation of Mitochondrial Function Test

1. **Levels of ATP** The level of ATP in cells is shown by ATP with excess magnesium added and with endogenous magnesium only. When excess magnesium is added and the result is for instance 1.37 (1.6–2.9 nmol/10^6), it indicates very low levels of ATP. At this point a supplementation with D-ribose is recommend, and the body uses D-ribose to make brand new ATP, as opposed to recycled ADP. By using up to three tea spoonfuls this equals to a daily amount of 15 g and the dose should be adjusted according to response. If a patient, however, has a very high cell free DNA, this may be caused by an inappropriate switch from efficient aerobic mitochondrial metabolism into inefficient anaerobic glycolysis with excessive production of lactic acid causing secondary cell damage. Indeed this often causes a symptom of fibromyalgia and D-ribose should be given as it has already been trialed in the treatment of fibromyalgia with excellent results [273]. D-ribose has a very short half-life and should be taken in small doses throughout the day in drinks (hot or cold). Interestingly caffeine enhances the effects of D-ribose so it is recommended to take it with

green tea, coffee, tea or whatever. It is worth supplementing with D-ribose even with low normal results because regularly there is a positive feedback from patients taking this supplement.

2. **Magnesium.** If endogenous magnesium shows a level of 0.74 (0.9–2.7 nmol/l) with a ratio of 0.54 (>0.65), this indicates a low magnesium status. Magnesium is a difficult mineral to replete and some people have to have it by injection. For a faster mode of action, at least a dose of 300 mg of orally magnesium daily is suggested (more if tolerated – up to 600 mg), together with magnesium by subcutaneous injection, e.g. using Evans 50% magnesium sulphate. One can give 2 ml on a weekly basis, since larger volume injections can be uncomfortable. When using ½ ml for patients to inject themselves every day, (using an insulin syringe) for 2 months then adjust the frequency of dose according to the clinical response many people end up injecting 2–3 times a week until they feel much better. These small volumes of injections are better tolerated and less inclined to leave injection lumps. Adding lignocaine often improves tolerance (0.05 ml lignocaine with 0.5 ml magnesium solution) giving the patients the prescription and demonstrating them how to do these injections. Low magnesium level is a real problem in patients with fatigue syndromes, since it is necessary for ATP to release its energy, it is necessary for oxidative phosphorylation and much of the resting energy goes to maintain the calcium magnesium ion pumps. That is to say, a low intracellular magnesium is both the cause and the symptom of mitochondrial failure. So there is a very clear indication to give magnesium by injection. The main problem can be injection lumps. To avoid these, wait for 2 min after the injection to allow capillary bleeding to stop. Then feel the injection site as a small "bump" of the injected magnesium can easily be felt, and massage the area gently until this "bump" disperses completely. However, if there is a bruise when using this technique, do not massage the site at all after the injection.

Epsom salts in the bath (a double handful) will improve magnesium status since magnesium is absorbed through the skin. The bath needs to be as warm as can be tolerated for at least 15 min –the longer the better. The 20 kg sack of Epsom salt can be purchased from garden centers, farm supply shops or health food centers.

3. **Oxidative phosphorylation.** Once the Krebs's Citric Acid Cycle and ADP to ATP conversion is going very slow at for instance 38.5% (normal range >60%), it indicates insufficient mitochondrial activity. To assess the efficiency with which ADP is converted to ATP, during the test an inhibitor is added and then removed to see how quickly ATP is reformed. Having added the inhibitor one expects levels of ATP and magnesium to drop below 0.3 – if this does not happen this suggests there is blocking of the active sites, an acceptable percentage is up to 14%; if however the result is 29.9%, this suggests that there is a significant block at the active sites (i.e. complexes I, II, III, IV and V) on the inner mitochondrial membranes. The most likely reason for this is toxic stress (i.e. from additives & preservatives in food) and one can explore the possible source of malfunction further by using microrespirometry measurements, which look at

7.4 Symptoms of Mitochondropathy of the CNS

oxidative phosphorylation in more detail. In order to identify the problem further it may be necessary to determine the

(a) **Vitamin B$_3$ levels**. Once the red cell NAD shows a mild B$_3$ deficiency as for instance 12.7 µg/ml (normal range 14–30) it can be considered a relevant negative result, because NAD is a functional test. While it reflects B$_3$ status, it also reflects the function of the Krebs's-citric acid cycle. The task of the Krebs cycle is to take energy from acetyl groups and convert it into NADH, which is then converted to NAD$_+$ in the process of driving the oxidative phosphorylation. In order to get normal levels of NAD one needs not only an adequate supply of B$_3$, but also a functioning Krebs's-citric acid cycle. So a low NAD$_+$ (among other things) may also imply poor acetyl L-carnitine levels. While most people can obtain all the NAD$_+$ they need from a combination of diet and a sufficiently high Vitamin B complex preparation, others seem to need much higher levels to achieve normal blood levels. It is recommended to start off with 500 mgs of niacinamide daily, a form of vitamin B$_3$ that is free from side effects. Use of pure niacin or nicotinamide is not recommended, as they may cause unpleasant flushing. Acetyl L-carnitine is normally present in mutton, lamb, beef and pork. If these foods are not consumed then it is recommended to take acetyl L-carnitine 2 g daily. Lamb contains about 5 g per kilo of acetyl L-carnitine so one needs to eat a lot! A small supplemental dose, geared to compensate for meat intake, may be necessary. Indeed, acetyl L-carnitine has been trialed in the treatment of CFS with positive results and studies show that carnitine supplementation can help to lower pain levels, boost the mental health of people with FMS, and lessens fatigue in those with CFS/FMS. Researchers also found that people with either condition tolerated carnitine well [274].

(b) **The Co-enzyme Q10,** also called ubiquinone, which reflects its presence in all tissues because it is present in all mitochondria with a normal range of .64 µmol/l (0.55–2.0), which however in case of CFS or any other mitochondropathy, should be in the upper range. This is because it is the most important antioxidant inside mitochondria and also a vital molecule in oxidative phosphorylation. Co-Q10 deficiency may also cause oxidative phosphorylation to go slow. The best clinical results are achieved if levels get up to 2.5 µmol/l or above. This seems to be necessary to accelerate mitochondrial action, to a point where the dose can be reduced according to clinical response. Such a regime has been worked out by the cardiologist, Dr. Sinatra, who (among D-ribose and l-carnitine) advocates Co-Q10 for treatment of patients with congestive heart failure secondary to mitochondrial failure, which are the results of mitochondrial myopathy [275]. The underlying pathology in these cardiomyopathies is the same as that in the fatigue syndrome. The problem in heart failure are primarily the heart muscles, in the fatigue syndrome, cells of different organs (especially in the CNS) are affected. It is suggested to start on 300 mg daily of Co-Q10 (the dose should be split into 100 mg three times daily) for three months then a maintenance dose of 100 mg daily. It should be outlined that Q10 shows very low water

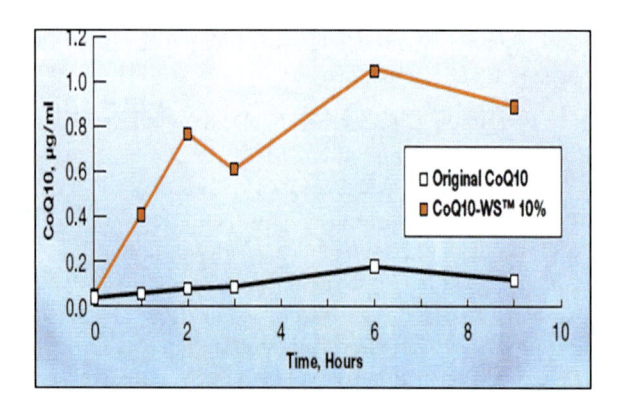

Fig. 7.48 Difference in blood plasma concentrations of Q10 (100 mg) in individuals taking either Q10 in a water-soluble form or as a regular lipid formulation. Note, Q10 when being administered as a water-soluble preparation (CoQ10 WS™ from blue California, manufacturer of botanical extracts) reflects an increase in bioavailability with tenfold higher plasma levels

solubility (Fig. 7.48). Therefore reabsorption via the intestine is very low, and special galenic formularies have been developed. For instance the Q10 nanotechnology from Kaneka company/Japan, results in a quasi 100% bioavailability.

7.4.2.4 Medium Chain Fatty Acids for Restoration of Mental Capacity

There is a further interesting peculiarity with respect to the energy supply to the brain which is different from that used by the rest of the body. Although the brain weighs just 2% of total body weight, in use it absorbs 20% of the body's oxygen requirement and 25% of its fuel needs! The number of mitochondria in the brain cannot simply explain this, since there are not enough. This indicates that there must be another energy-generating source. Brain cells are also very different from normal cells. They have a cell body, and very long tails – or dendrites, which communicate with other cells. To get an idea of the length of such a dendrite just imagine a nerve-cell body of the spinal cord that supplies one's toes would have the size of a football sitting on the desk, its tail would be in New York! These tails (dendrites) are too small to contain mitochondria, but it has been suggested that the energy supply comes directly from the myelin sheaths themselves [276]. They too can produce ATP and it is this sheath that supplies the energy for neurotransmission. Myelin sheaths are made up almost entirely of fats, so we need to look to oils and fats for improved energy supply to brain cells. This assumption is supported by the fascinating work of Dr. Mary Newport who demonstrated conclusively that coconut oils can cure dementia (see chapter on Alzheimer page…). The key ingredients are medium-chain triglycerides (MCT). This suggests that they are an essential part of this process, are an excellent energy

7.4 Symptoms of Mitochondropathy of the CNS

Fig. 7.49 The Polynesian Archipel a source for the coconuts with their important medium chain triglycerides, imperative for normal brain function

supply to the brain, which also likes to work with short-chain fatty acids or ketones. The hypothesis is that the myelin sheath membranes have to be of just the right consistency to allow ATP to be synthesized. If this were the case, it would explain how general anesthetics work – presently no one really knows! If general anesthetics upset the consistency of myelin-sheath cell membranes enough, then the energy supply to nerve cells would be switched off; the nerves would cease to work and not just resulting in a foggy-brain but leading into unconsciousness!

On the other hand one should not be too concerned about dietary fat and cholesterol causing arterial disease! And although statins are being routinely touted as necessary adjuncts to lower cholesterol levels and their benefits to coronary artery disease have been copiously documented and are incontrovertible, other have questioned their potential benefit especially in the older population where statins may even lead into advanced memory loss [277–279]. Also, the results of some studies of Polynesian peoples for whom coconut is the chief source of energy should put your mind at rest (Fig. 7.49). It even has been suggested (although controversial) that statins, by reducing the cholesterol that the brain loves, may contribute to the current epidemic of Alzheimer's disease. Certainly it is rare for CFS patients to tolerate statins, but nearly always they make them ill. Thus, adding coconut oil to the regime of anyone with symptoms of foggy brain to dementia is likely to be helpful, partly by improving fuel supply and partly by improving membrane function.

Oils for Treatment of Dementia

Humans evolved on the East Coast of Africa eating a diet rich in seafood. It is suggested that the high levels of oils, particularly DHA (Docosahexanoic acid), allowed the brain to develop fast, thus allowing humans to outstrip other mammals. So Homo Sapiens came to have bigger brains allowing intelligence to develop. Phospholipid exchange is a technique for supplying the correct proportion of fats and oils in a bioavailable form to replenish cell membranes and membranes within cells. Dr. Patricia Kane, who has seen remarkable clinical results, has pioneered this technique. She uses intravenous therapy, but good results are also possible with oral therapy. Together with patients various physicians throughout Europe are using this techniques of replenishment of cell membranes to get the best results. The underlying principles are:

1. The basic membrane structure is made up of phosphatidylcholine. The best source of this is lecithin (available as egg, soya or sunflower lecithin).
2. Membranes then need the right proportion of Ω-6 to Ω-3 oils, i.e. 4:1. Hemp oil is very close to this having a ratio at 3.8:1, where hemp oil is not the same as linseed oil! So, add in a small amount of sunflower oil, say 5%.
3. Small amount of Eskimo oil; not necessary if one already takes VegEPA®
4. The perfect fuel for brain cells is coconut oil, which is rich in medium chain triglycerides (MCT).
5. If there is poor digestion or poor gall bladder function for any reason then adding bile salts may be very helpful by further emulsifying fats and facilitating digestion and absorption.

Providing an abundance of clean oils helps to displace oils in the brain, which holds polluting heavy metals, pesticides, volatile organic compounds etc. That is to say, these "clean" fats will displace "dirty" fats and also help during detoxification. Hemp oil is near enough perfect with a ratio of 3.8–1 of Ω-6 to Ω-3. However there are combinations of other oils one can use in case allergy gets in the way! Ideally, all these oils should be organic.

Throughout life, the brain creates a million new connections every second! This means there is huge potential for healing and repair; it is simply a case of moving things in the right direction! But the brain has to have the optimum energy supply to allow this process to happen! Research done by Professor Caroline Pond has shown that the immune system, like the brain, is also fat-loving [280]. Wild animals, if they have an abundance of food available, will first deposit fat around lymph nodes where energy is needed for immune activity. This may explain why

7.4 Symptoms of Mitochondropathy of the CNS

people who are apple-shaped are more prone to heart disease compared to those pear-shaped types. Fat deposited around the gut indicates inflammation and inflammation results in arterial damage.

Humans evolved on the East Coast of Africa eating a diet rich in seafood. It is suggested that the high levels of oils, particularly DHA (Docosahexanoic acid), allowed the brain to develop fast, thus allowing humans to outstrip other mammals. So Homo Sapiens came to have bigger brains allowing intelligence to develop. There is a lot of research showing that essential fatty acids are indeed "essential" for normal brain function; so oils that would be helpful in addition to coconut oil would be Ω-3 (fish), Ω-6 (evening primrose) and Ω-9 (olive), together with lecithin (which is a phosphatidylcholine) i.e. the main component of all cell membranes. A suggested therapeutic regime to start off:

- **Lecithin** – one teaspoon (5 ml) twice daily, the raw material for the basic building block of membranes.
- **Coconut oil** – one dessert spoonful (10 ml) twice daily, the perfect fuel for brain cells.
- **Hemp oil** which has the right proportion of Ω-6 to Ω-3 (4–1), since it ensures membranes of perfect consistency – not too stiff, not too elastic.

Hemp oil is near enough perfect with a ratio of 4–1 of Ω-6 to Ω-3. However there are combinations of other oils one can use in case allergy develops with hemp oil! Ideally, all these oils should be organic. A very delicious way to take coconut oil is first to melt it in a warm place, then stir in organic cocoa powder (e.g. green and black), add a teaspoon of D-ribose, allow it to cool and go hard – and one has the perfect chocolate! One may vary the ingredients according to taste; but what is recommended is 1/3 pot (150 g) of coconut oil with 1/2 pot of cocoa powder (60 g) and a teaspoon of D-ribose! The rational for taking the proper oils and fats relates to the fact that throughout life, the brain creates a million new connections every second. This means there is a huge potential for healing and repair; it is simply a case of giving the optimal supplements for healing! However, the brain has to have the optimum energy supply to allow this process to happen. Research data from Caroline Pond has shown that the immune system, like the brain, is also fat-loving [281]. Wild animals, if they have a food overabundance, will first deposit fat around lymph nodes where energy is needed for immune activity. This may be extrapolated into the different shapes of people. Those who are apple-shaped are more prone to heart disease compared to those pear-shaped, since fat deposited around the gut indicates inflammatory mediators in fat cells (with the possibility of allergy or a fermenting gut) with inflammation in the endothelium resulting in arterial damage.

Alcohol Intolerance and Foggy Brain in CFS/FMS

Alcohol intolerance is almost universal in CFS but also in FMS patients. It could be partly explained by poor detoxification process, but clinically there seems to be a direct effect on the neuronal cells of the brain, similar to that of hypersensitivity. The physicochemical property of alcohol explains all. Alcohol readily dissolves fats and is excellent at changing the consistency of the myelin sheath cell of membranes. It works just like a general anesthetic; indeed, many CFS sufferers are intolerant to anesthetics. The explanation is that the more fat-dissolving a chemical, the greater its ability to produce a foggy brain in CFS. This could also apply to a range of prescription medication a knowledge, which clinically may become very useful when CSF patients use a medication resulting in an excessive sedation. People who are intolerant to alcohol need brain fats and oils as outlined above. And as we age, the intolerance to such agents increases, indicating a greater need for the correct fats and oils; as we grow older we become metabolically less efficient!

Treatment of the Foggy Brain in CSF and FMS Patients

How can a supply of fuel and oxygen be translated into ATP, the energy for brain cells to work ? The following facts should be taken into consideration when it comes to replenishment of energy to neuronal cells:

- **Coconut oils** likely to be very helpful – the brain works well on short chain fatty acids and ketones, which are in abundance in coconut oil. These must be pure cold-pressed organic virgin oils, which are semisolid at room temperature – not hydrogenated oils that are hard at room temperature.
- **Allergy** can certainly cause foggy brain. Wheat is a common cause of brain fog. Many religious groups will fast for several days in order to "clear the brain". Food allergy usually causes more than one symptom – often several.
- **Sleep** – are you getting enough good quality sleep? During sleep, healing and repair of impaired mitochondria takes place, and without these function of organs gradually declines. For instance, after the First World War a strain of Spanish flu swept through Europe killing 50 million people worldwide. Some people sustained neurological damage and for some this virus damaged their sleep centre in the brain. This meant they were unable to sleep at all. All those people were dead within 2 weeks. This was the first solid scientific evidence that sleep is more essential for life than food and water. And indeed, all living creatures require a regular "sleep" (especially REM sleep) during which time healing and repair takes place, a fact that even nowadays takes a major percentage in the healing process of any disease. Therefore it is essential to remind patients regularly to put as much effort into their sleep as in their diet. This is of importance since modern world has created a generation of laborer where sleep deprivation is accepted as something natural resulting in a chronic sleep-restricted state which can cause fatigue, daytime sleepiness, clumsiness, weight loss or even weight gain [282]. It adversely affects the brain and cognitive function [283] with an increased risk of fibromyalgia and a higher percentage of diabetes type 2 [284] (Fig. 7.50).

7.4 Symptoms of Mitochondropathy of the CNS

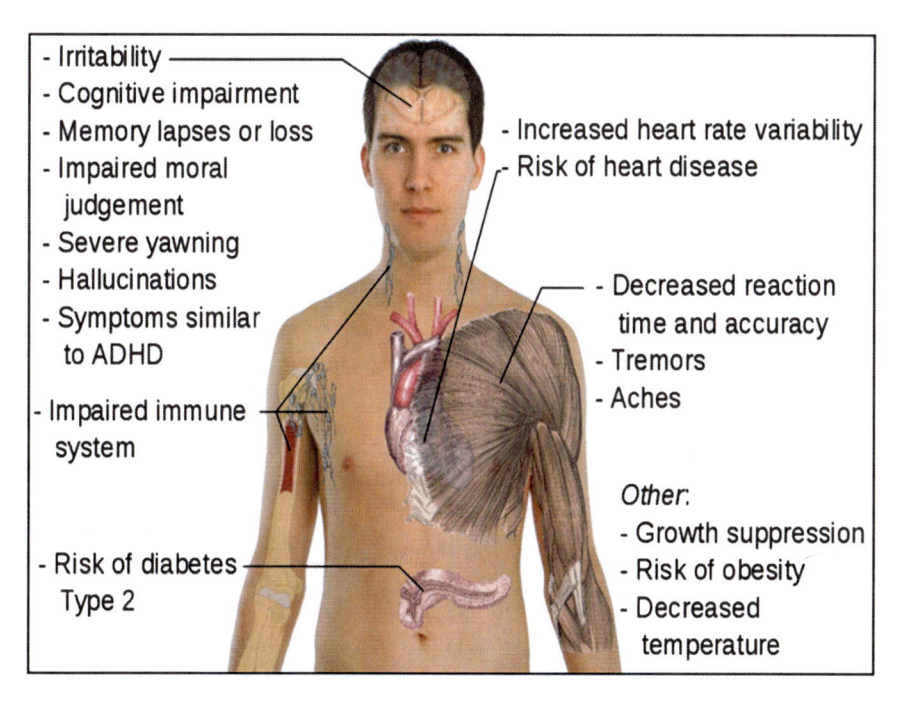

- Irritability
- Cognitive impairment
- Memory lapses or loss
- Impaired moral judgement
- Severe yawning
- Hallucinations
- Symptoms similar to ADHD
- Impaired immune system
- Risk of diabetes Type 2

- Increased heart rate variability
- Risk of heart disease

- Decreased reaction time and accuracy
- Tremors
- Aches

Other:
- Growth suppression
- Risk of obesity
- Decreased temperature

Fig. 7.50 Summary of deleterious effects of chronic sleep deprivation

Recognize the Sleep Wave Cycle

Actually sleep does not gradually come on us during the evening – it comes in waves. There is a sleep wave about every 90 min and you will get to sleep most efficiently if you learn to recognize and ride the sleep wave. Often there is a lesser one earlier in the evening when people drop off to sleep in front of the television, or they jump and make a cup of tea or coffee just to wake themselves up because "they are not ready to go to bed" although in reality they are.

Getting the Mind Conditioned for Sleep

Getting the physical things in place is the easy task. The hard part is getting your brain off to sleep. Since throughout life, the brain makes a million new connections every second!! This means it has a fantastic ability to learn new things. It also indicates that it is perfectly possibly to teach the brain to go off to sleep, it is simply a case of getting into the right mental state, since getting off to sleep is all about developing a conditioned reflex. The first historical examples of this are Pavlov's dogs. Pavlov, a Russian physiologist showed that when dogs eat food, they produce stomach acid. He then "conditioned" them by ringing a bell whilst they ate food. After 2 weeks of conditioning, he could make them produce stomach acid simply by ringing a bell. This of course is a completely useless conditioned response, but it demonstrates the brain can be trained to do anything. By applying this to an insomniac, firstly, one has to get into a mind-set, which does not involve the immediate past or immediate future. That is to say if one is thinking about reality then there is no chance of getting off to sleep. Using a hypnotic, which will get one off to sleep. By applying the two together for a period of few days or a week the brain is "conditioned" and learns to go off to sleep, and is put into that particular mindset. After this the drug is withdrawn, as it is no longer needed. However, things can break down during times of stress and a few days of drug intake may be required to reinforce the conditioned response. It is vital to use the correct "mind-set" every time the drug is used, or the conditioning is weakened. Although not being easy, it however allow one's mind to eliminate reality when one is trying to sleep which must be considered a complete self-indulgence. Treat your brain like a child! It is simply not allowed to free wheel.

Sleep Essential in CSF/FMS – Implementing Self-Hypnosis

Everyone has to work out his or her best mind-set. It could be a childhood dream, or recalling details of a journey or walk, or whatever. It is actually a sort of self-hypnosis. What you are trying to do is to "talk" to your subconscious. This can only be done with the imagination, not with the spoken language. We know that the hypnotic state is characterized by extreme responsiveness to suggestion. You can use this information for conditioning yourself into self-hypnosis. There is a standard procedure to follow as worked out by Dr. Myhill from the UK. Lie down in bed, ready for sleep initially with your eyes open while the room needs to be dark. Mentally give yourself the suggestion that your eyes are becoming heavy and tired. Give yourself the suggestion that as you count to ten your eyes will become very heavy and watery and that you will find it impossible to keep your eyelids open by the time you reach ten. If you find that you cannot keep them open and have to close them, then you are probably under self-hypnosis. At this point deepen the state by again slowly counting to ten. Between each count mentally give yourself suggestions that you are falling into a deep hypnotic state. Give yourself suggestions of relaxation. Try to reach a state where you feel you are about to fall asleep. Give yourself the suggestion that you are falling more deeply down into sleep. Some may get a very light feeling throughout the body while others may get a heavy feeling. As you work through the task, you gradually get the feeling and achieve proficiency. If the implementation of sleep hygiene does not work one should consider other causes of sleep deprivation.

The Essentials of a Bedtime Routine for a Good Night's Sleep

Being all creatures of habit, first of all it is essential is to have a regular bedtime hygiene.
- A regular pre-bedtime routine with the "alarm" going off at 9 p.m. at which point one should drop all activities and move into the bedtime routine.
- A regular bedtime at 9.30 p.m. the latest, perhaps earlier in winter.

- One should learn to recognize the sleep wave as they come every 90 min.
- The day needs the right balance of mental and physical activity and recovery, avoiding sleep deprivation by all means.
- One should not share the same room with a bedfellow who snores – this indicates the need for different rooms!
- A small carbohydrate snack just before bedtime (e.g. nuts, seeds) helps to prevent nocturnal hypoglycemia – often manifested by vivid dreams or sweating or waking up in the night. Hypoglycemia is a common reason for disturbed sleep.
- Perhaps restrict fluids in the evening if the night is disturbed by the need to urinate.
- No stimulants such as caffeine or adrenaline-inducing TV, no arguments with the partner, no phone calls or discussions of family matters before bedtime! Since caffeine has a long half-life – no intake after 4 pm.
- Dark room – the slightest chink of light landing on the skin will disturb the production of melatonin, the bodies own natural sleep hormone. Have thick curtains or blackouts to keep the bedroom dark; this is particularly important for children! Do not switch the light on or a clock watch for wake-up.
- A source of fresh, preferably cool air.
- A warm comfortable bed. We have been brainwashed into believing a hard bed is good for you and many people end up with sleepless nights on an uncomfortable bed. It is the shape of the bed that is important. It should be shaped to fit the body approximately and then being very soft to distribute the weight evenly thus avoiding pressure points. Tempura mattresses and, although expensive waterbeds can be helpful.
- If sweating disturbs the sleep then this is likely to be a symptom of low blood sugar.
- Another common cause of disturbed sleep is hyperventilation, which often causes vivid dreams or nightmares. This can now be tested for by measuring a Carbonic anhydrase in red blood cells (for further information see Acumen® Laboratories in the UK). A short acting benzodiazepine such as midazolam 5–10 mg at night reduces the sensitivity of the respiratory centre and acts as a sleep inducer. However, because of their habit-forming properties, any benzo should not be taken for more than a week!
- If sleep is disturbed by pain, one should not refrain from taking a mild OTC analgesic such as an NSAID or a low to medium potent opioid such as meptazinol (Meptid®). Avoid tramadol since this results in an activation of the fast beta waves in the EEG being contraproductive to reach the deep and restorative plane of NREM sleep [285]. If nothing is available then one must take whatever painkiller is at hand to control this pain because it simply worsens the lack of sleep.
- If one wakes up in the night with symptoms such as asthma, chest pain, shortness of breath, indigestion etc., this may suggest food allergy with withdrawal symptoms occurring during the small hours.
- Some people find any food disturbs sleep and they sleep best if they do not eat after 6 p.m.

- If one wakes up in the night, do not switch on the light, do not get up and pound around the house as there will be a lesser chance of dropping off again to sleep.
- Learn a "sleep routine" by training the subconscious mind to switch on the sleep button when being asked for!

> **The most likely cause of disturbed sleep in the night is hypoglycemia.**

It is critically important for the body to maintain blood sugar levels within a narrow range. If the blood sugar level falls too low, energy supply to all tissues, particularly the brain is impaired. However, if blood sugar levels rise too high, then this is very damaging to arteries and the long-term effect of arterial disease is heart disease and strokes. This is caused by sugar sticking to proteins and fats to make AGEs (Advanced Glycation End-products), which accelerate the ageing process. Normally, the liver controls blood sugar levels. It can create the sugar from glycogen stores inside the liver and releases sugar into the blood stream minute by minute in a carefully regulated way to cope with body demands, which may fluctuate from minute to minute. Excess sugar flooding into the system after a meal can be used up by muscles, but only so long as there is the need for his extra energy. Eating the wrong thing or not exercising upsets this system of control works perfectly well until it. Eating excessive sugar loads at one meal, or excessive refined carbohydrate, which is rapidly digested with a rise in blood plasma glucose, can suddenly overwhelm the muscle and the liver's normal control of blood sugar levels. This is because mankind devolved over millions of years eating a diet that was very low in sugar and had no refined carbohydrate. Control of blood sugar therefore largely occurred as a result of eating the *Cave Man Diet* and the fact that we were exercising vigorously, so any excessive sugar in the blood was quickly burned off. Nowadays the situation is different: we eat large amounts of sugar and refined carbohydrate and do not exercise enough in order to burn off this excessive sugar. The body therefore has to cope with this excessive sugar load by other mechanisms.

7.4.3 Situations Developing into Hypoglycemia

When food is digested, the sugars and other digestive products go straight from the gut in the portal veins to the liver, where they should all be mopped up by the liver and processed accordingly. If excessive sugar or refined carbohydrate overwhelms the liver, the sugar spills over into the systemic circulation. If not absorbed by muscle glycogen stores, high blood sugar results, which is extremely damaging to arteries. If one were exercising hard, this would be quickly burned off. However, if one is not, then other mechanisms of control are brought into play. The key player here is insulin, a hormone excreted by the pancreas. This is very good at bringing blood sugar levels down and it does so by shunting the sugar into fat. Indeed, this includes the "bad" cholesterol LDL. However, there is a rebound effect and blood sugars may well go down too low. Low blood sugar is also dangerous to the body because the energy supplied to all tissues is impaired. When the blood sugar is low, this is called "hypoglycemia". Subconsciously people quickly work out that eating more sugar alleviates these symptoms, but of course they invariably turn things into the wrong direction; the blood sugar level then goes high and one ends up on a roller-coaster ride of blood sugar level going up and down throughout the day. Ultimately, this leads to metabolic syndrome or syndrome X – a major cause of disability and death in Western societies, since it is the forerunner of diabetes, obesity, cardiovascular disease, degenerative conditions and cancer.

- **Fructose intolerance.** Fructose is fruit sugar generally perceived to be a healthy alternative to glucose. No problem if one is tolerant of fructose or if it is taken in small amounts, but intolerance of fructose or excessive intake can result in hypoglycemia. This is because the control mechanisms that apply to glucose are bypassed if the system is flooded with fructose. In fructose intolerance (aldolase type B deficiency), fructose-1-phosphate builds up because it inhibits glycogen phosphorylase, which is essential for the provision of glucose from glycogen, and it also inhibits fructose-16-biphosphatase is essential for provision of glucose from protein and fat. This combination can result in severe hypoglycemia because it means effectively the body cannot mobilize glucose from stores in the liver when blood sugar levels fall. This combination can lead to severe hypoglycemia.
- **Excessive fructose intake** will stress the same pathways, even if the enzyme works perfectly well. This is because the sugar stores in the liver cannot be mobilized and the liver uses up short chain fatty acids for the production of glucose in order to try to avoid hypoglycemia. Looking at short chain fatty acids in the blood and also measuring levels of fructose-6-phosphate, which gets induced in this situation, can measure this tendency. These three metabolic problems i.e. levels of short chain fatty acids, levels of fructose-6-phosphate and LDH isoenzyme (indicative for liver damage), can help diagnose such a problem. It is recommend that people avoid tropical fruits (high fructose) before going to bed, and go for berries which are low fructose but rich in good antioxidants! Failure to tackle hypoglycemia will result in diabetes, which is an inevitable consequence of Western diets and lifestyles, because on current figures not only 50% of the UK but of the whole European population will be diabetic by the year 2030.

Initial complications. The problem for the established hypoglycemic is that it may take many weeks or indeed months for the liver to regain full control of blood sugar and therefore the symptoms of hypoglycemia may persist for some time whilst the sufferer continues to avoid sugar and refined carbohydrate. This means that when you change your diet you will get withdrawal symptoms and it may take many weeks of a correct diet before these symptoms resolve. This type of addiction is very much like that which the smoker or the heavy drinker suffers from. With time the regime can be relaxed, but a return to excessive sugar and refined carbohydrate means the problem starts again. Finally, many sufferers of hypoglycemia may need something sweet to eat immediately before and during exercise, until the body learns to fully adapt. Once a so called "Cave Man Diet" is established, this often helps considerably with sleep, but in the meantime have a snack as a last thing at night (e.g. nuts and seeds with a small piece of fruit) and, if disturbed, maybe eat again in the night (for further information on *Cave Man Diet* see Sect. 8.5.3). Finally, medical researchers have come up and warn that carbohydrates are behind the obesity epidemic. Worse than most carbs, Robert Lustig of the University of California at San Francisco says that fructose, because of the insulin response it generates, is the metabolic equivalent of poison. Being an obesity expert who has worked extensively with children, Lustig says too much insulin triggered by fructose consumption (i.e. the fructose you get from drinking orange juice) interferes with the hormone leptin, which normally tells our brain when we've had enough to eat.

7.4.3.1 Fructose Consumption Leading into Obesity

Obesity is not the result of gluttony or lack of exercise, as the low-fat mafia would have you believe. Research by experts such as Lustig challenges federal dietary guidelines that recommend more consumption of carbohydrates, rather than proteins and fats, or that feeding kids lots of fruits is healthy. But if you must eat fruit, eat whole fruit, not juice, for the fiber it contains. Drinking a glass of juice is far worse than any fat you might eat. Fructose is a component of the two most popular sugars. One is table sugar — sucrose, the other is high-fructose corn syrup. High-fructose corn syrup has become ubiquitous in soft drinks and many other processed foods. Lustig's own groundbreaking studies more than a decade ago stimulated the development of his controversial ideas about metabolism and biological feedback in weight control [286]. One not-yet-popular idea is that, calorie for calorie, sugar causes more insulin resistance in the liver than other edibles. The pancreas then has to release more insulin to satisfy the liver's needs. It is believed, that high insulin levels, in turn, interfere with the brain's receipt of signals from a hormone called leptin, secreted by fat cells. It was also demonstrated that fructose generates greater insulin resistance than other foodstuffs, and that fructose calories, therefore, fail to blunt appetite in the same way as other foods. Eating stimulates secretion of insulin and leptin. The conventional view holds that insulin, like leptin, feeds back in the brain to limit food intake. Most people think insulin does the same thing as leptin, but Lustig thinks just the opposite. Over the past century, Americans have increased their fructose consumption from 15 g per day to 75 g per day or more. The trend accelerated about three decades ago, when cheap, easy-to-transport high-fructose corn syrup became widely available. In addition, much of processed food labeled "reduced fat" instead has sugar added to make it more palatable. But when it comes to harmful health effects, sugar is worse than fat, since consumption of either results in elevated levels of artery-clogging fats being made by the liver and deposited in the bloodstream. But fructose causes even further damage to the liver and to structural proteins of the body while

7.4 Symptoms of Mitochondropathy of the CNS

Fig. 7.51 Some of the light drinks which is high in fructose corn syrup (HFCS)

fomenting excessive caloric consumption. There are four simple guidelines for parents coping with kids who are too heavy [287].

- Get rid of every sugared liquid in the house (Fig. 7.51). Kids should drink only water and milk.
- Provide carbohydrates associated with fiber.
- Wait 20 min before serving second portions.
- Have kids buy their "screen time" minute-for-minute with physical activity.

Fructose is abundant in fruit. Fruit is fine, but we should think twice before drinking juice or feeding it to our kids. The fiber in whole fruit contributes to a sense of fullness. It is rare to see a child eat more than one orange, but it is common for kids to consume much more sugar and calories as orange juice.

Eating fiber also results in less carbohydrate being absorbed in the gut. In addition, fiber consumption allows the brain to receive a satiety signal sooner than it would otherwise, so we stop eating sooner. Also, in contrast to common belief, exercise burns only a modest amount of calories. But it does have other benefits, since exercise improves insulin sensitivity in skeletal muscle, lowers insulin levels in the bloodstream, reduces stress and, therefore, reduces stress-induced eating. And Lastly, exercise increases the metabolic rate.

Fructose like high fructose corn syrup is definitely a poison.

Fig. 7.52 Significant increase in nitric oxide metabolites following nutrition with fructose in rats as in the Western diet. Since NO metabolites are involved in hypertension, diabetes, cardiac insufficiency, but also in Alzheimer and Parkinson, such data suggest a close correlation (Adapted from [289])

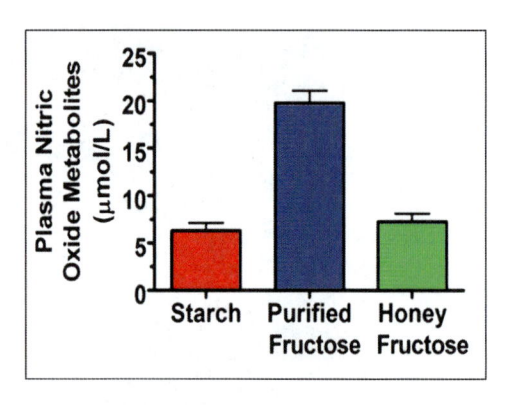

Not enough, that excess fructose consumption leads into obesity. Mercury cell chlor-alkali products are used to produce thousands of products including food ingredients such as citric acid, sodium benzoate, and you guessed right, it is also high in fructose corn syrup. High fructose corn syrup is used in food products to enhance shelf life. In a pilot study it was found that high fructose corn syrup contains mercury, a toxic metal historically used as an anti-microbial, which accumulates in the fatty tissues like the central nervous system. The samples from three different manufacturers were found to contain levels of mercury ranging from below a detection limit of 0.005–0.570 µg mercury per gram of high fructose corn syrup. Since average daily consumption of high fructose corn syrup is about 50 g per person in the United States, and with respect to total mercury exposure, it may be necessary to account for this source of mercury in the diet of children and sensitive populations [288].

And lastly the deleterious effect of fructose on NO-metabolism with NO-radicals has been demonstrated in experiments where other carbohydrates such as starch or honey practically had no effect (Fig. 7.52).

And now the bad news, since fructose intake has increased dramatically in recent decades and cellular uptake of glucose and fructose uses distinct transporters, fructose provides an alternative substrate to induce pancreatic cancer cell proliferation. In comparison with glucose, cancer cells metabolized the fructose in very different ways. In the case of fructose, the pancreatic cancer cells used the sugar in the transketolase-driven non-oxidative pentose phosphate pathway to generate nucleic acids, the building blocks of RNA and DNA (Fig. 7.53), which the cancer cells need to divide and proliferate. [290]. These findings show that cancer cells can readily metabolize fructose to increase proliferation. This have major significance for cancer patients given dietary refined fructose consumption, and indicate that efforts to reduce refined fructose intake or inhibit fructose-mediated actions may disrupt cancer growth (or even vice versa!).

7.4.3.2 Sugar and Fast Carbohydrates are Addictive

The problem is that people feel boosted by a high level of blood sugar. This is because they have a good energy supply to their muscles and brain – albeit short-term.

7.4 Symptoms of Mitochondropathy of the CNS

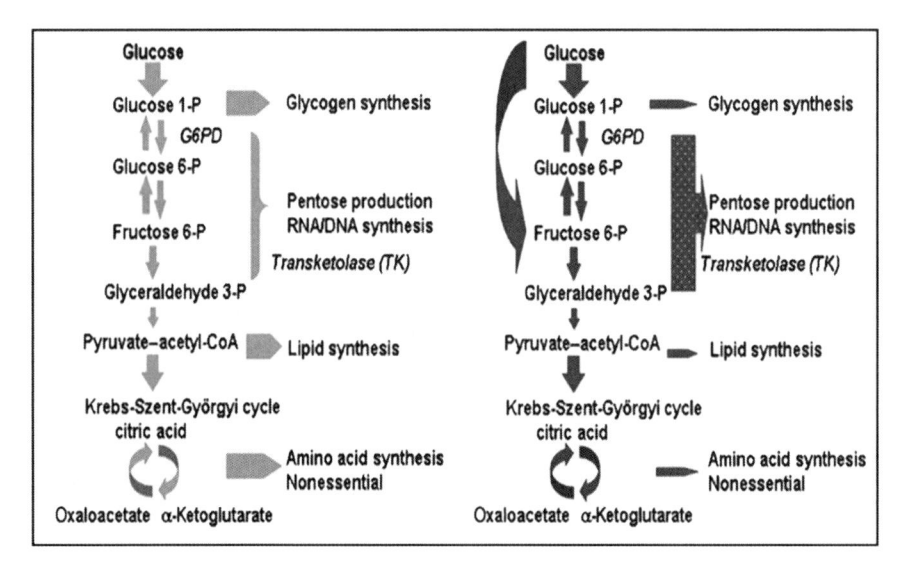

Fig. 7.53 Fructose induces transketolase flux promoting cancer growth. Transketolase is an enzyme of the pentose phosphate pathway in animals, and connects the pentose phosphate pathway to glycolysis, feeding excess sugar phosphates into the main carbohydrate metabolic pathways. Its presence is necessary for the production of NADPH

The problem arises when blood-sugar levels dive as a result of insulin being released and the energy supply to the brain and the body is suddenly impaired. This results in a whole host of symptoms: the brain-symptoms include difficulty in thinking clearly, feeling spaced out and dizzy, poor word finding ability, foggy brain and sometimes even blurred vision or tinnitus. The body-symptoms include suddenly feeling very weak and lethargic, feeling faint and slightly shaky, rumbling tummy and a craving for sweet things. Sufferers may look as if they are about to faint (and indeed often do) and have to sit down and rest. Eating something sweet can quickly alleviate the symptoms; if nothing is done, then the sufferer gradually recovers. These symptoms of hypoglycemia can be brought upon by missing a meal (or on the usual sweet snack top-up such as a sweet drink), by vigorous exercise or by alcohol (Fig. 7.54). Diabetics may become hypoglycemic if they use too much medication. Since the brain likes sugar, this is a reason why sugar can be considered an addiction forming substance [291]. Running on a high blood sugar allows the brain to function efficiently and also releases the happy neurotransmitters such as GABA and serotonin, which have a calming effect. We all recognize this because comfort-eating foods are carbohydrates. The second problem is that we have a "sugar-thermostat" for blood sugar (i.e. a system in the hypothalamus that constantly measures the actual blood-sugar level and compares it with the control value), which gets an upward set if the blood sugar consistently runs high. For instance, people with diabetes who consistently run a high blood sugar level, feel hypoglycemic if their blood sugars drop below 7 or 8 mmol/l. Whatever interventions one makes to control high blood sugar, this must be done slowly to assure that the *"sugar-thermostat"* is gradually reset.

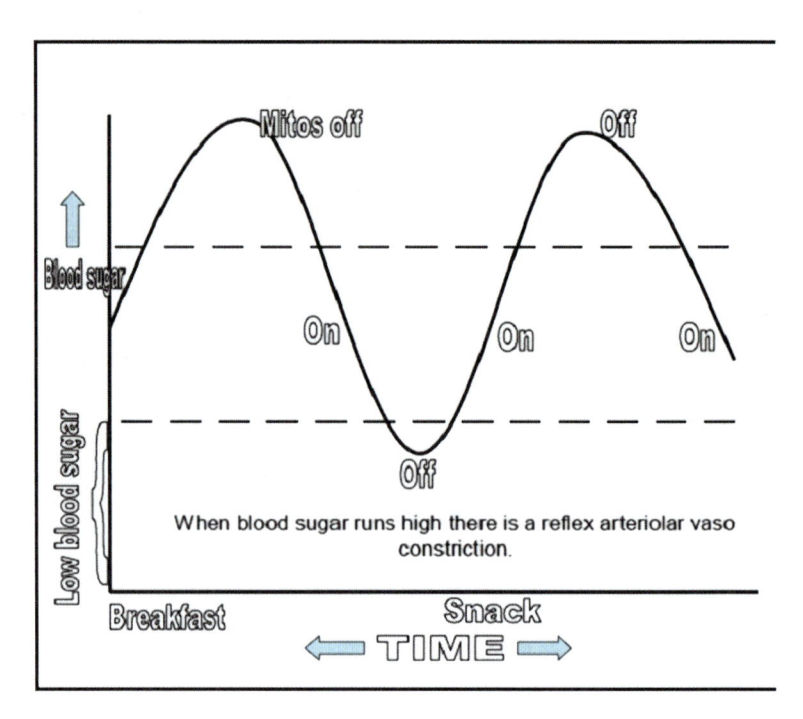

Fig. 7.54 The effect of the ups and downs of glucose blood levels with a secondary effect on mitochondrial activity

Treatment of Hypoglycemia and Carbohydrate Addiction

Treatment is to avoid all foods containing sugar and refined carbohydrate and take extra supplements. The problem for the established hypoglycemic is that it may take many weeks or even months for the liver to regain full control of blood sugar and therefore the symptoms of hypoglycemia may persist for some time while the sufferer continues to avoid sugar and refined carbohydrate. This means that when one changes the diet one will get withdrawal symptoms and it may take many weeks of a correct diet before these symptoms resolve. This type of addiction is very much like that which the smoker or the heavy drinker suffers from. In order to avoid the problem one needs to switch to a diet, which concentrates on eating proteins, fats and complex (i.e. slowly digestible) carbohydrates. Initially, it is suggested taking a high protein high fat diet, but include all vegetables (care with potato), nuts, seeds, etc. Fruit is permitted but rationed, since excessive amount of fruit juices or dried fruits contain too much fruit sugar for the liver to be able to deal with. It is also recommended taking high dose probiotics as an essential part of controlling low blood sugar. This is because probiotics ferment carbohydrates to short chain fatty acids, which have no effect on blood glucose level and are the preferred fuel of mitochondria. The best and cheapest way to do this is to brew your own. Probiotics also displace yeast, which worsen the hypoglycemia problem.

7.4 Symptoms of Mitochondropathy of the CNS

With time after the system has reset, the former tight food regime can be relaxed. However, any return to excessive sugar and refined carbohydrate means the problem starts again. Finally, many sufferers of hypoglycemia may need something sweet to eat immediately before and during exercise, until the body learns to fully adapt. Hypoglycemia is usually accompanied by micronutrient deficiencies, a major reason, why one should also take nutritional supplements. To tackle hypoglycemia one needs to do a diet based on foods with a low glycemic index. The glucose index (GI) is a measure of the ability of a food component to raise the blood glucose levels. Sugar (i.e. disaccharides) has arbitrarily been given a GI of 100. High GI foods are the grains (wheat, rye, oats, rice, etc.), root vegetables (potato, sweet potato, yam, parsnip), alcohol, sugars, and fruits, dried fruits and fruit juices. But expect to see withdrawal symptoms, which can persist for weeks. Hypoglycemia is not just about diet as the body has a very difficult balancing act with respect to blood sugar. If levels drop too low, then this will cause unconsciousness followed by death. On the other hand, if the blood sugar level goes too high, glucose will stick onto many other membranes and create advanced glycation end products. This effectively causes accelerated ageing. So the body undergoes a great deal of metabolic work in order to keep the blood sugar tightly controlled between about 3 mmols and 6 mmols/l. The mechanism to achieve such levels is delicately balanced and therefore there is great potential for things to get out of regulation. This is further complicated by the fact that the brain likes sugar. What, however, makes a blood glucose level go up? **"Diet"** is the most obvious answer, since fast reabsorbable carbohydrates are broken down into sugar, which increase the blood glucose level. It is because of this increase that food components have been given a measure being called the *"glycemic index"*. It is a measure of the ability of nutrients to raise the blood glucose level. This can be affected by many factors, not just the food itself. Foods that are cooked will be more rapidly digested and therefore have a higher glycemic index. Foods that are processed and refined such as flour are more rapidly digested and therefore have a high glycemic index. Carbohydrates that are rapidly broken up such as sugar and alcohol again are rapidly absorbed. Any carbohydrates that are consumed should therefore be unrefined complex carbohydrates which are slowly digested, if possible eaten raw, although this is obviously not palatable with some carbohydrates such as the potato. In addition, food should be slowly eaten. What causes insulin to be released is the rate at which the blood sugar level rises. A quick rise will produce a release of insulin, with an overhang for a long time, causing a subsequent hypoglycemia. Therefore, food should be eaten and chewed slowly, it should not be gobbled down, and it should be a mix of carbohydrates and high fiber foods, vegetables, meat and fat so that the absorption of carbohydrate is slowed.

It is easy to identify the carbohydrate addicts; they like their carbohydrates highly refined such as sugar, sweets, crisps, white bread, pasta and refined breakfast cereals and fruit juice. They tend to gobble their food. They are not content with a normal meal of meat and vegetables without the sweet sticky desert to follow!

Since alcohol, is regularly served in conjunction with the dinner it results in a symptom causing hypoglycemia with sleeplessness.

The Fermenting Gut, a Cause for Hypoglycemia

This is a major cause of hypoglycemia because sugars (or carbohydrates that are digested to sugars) are fermented to produce various alcohols, which destabilize blood sugar levels. This is because the human gut is almost unique amongst mammals, with an upper gut that is sterile, for digesting carnivorous food (like that of a dog or a cat) to deal with meat and fat, while the lower gut (the large bowel or the colon) is full of bacteria and is a fermenting, vegetarian gut (like that of a horse or a cow) to digest vegetables and fiber. From an evolutionary perspective this has been a highly successful strategy, because it allows Eskimos to live on fat and protein and other people to survive on pure vegan diets. The problems arise when the upper gut starts to ferment. The stomach, duodenum and small intestine should be free from microorganisms (bacteria, yeast and parasites). This is normally achieved by eating a *"Cave Man's Diet"*, having an acidic stomach which digests protein efficiently and kills the acid sensitive bugs, then followed by an alkali duodenum, which kills the alkali sensitive bugs with bicarbonate and finally the bile salts (which are also toxic to bugs) plus the pancreatic enzymes to further digest protein, fats and carbohydrates. The small intestine does more digesting and also absorbs the amino acids, fatty acids, glycerol and simple sugars that result from digestion. Bacteria flourish in the large bowel, where foods that cannot be digested upstream are then fermented to produce many substances highly beneficial to the body. This also generates heat to help keep us warm. The human body is made up of ten million cells; in our gut we have 100 million bacteria or more, i.e. ten times as many! Bacteria make up 60% of dry stool weight, there are over 500 different species, and 99% of these are from 30 to 40 different species. Since bacteria in the intestinal tract are so vital for well being (or even survival) any long-term problems in digestion should be evaluated in regard to the bacterial content in the gut. Metametrix laboratories in USA or the laboratories Biovis and Ganzimmune in Germany offer the *Microbial Ecology Profile*, which measures the amount of different bacteria in a stool sample. This tells us which bacteria are present and their numbers, with the most important ones, the bactericides E. coli, lactobacilli and the bifidobacteria.

- **Bacteroides (65%)**. They are by far the most abundant bacteria in the intestinal tract. It is bacteroides, which allow us to digest soluble fiber and make short chain fatty acids. This is the main source of food for the colonocytes, the cells lining the bowel and if this is low, then it will result in atrophy of the colon. Short chain fatty acids also protect us from hypoglycemia. Indeed, it is estimated that up to 540 Kcal per day may be generated – a very significant source of energy! They are essential for recycling of bile acids, so low levels of bile acids may indicate poor levels of bacteroides. They occupy the surface of the gut so preventing pathogenic species (such as salmonella, shigella and clostridia) from adhering and causing infection. There is no probiotic, which contains bacteroides. This is because it is also a potential pathogen! We just have to feed the gut with the right food (prebiotics) found in pulses, nuts, seeds and vegetables. Bacteroides are the most common bacteria in the colon, outnumbering E. Coli by

7.4 Symptoms of Mitochondropathy of the CNS

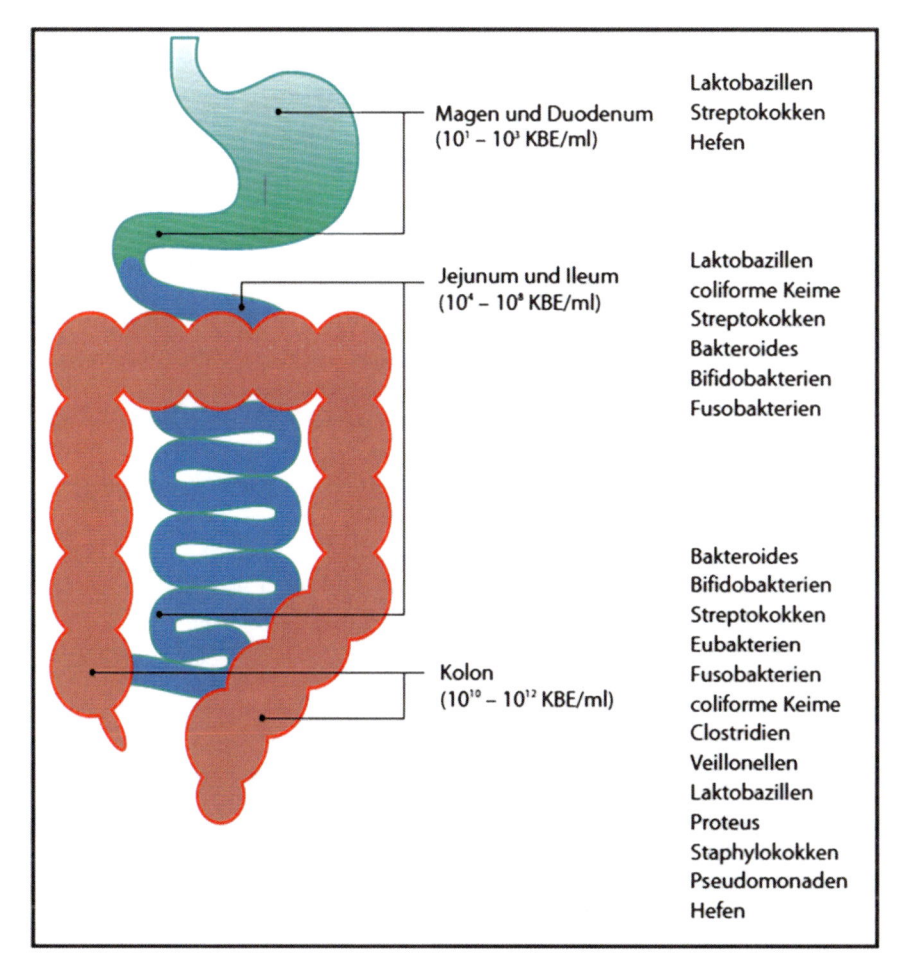

Magen und Duodenum ($10^1 - 10^3$ KBE/ml)
- Laktobazillen
- Streptokokken
- Hefen

Jejunum und Ileum ($10^4 - 10^8$ KBE/ml)
- Laktobazillen
- coliforme Keime
- Streptokokken
- Bakteroides
- Bifidobakterien
- Fusobakterien

Kolon ($10^{10} - 10^{12}$ KBE/ml)
- Bakteroides
- Bifidobakterien
- Streptokokken
- Eubakterien
- Fusobakterien
- coliforme Keime
- Clostridien
- Veillonellen
- Laktobazillen
- Proteus
- Staphylokokken
- Pseudomonaden
- Hefen

Fig. 7.55 Microbial colonization in different parts of the intestine with a total of up to 10^{14} microbial agents (100 billion, especially in the lower part) resulting in a weight of 1.5 kg (!) (Modified from [294, 295])

at least 100 to 1 (Fig. 7.55). Some species are pathogenic and are often found in necrotic tissue and in the blood after an infection.

- **Firmicutes (39%)**. Another phylum of bacteria in the gut, which are able to divide complex carbohydrates within the diet, thus providing additional glucose to the body, a potential disadvantage in patients with diabetes type 2. Often the relative proportion of bacteroides to firmicutes (norm 50:50) is decreased in obese people by comparison with lean people, and that this proportion increases with weight loss on two types of low-calorie diet [292]. Thus, obesity has a microbial component, which might even result in potential therapeutic implications.
- **E-coli (45%)**. One gram of stool should contain between 7 million and 90 million. E-coli ferments to produce folic acid, vitamin K_2 (which protects against

osteoporosis), Co-enzyme Q10 (essential for mitochondrial function), together with 3 amino acids, namely tyrosine and phenylalanine, the precursors of dopamine and where their lack results in low mood, and tryptophan. Tryptophan is a precursor of serotonin, which is responsible for gut motility. So, if there are low counts of E-coli, one can expect problems in all the above areas, i.e. osteoporosis and bone problems, mitochondrial function, low mood and poor gut motility. A study done in Germany with E-coli probiotics given for the treatment of constipation demonstrated a dramatic improvement from 1.6 motions a week to six, illustrating the effects on motility [293]. E-coli is contained in the probiotic agent Mutaflor® produced commercially in Germany.

- **Lactobacilli (9%)**. These ferment sugars to lactic acid, which provides an acid environment in the large bowel to protect against infection. Also highly protective against bowel cancer and is found in abundance in Kefir®, a fermented milk drink.
- **Bifidobacteria (32%)**. These assist digestion, protect against development of allergies and cancer. It is therefore necessary concentrate all efforts on the above good bacteria – once they are in the right amount other bacteria will not flourish and the numbers of the bad bacteria decline resulting in a well kept balance- Nature abhors a vacuum- empty or unfilled spaces are unnatural!
- **Streptococcus**. This ferments to produce large amounts of lactic acid. This gives a tendency to acidosis. Lactic acid is metabolized in the liver by lactate dehydrogenase, so high levels of this may indicate bowel overgrowth with streptococcus. Fermentation produces two isomers of lactic acid, namely L-lactate and D-lactate. It is D-lactate, which is the problem, since the body cannot metabolize this isomer; it accumulates in mitochondria and inhibits their activity.
- **Prevotella** (bacteroides in the upper gut). They ferment and produce hydrogen sulphide (H_2S). Hydrogen sulphide inhibits mitochondrial function by a direct action. So a positive hydrogen sulphide urine test shows there is a severe gut dysbiosis with overgrowth of prevotella secondary to undergrowth of the good bacteria (for further information on the hydrogen sulphide test see page 185, 197 ff)!

Problems could arise either from having the wrong colonization in the lower gut (large intestine), or having the wrong intestinal flora in the upper gut (stomach, duodenum, jejunum, small intestine). Especially, local mitochondropathy of the upper layer of the intestine means of xenobiotics in the diet, wrong bacteria colonization or a local inflammatory reaction due to allergy, these **enterocytes**, or **intestinal absorptive cells** (i.e. simple columnar epithelial cells found in the small intestines and colon) have to fulfill an important task. For instance, they have a glycocalyx surface that contains digestive enzymes and the microvilli on the apical surface increase surface area for the digestion while at the same time there is an active carrier system for molecules from the intestinal lumen. The cells also have a secretary role and in order to work efficiently energy has to be generated. This however, is not done by use of glucose, rather these cells obtain energy by glutamine metabolism for which they need the formation of ATP which is supplied by mitochondria within the enterocytes [296]. When these cells are depleted from ATP formation, there is an insufficient

7.4 Symptoms of Mitochondropathy of the CNS

synthesis of vitamin K, an impaired uptake of essential vitamins (especially of the B-type), a reduced active transport of fatty acids and micronutrients and an insufficient absorption of proteins. Especially in the elderly population these cell undergo reduced activity often resulting in malnutrition with a shortage in micronutrients, the reason why these people need supplementation.

7.4.3.3 Problems in Digestion: Bacteria and Yeast in the Upper Gut

In some patients there are bacteria, yeasts and possibly other parasites in the upper gut, which ferment the food components instead before they are being digested. When foods get fermented they produce all sorts of unwanted products, which have to be detoxified by the liver cytochrome P450 detox system but at the same time are also toxic to the enterocytes. These products include:

- **Alcohols** such as ethyl-, propyl-, butyl- and possibly methyl alcohol. These would be metabolized by stage one into acetaldehyde, propylaldehyde, butylaldehyde and possibly formaldehyde. Alcohol and acetaldehydes result in foggy brain, the "toxic brain", feeling of being "poisoned", mental clouding, etc. Alcohol also upsets blood sugar levels. This makes the sufferer crave for more sugar and refined carbohydrates, since the bacteria need more nutrition in the upper gut to ensure their own survival. This is a clever evolutionary trick used by the bacteria to ensure their own survival!
- **Noxious gases** such as hydrogen sulphide (H_2S), nitric oxide, ammonia and others. Hydrogen sulphide is known to inhibit mitochondria and block the oxygen carrying capacity of hemoglobin. It also greatly increases the toxicity of heavy metals.
- **Abnormal sugars such as D-lactate**. This right-handed sugar cannot be detoxified by the liver enzyme lactate dehydrogenase. If D-lactate is present, this points to a problem of gut fermentation, resulting in lactic acidosis. These patients typically have episodic metabolic acidosis, usually occurring after high carbohydrate meals with characteristic neurological abnormalities including confusion, cerebellar ataxia, slurred speech, and a loss of memory. In a review of 29 reported cases, for example, all patients exhibited some degree of altered mental status [297]. In the absence of ethanol intake they may complain or appear to be drunk. The veterinarian much better knows this phenomenon, since D-lactate is a recognized as a cause of neurological manifestations in cattle. There, the encephalopathy with liver failure can be treated only by antibiotics to wipe out unwelcome overgrowth of fermenting flora in the gut. D-lactate is fermented from sugars, including fruit sugars. This is a further reason to cut out sugar and fruit strictly from the diet as one molecule of sugar generates two molecules of D-lactate!

In theory, all the above toxins all have to be detoxified by the P450 cytochrome system, but in practice some of these can spill over into the systemic circulation with a simple poisoning effect and result in the production of free radicals with an inhibition of mitochondrial function. Since the products of fermentation add to the total chemical load they therefore may worsen any underlying poisoning (such as chronic organophosphate poisoning) by overwhelming the detoxifying defense system of the liver.

Causes for Upper Gut Fermentation

There are a number of causes, which result in upper intestinal fermentation

- Western lifestyles! For instance, a traditional Chinese diet does not include dairy products, gluten grains, alcohol and only very modest amounts of fruits!
- Failure to inoculate the gut at birth with the correct friendly bacteria (for further information see probiotics on page 178 ff)
- Eating the wrong diet:
  - high in sugar and refined carbohydrate,
  - low in vegetables, pulses, nuts and seeds,
  - low in other micronutrients.
- Poor digestion of food – as a result of hypochlorhydria and an insufficient pancreatic exocrine function.
- Modern medication (consider a medication of >4 different agents as contraproductive because of unpredictable interactions)
  - All antibiotics, which wipe out the flora of the intestine
  - The anticontraceptive pill and hormone replacement. therapy, both of which suppress the immune system and encourage yeast overgrowth.
  - Acid blockers such as antacids, H_2 blockers and proton pump inhibitors (PPI), which inhibit stomach acid production with hypochlorhydria. Allergies to food may also result as an additional problem.

Clinical Problems Related to Upper Gut Fermentation

- **Symptoms** – For example, Drs De Meileir and Butt, both from the University of Melbourne have shown that high levels of enterococcus are associated with symptoms of headache, arm pain, shoulder pain, myalgia, palpitations and sleep disturbance. High levels of streptococcus are associated with post exertional fatigue, photophobia, mind going blank, palpitations, dizziness, and faintness.
- **Malabsorption of micronutrients.**
- **When nutritional supplements make things worse.** This very likely can be explained by the fermenting gut. Instead of nourishing the body, the supplements nourish the wrong bacteria. This encourages their growth and therefore fermentation increases! This makes sense because from an evolutionary perspective mitochondria are derived from bacteria. So what mitochondria like bacteria will also thrive on.

7.4 Symptoms of Mitochondropathy of the CNS

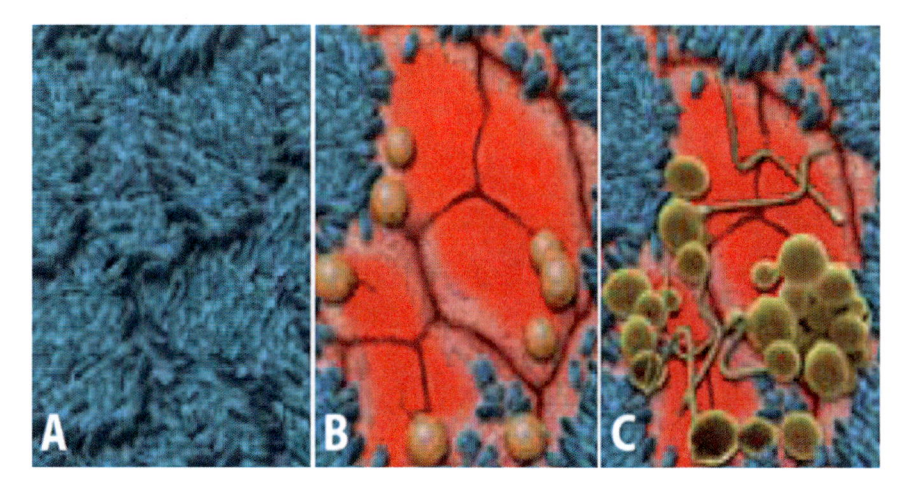

Fig. 7.56 Optimal versus wrong microflora in the intestine: (**a**). Intact resistance of the inner lining of the gut with optimal microbial colonization. (**b**). Areas with beginning gaps in intact colonization due to wrong strains. (**c**). Pathogenic microflora with Candida resulting in leaky gut with invasion of pathogens

- **Wind, gas, bloating** from physical distension.
- **Disturbances of normal gut movement** – constipation or diarrhea;
- **Susceptibility to infections.**
- **Leaky gut system**. With a leaky gut (Fig. 7.56), short chain polypeptides may leak into the blood stream and act as antigens or even worse as hormone mimics. For example, a strip of amino acids Ser-Tyr-Set-Met would mimic ACTH – the hormone that stimulates the adrenal gland. An eight amimo acid fragment could act like glycogen and thus depletes glycogen (sugar) stores in the liver.
- **Susceptibility to heavy metal poisoning** – because hydrogen sulphide binds to heavy metals rendering them "organic" instead of "inorganic" and therefore is much more likely to enter the body and bioaccumulation the body.
- **Prion disorders** – heavy metals in the wrong department can twist normal proteins and convert them into pathogenic prions, which are implicated in prion disorders.
- **Dental, gum and mouth problems** – one is likely to have similar bacteria in the mouth as the gut. If a person has a clean tongue and no dental plaque then he/she is likely to have a good gut flora.
- **Allergic reactions-** due to a leaky gut pathogens can now traverse through the inner lining to the enteric immune system where they act as antigens inducing a shift of the Th2 ->Th1 relation in immune cells (Fig. 7.57) which results in an overexpression of cellular reactions with (even distant) symptoms of allergy in mucous membranes (e.g. bronchial tree with asthma, sinuses with sinusitis) and the skin (e.g. neurodermatitis).

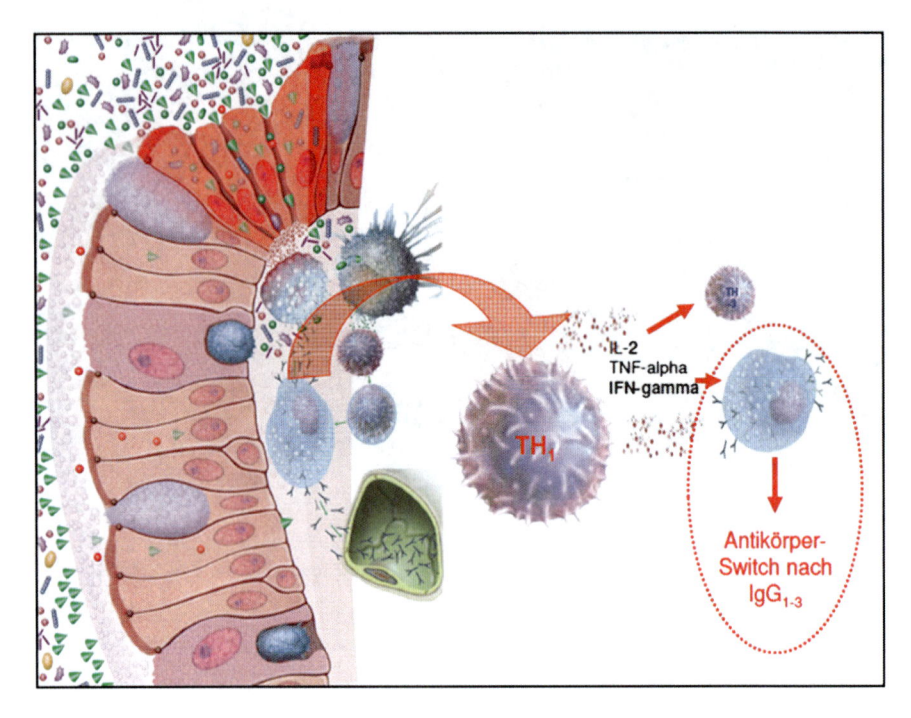

Fig. 7.57 The intestinal immune system with the activation of inflammation and an antibody switch to IgG_{1-3}, TNF-α, IL-2 and IFN-γ production (Th1-shift). Higher intake of dietary fiber, however, results in the growth of a bifiodogenic flora with short-chain fatty acid formation inducing macrophage production, which produce IL-10 thus activating Th3 cells, which are relevant for oral tolerance, immunosuppression, and an intact immune defense system (Modified from [298])

7.4.3.4 Tests for Bacteria in the Upper Gut Fermentation

- **The Gut fermentation profile** measures levels of alcohol in the blood. It also looks for short chain fatty acids, which are desirable products of fermentation by friendly bacteria in the large bowel.
- **Determining the type of bacteria in the intestine.** By using the contents of stool samples it does not tell us where the bacteria came from, however, it is something to start with. A Comprehensive Digestive Stool Analysis or a Comprehensive Digestive Stool Analysis with parasitological analysis is helpful (Table 7.7). The bacteria, which are thought to be the main fermenters, are enterococcus, streptococcus and prevotella (bacteria) and candida (yeast). Conversely, in the fermenting gut there may be low levels of bacteroides, lactobacilli and bifidobacteria. This is because, once there is fermentation upstream, there is little substrate for fermentation downstream! A more detailed test of the actual bugs present, which includes bacteroides, is a total microbial profile.
- **D-lactate** can be measured by a blood test following a carbohydrate meal. The snag is that postal samples are not completely reliable and in order to have the

7.4 Symptoms of Mitochondropathy of the CNS

Table 7.7 Among anaerobic bacteria, Prevotella is the most consistently overgrown bacteria in CFS/ME patients

Organisms	Control	ME patients	p-value
Bacteroides spp.	3.2×10^{11}	1.6×10^{11}	$p = 0.39$
Prevotella spp.	1.0×10^{8}	9.0×10^{9}	$p < 0.001$
Bifidobacterium spp.	6.0×10^{8}	5.5×10^{9}	$p = 0.001$
Lactobacillus spp.	2.7×10^{7}	1.8×10^{8}	$p = 0.002$

test, the patient either needs to go to a laboratory to have blood taken and processed straight away or he/she has to find a laboratory where they will spin and separate blood immediately and freeze it after which a frozen sample is sent to the analyzing lab.

- **Hydrogen sulphide (H_2S)** can be tested with an urine test. Normal gut fermentation produces hydrogen and methane which allows one to "light their own flatus", and is also odorless. With sulphate reducing bacteria are present in the gut, hydrogen sulphide is produced giving the rotten eggs smell and a positive hydrogen sulphide in urine test. This test is called the "Neurotoxic metabolite test" and can be ordered online directly from Protea Biopharma in Belgium (www.redlabs.com).

- **A good clinical anamnestic test** for upper gut fermentation is whether one produces wind or gas (belching, bloating, feeling full, noisy gut, etc.) after eating complex carbohydrates! The following reported symptoms, gathered from people with chronique fatigue syndrome, which are related to an ongoing absorption of antigens (antigenstress) and xenobiotics from the intestine result in different symptoms (Fig. 7.58).

- Headaches, "heavy head," "heavy-feeling headaches"
- Alternated periods of mental "fuzziness" and greater mental clarity.
- Feeling "muggy-headed" or "blah" or sick in the morning.
- Transient malaise, flu-like symptoms.
- Transiently increased fatigue, waxing and waning fatigue, feeling more tired and sluggish, weakness.
- Dizziness.
- Irritability.
- Sensation of "brain firing: bing, bong, bing, bong," "brain moving very fast".
- Depression, feeling overwhelmed, strong emotions.
- Greater need for "healing naps."
- Swollen or painful lymph nodes.
- Mild fevers
- Runny nose, low grade "sniffles," sneezing, coughing.
- Sore throat.
- Rashes.
- Itching.
- Increased perspiration, unusual smelling perspiration.
- "Metallic" taste in mouth.

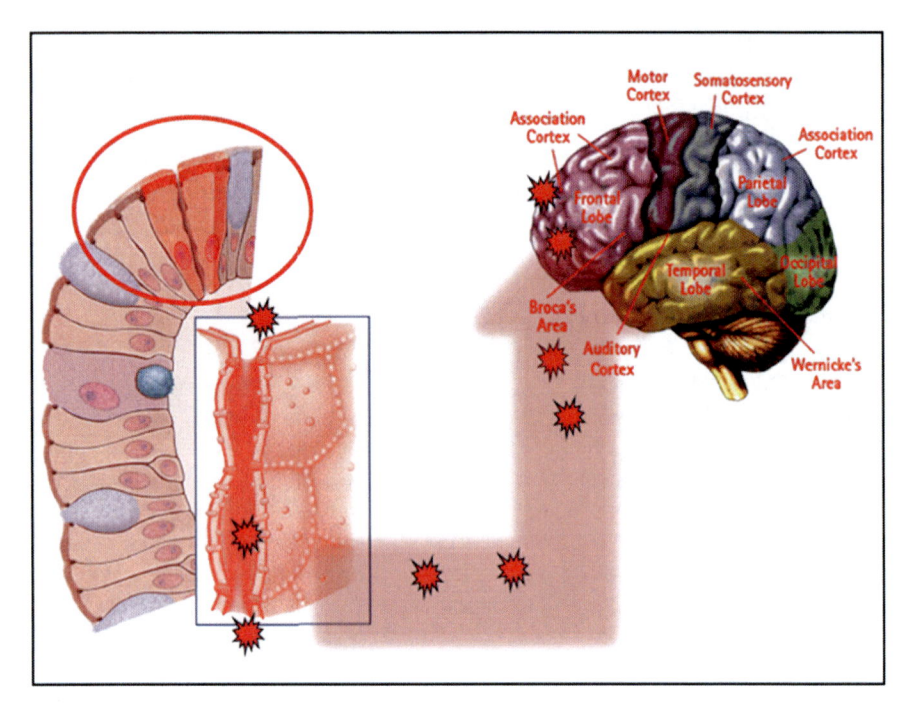

Fig. 7.58 Following migration of antigen material and xenobiotics through disrupted tight junction of the intestine, the immune system is activated, both of which result in distant effects on functions within the central nervous system

- Transient nausea, "sick to stomach"
- Abdominal cramping/pain.
- Increased bowel movements.
- Diarrhea, loose stools, urgency.
- Unusual color of stools, e.g. green.
- Temporarily increased urination
- Transiently increased thirst.
- Clear urine.
- Unusual smelling urine
- Transient increased muscle pain.

There is now good evidence that the central pathological lesion in CFS is mitochondrial failure. What is critical to the optimal function of mitochondria is a good redox state, that is to say the balance between free radical stress and the ability to cope with those free radicals, i.e. the body's antioxidant status. Free radicals damage mitochondria so they go slow, but the body has a system of antioxidants for protection against free radical stress. Once however there is an invasion of xenobiotics having entered the body via the intestine, than a dysbiotic gut can cause mitochondrial failure, which results in fatigue.

7.5 The Methylation Cycle-Hypothesis for Pathogenesis of Fatigue

Since chronic Fatigue Syndrome is a symptom, not a diagnosis, the name of the game is to identify the underlying causes. In fatigue syndromes there is no macropathology, but one sees micropathology, i.e. the problems are biochemical and occur at the molecular level. There are several cycles, which now have become known to be centrally important in causing fatigue. All these cycles interlink with each other like Olympic rings and getting one cycle going will drive another. The important cycles which are known to be major players include blood sugar oscillations, allergy problems, disturbed sleep cycles, mitochondrial function, an insufficient antioxidant status, an upregulated NO/OONO⁻ cycle, disturbed thyroid and adrenal hormones cycles and insufficient detoxification after exposure to toxic agents found in the food and the environment. In this scenario another new player, which interlinks with many of the above, has entered the scene, as the methylation cycle seems to maintain the ailment. The methylation cycle (also called the methionine cycle) is a major part of the biochemistry of sulfur and of methyl (CH_3) groups in the body. It is also tightly linked to folate metabolism and is one of the two biochemical processes in the human body that require vitamin B_{12} (the other being the methylmalonate pathway, which enables use of certain amino acids to provide energy to the cells). This cycle supplies methyl groups for a large number of methylation reactions, including those that methylate (and thus silence) DNA [299], and those involved in the synthesis of a wide variety of substances, including creatine, choline [300], carnitine [301], coenzyme Q10 [302], melatonin [301], and myelins basic protein [303]. Methylation is also used to metabolize the catecholamines dopamine, norepinephrine and epinephrine [301], to inactivate histamine [301], and to methylate phospholipids [304], promoting transmission of signals through membranes. The role of the methylation cycle in the sulfur metabolism is to supply sulfur-containing metabolites to form a variety of important substances, including cysteine, glutathione, taurine and sulfate, via its connection with the transsulfuration pathway [301]. Originally proposed as a cause for autism is that ineffective methylation is a major cause of chronic fatigue [305]. This hypothesis was further extended by other researchers [306] who suspect that a large subset of the CFS patients have a blocked methylation cycle. Therefore, the main goal of treatment must be to remove this block and get the methylation cycle back into normal operation. There are many possible reasons but those, which have been identified and in which methylation is essential are (Fig. 7.59):

- To produce vital molecules such as Co Q10 and carnitine.
- To switch on DNA and switching off DNA. Activating and deactivating genes by methylation achieve this. This is essential for gene expression and protein synthesis. Proteins of course make up the hormones, neurotransmitters, enzymes, and immune factors and are fundamental to good health. When viruses attack our bodies, they take over our own DNA in order to replicate themselves. If we can't switch DNA/RNA replication off then we will become more susceptible to viral infection.
- To produce myelin for the brain and nervous system.

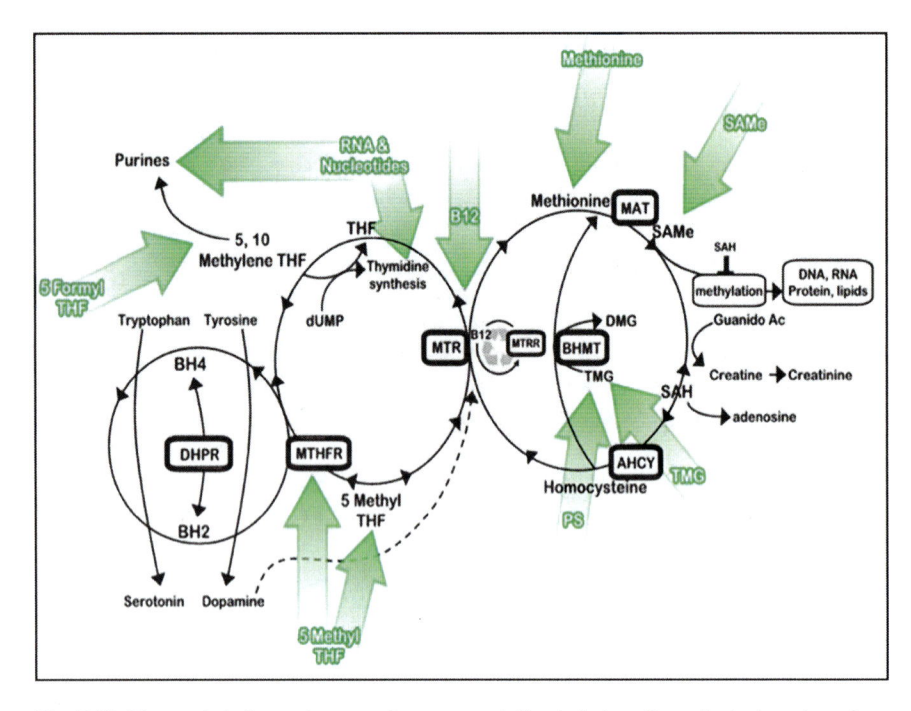

Fig. 7.59 The methylation pathway, acting as an on/off switch that allows the body to learn how to respond to environmental changes. It represents the only cellular pathway that affects both adaptability and structural integrity of the body. Like the simple water molecule, methyl groups are necessary for life

- To determine the rate of synthesis of glutathione which is essential for detoxification.
- To determine the rate of synthesis of glutathione, which is an essential antioxidant as glutathione-peroxidase. Furthermore oxidative stress blocks glutathione synthesis – being another vicious cycle!
- To control sulphur metabolism of the body, not just glutathione but also cysteine, taurine and sulphate. This is an important process for detoxification.
- As part of folic acid metabolism. This also switches on the synthesis of new DNA and RNA.
- For normal immune function. The methylation cycle is essential for cell mediated immune function and any blockade results in an insufficient and adequate response to any infection. This has been observed clinically because many patients report that once they get their B_{12} injection (an essential cofactor for methylation) it seems to protect them from getting an infection.

If the methylation cycle doesn't work the immune system malfunctions, the detoxification system malfunctions, the ability to heal and repair is reduced and the antioxidant system is not working properly.

7.5 The Methylation Cycle-Hypothesis for Pathogenesis of Fatigue

7.5.1 Biochemistry of the Methylation Cycle

There are four cornerstones to the methylation cycle and on each cornerstone sit four molecules namely homocysteine, methionine, S-adenosylmethionine (SAMe) and S-adenosylhomocysteine. Each of these molecules leads into the next one by means of enzymes. The important cofactors that allow this to happen are the B vitamins such as folic acid, vitamin B_{12} and vitamin B_6. In converting from S-adenosyl homocysteine into homocysteine, a methyl group is given up and this can be used to stick on to other molecules – hence the name, the methylation cycle. However, there is a particular biochemical problem. In order for the methylation cycle to work, these B vitamins have to be in their activated form, namely methylcobalamin, folic acid and pyridoxyl-5-phosphate. In order to get cobalamin into methylcobalamin, the methylation cycle has to be working. So if this cycle is completely inactive, the body can't make methyl cobalamin in order to get up and running again. Since this cycle is so fundamental to other biochemical cycles including trans-sulphuration and folate metabolism, it can't change the vitamin B_6, folic acid and cobalamin into the active forms necessary for the methylation cycles to work (Fig. 7.59). This means that in order to get this cycle up and running, initially one has to prime the pump with the activated vitamins, and hopefully once the methylation cycle is up and running again, it can function on the vitamins in their non-activated states.

Any pathology of the methylation pathway in the body, aside from Myelodysplastic syndromes (MDSs), i.e. hematologic disorders that frequently represent an intermediate disease stage before progression to acute myeloid leukemia (AML), has also ties to most major diseases including [307]:

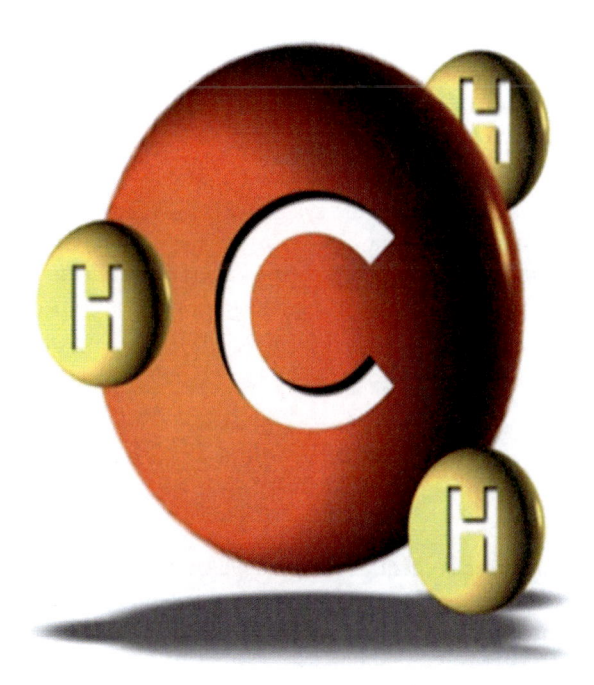

- Stroke
- Cancer
- Diabetes, and
- MS
- Alzheimer's disease
- ALS
- Heart disease
- Parkinson's disease
- Huntington's disease
- CFS/FMS
- Mitochondrial disease
- SLE, neural tube defects
- Miscarriages
- Down's syndrome
- Bipolar disorder
- Schizophrenia
- Repair of tissue damage
- Proper immune function
- The aging process
- Autism

7.5.2 Testing How Well the Methylation Cycle Works in the Body

There is no simple test to see how well the methylation cycle works. What we can do is measure levels of homocysteine and SAMe. If these are raised this would mirror a blockage in one part of the pathway. Indeed, a raised homocysteine, known to be a major risk factor for arterial disease, almost certainly represents blockages in the methylation cycle. However, one could have a normal homocysteine and normal SAMe but blockages elsewhere in the system, which would still impair the ability to methylate. And since there is no simple test available, one can measure urinary MMAs (a test for methylated B_{12}) and FIGLu (a test for methylated folate). These however, can only be done as part of Organic acids present in the urine metabolic analysis profile.

7.5.3 Block in the Methylation Cycle – Which Supplements to Take

Rich van Konynenburg has identified a package of micronutrients specifically to support the methylation cycle [306]. He recommends the activated form of vitamins. These are more expensive than the basic forms, but the idea behind this is that

7.5 The Methylation Cycle-Hypothesis for Pathogenesis of Fatigue

they are necessary in the short term to get the cycle working while in the long term they can exchange for the basic form. In addition to the basic three B vitamins Rich van Konynenburg has one or two other additional supplementations, which one may also choose to use. The package of supplements to support the methylation cycle needs to be taken in addition to everything else, i.e. the standard nutritional package (multivitamins, multiminerals, EFAs, vitamins C + D) and the mitochondrial rescue package (D-ribose, acetyl-L-carnitine, CoQ10, medium chain fatty acids). But the methylation package will change with time because as the methylation cycle starts to work again, it will start to stand on its own feet. Everyone will need a different supplementation depending on how poorly the cycle is working. One day we will have the biochemical tests to tailor make each package for each person, but until then I suggest the following regime for those sufferers who have been taking vitamin B_{12} in oral form, as either hydroxocobalamin or cyanocobalamin:

For 2 months a daily dose of

- Methylcobalamin 1,000 μg sublingually
- Methyltetrahydrofolate 800 μg
- Pyridoxal-5-phosphate 50 mg twice daily
- Glutathione 250 mg daily
- Phosphatidyl Serine 100 mg twice daily

If patients get better – fine! If symptoms get worse, it may be the reaction to the methylation package because it can be the cause of an acute detox reaction (see below). In such case one slows down the regime by taking smaller amounts of the supplements and increase the dose slowly. If symptoms are unchanged, the sublingual vitamin B_{12} should be exchanged for the injectable B_{12} i.e.:

- Daily subcutaneous injections methylcobalamin 1/2 ml (= 1,000 μg; this is a bit more expensive than cyanocobalamin). It is suggested that people start with this regime and although many do not fancy the idea of injections, they are easy to do and almost painless.
- Methyltetrahydrofolate 800 μg
- Pyridoxal-5-phosphate 100 mg (50 mg twice daily)
- Glutathione 250 mg daily (but only the reduced form)
- Phosphatidyl Serine 200 mgs (100 mg twice daily)

If the person gets better – fine! If the person gets worse – it may be the reaction. If symptoms are unchanged, add in:

- Trimethylglycine (also known as betaine hydrochloride) frequently used to increase stomach acid, take at meal times and be mindful because it may produce symptoms of acidity.
- Lecithin (phosphatidyl choline) and Phosphatidyl Ethanolamine.
- S-adenosyl methionine (SAMe) directly as a supplement 400 mg daily

For those sufferers who have already tried B_{12} by injection as either hydroxocobalamin or cyanocobalamin before starting the methylation cycle protocol, go straight on to injections of methylcobalamin. Once you are substantially better, then the regime can be relaxed. Once the patients change to a good methylator, methyl B$_{12}$, ActiFolate and glutathione can be tapered off. Injections could be swapped for oral supplements. However, do this slowly as some people need a small supplement in the long term in order to stay well.

- Methyltetrahydrofolate 800 µg (ActiFolate).
- Hydroxocobalamin 5,000 µg sublingually (or cyanocobalamin sublingually). It may be necessary for some people to continue with B$_{12}$ by injection to get the best effect (easy to self inject 1/2 ml daily – once there is improvement on methylcobalamin, then switch to the less expensive cyanocobalamin)
- Pyridoxyl-5-phosphate 50 mg
- Phosphatidyl Serine 200 mg (100 mg twice daily).

7.5.4 Problems Arise When Starting Treatment

Rich van Konynenburg has been in contact with patient and support groups who have gone through this regime. There seems to be two categories of effect [306]. Firstly, there is a quite rapid and profound improvement in some of the common symptoms, and secondly symptoms worsening or new symptoms arise, because in getting the methylation cycle going one suddenly is detoxified with increasing symptoms. The reason for this is that when the methylation cycle was not working the body was unable to detoxify properly and unable to produce cell mediated immune responses to get rid of chronic infections. Once the methylation cycle is up and running, the body is able to cope with respect to detoxification and cell mediated immune responses, which initially can make the person worse. The reasons for this are obvious – as soon as one start to detoxify chemicals and toxins are being mobilized and released into the blood stream, which makes people ill. Secondly, it is not the viruses or the chronic infections that make one ill; it is the immune reaction against them. Cell-mediated immune responses make one feel sick! Therefore it is really important to get into the detoxification regime smoothly, remembering that initially it can make things worse but one should interpret this as a good sign.

If one wants to explore the methylation pathways by lab tests, the best panel and the fastest service is offered by the Health Diagnostics and Research Institute in New Jersey, USA, where the panel costs approx. $300 US.

The following symptoms of CFS have been reported to be corrected in people with chronic fatigue on treatment. Adapted from [306]

- Improvement in sleep (though a few have reported increased difficulty in sleeping initially).
- Ending the need for of continues thyroid hormone supplementation.

- Termination of excessive urination and nighttime urination.
- Restoration of normal body temperature from lower values.
- Restoration of normal blood pressure from lower values.
- A more stable immune system fighting longstanding infections.
- Increase of energy and ability to carry on higher levels of activity without post-exertional fatigue or malaise. Termination of "crashing."
- Lifting of brain fog, increase in cognitive ability, return of memory.
- Relief from hypoglycemic symptoms.
- Improvement in alcohol tolerance
- Decrease in pain (although some have experienced a temporary increase in pain, as well as an increase in headache, presumably as a result of toxins being released from fatty tissues).
- Improvements in the patient's condition as noticed by friends and therapists.
- Not as much caregiving is needed by spouse and working out more balanced relationship in view of improved health and improved desire and ability to be assertive.
- Return of ability to read and retain what has been read.
- Return of ability to take a shower while standing up.
- Return of ability to sit up for longer times.
- Return of ability to drive for long distances.
- Improved tolerance for heat.
- Feeling unusually calm and relaxed.
- Feeling "more normal and part of the world."
- Ability to stop steroid hormone support without experiencing problems from doing it.
- Lowered sensation of continuously being under stress.
- Loss of excess weight.

7.5.5 Creating a Healthy Digestive Tract

The definitive treatment has yet to be established. There is a sharp learning curve; the key thing to realize, is to sterilize the upper gut and normalize digestion of food so that only the friendly bacteria can grow downstream. However, there are a few principles of treatment:

7.5.5.1 Recreate Ideal Environment for Bacteria in the Digestive Tract

- **Eat a diet with low fermentable substrates,** since it is sugar and all the refined carbohydrates which microbes most love and ferment. The diet needs to have low glycemic index and should be rich in raw or lightly cooked vegetables. This is because these foods contain a range of natural antimicrobials to inhibit bacterial overgrowth in the upper gut, together with many enzymes essential for their own digestion and fiber for fermentation in the large bowel by friendly bacteria into short chain fatty acids. When fruits are eaten, only berries are permitted!

- **Feed the bacteroides** – by eating pulses, nuts, seeds and vegetables, i.e. food, which is rich in probiotics. These provide food for bacteroides in the large bowel, which then ferment them to short chain fatty acids. This is the fuel for the cell inner lining of the gut, without that they atrophy. It is also the fuel which mitochondria can use when blood glucose levels (for example during sleep) fall low. The best sources are in pulses, vegetables, nuts and seeds.
- **Acid stomach and alkali duodenum** – An acid stomach helps to kill microbes, which are acid sensitive, while an alkali duodenum helps to kill microbes, which are alkali sensitive. A normally acid stomach would be pH 4 or less, and the duodenum would have a pH of 8 or above. Since the pH scale is a logarithmic scale a single change in these figures represent a 10,000-fold difference in acidity. It has been well known now for many decades that childhood asthma is associated with hypochlorhydria. Asthma in children tends to be caused by allergy to foods. If these foods are poorly digested, then they will be very much more antigenic and therefore very much more likely to switch on allergies and, therefore, asthma. Indeed, a study done in the 1930s showed that 80% of children with asthma also have hypochlorhydria. The two conditions are undoubtedly related. As the child's stomach matures and acid eventually is produced, then the asthma disappears. What often occurs with hypochlorhydric children is that they malabsorb their food and therefore tend to be underweight. So clinically it is unusual to see overweight kids with asthma – almost invariably they are thin children who wheeze. The treatment is as above, as well as trying to identify provoking foods. The commonest allergies to food are dairy products.

7.5.5.2 A Test for Hypochlorhydria by Using Salivary VEGF

One test for a non-acid stomach us a simple saliva test. The idea is that in general it is very difficult for the stomach to produce stomach acid. The normal acidity of blood is about pH 7, but the acidity of stomach acid can be as low as pH 1 – that means that hydrogen ions (which create the acidity) are a million times more concentrated in the stomach than in the bloodstream. So the stomach wall cells have a big task to do. Since for this job to be done, gastric parietal cells need a lot of energy from ATP to pump hydrogen ions from the inside of the parietal cell into the lumen of the stomach. The difficult part is stopping these hydrogen ions leaking back again intro the cells. This however, is achieved by the gastric parietal cells, which form a protective barrier between each other cell membrane tight junction to stop hydrogen ions from diffusing back. This coincides with the use of energy, which the body does not want to waste. And since the main inhibitor for the reabsorption via the cell membrane tight junctions is the *vascular endothelial growth factor* (VEGF), the more stomach acid is produced, the more VEGF will be released to inhibit the acid from being taken up by the gastric parietal cells. Therefore, parallel to the salivary VEGF levels is a proportionate amount of stomach acid is released. And although a huge amount of research has been done with respect to VEGF, most of which is related to high levels, the reverse is also true and low levels of VEGF would point towards hypochlorhydria. Taking proton pump inhibitors and possibly other acid blockers would invalidate this test. So in order to have an accurate result, these drugs need to be stopped 4 days prior to doing the test. Saliva for VEGF tests should be unstimulated, i.e. a sample is given at least 1 hour after a meal and at least 15 min after drinking soft drinks, tea or coffee. A break of 24 h after alcohol ingestion is needed AND a similar break after any proton-pump-inhibitor drugs are used and ideally any drug that interferes with stomach acidity. Put 1–2 ml of saliva into the blue topped trace element free tube supplied and post it to a laboratory that does the VEGF analysis (e.g. Acumen/England) to arrive on a working day.

There is no easy way to test for an alkali duodenum. Where there is hypochlorhydria, take additional acid with food (as ascorbic acid or betaine hydrochloride). The stomach normally takes 1–2 h to empty; at this point take magnesium carbonate 1–2 g, which neutralizes stomach acid and assists digestion in the duodenum.

- **Eat smaller meals** - lesser amounts of foods are easier to deal with. Anyone who has gobbled too much of a large meal will recognize the symptoms of a fermenting gut – fatigue, bloating, discomfort and, later, foul smelling winds!
- **Sterilize the upper gut** – the key here is to take something, which kills bugs in the upper gut but does not upset bacteria in the lower gut. This is why antibiotics are not ideal – they will also upset lower gut flora. Bacteria and yeast are greedy for micronutrients, especially minerals. Indeed, this may explain why some patients worsen when they take micronutrients – these simply feed the upper gut flora so they ferment harder. The following ways to tackle this problem:
  1. **Take high dose ascorbic acid** in between and during meals. The acid and the ascorbate both kill microbes. In the right dose they can sterilize the upper gut but since most is absorbed the lower gut is not affected. If very high doses of vitamin C are taken (<2,000 mg) this will spill over into the large bowel and cause diarrhea – this is considered a useful therapeutic option in getting rid of gut infections such as viral gastroenteritis.
  2. **Take minerals** through the skin by mixing them with aqueous cream. This may explain why the tiny amounts of magnesium injections are so effective!
  3. **Mix herbal tannins** (e.g. Viracin®), which chelate with minerals so they are not available for bacteria. This may explain why tea drinking is so popular – the tannin in the tea has the same effect. Also spicy foods kill microbes and may explain why many spices (onions, garlic, horseradish, ginger, black pepper, chili, leek, and curry) are getting an increased attention. Since gut fermentation is a common problem in people eating Western diets, subconsciously people have worked out that curry with a cup of tea to improve digestion makes us feel better!

A fast and inefficient propulsion of food means that there is less available for digestion and to be fermented downstream. This is why one may need:

- **Acid supplements:** Indeed, there may be a role for vitamin C as ascorbic acid. Ascorbic acid is an acid and improves the digestion of protein. It is also toxic to all microbes including bacteria, yeast and viruses as well as being an important anti-oxidant – indeed the overall receiver of electrons from free radicals. Humans, guinea pigs and fruit bats are the only mammal species which cannot make their own vitamin C and that is why they have to get a supplement via their diet. Scaling up from other mammals we should be consuming 2–6 g daily (a 100-fold more than the government RDA of 30 mg daily!). One could get the dose just right so that ascorbic acid when taken together with food sterilizes the upper gut, but is absorbed and thus has no effect on the lower gut. If one takes excessive vitamin C this will cause diarrhea as too much gets into the lower gut, kills off the good bacteria and then empties the gut completely!

- **Pancreatic enzymes.** The pancreas is a large gland, which lies behind the stomach and upper gut. It has two major functions of clinical importance – firstly it acts as an endocrine organ to produce insulin and other hormones essential for the control of blood sugar. Secondly it has an exocrine function to produce enzymes essential for the digestion of food. These enzymes include those to digest proteins, fats and starches and to work best they need an alkali environment. This alkali environment is provided by the liver, which produces bile containing bile salts and bicarbonate. When food is present in the duodenum and jejunum, the gall bladder contracts sending a bolus of bile salts and bicarbonate which meet up downstream with pancreatic enzymes to allow digestion to take place in the duodenum and jejunum.

7.5.5.3 Symptoms of Poor Pancreatic Exocrine Function

If the pancreas does not produce sufficient enzymes then foods will not be digested. This can lead to problems downstream. Firstly, foods may be fermented instead of being digested and this can produce the symptom of bloating due to wind, together with the production of metabolites such as various alcohols, hydrogen sulphide and other toxic compounds. Secondly, foods are not fully broken down so that they cannot be absorbed and this can result in malabsorption. In cases of severe pancreatic dysfunction it is obvious because the stools become greasy and fatty, foul smelling, bulky and difficult to flush away. Where there is malabsorption of fat, there will be malabsorption of essential Ω-3 and Ω-6 fatty acids plus there will be a malabsorption of fat-soluble vitamins such as vitamins A, D, E and K. If food components are poorly digested, it results in the presentation of large antigen molecules to the immune system, downstream in the gut, alerting the defense with a switch in inflammatory cytokine production (Th1-shift) followed by allergies. Therefore,

Poor digestion of food can be considered a risk factor for allergy

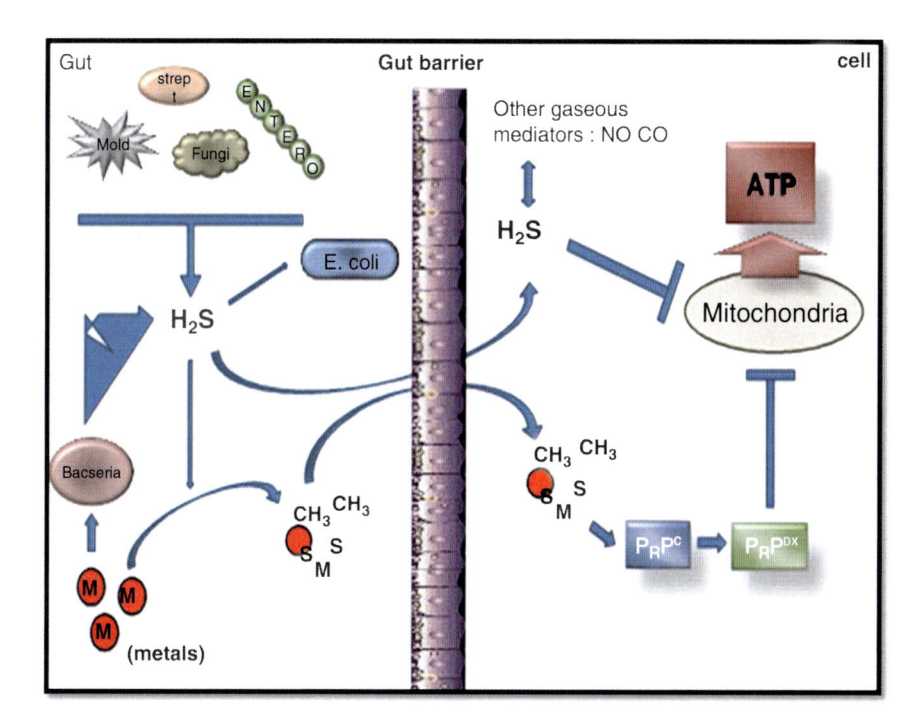

7.5 The Methylation Cycle-Hypothesis for Pathogenesis of Fatigue   197

Fig. 7.60 Exogenous hydrogen sulfite (H_2S) production by bacteria is a potent toxic for mitochondrial function since it directly inhibits enzymes involved in cellular energy production. It also interferes with oxygen transport by blocking hemoglobin. Enterococcus, Streptococcus and Prevotella are strong H_2S producers

7.5.5.4 Diagnosis/Treatment of Poor Pancreatic Function

A Comprehensive Digestive Stool Analysis (CDSA) looks at the different fats appearing in the stool and if these are raised, this points to an insufficient pancreatic function. In addition, fecal elastase is also a useful test for pancreatic function.

- If there is acute inflammation in the pancreas, amylase levels may be raised. In acute pancreatitis there is severe dull central abdominal pain, which typically is relieved when the sufferer leans forward – that is because the weight of the stomach is pulled off the pancreas. Conventional medicine tells us that a third of all pancreatitis cases are due to gallstones, another third is due to alcohol consumption and a the last third is unknown. This unknown category is almost certainly due to poor antioxidant status, particularly low levels of glutathione peroxidase. This problem most likely co-exists in all cases of pancreatitis and is in all probability the mechanism by which alcoholism causes pancreatitis!
- Also, Kenny de Meirleir developed a test, manufactured by Protea Biopharma, in which the presence of hydrogen sulphide (H_2S), produced in the intestines when bacteria come in contact with heavy metals is detected (Fig. 7.60). People with

CFS have been shown to have higher concentrations of intestinal bacteria than normal, which leads to higher levels of H_2S. H_2S is a gas present in minuscule concentrations in normal people but has toxic levels in CFS patients. The reasons for overproduction of bacteria can range from lactose intolerance to viral infection, mercury intoxication to stress. According to an article published by the De Meirleir and his team, H_2S causes intolerance to light and noise, a depressed immune system and a low white blood cell count. It also leads to retention of mercury by the body, which in turn produces cell death and damage to energy metabolism. The biggest effects, though, are produced on the central nervous system, explaining the main symptoms of CFS [308]. Protea Biopharma/Belgium has launched a bedside test, which measures abnormal high hydrogen sulfide in the urine, which is considered a neurotoxic metabolite (Fig. 7.60) affecting mitochondrial function and is commercialized on a fee-per-service basis by R.E.D Laboratories (www.redlabs.com).

Treatment of Poor Pancreatic Function

Correction of the above factors, in particular poor antioxidant status.

- Pancreatic enzymes, e.g. Kreon®/Creon® 10,000 contains phthalates and methacrylate and while in theory it should be perfect agent in case of insufficient pancreatic function, these toxins make it unsafe. One can buy pancreatic enzymes on the open market such as Polyzyme Forte, Spectrazyme, Digestaid, etc. In general 1–3 tablets with meals, depending on the size of the meal are recommended.
- A study done on patients with food allergy showed that the majority of them either had hypochlorhydria, or poor pancreatic function, or both [309]. What is interesting is that through correcting the hypochlorhydria with betaine hydrochloride and the poor pancreatic function with pancreatic enzymes, many symptoms due to food allergy were resolved. Supplementation is usually not for life – once the above factors are corrected the pancreas can recover and normal function is restored. A regular dose would be 1–3 capsules of Polyzyme Forte® (BioCare) with meals depending on the size of the meal. Be aware that many prescribable pancreatic enzymes contain toxic dimethicones or phthalates (e.g. Creon®).
- Magnesium carbonate as above
- Also, bile salts to emulsify fats (from 333 to 1,000 mg) and magnesium carbonate to provide an alkaline environment (500 mg to 2 g) depending on meal size should be added, given 60–90 min after the meal. Bile acids can be measured as part of a CDSA but one has to ask specifically for this test, as it is not a part of the basic package. Apparently prevotella, the bug shown by Kenny de Meirleir to be a major fermenter, is susceptible to bile acids. Increasing fats and oils in the diet will improve bile flow and my help flush out unwanted bacteria in the biliary tree.

7.5 The Methylation Cycle-Hypothesis for Pathogenesis of Fatigue

Improve the Lining of the Gut

- **Chewing gum.** The parotid salivary gland provides a rich source of endothelial growth factor (indeed, this is what John McLaren Howard measures in his hypochlorhydria test), which stimulates growth of the lining of the gut. Chew, because this stimulates flow of saliva using a sugar-free, additive-free gum!
- **Intake of red meat and soya bean protein** for vegetarians, as recommend by other scientists, although they consider this very much as being the "second best".
- Also, **glutamine** in a slow release capsule is recommended to nourish the gut lining and also to correct the antioxidant status (superoxide dismutase, Co-enzyme Q10 and glutathione peroxidase).
- Use of **probotics** since they nourish those bacteria (provotella, formicuta, bacteroides), which are necessary for normal gut function fighting off pathogens. And lastly,
- By all means, avoid the following medication
  - Antibiotics,
  - Anticontraceptives or using combined hormone replacement therapy (HRT),
  - Any acid blocking agents, especially proton pump inhibitors (PPI),
  - Antidepressants as they tend to result in a dry mouth.

So far, probiotics have not lived up to their full therapeutic potential. This very likely is because the single most important probiotic is, that bacteroides, a genus of gram-negative bacillus bacteria making up the most substantial portion of the mammalian gastrointestinal flora, where they play a fundamental role in processing of complex molecules to simpler ones in the host intestine. They cannot exist outside the human gut; since oxygen kills it within minutes with the main source of energy is polysaccharides from plant sources such as starch and glycogen, and structural polysaccharides such as cellulose and chitin. There is no probiotic on the market, which contains bacteroides, as the human acquires bacteroides at birth and retains those bacteria for the rest of his life. Kefir® is useful supplement, because it contains a combination of bacteria that produces a toxin which kills yeast. In the following there are some basic suggestion of how to improve the flora of the intestine being a necessary adjunct in the improvement of mitochondrial function.

Treatment of a Low E-coli Colonization

- **All the above** plus
- **Mutaflor®** – ½ a capsule daily for 2 days, then 1 capsule daily for 7 days then 2 capsules daily for 14 days and adjust the dose subsequently according to response.
- **Prebiotics like galactose**. It is vital that E-coli is given the prebiotic galactose (galactose oligosaccharides) mainly found in pulses and nuts. These will ferment in the large bowel to produce wind (hydrogen and carbon dioxide) – this only becomes foul smelling if other foods are poorly digested.

Treatment of Low Numbers of Bacteroides

- **All the above**, especially prebiotics
- Low numbers of bacteroides should build up easily with structural polysaccharides such as cellulose as found e.g. the apple. Again it has to be pointed out that special care is put on not eating preprocessed fruits with pesticides, fungicides or fruits that is genetically modified!
- A major problem would arise if there were no bacteroides. There is no probiotic, which contains bacteroides because it does not survive outside the human gut. One possibility would be to consider fecal bacteriotherapy, i.e. the use of fresh live actively fermenting bacteroides from another human gut. The main proponent of this therapy is Dr Thomas Borody an Australian gastroenterologist based in Sydney who is noted for his work in novel therapies for gastrointestinal disorders [310].

Treatment of High Streptococcus Levels

- All the above
- Strictly avoid glucose and fructose– each molecule of glucose and fructose will produce two molecules of lactic acid and create a marked tendency to acidosis. It is noteworthy that that Chinese people do not eat fruit at all, which may be a reason why they do have a low incidence of gastrointestinal problems unless they change to the so-called western diet. Only the wealthy Chinese do and then very occasionally. Glucose and fructose are potentially very damaging to the body because they get fermented by streptococcus into D-lactate. This is another mechanism by which sugars can result in foggy brain.
- **Erythromycin** 500 mg twice daily for 7 days or any such macrolide or even longer. Happily bacteroides, the most abundant probiotics are largely resistant to erythromycin!
- **Magnesium carbonate and pancreatic enzymes**. One can monitor any progress by looking at foggy brain symptoms – they should eventually disappear and when it does one can terminate Erythromycin intake, but continue with an E-coli probiotic.

Treatment of High Enterococcus

- All the above
- Take Kefir; the best results with this probiotic come from using live cultures. One can grow your own Kefir as it grows well at room temperature. Because dairy products are not evolutionarily correct foods, Kefir should be grown on non-dairy foods such as soya milk, rice milk or coconut milk. It contains bacteriocins, which inhibit streptococcus, but it also contains lactobacillus plantarum, which can ferment sugar to D-lactate, so strictly avoid sugar as this results in another

major problem! L. Plantarum has good anti-inflammatory action, which makes it a desirable probiotic in any inflammatory bowel disease.
- Prevotella species. There does not appear to be an antibiotic that Prevotella is sensitive to, but apparently it can be reduced by taking bile salts. These are also available on prescription as Ursodeoxycholic acid and it is suggested to take 150 mg with meals, perhaps more!
- **Neem** or *Azadirachta indica* is a fast-growing tree that can reach a height of 15–20 min from the mahogany family Meliaceae. All parts of the tree are said to have medicinal properties (seeds, leaves, flowers and bark) and are used for preparing many different medical preparations. Neem is extremely safe and has activity against upper gut fermenters (yeast and aerobic, oxygen loving bacteria) but no activity against desirable lower gut fermenters (anaerobic bacteria). It is recommended to take one 500 mg capsule of neem leaf daily initially and, if tolerated, gradually increase to 2 capsules with each meal. Neem is very inexpensive, a fact which should be considered when comparing the costs of any other medication- an important bonus!

7.5.5.5 Free radicals from the Liver Resulting in Mitochondropathy

Excessive free radical production, which cannot be dealt with by antioxidant reserves, will damage and switch off mitochondria. One would think that the largest source of free radicals comes from mitochondria themselves since here we have large amounts of glucose being oxidized in the presence of oxygen to produce energy with a large potential to produce free radicals, such as superoxide. While this is undoubtedly a major source, even greater than this is the liver P450 cytochrome system. Humans are able to eat a wider variety of foods than any other mammal because of this amazing detoxification system of enzymes. This has resulted in humans becoming the most successful mammal because they can occupy almost any ecological niche. When the P450 detox system is working well, then this has enormous evolutionary advantages. However, if things start to go wrong, excessive amounts of free radicals are produced with the potential to switch on a chronic fatigue syndrome. At this point it must be emphasized that a chronic fatigue syndrome is a protective adaptive response. If that person did not become acutely fatigued and succeeded in pushing on physically or mentally, then the excessive free radicals so generated would have the potential to cause enormous pathological damage. This is probably why we do not see wild animals with chronic fatigue syndromes. They simply push themselves to destruction because they have to survive.

The Liver P450 Detoxification, Source of Free Radicals

The liver is one of the most important organs in the body when it comes to detoxifying or getting rid of foreign substances or toxins, especially from the gut.

The liver plays a key role in most metabolic processes, especially detoxification. The liver detoxifies harmful substances by a complex series of chemical reactions. The role of these various enzyme activities in the liver is to convert *fat-soluble* toxins into *water-soluble* substances that can be excreted in the urine or the bile depending on the particular characteristics of the end product. Many of the toxic chemicals that enter the body are fat-soluble, which means they dissolve only in fatty or oily solutions and not in water. This makes them difficult for the body to excrete. Fat-soluble chemicals have a high affinity for fat tissues and cell membranes, which are composed of fatty acids and proteins. In these fatty tissues of the body, toxins may be stored for years, being released during times of exercise, stress or fasting. During the release of these toxins, several symptoms such as headaches, poor memory, stomach pain, nausea, fatigue, dizziness and palpitations can occur. There are two stages in liver detoxification. Stage one is an oxidation reaction to make molecules a bit more active in order that stage two can take place, in which another molecule is stuck on. This stacking on or conjugation allows the toxin to become less active and more water-soluble so it can be excreted via the urine. The conjugation process could be glucuronic acid (glucuronidation), amino acid, methyl groups (methylation) glutathione, sulphate group (sulfatation) and acetyl-CoA (acetylation).

There are many possible ways the liver P450 cytochrome system could be overwhelmed.

1. **Genetic**: some people simply have genetically poor detox ability. One example of this, of course, is Gilbert's syndrome, where conjugation with glucuronide (stage 2 detox) is lacking. There are two steps to detoxification: the first is an oxidation reaction, which may make some toxins more toxic! Many CFS sufferers have fast stage one and slow stage two metabolism, which means they have a P450 system which initially produces more rather than less toxic stress! So, for example, over 80% of Gilbert's sufferers complain of fatigue. One example is alcohol. This is metabolized initially into acetaldehyde, which is a toxic compound responsible for hangovers! Alcohol intolerance is almost universal in CFS sufferers.
2. An **acquired metabolic lesion** as a result of nutritional deficiency. For example, many of these P450 cytochrome enzymes are highly dependent on metal co-factors such as zinc, magnesium, or selenium, B vitamins and essential fatty acids.
3. **Toxins** produced from **normal metabolism** e.g. detoxifying neurotransmitters, products from immune activity, breakdown products from damaged tissues etc.
4. Overwhelming **toxins** from the **outside** world, such as persistent organic pollutants, or of course prescribed drug medication or social drugs of addiction (caffeine, alcohol, drugs, medication). This is part of the explanation why so many CFSs do not tolerate prescription medication. Other reasons are that many drugs inhibit mitochondria directly, or destabilize membranes in the brain resulting in poor energy delivery to brain cells. Patients refusing medication then get labeled as uncooperative and are dropped from medical care (Fig. 7.60).
5. Intoxicants arising as a result of fermentation from the upper gut.

There are genetic tests, for the state of the detoxification system in the liver, such as single nucleotide polymorphisms (or SNIPs) through Genova laboratories (www. genovadiagnostics.com) available. Genova also offers functional tests to look at stage one and stage two detoxification.

References

1. Nikić I et al (2011) A reversible form of axon damage in experimental autoimmune encephalomyelitis and multiple sclerosis. Nat Med 17:495–499
2. Wu G, Meininger CJ (2009) Nitric oxide and vascular insulin resistance. Biofactors 35:21–27
3. Toutouzas K et al (2008) Asymmetric dimethylarginine (ADMA) and other endogenous nitric oxide synthase (NOS) inhibitors as an important cause of vascular insulin resistance. Horm Metab Res 40:655–659
4. Mastrocola R et al (2005) Oxidative and nitrosative stress in brain mitochondria of diabetic rats. J Endocrinol 187:37–44
5. Dixon LJ et al (2005) Increased superoxide production in hypertensive patients with diabetes mellitus: role of nitric oxidesynthase. Am J Hypertens 18:839–843
6. Yudkin JS, Stehouver CDA (2005) "Vasocrine" signalling from perivascular fat: a mechanism linking insulin resistance to vascular disease. Lancet 365:1817–1820
7. See reference [4]
8. Yeh CH et al (2001) Requirement for p38 and p44/p42 mitogen-activated protein kinases in RAGE-mediated nuclear factor-kappaB transcriptional activation and cytokine secretion. Diabetes 50:1495–1504
9. Blau N et al (2010) Phenylketonuria and BH4 deficiencies, 1st edn. UNI-MED Science, Bremen
10. See reference [5]
11. Fayers KE et al (2003) Nitrate tolerance and the links with endothelial dysfunction and oxidative stress. Br J Clin Pharmacol 56:620–628
12. See reference [5]
13. Chen W et al (2010) Peroxynitrite induces destruction of the tetrahydrobiopterin and heme in endothelial nitric oxide synthase: transition from reversible to irreversible enzyme inhibition. Biochemistry 49:3129–3139
14. Fukuda Y et al (2002) Tetrahydrobiopterin restores endothelial function of coronary arteries in patients with hypercholesterolaemia. Heart 87:264–269
15. Maier W et al (2000) Tetrahydrobiopterin improves endothelial function in patients with coronary artery disease. J Cardiovasc Pharmacol 35:173–178
16. Shinozaki K et al (1999) Abnormal biopterin metabolism is a major cause of impaired endothelium-dependent relaxation through nitric oxide/O imbalance in insulin-resistant rat aorta. Diabetes 48:2437–2445
17. Verma S et al (2002) Novel cardioprotective effects of tetrahydrobiopterin after anoxia and reoxygenation: identifying cellular targets for pharmacologic manipulation. J Thorac Cardiovasc Surg 123:1074–1083
18. Do Carmo M et al (2004) Tetrahydrobiopterin improves endothelial dysfunction and vascular oxidative stress in microvessels of intrauterine undernourished rats. J Physiol 558:239–248
19. Landmesser U, Drexler H (2007) Endothelial function and hypertension. Curr Opin Cardiol 22:316–320
20. Bevers LM et al (2004) Tetrahydrobiopterin, but not L-arginine, decreases NO synthase uncoupling in cells expressing high levels of endothelial NO synthase. J Physiol 558: 239–248

21. Kawashima S, Yokoyama M (2004) Dysfunction of endothelial nitric oxide synthase and atherosclerosis. Arterioscler Thromb Vasc Biol 24:998–1005
22. Baker H et al (2006) Interleukin-2 enhances biopterins and catecholamine production during adoptive immunotherapy for various cancers. Cancer 64:1226–1231
23. Chen L et al (2010) Roles of tetrahydrobiopterin in promoting tumor angiogenesis. Am J Pathol 177:2671–2680
24. Naylor AM et al (2010) The tetrahydrobiopterin pathway and pain. Curr Opin Invest Drugs 11:19–30
25. Moens AL et al (2008) Reversal of cardiac hypertrophy and fibrosis from pressure overload by tetrahydrobiopterin (fig.). Efficacy of recoupling nitric oxide synthase as a therapeutic strategy. Circulation 117:2626–2636
26. See reference [4]
27. Wever RMF et al (1997) Tetrahydrobiopterin regulates superoxide and nitric oxide generation by recombinant endothelial nitric oxide synthase. Biochem Biophys Res Commun 237:340–344
28. Antoniades C et al (2006) 5-Methyltetrahydrofolate rapidly improves endothelial function and decreases superoxide production in human vessels-effects on vascular tetrahydrobiopterin availability and endothelial nitric oxide synthase coupling. Circulation 114:1193–1201
29. Kuzkaya N et al (1950) Interactions of peroxynitrite, tetrahydrobiopterin, ascorbic acid and thiols: implications for … The fate of uric acid in man. J Biol Chem 183:21–31
30. Chan E (2002) The role of complementary and alternative medicine in attention-deficit hyperactivity disorder. J Dev Behav Pediatr 23:S37–S45
31. Kremer H (2004) Die stille Revolution der Krebs- und AIDS-Medizin. Neue fundamentale Erkenntnisse über die tatsächlichen Krankheits- und Todesursachen bestätigen die Wirksamkeit der biologischen Ausgleichstherapie. Ehlers Verlag, Wolfratshausen
32. Duesberg P (1992) AIDS acquired by drug consumption and other noncontagious risk factors. Pharmacol Ther 55:201–277
33. Carr A, Cooper DA (2000) Adverse effects of antiretroviral therapy. Lancet 356:1423–1430
34. Brinkman K, Kakuda TN (2000) Mitochondrial toxicity of nucleoside analogue reverse transcriptase inhibitors: a looming obstacle for long-term antiretroviral therapy? Curr Opin Infect Dis 13:5–11
35. Lewis W, Dalakas MC (1995) Mitochondrial toxicity of antiviral drugs. Nat Med 1:417–422
36. Kakuda TN (2000) Pharmacology of nucleoside and nucleotide reverse transcriptase inhibitor-induced mitochondrial toxicity. Clin Ther 22:685–708
37. Brinkman K et al (1999) Mitochondrial toxicity induced by nucleoside-analogue reverse-transcriptase inhibitors is a key factor in the pathogenesis of antiretroviral-therapy-related lipodystrophy. Lancet 354:112–1115
38. Baumbach GL, Sigmund CD, Faraci FM (2004) Structure of cerebral arterioles in mice deficient in expression of the gene for endothelial nitric oxide synthase. Circ Res 95:822–829
39. Peters C (2006) Bestimmung von 3-Nitrotyrosin im Liquor als Hinweis fur nitrosativen Stress. Department of Pediatrics, Rheinisch-Westfälische Technische Hochschule, Aachen
40. Husmann M, Keller M, Barton M (2007) Atherosklerotische Gefässerkrankungen und Stickstoffmonoxid (NO):Die wachsende Bedeutung von hoher Lebenserwartung und Übergewicht für die Klinik. Schweiz Med Forum 7:1008–1011
41. Giulivi C et al (2010) Mitochondrial dysfunction in autism. JAMA 304:2389–2396
42. Szabó C, Módis K (2010) Pathophysiological roles of peroxynitrite in circulatory shock. Shock 34:4–14
43. Hartl D et al (2008) Dysregulation of innate immune receptors on neutrophils in chronic granulomatous disease. J Allergy Clin Immunol 121:375–382
44. Selley ML (2004) Increased (E)-4- hydroxy-2-nonenal and asymmetric dimethylarginine concentrations and decreased nitric oxide concentrations in the plasma of patients with major depression. J Affect Disord 80:249–256
45. Lamson DW, Plaza SM (2002) Mitochondrial factors in the pathogenesis of diabetes: a hypothesis for treatment -mitochondrial factor diabetes. Altern Med Rev 7:94–111

46. Kelley DH et al (2002) Dysfunction of mitochondria in human skeletal muscle-type 2 diabetes. Diabetes 51:2944–2950
47. Petersen KF et al (2004) Impaired mitochondrial activity in the insulin-resistant offspring of patients with type 2 diabetes. N Engl J Med 350:664–671
48. Rötig A et al (1997) Aconitase and mitochondrial iron sulphur protein deficiency in Friedreich ataxia. Nat Genet 17:215–217
49. Seznec H et al (2004) Idebenone delays the onset of cardiac functional alteration without correction of FE-S enzymes deficit in a mouse model for Friedreich ataxia. Human Mol Genet 13:1017–1024
50. Buyse G et al (2003) Idebenone treatment in Friedreich's ataxia. Neurological, cardiac, and biochemical monitoring. Neurology 60:1679–1681
51. Kuklinski B (2005) Zur Praxisrelevanz von nitrosativem stress. umwelt·medizin·gesellschaft 18:95–106
52. Kuklinski B (2008) Praxisrelevanz des nitrosativen Stresses, 1. Mitteilung: Diagnostik und Therapie neurologischer Erkrankungen. OM & Ernährung 124:F13–F30
53. Keil U (2005) Schlüsselfunktion der Mitochondrien in der Pathogenese der Alzheime Demenz. Department of Geriatrics, Johann Wolfgang Goethe Universität, Frankfurt
54. Jain A, Martensson J, Stole E (1991) Glutathione deficiency leads to mitochondrial damage in brain. Proc Natl Acad Sci USA 88:1913–1917
55. Lafon-Cazal M et al (1993) Nitric oxide, superoxide and peroxynitrite: pulative mediators of NMDA induced cell death in cerebellar granule cells. Neuropharmacology 32:1259–1266
56. Kidd PM (2005) Neurodegeneration from mitochondrial insufficiency, nutrients, stem cells, growth factors and prospects from brain rebuilding using integrative management. Altern Med Rev 10(4):268–293
57. Beal MF (2004) Mitochondrial dysfunction and oxidative damage in Alzheimer's and Parkinson's diseases and coenzyme Q10 as a potential treatment. J Bioenerg Biomembr USA 36:381–386
58. Milunsky A, Milunsky J (2010) Genetic disorders and the fetus: diagnosis, prevention and treatment, 6th edn. Wiley-Blackwell, Oxford
59. Chiou G (2001) Review: effects of nitric oxide on eye diseases and their treatment. J Ocul Pharmacol Ther 17:189–198
60. Der-Chong T et al (2002) Significant variation of the elevated nitric oxide levels in aqueous humor from patients with different types of glaucoma. Ophthalmology 216:346–350
61. Saccr S (2002) Nitric oxide as a mediator of glaucoma pathogenesis. Med Sci Monit 8:LE41–LE42
62. Boger RH (2004) Asymmetric dimethylarginine and endogenous inhibitor of nitric oxide synthase, explains the "L-arginine paradox" and acts as a novel cardiovascular risk factor. J Nutr 10(Suppl):2842–2847
63. Murray AJ, Edwards LM, Clarke K (2007) Mitochondria and heart failure. Curr Opin Clin Nutr Metab Care 10:704–711
64. Rosca MG et al (2008) Cardiac mitochondria in heart failure: decrease in respirasomes and oxidative phosphorylation. Cardiovasc Res 80:30–39
65. Abe K et al (1995) Induction of nitrotyrosine like immunoreactivity in the lower motor neuron of amyotrophic lateral sclerosis. Neurosci Lett 199:152–154
66. Weishaupt JH et al (2006) Reduced oxidative damage in ALS by high-dose enteral melatonin treatment. J Pineal Res Denmark 41:313–323
67. Jacob S et al (2002) Melatonin as a candidate compound for neuroprotection in amyotrophic lateral sclerosis (ALS): high tolerability of daily oral melatonin administration in ALS patients. J Pineal Res Denmark 33:186–187
68. Dupuisa L et al (1994) Mitochondria in amyotrophic lateral sclerosis: a trigger and a target. Neurodegener Dis 1:245–254
69. Fabig KR (2002) Die Auslösung chemikalienassoziierter Symptome und Befunde der NAT2, GST M1 und GST T1 bei 603 Personen. Umweltmed Forsch Prax 7:226–227

70. Garbe TR, Yukawa H (2001) Common solvent toxicity, antioxidation of respiratory redox-cyclers enforced by membrane derangement. Z Naturforsch 56:483–491
71. Wiesmüller GA et al (2004) Genetische Disposition und multiple chemische Sensitivität (MCS): Stand des Wissens und Konsequenz für die molekulargenetische Diagnostik. Umweltmed Forsch Prax 9:275–280
72. McKeown-Eyssen G et al (2004) Case-control study of genotypes in multiple chemical sensitivity: CYP 2D6, NAT1, NAT2, PON1, PON2 and MTMFR. Int J Epidemiol 33:1–8
73. Pall ML (2002) NMDA-sensitization and stimulation by peroxynitrite, nitric oxide and organic solvents as the mechanism of chemical sensitivity in multiple chemical sensitivity. FASEB J 16:1407–1417
74. Sorg BA, Prasad BM (1887) Potential role of stress and sensitization in the development and expression of multiple chemical sensitivity. Environ Health Perspect 105:467–471
75. Drapier JC, Hibbs JB (1989) Murine activated macrophages inhibit aconitase in tumor cells. Inhibition involves the iron-sulfur prosthetic group and is reversible. J Clin Invest 78:790–794
76. Granger DL, Lehninger AL (1982) Sites of inhibition of mitochondrial electron transport in macrophage-injured neoplastic cells. J Cell Biol 95:527–531
77. Carew JS, Huang P (2002) Mitochondrial defects in cancer. Mol Cancer 1:9–22
78. Fosslien E (2008) Cancer morphogenesis: role of mitochondrial failure. Ann Clin Lab Sci 38:307–330
79. Zhou L et al (2010) Emergent ventricular fibrillation caused by regional mitochondrial depolarization in cardiac muscle. Circulation 122:A2090
80. Fosslien E (2003) Mitochondrial medicine – cardiomyopathy caused by defective oxidative phosphorylation. Ann Clin Lab Sci 33:371–395
81. Song Y et al (2007) Diabetic cardiomyopathy in OVE26 mice shows mitochondrial ROS production and divergence between in vivo and in vitro contractility. Rev Diabet Stud 4:159–168
82. Marin-Garcia J, Ananthakrishnan R, Goldnetkal R (1998) Hypertrophic cardiomyopathy with mitochondrial DNA depletion and respiratory enzyme defect. J Pediatr Cardiol 19:266–268
83. Hamza M et al (2010) Nitric oxide is negatively correlated to pain during acute inflammation. Mol Pain 6:55–76
84. Celerier E et al (2006) Opioid-induced hyperalgesia in a murine model of postoperative pain: role of nitric oxide generated from the inducible nitric oxide synthase. Anesthesiology 104:546–555
85. Muscoli C et al (2007) Therapeutic manipulation of peroxynitrite attenuates the development of opiate-induced antinociceptive tolerance in mice. J Clin Invest 117:3530–3539
86. Babey AM et al (1994) Nitric oxide and opioid tolerance. Neuropharmacology 33:1463–1470
87. Toda N et al (2009) Modulation of opioid actions by nitric oxide signaling. Anesthesiology 110:166–181
88. Kodo H et al (2000) Dietary zinc-deficiency decreases glutathione-S-transferases expression in the rat olfactory epithelium. J Nutr 130:38–44
89. Massaad CA, Eric Klann E (2011) Reactive oxygen species in the regulation of synaptic plasticity and memory. Antioxid Redox Signal 14(10):2013–2054
90. Pall ML, Bedient SA (2007) The NO/ONOO- cycle as the etiological mechanism of tinnitus. Int Tinnitus J 13:99–104
91. See reference [46]
92. Kuklinski B (2007) Nahrungsfett, metabolisches Syndrom, mitochondriale Zytopathie. OM & Ernährung 120:F63–F69
93. Kalman B (2006) Role of mitochondria in multiple sclerosis. Curr Neurol Neurosci Rep 6:244–252
94. Liu JS et al (2001) Expression of inducible nitric oxide synthase and nitrotyrosine in multiple sclerosis lesions. Am J Pathol 158:2057–2061
95. Costa B et al (2002) Therapeutic effect of the endogenous fatty acid amide, palmitoylethanolamide, in rat acute inflammation: inhibition of nitric oxide and cyclo-oxygenase systems. Br J Pharmacol 137:413–420

96. Finsterer J, Gelpi E (2006) Mitochondrial disorder aggravated by propranolol. South Med J 99:768–771
97. Mais AH (2006) The role of nitric oxide in chondrocyte models of osteoarthritis. Department of Biology, University of Konstanz, Konstanz
98. Liu JT et al (2010) Mitochondrial function is altered in articular chondrocytes of an endemic osteoarthritis, Kashin-Beck disease. Osteoarthritis Cartilage 18:1218–1226
99. Blanco FJ, López-Armada MJ, Maneiro E (2004) Mitochondrial dysfunction in osteoarthritis. Mitochondrion 4:715–728
100. Gaist D et al (2002) Statins and risk of polyneuropathy. Neurology 58:1321–1322
101. Guo FH et al (1995) Continuous nitric oxide synthesis by inducible nitric oxide synthase in normal human airway epithelium in vivo. Proc Natl Acad Sci USA 99:7809–7813
102. Knepler J et al (2001) Peroxynitrite causes endothelial cell monolayer barrier dysfunction. Am J Physiol Lung Cell Mol Physiol 281:C1064–C1075
103. Lee DM, Schur PH (2003) Clinical utility of the Anti-CC Passay in patients with rheumatic disease. Ann Rheum Dis 62:870–874
104. Gross WL, Moosig F, Lamprecht P (2009) Anticitrullinierte protein/peptid-Antikörper bei rheumatoider Arthritis. Dt Ärzteblatt 106:157–158
105. Pettersson A, Hedner T, Milsom J (1998) Increased circulating concentrations of asymmetric dimethyl arginine (ADMA), an endogenous inhibitor of nitric oxide synthesis, in preeclampsia. Acta Obstet Gynaecol Scand 77:808–813
106. Lacza Z et al (2001) Mitochondrial nitric oxide synthase is constitutively active and is functionally upregulated in hypoxia. Free Radic Biol Med 31:1609–1615
107. Jaksch M (2004) Mitochondriale DNA-mutationen. In: Sperl W, Freisinger P (eds) Mitochondriale Encephalopathien im Kindesalter. SPS-Verlagsgesellschaft, Heilbronn, pp 59–67
108. Rothman S (1983) Synaptic activity mediates death of hypoxic neurons. Science 220:536–538
109. Soderkvist P et al (1996) Glutathione-s-transferase M1 null genotype as a risk modifier for solvent-induced chronic toxic encephalopathy. Scand J Work Environ Health 22:360–363
110. Brown G (2004) Pyruvatdehydrogenasekomplex - Defekte – ein Überblick. In: Sperl W, Freisinger P (eds) Mitochondriale Encephalomyopathien im Kindesalter. APS-Edition, Heilbronn, pp 32–46
111. Mishra O et al (2000) Hypoxia-induced generation of nitric oxide free radicals in cerebral cortex of newborn guinea pigs. Neurochem Res 25:1559–1565
112. Kuklinski B (2006) Das HWS trauma. Aurum-Verlag, Bielefeld
113. Bielicki JK, Forte TM (1999) Evidence that lipid hydroperoxidases inhibit plasma lecithin: cholesterol acyltransferase activity. J Lipid Res 40:948–954
114. Salvemini D, Billiar TR, Vodowatz Y (2001) Nitric oxide and inflammation. Birkhäuser-Verlag, Basel
115. Takemura S et al (1999) Hepatic cytochrome P450 is directly inactivated by nitric oxide, not by inflammatory cytokines, in the early phase of endotoxemia. J Hepatol 30:1035–1044
116. Ince Y (2008) Untersuchungen zur intrazellulären Zinkhomöostase: Effekt von nitrosativem Stress und Hitze-Schock. Dept of Biochemistry and Molecular Biology, Heinrich- Heine-Universität Düsseldorf, Düsseldorf
117. Menegon A (1998) Parkinson's disease, pesticides and glutathion transferase polymorphisms. Lancet 362:1344–1346
118. Ozawa T et al (1991) Patients with idiopathic cardiomyopathy belong to the same mitochondrial DNA gene family of Parkinson's disease and mitochondrial encephalomyopathy. Biochem Biophys Res Comm 177:518–525
119. Birkmayer JGD, Vrecko C, Volc D (1993) Nicotinamid adenin dinucleotide (NADH) – a new therapeutic approach to Parkinson's disease. Comparison of oral and parenteral application. Acta Neurol Scand 87(Suppl 146):32–35
120. Shults CW et al (2002) Effects of coenzyme Q10 in early Parkinson disease: evidence of slowing of the functional decline. Arch Neurol 59:1541–1550

121. Novelli A et al (1988) Glutamate becomes neurotoxic via the N-methyl-D-aspertate receptor when intracellular energy levels are reduced. Brain Res 451:205–212
122. McMahon SB, Lewin GR, Wall PD (1993) Central hyperexcitability triggered by noxious inputs. Curr Opin Neurobiol 3:602–610
123. Joseph EK, Levine JD (2006) Mitochondrial electron transport in models of neuropathic and inflammatory pain. Pain 121:105–114
124. Bonkowsky HL et al (1975) Porphyrin synthesis and mitochondrial respiration in acute intermittent porphyria: studies using cultured human fibroblasts. J Lab Clin Med 85:93–102
125. Ferrer MD et al (2010) Impaired lymphocyte mitochondrial antioxidant defences in variegate porphyria are accompanied by more inducible reactive oxygen species production and DNA damage. Br J Haematol 149:759–767
126. Cordero MD et al (1998) Mitochondrial dysfunction and mitophagy activation in blood mononuclear cells of fibromyalgia patients: implications in the pathogenesis of the disease. Morphologic aspects of fibromyalgia. Z Rheumatol 57:47–51
127. Cordero MD et al (2010) Mitochondrial dysfunction and mitophagy activation in blood mononuclear cells of fibromyalgia patients: implications in the pathogenesis of the disease. Arthritis Res Ther 12:R17
128. Pall ML (2006) The NO/ONOO- cycle as the cause of fibromyalgia and related illnesses: etiology, explanation and effective therapy. In: Pederson JA (ed) New research in fibromyalgia. Nova Biomedical Publishers, Inc, New York, pp 39–59
129. Bell DS (2007) Cellular hypoxia and neuro-immune fatigue. CeWingSpan Press Inc, Livermore
130. Maloney EM et al (2010) Chronic fatigue syndrome is associated with metabolic syndrome: results from a case-control study in Georgia. Metabolism: clinical and experimental. Psychosomatic Med 59:1351–13579
131. Lakhan SE, Kirchgessner A (2010) Gut inflammation in chronic fatigue syndrome. Nutr Metab 7:79–89
132. Müller-Höcker J, Jacob U, Seibel P (1998) Hashimoto thyroiditis is associated with defects of cytochrome-c oxidase in oxyphil Askanazy cells and with the common deletion (4,977) of mitochondrial DNA. Ultrastruct Pathol 22:91–100
133. Vallas M et al (1994) Prevalence of antimitochondrial antibodies in women with Hashimoto's thyroiditis. Press Med 23:1117–1120
134. Williams D, Geraci A, Simpson DM (2001) AIDS and AIDS-treatment neuropathies. Curr Neurol Neurosci Rep 1:533–538
135. Teodor A, Teodor D, Luca V (2004) Side effects of antiretroviral therapy. Rev Med Chir Soc Med Nat Iasi 108:23–26
136. Montessori V et al (2004) Adverse effects of antiretroviral therapy for HIV infection. CMAJ 170:229–238
137. Heales SJ et al (1999) Nitric oxide, mitochondria and neurological disease. Biochim Biophys Acta 1410:215–228
138. Stewart VC, Heales SJR (2003) Nitric oxide-induced mitochondrial dysfunction: implications for neurodegeneration. Free Radic Biol Med 34:287–303
139. Hashiatni H, Lang R, Suzuki H (2010) Role of perinuclear mitochondria in the spatiotemporal dynamics of spontaneous Ca2+ waves in interstitial cells of Cajal-like cells of the rabbit urethra. Br J Pharmacol 161:680–694
140. Renström Koskela LNP, Wiklund J (2007) Nitric oxide in the painful bladder/interstitial cystitis. Urol Urogynäkol 14:18–19
141. Kovalenko AN, Loganovsky KN (2001) The chernobyl catastrophe-consequences on human health. Ukrainian Med J 6:XI–XII
142. Castro L et al (1998) Nitric oxide and peroxynitrite dependent aconitase inactivation and iron-regulatory protein. I. activation in mammalian fibroblasts. Arch Biochem Biophys 369:215–224
143. Ashton E et al (2011) Why did high-dose rosuvastatin not improve cardiac remodeling in chronic heart failure? Mechanistic insights from the UNIVERSE study. Int J Cardiol 146:404–407

144. Wagstaff LR et al (2003) Statins associated memory loss: analysis of 60 case reports and review of the literature. Pharmacotherapy 23:871–880
145. Norman C et al (2004) Salicylic acid is an uncoupler agent that inhibits ATP synthesis by dissociating it from the electron transport system at one or more of the phosphorylation sites and inhibitor of mitochondrial electron transport. Plant Physiol 134:492–501
146. Blanche S et al (1999) Persistent mitochondrial dysfunction and perinatal exposure to antiretroviral nucleoside analogues. Lancet 354:1084–1089
147. Lebrecht D et al (2003) Time-dependent and tissue-specific accumulation of mtDNA and respiratory chain defects in chronic doxorubicin cardiomyopathy. Circulation 108:2423–2429
148. Brunmair B et al (2004) Thiazolidinediones, like metformin, inhibit respiratory complex I: a common mechanism contributing to their antidiabetic actions? Diabetes 53:1052–1059
149. Almotrefi AA, Dzimiri N (1992) Effects of beta-adrenoceptor blockers on mitochondrial ATPase activity in guinea pig heart preparations. Eur J Pharmacol 215:231–236
150. Wei YH et al (1985) Inhibition of the mitochondrial Mg2+−ATPase by propranolol. Biochem Pharmacol 34:911–917
151. Cocco T et al (2002) The antihypertensive drug carvedilol inhibits the activity of mitochondrial NADH-ubiquinone oxidoreductase. J Bioenerg Biomembr 34:251–258
152. Zaiton Z et al (1993) The effects of propranolol on skeletal muscle contraction, lipid peroxidation products and antioxidant activity in experimental hyperthyroidism. Gen Pharmacol 24:195–199
153. Kuncl RW, Meltzer HY (1979) Beta adrenergic-mediated myofibrillar disruption and enzyme efflux in an experimental myopathy related to isometric muscle activity. Exp Mol Pathol 31:113–123
154. Forfar JC, Brown GJ, Cull RE (1979) Proximal myopathy during beta-blockade. Br Med J 2:1331–1332
155. Goli AK et al (2002) Simvastatin-induced lactic acidosis: a rare adverse reaction? Clin Pharmacol Ther 72:461–464
156. Petersen KF et al (2004) Impaired mitochondrial activity in the insulin-resistant offspring of patient with type 2 diabetes. N Engl J Med 123:664–671
157. Vallance P et al (1992) Endogenous dimethyl-arginine as an inhibitor of nitric oxide synthesis. J Cardiovasc Pharmacol Ther 20:S60–S62
158. Böger RH et al (1998) Asymmetric dimethylarginine: a novel risk factor for endothelial dysfunction. Its role in hypercholesterolemia. Circulation 98:1842–1847
159. Schulman SP et al (2006) L-arginine therapy in acute myocardial infarction-The Vascular Interaction With Age in Myocardial Infarction (VINTAGE MI) randomized clinical trial. JAMA 295:58–64
160. Chan K et al (2005) Drug-induced mitochondrial toxicity. Expert Opin Drug Metab Toxicol 1:655–669
161. Xia T, Li N, Nel AE (2009) Potential health impact of nanoparticles. Annu Rev Public Health 30:137–150
162. Yazdi AS et al (2010) Nanoparticles activate the NLR pyrin domain containing 3 (Nlrp3) inflammasome and cause pulmonary inflammation through release of IL-1α and IL-1β. PNAS 107:19449–19454
163. Vasile B et al (2003) The pathophysiology of propofol infusion syndrome: a simple name for a complex syndrome. Intensive Care Med 29:1417–1425
164. Pryor WA, Squadrito GL (1995) The chemistry of peroxynitrite: a product from the reaction of nitric oxide with superoxide. Am J Physiol Lung Cell Mol Physiol 268:699–722
165. Oney JW et al (1996) Increasing brain tumor rates: is there a link to aspartame? J Neuropathol Exp Neurol 55:1115–1123
166. Academies TN (2005) 8. Dietary fats: total fat and fatty acids. In: FNB (ed) Dietary reference intakes for energy, carbohydrate, fiber, fat, fatty acids, cholesterol, protein, and amino acids (macronutrients). The National Academies Press, Wagington, DC, pp 422–541
167. Staff MC (2009) Trans fat is double trouble for your heart health

168. HPSC Group (2005) The effects of cholesterol lowering with simvastatin on cause-specific mortality and on cancer incidence in 20,536 high-risk people: a randomised placebo-controlled trial. BMC Med 3:1–21
169. Dworkin RH, Fields HL (2005) Fibromyalgia from the perspective of neuropathic pain. J Rheumatol 75(Suppl):1–5
170. Levine JD, Alessandri-Haber N (2007) TRP channels: targets for the relief of pain. Biochim Biophys Acta 1772:989–1003
171. See reference [85]
172. Foxton RH, Land JM, Heales SJR (2010) Tetrahydrobiopterin availability in Parkinson's and Alzheimer's disease; potential pathogenic mechanisms. Neurochem Res 32:751–756
173. Dissing IC et al (1989) Tetrahydrobiopterin and Parkinson's disease. Acta Neurol Scand 79:493–499
174. Tsai LY (1999) Psychopharmacology in autism. Psychosom Med 61:651–665
175. Curtius HC, Müldner H, Niederwieser A (1982) Tetrahydrobiopterin: efficacy in endogenous depression and Parkinson's disease. J Neural Transm 55:301–308
176. Bonafé L et al (2001) Diagnosis of dopa-responsive dystonia and other tetrahydrobiopterin disorders by the study of biopterin metabolism in fibroblasts. Clin Chem 47:477–485
177. Longo N (2009) Disorders of biopterin metabolism. J Inherit Metab Dis 32:333–342
178. Willis WD (2001) Role of neurotransmitters in sensitization of pain responses. Ann N Y Acad Sci 933:175–184
179. Garbem TR, Yukama H (2001) Common solvent toxicity: auto oxidation of respiratory redox-cyclers enforced by membrane derangement. Z Naturforsch 56:483–491
180. Schultz JB et al (1997) The role of mitochondrial dysfunction and neuronal nitric oxide in animal models of neurodegenerative diseases. Mol Cell Biochem 174:171–184
181. Doble A (1999) NMDA and neurogenative conditions (reviews). Pharmacol Ther 81:163–221
182. Dawson VL, Dawson TM (1996) Nitric oxide neruotoxicity. J Chem Neuroanat 10:179–190
183. Albenzi BC (2001) Models of brain injury and alterations in synaptic neuroplasticity. J Neurosci Res 65:279–283
184. Mayhan WG (2000) Nitric oxide donor-induced increase in permeability of the blood-brain barrier. Brain Res 866:101–108
185. Beckman JS, Crow JP (1993) Pathologic implications of nitric oxide, superoxide and peroxynitrite formation. Biochem Soc Trans 21:330–333
186. Khatsenko OG et al (1993) Nitric oxide is a mediator of the decrease in cytochrome p450-dependent metabolism caused by immunostimulants. Proc Natl Acad Sci USA 90:11147–11151
187. Ziem GE (1999) Profile of patients with chemical injury and sensitivity. Int J Toxicol 18:401–409
188. Meggs W et al (1996) Prevalence and nature of allergy and chemical sensitivity in a general population. Arch Environ Health 51:275–282
189. Kailin E, Hastings A (1966) Cerebral disturbances from small amounts of DDT; a controlled study of MCS patients. Med Ann DC 35:519–524
190. Meggs W (1993) Neurogenic inflammation and sensitivity to environmental chemicals. Environ Health Perspect 101:234–238
191. Callender T et al (1993) Three-dimensional brain metabolic imaging in patients with toxic encephalopathy. Environ Res 60:295–319
192. Callender T, Morrow L, Subramanian K (1993) Evaluation of chronic neurological sequelae after acute pesticide exposure using SPECT brain scans. J Toxicol Environ Health 41:275–284
193. Ferdinandy P et al (2000) Peroxynitrite is a major contributor to cytokine-induced myocardial contractile failure. Circ Res 87:241–247
194. Russo A et al (2006) Bioflavonoids as antiradicals, antioxidants and DNA cleavage protectors. Cell Biol Toxicol 16:91–98
195. Vanella A et al (2000) L-propiony L-carnitine as superoxide scavenger, antioxidant, and DNA cleavage protector. Cell Biol Toxicol 16:99–104

196. Yamadaa J et al (2003) Cell permeable ROS scavengers, tiron and tempol, rescue PC12 cell death caused by pyrogallol or hypoxia/reoxygenation. Neurosci Res 45:1–8
197. Siegel D et al (2004) NAD(P)H:quinone oxidoreductase 1: role as a superoxide scavenger. Mol Pharmacol 65:1238–1247
198. Zhu H et al (2007) The highly expressed and inducible endogenous NAD(P)H:quinone oxidoreductase 1 in cardiovascular cells acts as a potential superoxide scavenger. Cardiovasc Toxicol 7:202–211
199. Kishida KT, Klann E (2007) Sources and targets of reactive oxygen species in synaptic plasticity and memory. Antioxid Redox Signal 9:233–244
200. Cole S, Vassar R (2007) The Alzheimer's disease β-secretase enzyme, BACE1. Mol Neurodegener 2:22–31
201. Li R (2004) Amyloid β peptide load is correlated with increased β-secretase activity in sporadic Alzheimer's disease patients. Proc Natl Acad Sci USA 101:3632–3637
202. Yang L-B (2003) Elevated β-secretase expression and enzymatic activity detected in sporadic Alzheimer disease. Nat Med 9:3–4
203. Vassar R (2004) BACE1: the β-secretase enzyme in Alzheimer's disease. J Mol Neurosci 23:105–113
204. Andrekopoulos C et al (2004) Bicarbonate enhances alpha-synuclein oligomerization and nitration: Intermediacy of carbonate radical anion and nitrogen dioxide radical. Biochem J 378:435–447
205. Gao H-M et al (2008) Neuroinflammation and oxidation/nitration of α-synuclein linked to dopaminergic neurodegeneration. J Neurosci 28:7687–7698
206. Reynolds MR, Berry RW, Binder LI (2005) Site-specific nitration and oxidative dityrosine bridging of the tau protein by peroxynitrite: implications for Alzheimer's disease. Biochemistry 44:1690–1700
207. Reynolds MR et al (2006) Peroxynitrite mediated tau modifications stabilize preformed filaments and destabilize microtubules through distinct mechanisms. Biochemistry 45:4314–4326
208. Reynolds MR et al (2006) Tau nitration occurs at tyrosine 29 in the fibrillar lesions of Alzheimer's disease and other tauopathies. J Neurosci 26:10636–10645
209. Etévez AG et al (1998) Nitric oxide and superoxide contribute to motor neuron apoptosis induced by trophic factor deprivation. J Neurosci 18:923–931
210. Sanelli T, Strong MJ (2007) Loss of nitric oxide-mediated down-regulation of NMDA receptors in neurofilament. Free Radic Biol Med 42:143–151
211. Jaiswal MK et al (2009) Impairment of mitochondrial calcium handling in a mtSOD1 cell culture model of motoneuron disease. BMC Neurosci 10:64–79
212. Lucas DR, Newhouse JP (1957) The toxic effect of sodium L-glutamate on the inner layers of the retina. AMA Arch Ophthalmol 58:193–201
213. Olney JW (1969) Brain lesions, obesity, and other disturbances in mice treated with monosodium glutamate. Science 164:719–721
214. Hughes JR (2009) Alcohol withdrawal seizures. Epilepsy Behav 15:92–97
215. Manev H et al (1989) Delayed increase of Ca2+ influx elicited by glutamate: role in neuronal death. Mol Pharmacol 36:106–112
216. Kim AH, Kerchner GA, Choi DW (2002) Blocking excitotoxicity. In: Marcoux FW, Choi DW (eds) CNS neuroprotection. Springer, New York, pp 3–36
217. Camacho A, Massieu L (2006) Role of glutamate transporters in the clearance and release of glutamate during ischemia and its relation to neuronal death. Arch Med Res 37:11–18
218. Fujikawa DG (2005) Prolonged seizures and cellular injury: understanding the connection. Epilepsy Behav 7:S3–S11
219. Temple MD, O'Leary DM, Faden AI (2001) The role of glutamate receptors in the pathophysiology of traumatic CNS injury. In: Miller LP, Hayes RL, Newcomb JK (eds) Head trauma: basic, preclinical, and clinical directions. Wiley, New York, pp 87–113
220. Clements JD et al (1992) The time course of glutamate in the synaptic cleft. Science 258:1498–1501

221. Kerr JF, Wyllie AH, Currie AR (1972) Apoptosis: a basic biological phenomenon with wide-ranging implications in tissue kinetics. Br J Cancer 26:239–257
222. Hinkle JL, Bowman L (2003) Neuroprotection for ischemic stroke. J Neurosci Nurs 35:114–118
223. Stavrovskaya IG, Kristal BS (2005) The powerhouse takes control of the cell: is the mitochondrial permeability transition a viable therapeutic target against neuronal dysfunction and death? Free Radic Biol Med 38:687–697
224. Siegel GJ et al (1999) Basic neurochemistry: molecular, cellular, and medical aspects, 6th edn. Lippincott, Williams & Wilkins, Philadelphia
225. Zhang X et al (2005) Bench-to-bedside review: Apoptosis/programmed cell death triggered by traumatic brain injury. Crit Care Med 9:66–75
226. Hardingham GE, Fukunaga Y, Bading H (2002) Extrasynaptic NMDARs oppose synaptic NMDARs by triggering CREB shut-off and cell death pathways. Nat Neurosci 5:405–414
227. Acheson A et al (1995) A BDNF autocrine loop in adult sensory neurons prevents cell death. Nature 374:450–453
228. Huang EJ, Reichardt LF (2001) Neurotrophins: roles in neuronal development and function. Annu Rev Neurosci 24:677–736
229. Bekinschtein P et al (2008) BDNF is essential to promote persistence of long-term memory storage. Proc Natl Acad Sci USA 105:211–216
230. Bell IR et al (1999) Neural sensitization model for multiple chemical sensitivity: overview of theory and empirical evidence. Toxicol Health 15:295–304
231. Bell IR, Baldwin CM, Schwartz GE (2001) Sensitization studies in chemically intolerant individuals: implications for individual difference research. Ann N Y Acad Sci 933:38–47
232. Ballantyne B, Marrs TC, Syversen T (2009) Chapter XX. Multiple chemical sensitivity: toxicological questions and mechanisms. In: General and applied toxicology. Wiley, London
233. Meggs WJ (1997) Hypothesis for induction and propagation of chemical sensitivity based on biopsy studies. Environ Health Perspect 105:473–478
234. Heuser G, Wojdani A, Heuser S (1992) Diagnostic markers of multiple chemical sensitivity. In: Commission on Life Sciences (ed) Multiple chemical sensitivities: addendum to biologic markers in immunotoxicology. National Academy Press, Washington, DC
235. Pall ML, Anderson JH (2004) The vanilloid receptor as a putative target of diverse chemicals in multiple chemical sensitivity. Arch Environ Health 59:363–372
236. Hu CL, Xiang JZ, Hu FF (2008) Vanilloid receptor TRPV1, sensory C-fibers, and activation of adventitial mast cells. A novel mechanism involved in adventitial inflammation. Med Hypotheses 71:102–103
237. Ashford NA, Miller CS (1998) Chemical exposures: low levels and high stakes. Van Nostrand Reinhold/Wiley, New York
238. Bell IR et al (1998) Differential resting quantitative electroencephalographic alpha patterns in women with environmental chemical intolerance, depressives, and normals. Biol Psychiatry 43:376–388
239. Kimata H et al (1991) Nerve growth factor specifically induces human IgG4 production. Eur J Immunol 21:137–141
240. Johansson A, Millqvist E, Bende M (2010) Relationship of airway sensory hyperreactivity to asthma and psychiatric morbidity. Ann Allergy Asthma Immunol 105:20–23
241. Saito M et al (2005) Symptom profile of multiple chemical sensitivity in actual life. Psychosom Med 67:318–325
242. Joffres MR, Sampalli T, Fox RA (2004) Physiologic and symptomatic responses to low-level substances in individuals with and without chemical sensitivities: a randomized controlled blinded pilot booth study. Health Perspect 113:1178–1183
243. Peckerman A et al (2003) Abnormal impedance cardiography predicts symptom severity in chronic fatigue syndrome. Am J Med Sci 326(2):55–60
244. Agius LM (2009) Hypothesis and dynamics in the pathogenesis of neurodegenerative disorders. Bentham Science Publishers, Sharjah

245. Biswas S et al (1997) Selective inhibition of mitochondrial respiration and glycolysis in human leukaemic leucocytes by methylglyoxal. Biochem J 323:343–348
246. Jones OAH, Maguire ML, Griffin JL (2008) Environmental pollution and diabetes: a neglected association. Lancet 371:287–288
247. Naughton DP, Petróczi A (2008) Heavy metal ions in wines: meta-analysis of target hazard quotients reveal health risks. Chem Cent J 2:22–29
248. Legret M, Pagotto C (2006) Heavy metal deposition and soil pollution along two major rural highways. Environ Technol 27:247–254
249. O'Connor JC, Chapin RE (2003) Critical evaluation of observed adverse effects of endocrine active substances on reproduction and development, the immune system, and the nervous system. Pure Appl Chem 75:2099–2123
250. Okada H et al (2008) Direct evidence revealing structural elements essential for the high binding ability of bisphenol A to human estrogen-related receptor-gamma. Environ Health Perspect 116:32–38
251. vom Saal FS, Myers JP (2008) Bisphenol A and risk of metabolic disorders. JAMA 300:1353–1355
252. La Du BN (1992) Human serum paraoxonase/arylesterase. In: Kalow W (ed) Pharmacogenetics of drug metabolism. Pergamon Press, New York, pp 51–91
253. Wolfe F et al (1990) The American College of Rheumatology 1990 criteria for the classification of fibromyalgia: report of the multicenter criteria committee. Arthritis Rheum 33:160–172
254. Wolfe F (1989) Fibromyalgia: the clinical syndrome. Rheum Dis Clin North Am 15:1–18
255. Clauw DJ et al (1997) The relationship between fibromyalgia and interstitial cystitis. J Psychiatr Res 31:125–131
256. Simms RW, Goldenberg DL (1988) Symptoms mimicking neurologic disorders in fibromyalgia syndrome. J Rheumatol Suppl 15:1271–1273
257. Glass JM (2006) Cognitive dysfunction in fibromyalgia and chronic fatigue syndrome: new trends and future directions. Curr Rheumatol Rep 8:425–429
258. Buskila D, Cohen H (2007) Comorbidity of fibromyalgia and psychiatric disorders. Curr Pain Headache Rep 11:333–338
259. Schweinhardt P, Sauro KM, Bushnell MC (2008) Fibromyalgia: a disorder of the brain? Neuroscientist 14:415–421
260. Bagis S et al (2005) Free radicals and antioxidants in primary fibromyalgia: an oxidative stress disorder? Rheumatol Int 25:188–190
261. Ozgocmen S et al (2006) Current concepts in the pathophysiology of fibromyalgia: the potential role of oxidative stress and nitric oxide. Rheumatol Int 26:585–597
262. Cordero MD et al (2009) Coenzyme Q10 distribution in blood is altered in patients with fibromyalgia. Clin Biochem 42:732–735
263. See reference [260]
264. Ozgocmen S et al (2006) Antioxidant status, lipid peroxidation and nitric oxide in fibromyalgia: etiologic and therapeutic concerns. Rheumatol Int 26:598–603
265. Teitelbaum JE, Johnson C, St Cyr J (2006) The use of D-ribose in chronic fatigue syndrome and fibromyalgia: a pilot study. J Altern Complement Med 12:857–862
266. Zhao X et al (1995) Oxygen free radicals may be involved in the pathogenesis of fibromyalgia. J Musculoske Pain 3:111
267. Eisinger J et al (2002) Lipid and protein peroxidations in fibromyalgia. Myalgies Int 3:37–42
268. Light AR et al (2008) Moderate exercise increases expression for sensory, adrenergic and immune genes in chronic fatigue syndrome patients, but not in normal subjects. J Pain 10:1099–1112
269. Akbar A et al (2010) Expression of the TRPV1 receptor differs in quiescent inflammatory bowel disease with or without abdominal pain. Gut 59:767–774
270. Park S et al (2008) Pregabalin and gabapentin inhibit substance P-induced NF-kappaB activation in neuroblastoma and glioma cells. J Cell Biochem 105:414–423

271. Goettl VM et al (2002) Reduced basal release of serotonin from the ventrobasal thalamus of the rat in a model of neuropathic pain. Pain 99:359–367
272. Pore RS (1984) Detoxification of chlordecone poisoned rats with chlorella and chlorella derived sporopollenin. Drug Chem Toxicol 7:57–71
273. See reference [265]
274. Chambers D et al (2006) Interventions for the treatment, management and rehabilitation of patients with chronic fatigue syndrome/myalgic encephalomyelitis: an updated systematic review. J R Soc Med 99:506–520
275. Sinatra ST (1996) Heartbreak & heart disease: a mind/body prescription for healing the heart. Keats Publishing, New Canaan
276. Raveraa S et al (2009) Evidence for aerobic ATP synthesis in isolated myelin vesicles. Int J Biochem Cell Biol 41:1581–1591
277. Orsi A, Sherman O, Woldeselassie Z (2001) Simvastatin-associated memory loss. Pharmacotherapy 21:767–769
278. See reference [144]
279. King DS et al (2003) Cognitive impairment associated with atorvastatin and simvastatin. Pharmacotherapy 23:1663–1667
280. Pond CM (2009) Paracrine provision of lipids in the immune system. Curr Immunol Rev 5:150–160
281. Pond C (1987) Fat and figures. New Sci 4:62–66
282. Taher S et al (2004) Short sleep duration is associated with reduced leptin, elevated ghrelin, and increased body mass index. PLoS Med 1:e62
283. Alhola P, Polo-Kantola P (2007) Sleep deprivation: impact on cognitive performance. Neuropsychiatr Dis Treat 5:553–567
284. Gottlieb DJ et al (2005) Association of sleep time with diabetes mellitus and impaired glucose tolerance. Arch Intern Med 165:863–867
285. Freye E et al (1998) The opioid tramadol demonstrates excitatory properties of non-opioid character - a preclinical study using alfentanil as a comparison. Schmerz 12(1):19–24
286. Bibbins-Domingo K et al (2007) Adolescent overweight and future adult coronary heart disease. New Engl J Med 357:2371–2379
287. Mietus-Snyder ML, Lustig RL (2008) Childhood obesity: adrift in the "limbic triangle". Ann Rev Med 59:147–162
288. Dufault R et al (2009) Mercury from chlor-alkali plants: measured concentrations in food product sugar. Environ Health 8:2–8
289. Busserolles J et al (2002) Substituting honey for refined carbohydrates protects rats from hypertriglyceridemic and prooxidative effects of fructose. J Nutr 132:3379–3382
290. Huang D et al (2010) Fructose induces transketolase flux to promote pancreatic cancer growth. Cancer Res 70:6368–6376
291. Teitelbaum J (2010) Beat sugar addiction now! the cutting-edge program that cures your type of sugar addiction and puts you on the road to feeling great - and losing weight. Fair Winds Press, Beverly
292. Ley RE et al (2006) Microbial ecology: human gut microbes associated with obesity. Nature 444:1022–1023
293. Krammer H, Schlieger F, Singer MV (2005) Therapeutic options of chronic constipation [Article in German]. Internist 46:1331–1338
294. Kullak K (1997) Bedeutung der Darmflora für den Menschen. Medizin und Ernährung 6(Suppl):56–59
295. Haenel H, Bendig J (1975) Intestinal flora in health and disease. Prog Food Nutr Sci 1:21–64
296. McCauley R, Kong S-E, Hall J (1998) Review: glutamine and nucleotide metabolism within enterocytes. J Parenter Enteral Nutr 22:105–111
297. Hingorani AD, Chan NN (2001) D-lactate encephalopathy. Lancet 358:1814
298. Hadis U et al (2011) Intestinal tolerance requires Gut homing and expansion of FoxP3 regulatory T cells in the lamina propria. Immunity 34:237–246
299. Brenner C, Fuks F (2006) DNA methyltransferases: facts, clues, mysteries. Curr Top Microbiol Immunol 301:45–66

References

300. Brosnan JT et al (2004) Methylation demand: a key determinant of homocysteine metabolism. Acta Biochim Pol 51:405–413
301. Zaghloul AA et al (2002) Bioavailability assessment of oral coenzyme Q10 formulations in dogs. Drug Dev Ind Pharm 28:1195–1200
302. Jonassen T, Clarke CF (2000) Isolation and functional expression of human COQ3, a gene encoding a methyltransferase required for ubiquinone biosynthesis. J Biol Chem 275:12381–12387
303. Kim YI et al (1997) NMDA receptors are important for both mechanical and thermal allodynia from peripheral nerve injury in rats. Neuroreport 8:2149–2153
304. Hirata F, Axelrod J (1980) Phospholipid methylation and biological signal transmission. Science 209:1082–1090
305. Kunz-Schughart LA et al (2004) The use of 3-D cultures for high-throughput screening: the multicellular spheroid model. J Biomol Screen 9(4):273–285
306. Van Konynenburg RA (2004) Is glutathione depletion an important part of the pathogenesis of chronic fatigue syndrome? In: Seventh international AACFS conference, Madison
307. Jiang Y et al (2009) Aberrant DNA methylation is a dominant mechanism in MDS progression to AML. Blood 113:1315–1325
308. Lemle MD (2009) Hypothesis: chronic fatigue syndrome is caused by dysregulation of hydrogen sulfite metabolism. Med Hypotheses 72:108–109
309. Oelgoetz AW, Oelgoetz PA, Wittenkind J (1935) The treatment of food allergy and indigestion of pancreatic origin with pancreatic enzymes. Am J Dig Dis Nutr 2:422–426
310. Grehan MJ et al (2010) Durable alteration of the colonic microbiota by the administration of donor fecal flora. J Clin Gastroenterol 44:551–561

Chapter 8
Points to Consider in Therapy of Mitochondropathy

8.1 Phospholipids Restoring Cellular Function

The cell membrane is a bilipid layer of two fatty acid tails facing each other. The outer layer of the cell contains mostly the phospholipids phosphatidylcholine and sphingomyelin, while the inner layer contains predominantly phosphatidylserine, phosphatidylinositol, and phosphatidylethanolamine. In the inner layer of the membrane are over 100 different complex protein molecules (Fig. 8.1). This inner layer is quite impermeable compared to the outer layer of the membrane, which has a high degree of permeability and fluidity. The two fatty acid tails of the membrane are in a constant state of intense movement, vibrating at millions of times a second! The cell membrane is the gateway to the cell. It is absolutely vibrant with activity. The amount and type of long-chain fatty acids in the diet tremendously alter the composition of the membrane. For example, the types of fatty acids consumed and incorporated into the cell membranes affect many functions dependent upon the membrane such as:

1. fluidity,
2. transport through ion channels,
3. the binding of hormones,
4. regulation of receptors,
5. signaling, and
6. the production of the communication chemicals, called eicosanoids.

Incorrect amounts and/or types of long chain fatty acids can likewise lead to a series of consequences cascading toward serious health problems. When the membrane is compromised with too many trans fats, found in margarine and most prepared baked goods, and too high a percentage of cholesterol, the structure of the cell membrane becomes more rigid, and the actual shape of the cell is changed (Fig. 8.1). These less flexible membranes also have less fluidity. Proper fluidity is needed for the ease of transport of nutrients into the cell, as well as the equally important function of expulsion of waste materials out of the cell.

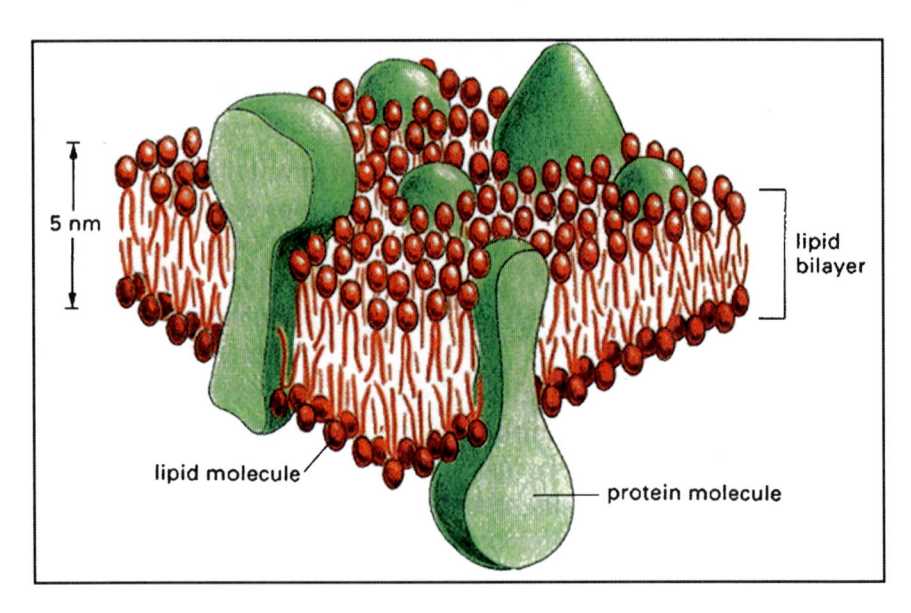

Fig. 8.1 Cell boundary is cell membrane – a phospholipid bilayer (double-layer). It is composed mainly of phospholipid molecules, that have a hydrophobic (fatty acids) part and a hydrophilic part. The molecules form a bilayer so that all the hydrophobic tails of the molecules are inside the bilayer and the hydrophilic heads are facing outside to make contact with water

The late Dr. Hans A. Nieper of Germany, focused much of his innovative work on restoring the health of the membrane through the use of the naturally occurring phospholipid metabolite called calcium EAP (phosphoric acid mono-{2-aminoethyl}-ester). Dr. Nieper believed that inadequate amounts of this phospholipid in the intracellular membrane were the cause of many illnesses. He utilized both an oral and an intravenous form of this phospholipid complexed with the minerals calcium, magnesium and/or potassium. He found it to improve the condition of many diseases: multiple sclerosis, diabetic nephropathy, diabetic retinopathy, immune thrombocytopathy, asthma, degenerative lung and small vessel diseases, as well as many other health disorders. This age-related change in the lipid composition of heart muscle cells of animals has been positively altered with injections of PC. The exciting results of animal studies have led to the administration of intravenous infusions of essential phospholipids in the form of PC to patients exhibiting various cardiovascular problems. One of the most challenging difficulties we face today is that of detoxification of the brain and CNS after exposure to neurotoxins from infection, toxic mold, chemicals, and heavy metals. Polymorphism of methylation enzymes may occur after toxic exposure, impacting DNA expression, detoxification, neurotransmitter balance, myelination, growth and rejuvenation in both adult and pediatric populations. Once the toxic burden becomes overwhelming as the patient's methylation and its EFA status deteriorates, all other cellular functions are compromised. The PK Protocol™ and Detoxx™ System, developed by Dr. Patricia Kane, provides health care professionals safe and efficient detoxification methods through targeted supplementation and intravenous therapy, by supporting membrane phospholipids, methylation and sulfation which are crucial for positive patient outcomes [1]. The approach is to detoxify from a cell membrane perspective, with respect to essential fatty acid metabolism, establishing phospholipid stability. Thus the health of the cell, the body and brain are stabilized. The PK Protocol™ is a new, clinically proven method to reach the systemic nature of neurotoxic and toxic syndromes and is applicable for patients presenting refractory heavy metal burdens, microbial infections, toxic mold exposure, CFIDS, Lyme, Fibromyalgia, MS, IBS, Autism, Depression, ALS, Psychosis, Stroke, Environmental Illness, Mood Disorders, Hypercoagulation, Parkinson's, Cardiovascular Disease, Hepatitis C and to increase longevity. Alternative physicians have found that besides the general improvement of one's health, intravenous treatments given with a PC product called Plaquex® (www.plaquex.ch), may help some cardiovascular patients avoid bypass surgery, angioplasty or diabetic amputations.

The Plaquex® product contains:

2,500 mg of essential phospholipids (70% phosphatidylcholine),
1,250 mg of deoxycholic acid,
10 mg vitamin E,
10 mg nicotinic acid,
10 mg adenosine triphosphate,
450 mg benzyl alcohol, and
120 mg of ethanol.

The Plaquex® is administered slowly intravenously over at least 90 min. Another PC product for intravenous application is LipoStabil® from Rhone-Poulanc/France with an oral formulation called extracted pure phospholipid (EPL) from the same company both of which are available worldwide.

- A series of 20–40 of these infusions has been documented to affect human ailments (especially Parkinson, Alzheimer and ALS) in several positive ways. The infusions give quicker results, and may be the preferred choice for people who are at significant cardiac risk. Dr. Kane has found through examination of the fatty acid profile in red blood cells that patients with neurodegenerative illnesses often have what are called renegade and odd chain fatty acids in their membranes that are the result of various biotoxins. These renegade fatty acids and odd chain fatty acids can disrupt the very structure of the membranes and consequently also affect function.
- The first and primary goal in the protocol is to facilitate membrane metabolism of these renegade fatty acids, thereby cleaning up the membrane, so to speak, and restoring it with appropriate fatty acids and phospholipids. Phospholipid exchange is a technique for supplying the correct proportion of fats and oils in a

bioavailability form to replenish cell membranes and membranes within cells. Dr. Patricia Kane, who has seen remarkable clinical results, has pioneered this technique. She uses intravenous therapy, but good results are also possible with oral therapy [1]. The underlying principle is based on the following:

1. The basic membrane structure is made up of phosphatidylcholine. The best source for replenishment of this structure is lecithin, available as egg, soya or sunflower lecithin.
2. Membranes need the right proportion of Ω-6 to Ω-3 oils, i.e. 4:1. Hemp oil is very close to this ratio at 3.8:1. Since hemp oil is not the same as flax seed oil, one adds a small amount of sunflower oil, say 5%.
3. Small amount of Eskimo oil, not being necessary if one already takes VegEPA®.
4. The perfect fuel for brain cells is coconut oil, which is rich in medium chain triglycerides (MCT).
5. If there is poor digestion or poor gall bladder function for any reason then adding bile salts may be very helpful by further emulsifying fats and facilitating digestion and absorption.

Providing an abundance of clean oils helps to displace oils in the brain, which hold polluting heavy metals, pesticides, volatile organic compounds (VOCs) etc. That is to say, these "clean" fats will displace "dirty" fats and also help detoxification. This protocol also utilized liberal amounts of ground flaxseeds, and balanced EFA oil, which contains a ratio of 4:1 of Ω-6 to Ω-3, which has been determined by recent research to be the proper ratio. The injections of PC help to restore the membranes of the cells throughout the body as well as being vital to the liver's ability to detoxify. The PK Protocol increases the metabolism from the membranes of the renegade fatty acids, heavy metals and biotoxins, and the bloodstream transports these compounds to the liver to be emptied into the bile. In the liver, these toxins bind with the fats in the bile and are dumped into the intestine, but some toxins end up being reabsorbed from the intestine and deposited back in the liver again. This cycle of reabsorption must be broken in order for the body to be relieved of its burden of toxins. The detox portions of the PK Protocol emphasizes the role of bile in detoxification and the necessity to increase bile production as well as toxin binding in the intestine with supplements of chlorella for complete removal from the body.

- In order to accomplish this, the PK Protocol uses supplementation of bile salts, as well as fat digesting enzymes. A special product is used to assist in detoxification: butyrate salts of calcium and magnesium, which are short chain fatty acids that are the number one fuel for the cells that line the colon. Butyrate fatty acids have been demonstrated to reduce ammonia levels, which are raised in some patients with poor liver detoxification abilities, contributing to brain fog. Besides providing fuel for the colon mucosal cells, it has been suggested that butyrate can help to clear renegade fatty acids from the liver.
- To facilitate greater detoxification of these odd chain and extra long fatty acids from the body, there are several special detox protocols used. A liver flush uses a special group of supplements for a few weeks that prepare the body for "the big plunge" of liver detoxification. They also suggest something called

"the intestinal peel", and the "butyrate detox enema"; all protocols that definitely require some professional health care guidance.
- This detoxification protocol works at building and healing the membrane, while simultaneously emphasizing detoxification. Liquid trace minerals and an oral electrolyte concentrate are also an important part of the protocol, as well as liberal amounts of water.
- Raising the level of glutathione, both through injection and/or diet, is also part of the protocol. Glutathione facilitates the removal of heavy metals like mercury, as well as other neurotoxins, and is of primary importance in one phase of liver detoxification. To raise levels of glutathione through oral means, the protocol uses supplementation with undenatured, bioactive whey protein, as well as selenium, lipoic acid, milk thistle and other supplements.

8.1.1 Switching Off the Sugar Addiction Gene

In the evolutionary sense, human mankind evolved eating so called Cave Man Diet. But every so often there would be time with an excess of carbohydrates available, perhaps the banana tree that ripens. At former times the only way primitive mankind could take advantage of this would be by eating a lot ! Also, because bananas do not store well and there was the risk of someone else eating them. Once starting to eat, the carbohydrate addiction gene is switched on and the individual would go on eating until there is none left. After engulfing all bananas this subject would gain weight, which would give him energy for other tasks such as building a house, fighting the neighbors, or protecting the wife from another male! This carbohydrate gene is switched on when just 3% of the diet contains sugar. The trouble with modern Western diets is they are high in sugar and refined carbohydrate and so the sugar addiction gene is permanently switched on. When however, the sugar intake is low, the gene is switched off and there is a stop in craving. Any concept of sugar addiction is complicated by a lack of consensus on the actual definition of addiction. There has been a reference to the idea of sugar addiction in the popular literature for a number of years. In 1998, Kathleen DesMaisons [2] outlined the concept of sugar addiction as a measurable physiological state caused by activation of mu-opioid receptors in the brain. Her work extracted data from studies done by Blass [3], showing that sugar acted as an analgesic drug whose effects could be blocked by a morphine antagonist. Acting on years of anecdotal evidence from her work in the field of addiction, DesMaisons noted that dependence on sugar followed the same track outlined in the DSM IV for other drugs of abuse such as opioids, amphetamines or even cocaine. Since that time, a growing body of laboratory evidence [4] has corroborated DesMaisons hypothesis. Bart Hoebel at Princeton began showing the neurochemical effects of sugar, outlining that sugar might serve as a gate drug for other addictable agents. He also stated that sugar affects are similar to opioids and dopamine in the brain, and thus might be expected to have addictive potential. The references "bingeing," "withdrawal," "craving" and "cross-sensitization" are each given operational definitions and demonstrated behaviorally with sugar bingeing as the reinforcer. These behaviors are then related to neurochemical changes in the brain that also occur with addictive drugs. Neural adaptations include changes in dopamine and opioid receptor binding, enkephalin mRNA expression and dopamine and acetylcholine release in the nucleus accumbens. Recent behavioral tests in rats further back the idea of an overlap between sweets and drugs. Drug addiction often includes three steps. A person will increase his intake of the drug, experience withdrawal symptoms when access to the drug is cut off and then face an urge to relapse back into drug use. Rats on sugar have similar experiences. Researchers withheld food for 12 h and then gave rats food plus sugar water. This created a cycle of binging where the animals increased their daily sugar intake until it doubled. When researchers either stopped the diet or administered an opioid blocker the rats showed signs common to drug withdrawal, such as teeth chattering and wet dog shaking. Early findings also indicate signs of relapse. Rats weaned off sugar repeatedly pressed a lever that previously dispensed the sweet solution.

8.1.2 Disturbed Sleep, a Common Symptom of Hypoglycemia

When blood glucose level falls for any reason, glycogen stores in the liver are mobilized for restoration. Another rapid and very effective way in which the body repletes the low glucose is by conversion of short chain fatty acids to glucose. In a healthy person on a good balanced diet the only time this is of importance is during the night because of the long break between food intake. Short chain fatty acids are used to restore circulating glucose and prevent a fall below the person's usual fasting glucose level. Short chain fatty acids are made in the gut by bacteria fermenting fibers (and starch that escapes small intestinal digestion). The production is maximized at about 3 h after food intake. That is to say, short chain fatty acids are highly protective against the dips we see in blood sugar. Therefore, the key symptom of hypoglycemia is disturbed sleep. This occurs typically at 2–3 a.m., when the blood sugar level falls and there are insufficient short chain fatty acids to maintain the glucose level. Low blood sugar is potentially serious to the brain, which can only survive on sugar and, therefore, there is an adrenalin reaction to bring the blood sugar back to normal level, however at the same time by waking up the sleeper. In this respect alcohol when consumed in the evening results in a common symptom of hypoglycemia and sleeplessness. And although, initially alcohol helps one to go to sleep, he then wakes up in the small hours with a rebound hypoglycemia.

8.1.3 Insulin a Stress Hormone Activated by High Blood Glucose Levels

There is a final twist to the hypoglycemia tale, which complicates the situation further. When someone becomes stressed for whatever reason, there is a release of stress hormones in order to allow to cope with that stressful situation. Insulin is such a stress hormone and has the effect of increasing the uptake of glucose into the cells. This produces a drop in blood glucose levels and also results in hypoglycemia. Therefore, hypoglycemia can be both a cause of stress and the result of stress, just another one of those vicious cycles that are so often seen in disease states.

8.1.4 Monitoring Symptoms of Hypoglycemia

Hypoglycemic symptoms and manifestations can be divided into those produced by the counter regulatory hormones (epinephrine/ adrenaline and glucagon) triggered by the falling glucose level, and the neuroglycopenic effects produced by a deprived brain glucose level.

- Adrenergic manifestations of a low glucose level
  Shakiness, anxiety, nervousness, palpitations, tachycardia sweating, feeling of warmth, pallor, coldness, clamminess, dilated pupils (mydriasis), feeling of numbness, "pins and needles" (paresthesia)
- Glucagon manifestations of a low glucose level
  Hunger, borborygmus (stomach growling), nausea, vomiting, abdominal discomfort headache.
- Neuroglycopenic manifestations of a low glucose level
  Abnormal mentation, impaired judgment with nonspecific dysphoria, anxiety, moodiness, depression, crying, negativism, irritability, belligerence, combativeness, rage, personality change, emotional lability, fatigue, weakness, apathy, lethargy, daydreaming, and finally sleep. In addition confusion, amnesia, dizziness, delirium, staring, "glassy" look, blurred vision, double vision followed by automatic behavior also known as automatism with difficulty speaking, or a slurred speech may be observed. Once the glucose level reaches very low levels ataxia, incoordination, sometimes mistaken for "drunkenness", and even focal or general motor deficit, paralysis, may be observed leading into hemiparesis, headache stupor, with coma, abnormal breathing or even generalized or focal seizures.

8.1.5 Longterm Treatment of Hypoglycemia

- **Low glycemic index diet.** Treatment is to avoid all foods containing sugar and refined carbohydrate, especially glucose and fructose. One needs to switch to a diet, which concentrates on eating proteins, fats and complex (and therefore slowly digestible) carbohydrates. Initially it was suggested to use a high protein high fat diet, but include all vegetables (being easy on potatoes), nuts, seeds, etc. Fruit is permitted but rationed, since excessive amount of fruit juices or dried fruits contain too much fructose for the liver to be able to deal with. Ii is suggested to take one piece of fruit at mealtimes.

- **Nutritional supplements** –It has become increasingly apparent that the long established Recommended Daily Allowance (RDAs) of nutritional supplements are now outdated. Either the levels have been incorrectly set, or they do not represent levels for optimum health, or they do not permit the necessary latitude for individual biochemical differences or when being used in a disease state. Increasingly, they are not relevant to people leading modern Western lifestyle and eating a Western diet. It is therefore time for a reset of the RDAs.

8.1.6 Resetting the Recommended Daily Allowance (RDA)

Lydia J. Roberts, Hazel Stiebeling, developed the RDA during World War II and Helen S. Mitchell, all part of a committee established by the U.S. National Academy of Sciences to investigate issues of nutrition that might "affect national defense". The committee was renamed the Food and Nutrition Board in 1941, after which they began to deliberate on a set of recommendations of a standard daily allowance for each type of nutrient. The standards would be used for nutrition recommendations for the armed forces, for civilians, and for overseas population who might need food relief. Roberts, Stiebeling, and Mitchell surveyed all available data, created a tentative set of allowances for "energy and eight nutrients", and submitted them to experts for review. The final set of guidelines, called RDAs for Recommended Dietary Allowances, were accepted in 1941. The allowances were meant to provide superior nutrition for civilians and military personnel, so they included a "margin of safety." Because of food rationing during the war, the food guides created by government agencies to direct citizens' nutritional intake also took food availability into account. The Food and Nutrition Board subsequently revised the RDAs every 5–10 years. In the early 1950s, USDA nutritionists made a new set of guidelines that also included the number of servings of each food group to make it easier for people to receive their RDAs of each nutrient. Even still, the RDAs are controversial in nutrition circles because the daily maximum for some nutrients, like sodium are higher in the U.S. than in other parts of the developed world, and are far above established safe minimum. For instance, the National Research Council has found that 500 mg of sodium per day (1/4 of a teaspoon) is a safe level. In Great Britain, the daily allowance for salt is 6 g (approximately 1 teaspoon, about the upper limit in the U.S.), but is still considered "too high." Presently, there is a need to reset the RDA simply because research has demonstrated an essential need for supplementation and because the previously published RDAs are much too low in order to have a beneficial effect on health and well-being. There are a number of reasons why these RDAs presently are of no value or even are misleading the consumer:

1. Modern agricultural techniques
    - The vast majority of our crops are annual crops, so that only the upper few inches of soil are exploited. This soil has rapidly become depleted of minerals.
    - Traditional systems of crop rotation have been abandoned in favor of monocultures. This process increased the need for pesticide and herbicides.

- Human dung is not recycled back onto the growing fields; therefore, there is a net depletion of minerals.
- The use of nitrogen fertilizers and pesticides reduces the humus content (mycorrhiza) of soil; therefore, plants will malabsorb minerals.
- Pesticides, such as glyphosphate, work by chelating minerals in the soil, thereby reducing mineral availability to plants.
- Modern plant varieties are often developed to suit the palate. Fruit, for example, is cultivated to increase sugar content – increasing fructose consumption, a major cause of the "metabolic syndrome".
- Genetically modified crops is largely done to develop pesticide resistant strains rather than for improving micronutrient density. As a result, more pesticides are used exacerbating the above problems.
- Foods may be bred or even genetically modified for the sake of "keeping quality" but at the expense of a good micronutrient content. Indeed, people often comment how tasteless modern varieties of agriculture products are, and this loss of flavor reflects the declining levels of micronutrients.

All the above issues result in food crops that are deficient in micronutrients. Plants grown on mineral-deficient soils will not be able to synthesize other micronutrients being essential to their health and the health of consumers downstream (farm animals and humans). In practice, this means the plants themselves are more susceptible to diseases, such as fungal infestation, and as a result they are more heavily sprayed with fungicides for counteraction resulting in a vicious cycle. Consumers downstream will be rendered deficient in not just minerals but more important in all vitamins, essential fatty acids and many other constituents essential for healthy life which they can only get from eating plant material. What is left is that they have to use supplements

2. Modern food processing and storage-what the industry does not tell us

For reasons of convenience we have adopted food-processing techniques, which further deplete the micronutrient content of food. All the following processes have this effect:

- Food storage – it may be many months between harvest and consumption. Storage inevitably results in micronutrient depletion.
- Food processing – whole fruits are acceptable, but fruit juices are not because of their high content in fructose. The potato is a good food until crisped! Food processing often results in production of hydrogenated and transfats, which are antinutrients leading to advanced arteriosclerosis.
- High temperature cooking results in the production of transfats and lipid peroxides, all of which are directly toxic in their own right.
- Cooking to serving times – it may be some time between cooking and eating, being a particular problem with ready meals, take-away foods, and precooked meals. Many times the latter are reheated in the microwave oven, another unnecessary modern device in the kitchen, which devitalizes the food further resulting in empty calories.

3. The choices of food and the chance of getting addicted

   Western diets are addictive with respect to sugar, refined carbohydrates, allergens (particularly dairy and wheat), caffeine, alcohol, chocolate and, of course, nicotine. This creates medical problems for the following reasons:

   - Human mankind consumes these micronutrient poor foods to the exclusion of micronutrient dense foods.
   - The above consumable goods require additional micronutrients to compensate for their elimination from the body.
   - Many of these consumable goods are diuretic (caffeine, tea, etc).
   - Eating potential addictive food means one does not get the normal cues that tell us when to stop eating. Western people eat more food than they need and thus there is a tendency to weight gain.
   - Fast food chains and snacks. These are now accepted parts of modern Western lifestyle, but lead to eating refined and processed food.
   - Whole-foods are less efficiently and only slowly digested. In contrast, refined foods are rapidly absorbed, resulting in a rapidly climbing blood sugar and insulin levels, which eventually leads into a metabolic syndrome.

   All these social accepted eating behaviors lead into some kind of addiction, they are anti-nutrient, they increase the need for micronutrients, and they encourage overconsumption of micronutrient-deficient food.

4. Poor education and ill-advised beliefs

   Many people make food or lifestyle choices as the mistaken belief they are doing the healthiest thing. Modern advertising campaigns are highly influential, are often aimed at children, and encourage poor food choices through subliminal association of ideas.

   - Many people still believe that saturated fats are unhealthy and believe that low-fat diets are healthy. By looking at the number of fat-reduced milk and other foodstuffs one should realize the insanity of this approach, since the body needs fats as an essential part for health and well-being (see Sect. 8.1).
   - Foods are highly colored with synthetic colorings just to make them more attractive and desirable.
   - Sunshine is the only substantial source of vitamin D. Vitamin D deficiency is pandemic because of western climates, but largely due to ill-founded advice to avoid sunshine and use sun blockers.
   - Dairy products are believed to be healthy options, but for many people this is not true. The evolutionarily correct diet is free from dairy products.

   As a consequence of western life style, one needs to eat less than our physically active ancestors. But lesser amounts of food carry lesser amounts of micronutrients therefore supplementation is mandatory.

8.2 Toxic Stress in Western World (Xenobiotics)

The term **xenobiotic** is related to a chemical which is found in an organism but which is not normally produced or expected to be present in it. It can also cover substances, which are present in much higher concentrations than are usual. Specifically, drugs such as antibiotics are xenobiotics in humans because the human body does not produce them itself, nor are they part of a normal diet. Natural compounds can also become xenobiotics if they are taken up by another organism, such as the uptake of natural human hormones by fish found downstream of sewage treatment plant outfalls, or the chemical defenses produced by some organisms as protection against predators. However, the term **xenobiotics** is very often used in the context of pollutants such as dioxins and polychlorinated biphenyls and their effect on the biota, because xenobiotics are understood as substances foreign to an entire biological system, i.e. artificial substances, which did not exist in nature before their synthesis by humans. In the context, heavy or toxic metals can also be considered as xenobiotics. They are stable trace elements, which cannot be metabolized by the body and bioaccumulate in soft tissues (and bone), passing up the food chain to humans. These toxins are an integral byproduct of industrial society and are unavoidable in our modern-day life (Fig. 8.2). These include: mercury, lead, nickel, arsenic, cadmium, aluminum, platinum, the metallic form of copper, and others. These heavy metals have no function in the body and can be highly toxic, negatively influencing the body's metabolic processes, and impairing multiple physiological systems throughout the body. Toxic metal burden can induce impairment and dysfunction in the cardiovascular, gastrointestinal, immune, reproductive, urinary, endocrine, central and peripheral nervous systems, detoxification (colon, liver, kidneys, skin), energy production and enzymatic pathways. Toxic heavy metals can negatively affect energy levels, memory, circulation, blood pressure, and cholesterol and triglyceride levels. If unrecognized or inappropriately treated, toxicity can result in significant illness, reduced quality of life, and possibly death. Symptoms of chronic exposure are very similar to symptoms of other medical conditions, and often develop slowly over months or even years, so their cause can be easily missed. Heavy metal toxicity is a significant global health issue that can cause health problems of varying degrees in multiple systems.

The two most common heavy metals that cause adverse effects are mercury and lead. Addressing heavy metals as part of an integrative medical plan can aid in the prevention and treatment of multiple ailments because heavy metals bind to Krebs-cycle enzymes, involved in mitochondrial function and can also upregulate the body's immune system response, which in turn can result in more cellular inflammation, increased immune system free radical activity and an impetus for increased nitric oxide production. Xenobiotic chemicals (pesticides, volatile organic compounds and heavy metals) all need to be detoxified and excreted from the body, but this process is highly demanding of micronutrients. Our increasing xenobiotic load increases micronutrient requirements. Furthermore, xenobiotics are directly antinutrient. For example, nickel increases our requirement for zinc, fluoride increases our need for iodine, and mercury increases our need for selenium.

Fig. 8.2 The heavy metals contamination of ocean fish is so bad that even the FDA warns expectant mothers to avoid eating fish. In this comic artist Dan Berger imagined the heavy metal content in fish tissue might be so bad that you could actually catch fish using powerful magnets. Being an exaggeration, in reality, if the heavy metals in the fish were strong enough to respond to magnets in this way, the fish could not possibly survive. Thanks to continued pollution by industrialized nations, that may be the ultimate fate of ocean fish. As NaturalNews reported on November 3, 2006, the world can expect to see a global collapse of seafood due to over fishing and the destruction of ocean ecosystems, mostly by synthetic chemicals flushed down toilets or emptied into rivers and streams, which ultimately empty into the ocean (Source: www.NaturalNews.com)

8.2.1 Origin of Xenobiotics Resulting in a Toxic Load

1. Contamination of our food and water in ways as outlined above.
2. Contamination by packaging – many products are wrapped in carbon boxes from recycled colored paper (containing petrol residues), sometimes cooked, in plastics with phthalates all leeching into the food (see Sect. 7.2).
3. Contaminations of the environment by persistent organic pollutants from agricultural industry, heavy metal-chemical industry, salting of roads in the winter, fine and exhaust particles from road traffic, mines and waste disposal.
4. Cosmetics often contain volatile organic chemicals and/or heavy metals such as nickel and aluminum – this increases the toxic load and thereby the requirements for micronutrients to detoxify them.
5. Jewellery and piercings increase exposure to toxic metals.

8.2 Toxic Stress in Western World (Xenobiotics)

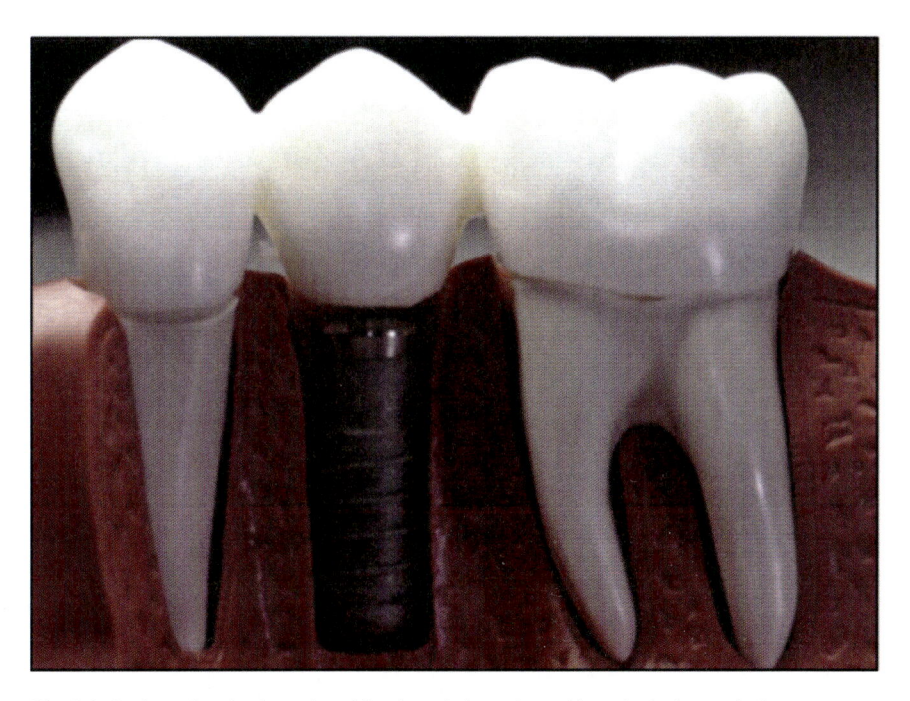

Fig. 8.3 Ionic molecules from dental implants being released into the body regularly can present a source of heavy metal poisoning which affects the immune system

6. Prescription medication – a good example is the malabsorption induced by proton pump inhibitors (PPI) and other acid blockers, resulting in an increased risk of osteoporosis. It is noteworthy to point out, that most drug side effects result from micronutrient deficiencies.
7. The vaccinations of otherwise healthy people, including children, which often contain heavy metals (mercury as an enhancer) together with live or attenuated viruses. Vaccinations have an immune disrupting potential and the potential to switch on to novel disease processes.
8. Dental work and surgical prostheses regularly involve the use of heavy metals such as mercury, palladium, titanium, nickel, gold and silicones, all of which may be toxic either directly or through their potential to disrupt the immune system (Fig. 8.3).
9. Last but not least, there is the issue with using recycled polyester in waste disposals (Fig. 8.4). It looks like the plastic bottle is here to stay, despite publicity about bisphenol A and other chemicals that may leach into liquids inside the bottle. Plastic bottles (which had been used for some kind of consumer product) are the feedstock for what is known as "post consumer recycled polyester". Recycled polyester, also called rPET, is now accepted as a "sustainable" product in the textile market. In textiles, most of what passes for "sustainable" claimed by manufacturers has some sort of recycled polyester in the mix, because it's a message that can be easily understood by consumers – and polyester is much cheaper than natural fibers.

Fig. 8.4 Not a single piece of plastic in this photograph was moved, placed, manipulated, arranged, or altered in any way. The image depicts the actual stomach contents of a baby bird in one of the world's most remote marine sanctuaries, more than 2,000 miles from the nearest continent (Source: national geographic)

8.2.2 Risk Factor for Alzheimer/Parkinson – Toxins Affecting Mitochondrial Function

While there is a small percentage of risk factors from inside the body where gut fermentation of foods produce alcohol, D-lactate, hydrogen sulphide and other toxins, the majority of threat comes from the outside with the kind of food (either processed, refined or contaminated) being consumed. Special attention should be paid:

- Caffeine in the short term is a mental stimulant. This can be helpful if you have to "perform", so long as you can rest and recover afterwards. If you are having more than three cups a day (tea, coffee, coca cola), then it is probably having an overall deleterious effect.
- As a result of any detoxification regime, there is a mobilization of toxins, which previously have been stored in fat tissue, now enters the circulation and induces a central nervous effect.

- Consumption of alcohol results in neurotoxicity since this beverage can be considered a neurotoxin. If one drinks more than one glass of wine daily, thiamine deficiency is likely to develop. Thiamine is essential for normal brain function.
- Prescription drugs (i.e. antidepressants, opioids, antiepileptics, etc.) often have profound effects on the brain, especially in older people and the very young (see Sect. 7.2 on page 88 ff).
- Chemical poisoning presents the greatest imperilment in the daily nutrition with all the additives not just to make it look better but more to prolong its shelf life (see chapter on chemical poisons and toxins on page 94 ff). The more one looks for them, the more one finds, especially in food chain being offered at the various large super markets. One just should just check the small writing on the label with all these E-numbers!

Especially when consumed over years, there is an accumulation of toxic residues, which not only may affect the digestive system, the liver or the pancreas. These substances have the tendency to be toxic for mitochondria suggesting that the steadily rising number of patients with Alzheimer is a home made epidemic disease. And although it is known that having a family history of Alzheimer's disease (AD) means that one has a higher risk of developing the disease, progressive brain is an age-related and irreversible brain disorder that occurs gradually and results in memory loss, behavior and personality changes, and a decline in thinking abilities. These losses are related to the breakdown of the connections between nerve cells in the brain and the eventual death of many of these cells. About 3% of men and women aged 65–74 have AD, and nearly half of those age 85 and older may have the disease. It is important to note, however, that Alzheimer is not a normal part of aging and therefore one should also consider other causes (Table 8.1), especially since it has been demonstrated that Alzheimer is connected with a loss in mitochondrial function (Ref. [172] in Chap. 7).

The detrimental role of a high carbohydrate (especially fructose) diet very likely is another major cause for the rapid increase in Alzheimer.

A crucial role is played by astrocytes, the glial cells that supply cholesterol and fats to neurons. A consequence of excess exposure to glucose and oxidizing agents may lead to an initial damage due to glycation and oxidation of ApoE in astrocytes. The subsequent cascade begins with defects in the transmission of neural signals,

Table 8.1 Summary of chemicals involved in the initiation of neurotoxicity

Risk factors for Alzheimer	Discussed mechanisms
Head trauma	Increase in NMDA activity
Alumium intake	Inhibition of BH_4 reductase leading to deficiency in BH4
Genetic origin carrying the apolipoprotein (Apo) ε4 allele with higher susceptibility to perioperative neural insults [6]	Activation of pro-oxidative properties; stimulation of NF-κB activity
Exposure to electro-magnetic waves	Increase in NADH oxidase with increase in superoxide production
Lead exposure	Inactivation of glutamate transporter system with intracellular glutamate accumulation followed by increased NMDA activity
Exposure to organic solvents	Stimulation of the TRPV1 and TRPV2 leading into increased NMDA activity
Exposure to organophosporus, a pesticide	Inhibition of GABA followed by increased NMDA activity
Exposure to organochloride, a pesticide	Inhibition of GABA leading into an increased NMDA activity
Exposure to 2,4-Dichlorophenoxyacetic acid (2,4-D) a common systemic pesticide used in the control of broadleaf weed	Dysfunction in energy production within mitochondria
Exposure to the neurotoxic non-protein amino acid β-methylamino-L-alanine (BMAA)	Activation of excessive NMDA-activity
Intake of MPTP (1-Methyl-4-phenyl-1,2,3,6-tetrahydro-pyridine) a neurotoxin and a toxic by-product in synthesis of the designer opioid MPPP (1-Methyl-4-phenyl-4-propion-oxy-piperidine) which acts similar to heroin	Block of energy production in complex I of mitochondria, followed by excessive superoxide production
Exposure to rotenone, an insecticide	Block of energy production in complex 1 of mitochondria, followed by excessive superoxide production
Exposure to maneb, a pesticide used as an antifungal	Inhibits the glutamate transporter system followed by increase of NMDA activity

progressing to mitochondrial dysfunction, insulin resistance, and the increased synthesis of Aβ. The dual roles of Aβ are to stand in for cholesterol and to redirect energy management from the mitochondria towards alternative cytoplasmic solutions, in order to reduce potential oxidative and glycation damage to active proteins and membranes. As such problems escalate, a neuron is eventually unable to function in its intended role (neurotransmission) without exposing itself and neighboring neurons and glial cells to dangerously high levels of oxidative agents. The microglia will then program the defective neuron for cell death. By this time, the proteasome and the lysosome systems have usually been so irreversibly damaged by oxidative exposure and insufficient ATP generation that the cell undergoes apoptosis. A legacy of complex protein debris is left in place.

8.2.3 Glycation End-Products Impede Supply of Cholesterol and Fats to Neurons

A diet high in processed carbohydrates and low in fats results in a rapid rise in blood glucose levels following meals. Over time, this may lead to insulin resistance and diabetes. Serum proteins that are exposed to high levels of glucose become impaired due to a process called glycation, resulting in the appearance of a diverse group of modified proteins known collectively as advanced glycation endproducts (AGEs). Fructose, a sugar increasingly in use in processed foods, also because of economical constraints, is estimated to be ten times more reactive than glucose in inducing glycation [7]. With the widespread availability of high fructose corn syrup as a sweetening agent, the Western diet is associated with much higher risk of AGE damage to proteins. Hemoglobin A1c, the protein whose serum level is the standard blood test for diabetes, is a prototypical AGE product. Fructose is by far more damaging than glucose as a reducing agent, leading to diverse AGE products. This is also reflected in results with a medium chain, trans fat with high fructose diet leading to obesity, nonalcoholic steatohepatitis and oxidative stress [8]. For example, an experiment involving feeding rats controlled diets containing either fructose, glucose, or sucrose demonstrated that the fructose-fed rats were worse off on many indicators of glycation damage [7]. This was supported by a study of Seneff and coworkers [9] who conclusively demonstrated that

- The amyloid-β present in Alzheimer's plaque may not be causal, since drug-induced suppression of its synthesis led to further cognitive decline in the controlled studies performed so far.
- Researchers have identified mitochondrial dysfunction and brain insulin resistance as early indicators of Alzheimer's disease (Ref. [57] in Chap. 7).
- ApoE-4 is a risk factor for Alzheimer's disease, and ApoE is involved in the transport of cholesterol and fats, which are essential for signal transduction and protection from oxidative damage.
- The cerebrospinal fluid of Alzheimer's brains is deficient in fats and cholesterol.
- Advanced glycation end products (AGEs) are present in significant amounts in Alzheimer's brains.
- Fructose, an increasingly pervasive sweetening agent, is ten times as reactive as glucose in inducing AGEs.
- Astrocytes play an important role in providing fat and cholesterol to neurons.
- Glycation damage interferes with the LDL-mediated delivery of fats and cholesterol to astrocytes, and therefore, indirectly, to neurons.
- ApoE induces synthesis of Aβ when lipid supply is deficient.

- Aβ redirects neuron metabolism towards other substrates besides glucose, by interfering with glucose and oxygen supply and increasing bioavailability of lactate and ketone bodies.
- Synthesis of the neurotransmitter, glutamate, is increased when cholesterol is deficient, and glutamate is a potent oxidizing agent.
- Over time, neurons become severely damaged due to chronic exposure to glucose and oxidizing agents, and are programmed for apoptosis due to highly impaired function.
- Once sufficiently many neurons are destroyed, cognitive decline is manifested.
- Simple dietary modification, towards fewer highly-processed carbohydrates and relatively more fats and cholesterol, is likely a be protective measure against Alzheimer's disease
- Aβ synthesis may represent a protective mechanism that redirects metabolism away from glucose and away from the mitochondria, in order to decrease the rate of further decline due to glycation and oxidative damage. Increased production of glutamate is necessary to maintain signal transport in the face of cholesterol deficiency, yet glutamate causes further oxidative damage.

8.2.3.1 Glycation- Key Factor in Triggering Cellular Peroxynitrite

All the evidence indicates that most of the cytotoxicity attributed to NO is rather due to peroxynitrite, produced from the diffusion-controlled reaction between NO and another free radical, the superoxide anion. As demonstrated, this results in glutamate cytotoxicity, which contributes to neuronal degeneration in many central nervous system diseases, such as epilepsy and ischemia. It was previously reported that a high-fat and low-carbohydrate diet, the ketogenic diet (KD), protects against kainic acid-induced hippocampal cell death in mice, and it was hypothesized, based on these findings that ketosis resulting from KD might inhibit glutamate cytotoxicity, resulting in inhibition of hippocampal neuronal cell death [10]. In this sense peroxynitrite interacts with lipids, DNA, and proteins via direct oxidative reactions or via indirect, radical-mediated mechanisms exacerbating neuronal cell death. These reactions trigger cellular responses ranging from subtle modulations of cell signaling to overwhelming oxidative injury, committing cells to necrosis or apoptosis. In vivo, peroxynitrite generation represents a crucial pathogenic mechanism in conditions such as stroke, myocardial infarction, chronic heart failure, diabetes, circulatory shock, chronic inflammatory diseases, cancer, and neurodegenerative disorders. Hence, novel pharmacological strategies aimed at removing peroxynitrite might represent powerful therapeutic tools for the future and recent work suggests

8.2 Toxic Stress in Western World (Xenobiotics)

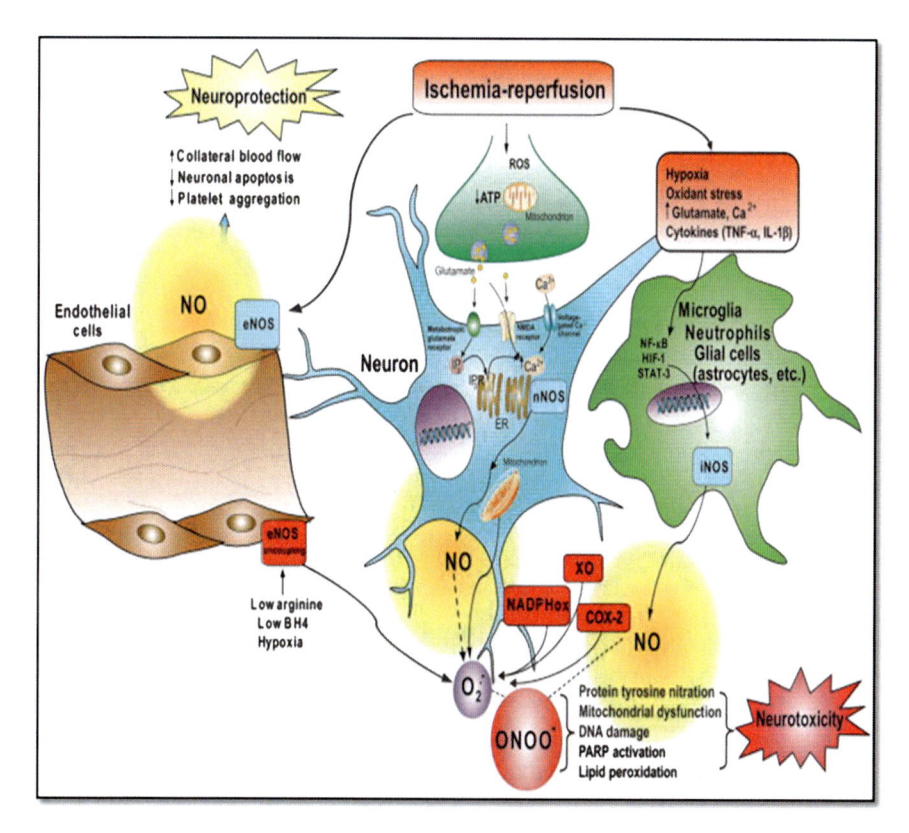

Fig. 8.5 Roles of NO and peroxynitrite in the pathophysiology of neurotoxicity

the detrimental role of a high carbohydrate diet [9]. A first step in the pathophysiology of the disease is represented by advanced glycation (sometimes called non-enzymatic glycosylation) end products in crucial plasma proteins concerned with fat, cholesterol, and oxygen transport, where gyration is the enzymatic process that attaches glycols to proteins, lipids, or other organic molecules. This leads to cholesterol deficiency in neurons, which significantly impairs their ability to function. As more and more damage is incurred by the cell membranes, without sufficient replenishment of the supplies of fats and cholesterol to repair them, an increase in ion leakage across all membranes leads to further depletion of ATP and further exposure to pathogens and oxidative damage (Fig. 8.5). Over time, the accumulated depletion of ATP leads to liposomal dysfunction, likely because of an inability to maintain a sufficiently acidic pH for the digestive enzymes to work properly. In parallel, the further gyration and oxidation of Aβ convert it into an algometric complex with both decreased function and reduced susceptibility to degradation by the lissome. Once a sufficient percentage of the cell membrane has been compromised due to oxidative damage, it incurs rapid calcium influx and subsequent apoptosis. The protein plaques and tangles of

AD are unrecyclable debris that remains in place after cell death. Thus the present knowledge on AD can be summarized as follows:

- The amyloid-β present in Alzheimer's plaque may not be causal, since drug-induced suppression of its synthesis led to further cognitive decline in the controlled studies performed so far.
- Researchers have identified mitochondrial dysfunction and brain insulin resistance as early indicators of Alzheimer's disease.
- ApoE-4 is a risk factor for Alzheimer's disease, and ApoE is involved in the transport of cholesterol and fats, which are essential for signal transduction and protection from oxidative damage.
- The cerebrospinal fluid of Alzheimer's brains is deficient in fats and cholesterol.
- Advanced glycation end products (AGEs) are present in significant amounts in Alzheimer's brains.
- Fructose, an increasingly pervasive sweetening agent, is ten times as reactive as glucose in inducing AGEs.
- Astrocytes play an important role in providing fat and cholesterol to neurons.
- Glycation damage interferes with the LDL-mediated delivery of fats and cholesterol to astrocytes, and therefore, indirectly, to neurons.
- ApoE induces synthesis of Aβ when lipid supply is deficient.
- Aβ redirects neuron metabolism towards other substrates besides glucose, by interfering with glucose and oxygen supply and increasing bioavailability of lactate and ketone bodies.
- Synthesis of the neurotransmitter, glutamate, is increased when cholesterol is deficient, and glutamate is a potent oxidizing agent.
- Over time, neurons become severely damaged due to chronic exposure to glucose and oxidizing agents, and are programmed for apoptosis due to highly impaired function.
- Once sufficiently many neurons are destroyed, cognitive decline is clinically manifested.
- Simple dietary modification, towards fewer or even total elimination of highly processed carbohydrates and relatively more fats and cholesterol, is likely a protective measure against Alzheimer's disease.

Brain ischemia and reperfusion leads to transient stimulation of the activity of endothelial NO synthase (eNOS), resulting in brief increases in endothelial NO generation, associated with neuroprotective actions in stroke. In parallel, ischemic energy depletion and oxidant (ROS) production triggers the release of glutamate, which results in neuronal calcium overload from extracellular (activation of calcium channels) and intracellular (phosphoinositol-3-kinase-endoplasmic reticulum signaling) sources. Calcium overload results in prolonged synthesis of NO, due to stimulated activity of the neuronal isoform of NO synthase (nNOS). Enhanced NO generation also depends on the induced expression of inducible NOS (iNOS) in various types of reactive inflammatory cells, upon the activation of several cell signaling pathways (HIF-1, STAT-3, and NF-κB) in response to hypoxia, cytokines, oxidants, and glutamate. During the same period of time, superoxide production is

8.2 Toxic Stress in Western World (Xenobiotics)

enhanced due to uncoupling of eNOS, mitochondrial dysfunction, and the stimulated activity of NADPH oxidase, xanthine oxidase, and cyclooxygnease-2 (COX-2). Formation of peroxynitrite is then markedly favored, damaging lipids, proteins, DNA, and triggering the activation of poly(ADP-ribose) polymerase (PARP), all of which contribute significantly to neurotoxicity in stroke (Ref. [15] in Chap. 5).

8.2.3.2 New Concepts in Therapeutic Approach of Alzheimer

So far there are three major lines of drugs have been developed or are under development for the treatment of Alzheimer Disease (AD): cholinergic drugs (mainly cholinesterase inhibitors), anti-beta-amyloid drugs, estrogens and anti-inflammatories. Although it is generally accepted, that with such treatment option there is no reverse of the disease and at best only a halt in the progression can be achieved, combination of cholinesterase inhibitors with estrogens, anti-oxidants and anti-inflammatories may represent a further improvement of the therapy. From the economical point of view, treatment with cholinesterase inhibitors is not cost neutral and the various Alzheimer's drugs, like Aricept®, have proven disappointing, with little real benefit and often distressing side effects. Dr. Newport from Florida discovered that with Alzheimer's disease, certain brain cells may have difficulty utilizing glucose made from the carbohydrates we eat, being the brain's principal source of energy. Without fuel, these precious neurons may begin to die. There is an alternative energy source for brain cells – fats known as ketones. If deprived of carbohydrates, the body produces ketones naturally. But this is the hard way to do it – who wants to cut carbohydrates out of the diet completely? Another way to produce ketones is by consuming oils that have medium-chain triglycerides (MCT). When MCT oil is digested, the liver converts it into ketones. In the first few weeks of life, ketones provide about 25% of the energy newborn babies need to survive. Dr. Newport learned that the ingredient in a drug trial, which was showing so much promise, was simply MCT oil derived from coconut oil or palm kernel oil, and that a dose of 20 g (about 20 ml or 4 teaspoons) was used to produce these results. When MCT oil is metabolized, the ketones which the body creates may, according to the latest research, not only protect against the incidence of Alzheimer's, but may actually reverse it [11]. By addressing the significant decrease in cerebral glucose in Alzheimer's patients is essential in combating this devastating disease, strategies has been to supplement the brain's normal glucose supply with ketone bodies. A company's data was drawn from two clinical studies, which examined the cognitive effects of induced ketosis. In both acute and chronic dosing, AC-1202 (Axona®) significantly induced ketosis 2 h after administration. Further analysis of the studies revealed that patients administered Axona who lacked the epsilon four variant of the APOE gene (E4(−)), demonstrated significant improvement from baseline values in the Alzheimer's Disease Assessment Scale-Cognitive (ADAS-Cog) and improvement compared to placebo. In the population of patients who were both APOE4(−) and dosage compliant, more pronounced improvements in ADAS-cog scores were observed at each assessment time point (Day 45 and Day 90). At Day 45 the improvement in

| DAY BEFORE COCONUT OIL | 14 DAYS ON COCONUT OIL | 37 DAYS ON COCONUT OIL |

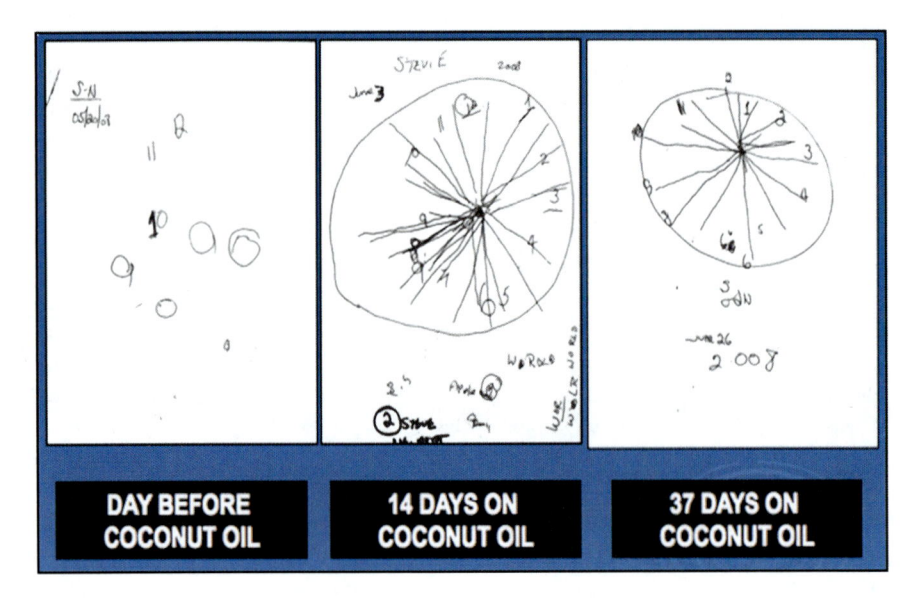

Fig. 8.6 Representative example of the simple and typical test for diagnosing Alzheimer (drawing the clock) demonstrating a deficit in cognitive function in a patient before and after taking a mixture of coconut oil and medium chain fatty acids (MCFA). Note the steady increase in performance. By courtesy of Dr. Newport [14]

ADAS-cog score relative to placebo was 6.26 ($p=0.001$), while at Day 90 the difference was 5.33 ($p=0.006$). These improvements were remarkable in that they were seen in patients who were already taking prescription Alzheimer medications [12]. Moreover, this may also be a potential treatment for Parkinson's disease, Huntington's disease, multiple sclerosis and amyotrophic lateral sclerosis (ALS or Lou Gehrig's disease), drug-resistant epilepsy, brittle type I diabetes, and type II (insulin-resistant) diabetes. In addition, in an open label trial with coconut oil and MCT supplementation, Dr. Newport presented data of anecdotal reports from caregivers of 47 persons with dementia showed improvement in the vast majority, many in aspects of human life other than memory and cognition. The positive responses are presumably due to metabolism of medium chain triglycerides to ketone bodies for use by neurons as an alternative fuel in cells with decreased ability to transport glucose, thereby improving neuron function and viability [13]. In one selective case Dr. Newport started giving the coconut oil twice a day. At this point, the patient could barely remember how to draw a clock (Fig. 8.6). Two weeks after adding coconut oil to his diet, his drawing improved. After 37 days, the drawing gained even more clarity (Fig. 8.6). The oil seemed to "lift the fog," and in the first 60 days, with remarkable changes concerning alertness and mood, being talkative, and making jokes although the gait was "still a little weird," but the tremor was no longer very noticeable. He was able to concentrate on things that he wanted to do around the house and in the yard and stay on task, whereas before coconut oil the patient was easily distractible and rarely accomplished anything unless he was

8.2 Toxic Stress in Western World (Xenobiotics)

Table 8.2 Summary of fats and oils with short and medium chain fatty acids

• Butter	11.5%
• Coconut oil	59%
• Palm kernel oil	54%
• Babassu oil	55% (Amazon palm)
• Ucuhuba butter	13% (Amazon tree seed)
• Nutmeg butter	3.1%
• Sheanut oil	1.7%

directly supervised. Over the next year, the dementia continued to reverse itself: the patient is able to run again, his reading comprehension has improved dramatically, and his short-term memory is improving – he often brings up events that happened days to weeks earlier and relays telephone conversations with accurate detail. A recent MRI shows that the brain atrophy has been completely halted.

Such promising therapeutic benefits were further substantiated in changes of patients with dementia and cognitive function between 4 and 10 months of therapy with coconut oil and MCFA (Table 8.2). In her personal experience with her husband changes, Dr. Newport has noticed the following changes over time, after implementing the intake of coconut oil and MCT:

- Conversational skills continued to improve
- Reading comprehension improved
- Short term and recent memory improved
- Stop having episodes of near syncope
- No longer depressed – says, "I have my life back."
- Wanted to do more with his life – volunteers in hospital warehouse
- Vacuums, cuts grass and weeds gardens again, instead of taking equipment apart!

Short Chain Fatty acids: The bottom message of these changes seem too support the notion that if the brain runs short of fats and ketones, it can swap to short chain fatty acids, which come from the large bowel fermenting soluble fiber and can provide up to 500 kcal a day. The trouble with sugar, it is a short-term fuel, like running on a reserve tank. The brain is constantly assessing the fullness of the tank and if the tank starts to run low, the brain stimulates the release of adrenaline – this will bring blood sugar up for the brain but one then suffers from the adrenaline effects while at the same time glucose is only be metabolized for a short period. If however this metabolisation pathway is impaired as in mitochondropathy, then there is a lack in the formation of energy, the cell deteriorates in function and finally dies off leaving a "scar" of amyloid-β. The brain's preferred fuels are ketones, which the liver synthesizes from medium chain fatty acids. The best source of these is coconut oil. This fuel source is much more constant than glucose and highly protective against hypoglycemia. Coconut oil 10–20 ml twice daily is often very helpful (Fig. 8.7) and the brain will just love it ! For additional information on nutrition for the brain see Sect. 7.4.2.

By having ketone bodies as an additional source of fuel, the dependence on glucose is reduced. But another effect that may be more important than this is the availability of high-quality fats to improve the condition of the myelin sheath. This idea was spurred by experiments done on human Alzheimer's patients. A placebo-controlled

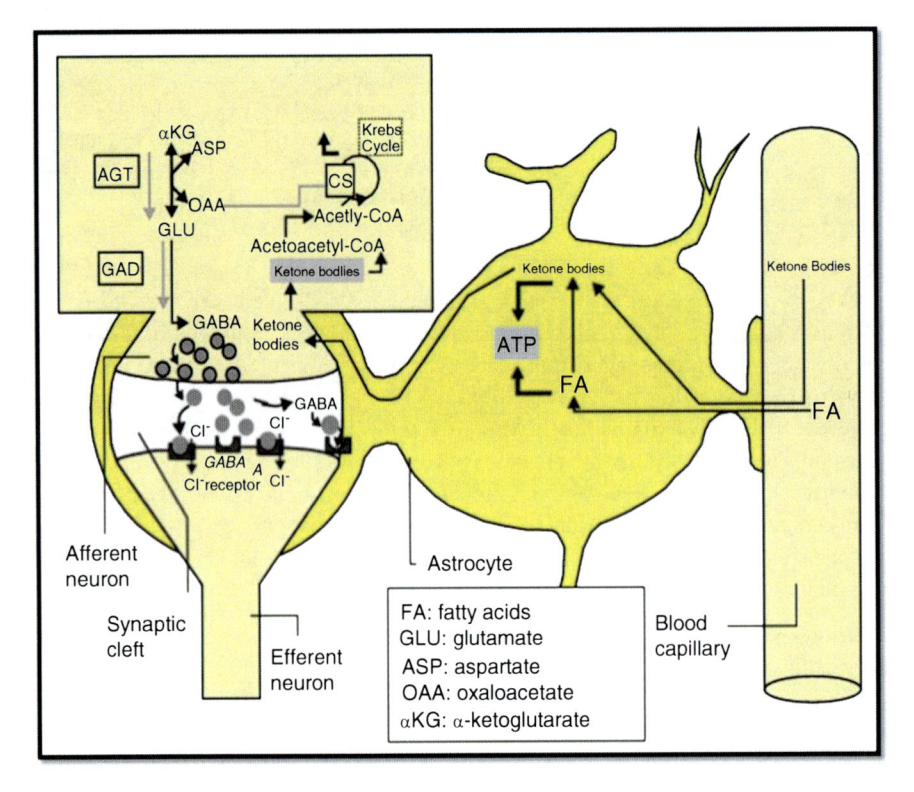

Fig. 8.7 One of the promising new treatment paradigms for Alzheimer's is to have the patient switch to an extremely high fat, low carb diet, a so-called "ketogenic" diet. The name comes from the fact that the metabolism of dietary fats produces "ketone bodies" as a by-product, which are a very useful resource for metabolism in the brain. It is becoming increasingly clear that defective glucose metabolism in the brain (so-called "type-3 diabetes") is an early characteristic of Alzheimer's. Ketone bodies, whether they enter the astrocyte directly or are produced in the astrocyte itself by breaking down fats, can be delivered to adjacent neurons

2004 study [15] of the effect of dietary fat enrichment on Alzheimer's is especially informative, because it uncovered a significant difference in effectiveness for the fat-enrichment for subjects who did not have the apoE-4 allele as compared with those who did. The experimental test group was given a supplemental drink containing emulsified medium chain triglycerides, found in high concentration in coconut oil. The subjects without the apoE-4 allele showed a significant improvement in score on a standard test for Alzheimer's, whereas those with the apoE-4 allele did not. This is a strong indicator that the benefit may have to do with an increase in uptake by the astrocyte of these high-quality fats, something that the subjects with the apoE-4 allele are unable to accomplish due to the defective IDL and LDL transport mechanisms.

Prebiotics, as soluble fibre in vegetables especially pulses, vegetables, nuts and seeds or as fructo-oligosaccharides. Prebiotics feed bacteroides in the large bowel, which ferment to produce short chain fatty acids – when blood sugar levels fall mitochondria happily swap to short-chain fatty acids (SCFAs) as a fuel source. Indeed over 500 kcal a day can be generated in this way. Short chain fatty acids (SCFAs) help to prevent hypoglycemia especially during sleep.

Probiotics – the present view is that probiotics have not lived up to their full potential. However they may be useful by displacing unfriendly bacteria or yeast. Kefir for example produces a toxin that kills yeast in the gut. The best and cheapest way to do this is to brew your own – In a normal situation free from antiseptics, antibiotics, high-carbohydrate diets, bottle feeding, hormones and other such paraphernalia of modern western life, the gut flora is safe. Babies start life in mother's womb with a sterile gut. During the process of birth, they become inoculated with bacteria from the birth canal and the perineum. These bacteria are largely bacteroides, which cannot survive for more than a few minutes outside the human gut. This inoculation is enhanced through breast-feeding because the first milk, namely colostrum, is highly desirable substrate for these bacteria to flourish. We now know that this is an essential part of immune programming. Indeed 90% of the immune system is gut associated. These essential probiotics programme the immune system so that they accept them and learn what is beneficial. A healthy gut flora therefore is highly protective against invasion of the intestine by other strains of bacteria or viruses. The problem, however, is there is no probiotic on the market that supplies bacteroides for the above reasons. If we eat probiotics that have been artificially cultured, for a short while the levels of these probiotics in the gut do increase. However, as soon as we stop eating them, the levels will taper off until they finally disappear. For bacteria to be accepted into the normal gut and remain, they have to be programmed first through somebody else's gut (in this case the mother's). So, when it comes to repleting gut flora, there are two ways there are two ways how this can be one- either one can take probiotics regularly (and the cheapest way to do this is to grow your own probiotics) or to take bacteroides directly. Indeed, this latter technique is well established in the treatment of Clostridium Difficile, a normally fatal gastroenteritis in humans. Dr Thomas Borody has developed these ideas further using fecal bacteriotherapy, which can provide a permanent cure in cases of ulcerative colitis, severe constipation, clostridium difficult infections and pseudomembranous colitis. The reason this technique works so well is because the most abundant bacteria in the large bowel, bacteroides, cannot survive outside the human gut and cannot be given by any other route. The gut flora is extremely stable and difficult to change. Therefore if one is going to take probiotics, they have to be taken for a long term. Many preparations on the market are ineffective. Those found to be most effective are milk ferments and live yoghurts where the product is freshly made. Keeping bacteria alive is difficult and it is not too surprising that they do not survive dehydration and storage at room temperature. So the best chance of eating live viable bacteria is to buy live yoghurts or drinks. These can be easily grown at home, just as one would make homemade yoghurt. If there is no easy grow from a culture, then this suggests that the culture is not active, and it is a good test of what is viable and what is not. In the meantime, the best one can do is to grow its own probiotics since this is a cheap and efficient way of sorting the situation out as follows:

8.2.4 Growing your Own Probiotics

The idea here is to take a substrate on which to grow the bugs and to which one is not allergic and make your own culture. This means one can swallow high dose probiotics, which are alive and active (so much better able to colonies the gut) and they can be eaten regularly throughout the day, being a cheap and delicious method. It also means that on what ever you grow the culture, the sugar is fermented out of it and so this provides a good low glycemic index food. This inhibits fermentation by yeasts. Furthermore, probiotics convert sugars and starches in the gut into short chain fatty acids, which are the preferred fuel for mitochondria. Therefore, anyone with a tendency to hypoglycemia will find their symptoms greatly reduced. Even for normally healthy people probiotics will stabilize blood sugar levels and reduce risk of obesity, diabetes, Syndrome X, heart disease, PCOS, cancer and all those problems arising from a hypoglycemic tendency. The idea of using fermented foods is very popular in many human societies and is associated with long and good health!

The sort of problems one expects to see in people with abnormal gut flora result clinically from the fermentation of sugars and starches by yeasts, which form alcohol and gas. They are:

- Gut symptoms – irritable bowel syndrome (alternating constipation or diarrhea, wind gas, pain), stools like pellets, foul smelling offensive wind, indigestion, poor digestion, constipation;
- Tendency to low blood sugar with carbohydrate craving;
- Tendency arising from Candida, such as thrush, skin yeast infections. Kefir is useful because it creates a toxin that kills yeast cells directly.
- Tendency to develop allergies to foods;
- Leaky gut (positive PEG test).

In theory any probiotics on the market can be used to start the culture going but in practice many of the dried preparations are inactive. You could try starting with plain live yoghurt, but the bacteria in yoghurt may be chosen for its ability to make tasty yoghurt rather than what is good for your gut, therefore Kefir, being a useful probiotic because it is rich in lactobacillus, which keeps the lower gut slightly acidic, displaces unfriendly bacteria, is directly toxic to yeast and is anti-inflammatory to the gut, is recommend.

8.2.5 Kefir for Growing Good Bacteria in the Intestinal Tract

You can grow you own Kefir and it goes well at room temperature. When you are a dairy allergic you also can use soya milk but it also grows on rice milk or coconut milk and who knows what else! To start off with 1 L of soya milk in a jug, add the Kefir sachet and within about 12–24 h it has gone semi solid. Then keep in the fridge, where it ferments further. This slower fermentation seems to improve the texture and flavor. However, it can be used at once as a substitute in any situation where you would otherwise use cream or custard. Once the kefir is down to nearly the bottom, add another liter of soya milk, stir it in and away you go again. I don't even bother to wash up the jug – the slightly hard yellow bits on the edge I just stir in to restart the brew. This way a sachet of Kefir lasts for lifetime! Another idea is the possibility of adding vitamins and minerals to the culture. This is because that they may be incorporated into the bacteria and thereby enhance the absorption of micronutrients. You could try this if you do not tolerate supplements well. The use of probiotics is an established practice in animal welfare and probiotics are actively marketed to the industry for this very reason. Furthermore, it is interesting to know that probiotics are routinely used in the pig industry to prevent post-weaning diarrhea. Anyone who has to take antibiotics for any reason should take these cultures as a routine to prevent "super-infection" with undesirable pathogens. These cultures are also an essential part of decolonizing the intestine following a triple eradication therapy.

The importance of different bacteria in the gut for well-being and health:

1. In acute gastroenteritis one should always use probiotics as a routine.
2. When antibiotics are prescribed then probiotics again should be given as a routine.
3. Irritable bowel syndrome seems to respond best to Bifido bacteria and also saccharomyces boulardii.
4. Giving prebiotics such as fructo-oligosaccharides 5 g may enhance the effect of probiotics. In eczema the best bacteria are lactobacillus rhamnosus and lactobacillus reuteri and lactobacillus GG.
5. VSL3 (a patented probiotic preparation of live freeze-dried lactic acid bacteria) is a good combination probiotic for all round use. In inflammatory bowel disease the best bacteria are bifidus longum, combined with 6 g of probiotics.
6. A combination of lactobacillus plantarum and lactobacillus paracasei, combined with pectin, fructo-oligosaccharides, inulin and resistant starch is recommended.
7. Many patients who are allergic to dairy products and soya cannot make their own ferments. However, Nutramigen Company now produces baby milk that has probiotics already added. Currently this is available in several countries within Europe.
8. Fermentation in the gut. A fermenting gut produces alcohol that further destabilizes blood sugar levels (for further information see page 182 ff).
9. Additional supplements – niacinamide and chromium are particularly helpful. I recommend taking a high dose for 2 months. Both these supplements have a profound effect on blood sugar levels to stabilize them but sometimes have to be given in high doses initially to kick-start the necessary mechanisms. By this I mean niacinamide 500 mg, three times daily at mealtimes and possibly double this dose. Rarely, niacinamide in these doses can upset liver enzymes but this is accompanied by nausea – so if one feels this symptom, reduce the dose to 500 mgs daily. Niacinamide is a really interesting vitamin – it shares the same action like the benzodiazepine diazepam (i.e. Valium®) by producing a calming effect

that, however, is not addictive. It is suspected that it works at the GABA receptor where it induces a benzodiazepine-like action [16].
10. Allergies to foods – this can certainly cause hypoglycemia – the three top allergens are grains, dairy products and yeast. But a person can be allergic to any kind of food!

8.3 Mitochondria Involved in Mutation to Cancer Cells?

During oxidative phosphorylation at complex V there is a release of intramitochondrial ROS, which immediately is inactivated by GSH (glutathion), the bodys own ROS-scavenging system. Once however, there is an abundance of ROS formation, a lack of GSH for the scavenging process, or the formation of excessive NOS via NOS synthase uncoupling, all this results in damage to mitochondria. If however >80% of mitochondria are damaged, the lack in ATP formation becomes obvious and the cell switches into the backup system, generating and producing ATP within the cytosol via anaerobic glycosylation using primarily glucose as an energy source. This process is less effective than oxidative breakdown, since only 1/9th of ATP is generated when compared to oxidative mitochondrial glycolysis. The continuing deficit in energy results in reduced activity with brain fog, accompanied by an excess formation of lactate. And although such individuals try to compensate this loss by an increase of carbohydrate intake (giving glucose to the starving cell), this eventually will only perpetuate the energy deficit with increased craving leading only into a further boost of body fat. If the stress is high enough such changes can be observed in patients with CFS and FMS, with the formation of lactic acid during metabolization. Such step in the metabolic generation of energy is regularly observed within cancer cells, and according to Warburg's five hypotheses, these cells:

1. **Don't undergo apoptosis (programmed cell death).**
2. **Derive their energy from anaerobic metabolisation.**
3. **Continuously replicate.**
4. **Loose the master plan to function with other cells.**
5. **Invade other cellular structures setting metastases.**

The alternative hypothesis advocates that cancer, malignant growth, and tumor growth are caused by mitochondrial dysfunction in which energy is mainly generated (i.e. adenosine triphosphate or ATP) by non-oxidative breakdown of glucose, a process called glycolysis. This is in contrast to "healthy" cells, which mainly generate energy from oxidative breakdown of pyruvate. Pyruvate is an end product of glycolysis, and is oxidized within the mitochondria. Hence and according to Warburg, cancer should be interpreted as a mitochondrial dysfunction. The concept that cancer

8.3 Mitochondria Involved in Mutation to Cancer Cells?

cells switch to glycolysis has become widely accepted, even if it is not seen as the *cause* of cancer, and some researcher even suggest that the Warburg phenomenon could be used to develop anticancer drugs. Parallel with this malfunction of mitochondria in cancerous cells there is a deficit in glutathione, the natural scavenger system for elimination of ROS and NOS. In order to survive the cell now switches to lower step in the evolutionary ladder in order to survive and generate energy. And since glucose no longer can be utilized within mitochondria using the oxidative phosphorylation pathway, now the cytosol takes over generating necessary energy by fermentation of glucose. During this process an archaic program is activated at which apoptosis is blocked (no programmed cell death), the cell continuously replicates, and the cell looses its differentiation (the three postulates for cancerous cells). In contrast to oxidative phosphorylation, only 1/9th of the energy is now being generated by fermentation within the cancer cells, a process, which needs additional substitutes for the loss in energy production. Therefore the compensatory degradation of amino acids from the musculature, channeled into the Krebs cycle, is initiated leading to an excessive loss of body weight in the terminal phase of cancer.

The fundamental findings of cellular biology in cancer cells can be summarized as follows:

1. Between the genomes of mitochondria and the nucleus there is a two-way interconnection.
2. Once there is a functional (and not a structural) failure of this two-way interconnection the cell no longer resorts to the highly differentiated activity of mitochondrial function of oxidative phosphorylation for the generation of ATP.
3. The cause for such failure is the increasingly malfunction of mitochondria, being the universal generator of >90% of all energy dependent biochemical processes within the body.
4. Because the usual production of ATP in a normal individual already amounts to 70 kg (the size of his body weight!) and lasts only for 3 s, it is therefore conceivable that the cancer cell had to switch another way in order to generate sufficient energy.
5. The archaic, non oxygen dependant (anaerobic) but less efficient way in generating energy goes in hand with a 20-fold higher need in glucose to make up for this deficit, however at the cost of other organs.
6. In cancerous cells especially complex IV of mitochondria is impaired which has the task to transfer electrons to molecular oxygen thus resulting in the reduction to water. Such a scenario has been demonstrated in cancer cells by researchers who, however, were unable to give a sound explanation for such phenomenon.
7. Already in 2002 researchers at the university of Helsinki in preclinical data and clinical trials together with microscopic evidence conclusively demonstrated that transformation into a cancer cell is associated with a loss of control of mitochondria during cellular mitosis.
8. Further evidence for the failure of mitochondria in cancer cells was given by researchers from the Anderson Cancer Research Center in Houston/Texas in 2003 using curcumin as a remedy to stop growth of tumor cells. They demonstrated that curucmin (the active ingredient of the plant curcuma longa)

effectively blocked all signal pathways in tumor cells. An explanation for such effect may be that curcumine absorbs the purple spectrum of visible light, i.e. the very same wave length of 415 nm which is being used by the electron transporter molecule cytochrome c in complex IV and which is degraded by the enzyme hemoxygenase within cancer cells. Thus, any deficit of action of the enzyme results in a short cut between complex III and IV within the electron transport chain of mitochondria, an otherwise necessary step in the formation of sufficient ATP. Curcumin therefore seems to be able to bridge this interruption in the chain leading to reconstitution of the normal biochemical pathways within impaired mitochondria, which now regain their ability for oxidative phosphorylation, and the synthesis of ATP.

Clinical trials with curcumin in Phase II trials suggest that the anticancer activity of curcumin is likely to be of value for chemoprevention and anticancer treatment. Curcumin, a compound in the spice turmeric, temporarily stopped advanced pancreatic cancer growth in two patients and substantially reduced the size of a tumor in another patient, according to a small study published in the Journal of Clinical Cancer Research. Of concern before starting the study was the fact that curcumin normally is poorly absorbed (low bioavailability), suggesting that only low levels get into the bloodstream after the capsule form has been taken by mouth. The fact that low levels of curcumin resulted in a benefit in this study, even in a small number of patients, suggests that if one could find a better way to get the agent to the tumor, one would see a greater response of induction of enzymes involved in the detoxification of the electrophilic products of lipid peroxidation that may contribute to the anti-inflammatory and anti-cancer activities of curcumin [17]. Curcuma's potential to regulate the inflammatory response through inhibition of pro-inflammatory mediators and the NF-κB signaling pathways; it is also able to induce pro-apoptotic proteins. The latter has been demonstrated by a preclinical study where the agent was capable of suppressing malignant glioma growth in vitro and in vivo [18]. Also, in colorectal cancer cell lines [19], curcumin potentiated the growth inhibitory effect of celecoxib (Celebrex®) by shifting the dose-response curve to the left. Such synergistic growth inhibitory effect is mediated through a mechanism that probably involves inhibition of the COX-2 pathway and may involve other non-COX-2 pathways, effects that are of clinical importance because they were achieved in the serum of patients receiving standard anti-inflammatory or antineoplastic dosages of celecoxib. All such data suggest that the pharmacologically safe agent curcumin holds a big promise for future clinical application in cancer therapy.

Aside from curcumin, orepigallocatechin-3-gallate (EGCG), a polyphenol antioxidant present in green and black teas, decreases inflammation and induces cell cycle arrest and apoptosis in vitro [20]. Also N-Acetyl-cysteine has been successfully given to patients not in a conventional chemoprevention protocol, but rather as a chemopreventive agent in the context of acute UV exposure [21]. Such data corroborate the common notion that oxidative stress is involved in cancer and agents targeting as scavengers have a potential therapeutic implication in cancer therapy and should be tested in a wider patient population.

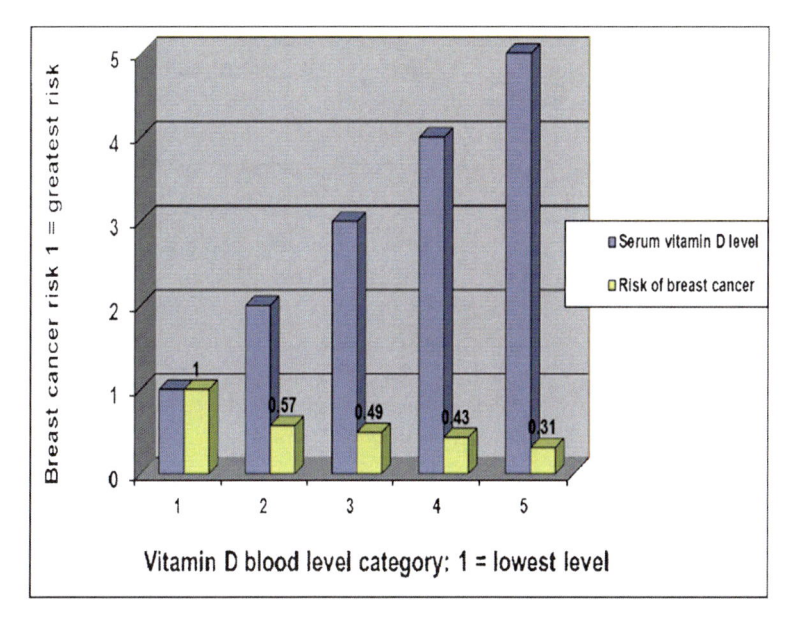

8.3 Mitochondria Involved in Mutation to Cancer Cells?

Fig. 8.8 Serum vitamin D_3 levels and risk of breast cancer. Adapted from [5]

Additional trace elements to take are as follows:

List of recommended daily trace mineral supplements with dosages

- **Calcium (as calcium chloride) – 60 mgs**
- **Magnesium (as magnesium chloride) – 300 mgs**
- **Potassium (as potassium chloride) – 40 mgs**
- **Zinc (as zinc chloride) – 6 mgs**
- **Iron (as ferric ammonium chloride) – 3 mgs**
- **Boron (as sodium borate) – 2 mgs**
- **Iodine (as potassium iodate) – 0.3 mg**
- **Copper (as copper sulphate) – 0.2 mg**
- **Manganese (as manganese chloride) – 0.2 mgs**
- **Molybdenum (as sodium molybdate) – 40 µg**
- **Selenium (as sodium selenate) – 40 µg**
- **Chromium (as chromium chloride) 40 µg, but most of all consider**
- **Vitamin B_{12} – 1000 µg plus**
- **Vitamin D (as cholecalciferol) – 1,000–2,000 IU as necessary**

Sunshine. Sunshine is necessary for synthesis of vitamin D_3. Vitamin D deficiency is extremely common and partly responsible for our epidemics of immune disorders (allergy and autoimmunity), osteoporosis, cancer (Fig. 8.8) and heart disease!

Vitamin D is extremely safe and should be taken at the rate of at least 1,000 IU with a maximum daily dose of 5,000 IU of vitamin D_3. Sunshine is the best way to increase vitamin D_3 levels. Thirty minutes of good sunshine on skin exposed by swimming costumes will give us about 5,000 IU of D_3. Since it is impossible to get adequate vitamin D_3 from sunshine in the northern parts of Europe or the US, hence there is the need for a supplementation.

Salt. If one eats not processed foods, then there is the need for some salt. Use sea salt, which also contains very rare trace elements that are also likely to be essential for normal metabolism. It is suggested to take 1/4 tsp daily on the food. The above nutritional supplements is designed for patients but can, of course, be implemented by anyone. Otherwise, the amounts of nutrients in the above list serves as a guide in putting together the own nutritional regime based on good quality multivitamin and multimineral preparations, essential fatty acids with vitamins C and D_3.

8.3.1 Exogenic Factors Rising Nitric Oxide, Blocking Endogenous Synthesis

1. In the majority of cases, any increase in the level of NO within the body stems from outside sources. This is, because most of the daily food we consume is contaminated with additives, which after metabolization results in the liberation of NO within the body. For instance nitrates are being used exceedingly used in fertilization, the groundwater we drink is contaminated with nitrates from fertilized acres, canned food is preserved with nitrates and even the corned beef or the salted sausage is preserved with nitrates.
2. Also, most of all the regular medication taken by patients such as nitroglycerine, amylnitrate or other slow release coronary dilators prescribed in coronary artery disease, results in a release of NO.
3. On the other hand all antibiotics destroy those bacteria in the gut that are necessary for the formation of sulfur containing compounds which are mandatory for detoxification of excess NO and nitrosated amino acids.
4. In addition, cortisone and all cholesterol-lowering agents (e.g. statins), as well as all peripherally acting analgesics (e.g. ibuprofen) or the daily intake of contaminated food with heavy metals (which according to a recent analysis was even found in the majority of different white and red wine brands (Ref. [247] in Chap. 7), inhibit the formation of coenzyme Q10, a necessary constituent within the respiratory chain of mitochondria.
5. Aside, all chemotherapeutics can be considered toxic for mitochondria, especially when taken over a long period of time (>7 days).
6. Any deficit in the formation of the bodies own antioxidative system essential for detoxification (e.g. SOD) result in the formation of toxic peroxides such as $ONOO^-$, which induces malfunction of mitochondria (Table 8.3). Because NO has a higher affinity to superoxide, as opposed to the detoxifying superoxide

8.3 Mitochondria Involved in Mutation to Cancer Cells?

Table 8.3 Summary of additives to daily food or exposure to agents resulting in nitrosative stress with damage to mitochondria with the likelihood of leading into chronic ailment and even cancer

- Xenobiotika exposure in solvents, or exposure to halogenated aromatic hydrocarbons
- All fungicides, pesticides, or heavy metals in food
- Nitric vapors
- Nitrates for conservation in sausages and meat products
- All NO-releasing agents *such as* isosorbiddinitrate *and* isosorbidmononitrate used in angina
- High blood pressure lowering drugs (i.e) angiotensin-converting enzyme (ACE) inhibitors, angiotensin (AT) receptor-blockers
- Cholesterol lowering agents such as statins, fibrates
- Antidiabetics such as metformin, or insulin sanitizers
- Potency boosters such as Viagra
- Anti-arrhythmic agents
- Antacids (proton pump inhibitors; PPI)
- Non-steroidal anti-inflammatory drugs (NSAIDs); acetaminophen
- All antibiotics
- All irradiation used in medicine and exposure following a nuclear catastrophe
- All kinds of chemotherapy
- Chronic psychostress
- Massive infections, and/or inoculation
- Nicotine abuse in smoking
- Solubiliser/emulsifiers (i.e. poloxamers) added to soft drinks
- All kinds of preservatives in food
- Food additives such as monosodium glutamate
- Dyestuffs in food
- Stabilizers in food or drinks
- Artificial coloring and sweetening of soft drinks

dismutase (SOD) within mitochondria (manganese-dependant) or the cellular matrix (cupper-, zinc-dependant) superoxide dismutase, any increased formation of $NO°$ and O_2^- results in the deadly cocktail peroxynitrite (Fig. 8.9), irreversibly damaging the respiratory chain in mitochondrial function and inhibiting the formation of mitochondrial superoxide dismutase [22].

As a result mitochrondropathy becomes evident, during which an insufficient production of the high-energy substrate ATP with a low in energy of the affected cellular formation. In order to survive the cell now switches into an emergency program with anaerobic glycolysis. In this sequence of metabolic effects, especially peroxynitrite is highly toxic, oxidizing and combines with aromatic amino acids such as tryptophan, serotonin, phenylalanine, or tyrosine. These nitrosated amino acids such as nitrotyrosine can be detected in the plasma similar to another side-product in NO-formation in the urine, citrulline. By combining with thyroxine or catecholamines new chemical structures are being formed, which when detected by the immune system, are identified as antigens with the formation of antibodies, leading into an autoimmune reaction and the beginning of an autoimmune disease. A typical example of such an antigen-antibody reaction is the autoimmune-induced disease Hashimoto thyreoiditis.

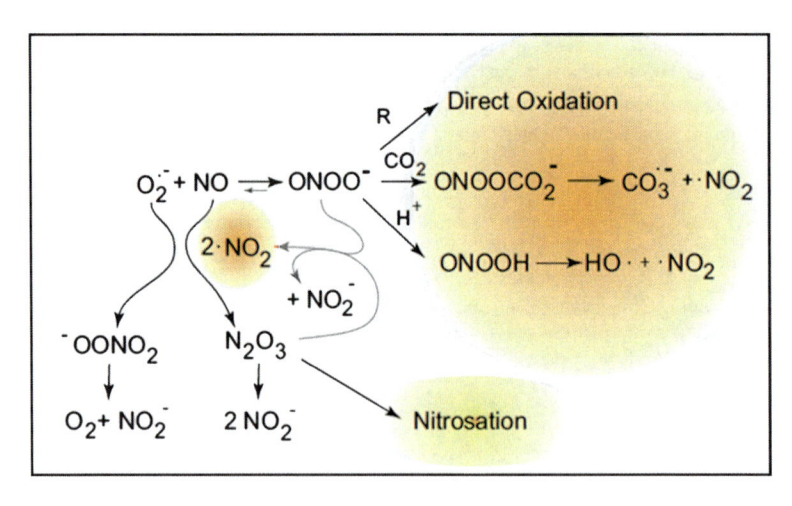

Fig. 8.9 The interplay of nitric oxide, superoxide, peroxynitrite, and nitrogen dioxide. When nitric oxide and superoxide are both present, they may also react with nitrogen dioxide to form N_2O_3 and peroxynitrate. Peroxynitrate decomposes to give nitrite and oxygen, while N_2O_3 can react with thiols to give nitrosothiols or with hydroxide anion to give nitrite. Adapted from (Ref. [15] in Chap. 5)

Table 8.4 Summary of stressors implicated in the initiation of different ailments leading into mitochondropathy

Chronic fatigue syndrome	Viral infection, bacterial infection, organ phosphorus pesticide exposure, carbon monoxide exposure, ciguatoxin poisoning, physical trauma, severe psychological stress, toxoplasmosis (protozoan) infection, ionizing radiation exposure
Multiple chemical sensitivity	Volatile organic solvent exposure, organophosphorus/carbamate pesticide exposure, organochlorine pesticide exposure, pyrethroid exposure, mercury exposure, carbon monoxide exposure, hydrogen sulfide exposure
Fibromyalgia	Physical trauma (particularly head and neck trauma), viral infection, bacterial infection, severe psychological stress, pre-existing autoimmune disease
Post-traumatic stress disorder	Severe psychological stress, physical (head) trauma

These stressors as outlined in the Table 8.4 are those most commonly associated with that specific illness. It should be noted that the majority of such stressors are involved in the initiation of more than only one illness. There are up to 17 diverse stressors associated in initiating chronic ailments, and one may ask, how these stressors are able to have such an impact on health ? It had been demonstrated, that all these stressors results in an increase in nitric oxide (NO) levels. Each of them has been experimentally demonstrated to increase the level of nitric oxide, or to stimulate a process, which by itself is known to increase nitric oxide. There is a striking common response and one may argue how the increase in nitric oxide might lead to chronic illness? This is because nitric oxide primarily acts through its oxidant product peroxynitrite (ONOO⁻), initiating a vicious biochemical cycle that is responsible for chronic ailments. While there is an initial cause of the illness via a short-term stressor(s), which starts this vicious cycle, being the initial spark for this cycle responsible for leading into mitochondropathy and a chronic phase of illness. This cycle is called the NO/ONOO⁻ cycle reflecting the elements nitric oxide (NO) and peroxynitrite (ONOO⁻), which are the main culprits in initiating chronic diseases (Fig. 6.1).

8.4 Avoiding Inflammation, Mediator of Nitrosative Stress

The prevalence of chronic conditions associated with inflammation is significant in the western world. It seems that nearly everyone is afflicted with a chronic inflammatory disorder of some sort or knows someone who is. Inflammation can develop across a wide range of common conditions that include psoriasis, eczema, rheumatoid arthritis, allergies, asthma, headaches, inflammatory bowel disease and many others. The following inflammatory disorders in one way or another induce the release of NO:

- Dermatitis
- Psoriasis
- Eczema
- Chronic Prostatitis
- Vasculitis
- Asthma
- Allergies
- Atherosclerosis
- Headaches
- Migraines
- Autoimmune Diseases:
  - Rheumatoid Arthritis
  - Inflammatory Bowel Disease
  - Cohn,s Disease
  - Colitis
  - Celiac Disease

Inflammation, from the Latin word "inflamatio" meaning to set on fire, describes the immune response that the body provokes when they encounter harmful stimuli. The body responds to these intruders and seeks to defend and heal our compromised tissue by stimulating a complex inflammatory response (characterized by swelling) that is designed to accomplish these goals. In chronic conditions, the inflammatory response is extended for days, months or years and involves the concurrent destruction and healing of body tissue and associated pain. Of significance are macrophage cells that are sent to the location of chronic inflammation to help out. Macrophage cells are very powerful white blood cells that help protect the body from harmful agents by engulfing them and releasing toxins in defense. These same protective toxins, however, damage our own body tissues in periods of chronic inflammation and often keep them inflamed, irritated, and painful with all the traditional signs of inflammation:

- Redness
- Heat
- Swelling
- Pain
- Loss of function

While in acute inflammation mast cells are the key players, which release TNF-alpha and histamine, the inflammatory response to tissue damage is of great value, since inflammation

- isolates the damaged area,
- mobilizes effector cells and molecules to the site, and – in the late stages,
- promotes healing, thus protecting the body from further harm.

8.4 Avoiding Inflammation, Mediator of Nitrosative Stress

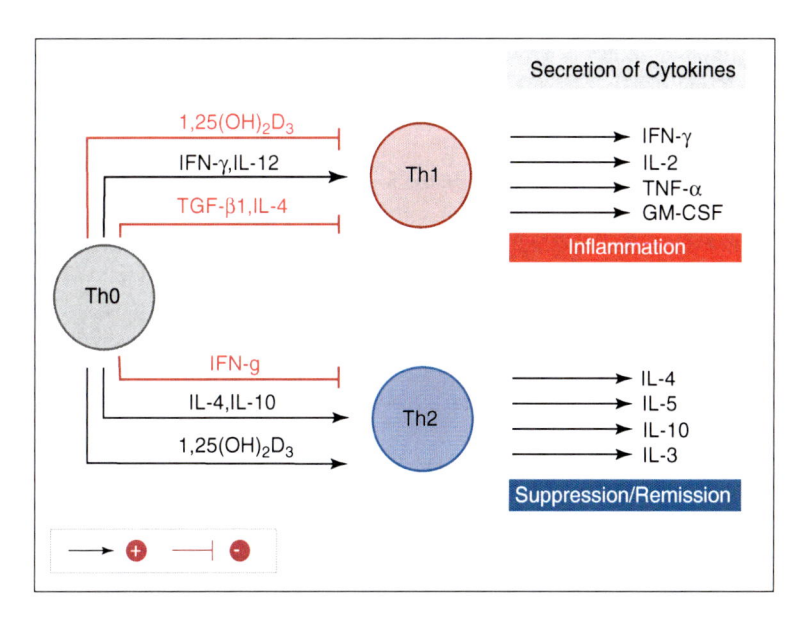

Fig. 8.10 By using different antioxidants the dysbalance within the immune Th1/Th2 system can be readjusted in order to fight off intracellular pathogens and reverse the inflammatory response

Contrary, in chronic inflammation, the inflammatory response is out of proportion to the threat it is faced with or is directed against inappropriate targets. In this case, the result can be more damage to the body than the pathogen itself would have produced. Thus, in many of these cases, the problem is made worse by the formation of antibodies against

- self antigens or
- persistent antigens from smoldering infections resulting in a dysbalance within the Th1/Th2-system (Fig. 8.10).

Those who suffer from chronic inflammatory disorders are well aware of how debilitating these conditions can be. Aside from antibiotics often, corticosteroids are prescribed to reduce inflammation in an inhaler, a topical cream, or as an oral pill form. Since both remedies especially when given over a long period of time, result in marked side effects, it is important to assess the benefits and risks of steroidal anti-inflammatories against NSAIDs (non-steroidal anti-inflammatories) to find the best option. However, as visualized in the above figure, re-establishing a balance in the Th1-dominance can be achieved by means of micronutrient therapy using antioxidant supplements such a Vitamin D_3, CoQ10, L-carnitine, NADH, magnesium, Vitamin B_{12} and the wide array of vitamin E tocopherols and tocotrienols in conjunction with natural vitamin C. Especially vitamin D_3 has been shown to activate the release of the transforming growth factor ß (TGF-1ß) being a natural inhibitor of Th1-mediated inflammatory response, and thus can be considered an anti-inflammatory cytokine [23]. It also should be noted, that once antibiotics are in use they immediately should be given in conjunction with probiotics to reassure that normal gut bacteria is not wiped out resulting in an overgrowth of a resistant strain which later is the cause for the intestinal bowel syndrome (IBS).

Aside from a local inflammatory processes, generalized inflammation as being observed in sepsis should also be considered as a major source in the generation of nitric oxide and peroxynitrite which in the long run compromise organs resulting in a multiorgan failure. A representative example of a patient with meningococcal sepsis and an acquired MRSA (methicillin-resistant staphylococcus aureus) infection, who was artificially ventilated (BIPAP) in the ICU with his blood pressure being only maintained with inotropic agents (dopamine, dobutmaine) and who because of intracerebral edema was given conventional antiepileptics for abortion of seizures. Aside from the regular antacid medication with PPI in conjunction with high dose antibiotics, such a patient in general has a very low survival index. Because it was felt that multiorgan failure could be reversed by inhibiting the devastating effect of ROS and NOS being expressed in the advanced course of sepsis in addition, he was given high dose vitamin supplementation (especially of the B-type), high dose vitamin C (up to 2 g/day), NAC (up to 2 g/day), CoQ10 in doses of 500 mg/day and magnesium (500–1,000 mg/day) together with an amino acid infusions and a short-chain fatty acids supplementation (Lipostabil®) for nutritional purposes. This was topped with an extract of berries, which in previous studies had demonstrated potent antioxidative scavenging capacity given via a nasogastral tube. The effect of this combined therapy is outlined below depicting various variables as indicators of oxidative stress and malnutrition in tissue and depressed mitochondrial activity (Fig. 8.11).

Chronic inflammation activates iNOS with local excess of NO-production followed by mitochondropathy.

8.4.1 Inflammation via NF-κB Activation

Any increase in ROS and NOS aside from TNF-alpha and IL-1, IL-6, are also known to induce NF-κB in the immune system resulting in an imbalance of the TH1/TH2 immune system and an activation of the inflammatory cascade (Fig. 8.12).

Mitochondropathy is especially upregulated when patients with chronic inflammation in addition demonstrate:

- A regular intake of a medication for lowering cholesterol levels
- A regular intake of medication with vasodilatory properties
- The regular use of amylnitrite (popper as used by addicts) because of their euphoric state. Amylnitrite, in common with other alkyl nitrites [25], is a potent vasodilator (*i.e.* expanding blood vessels via NO release, resulting in lowering of the blood pressure). Alkyl nitrites function as a source of nitric oxide, which signals for relaxation of the involuntary muscles. Physical effects include decrease

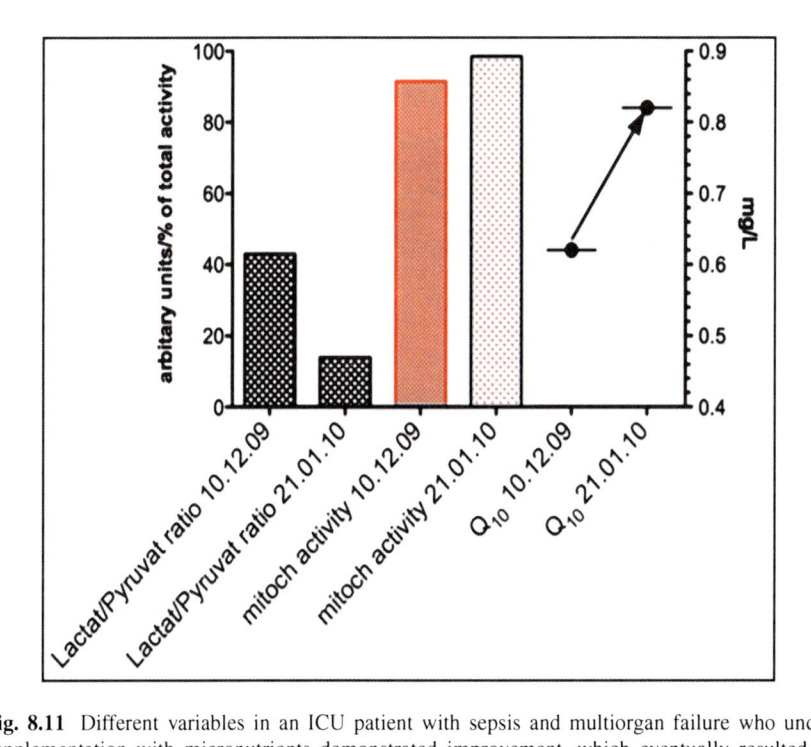

Fig. 8.11 Different variables in an ICU patient with sepsis and multiorgan failure who under supplementation with micronutrients demonstrated improvement, which eventually resulted in clinical recovery over a period of 1 month. He could be weaned off from the respirator and his cardiac function improved being reflected in the increase in his mitochondrial activity while the previously low Q10 level (*right side*) rose to higher levels

in blood pressure, headache, flushing of the face, increased heart rate, dizziness, and relaxation of involuntary muscles, especially the blood vessel walls and the anal sphincter.
- Digestion of food contaminated with chemicals for preservation while not taking the standard recommendations of nutritional supplements.
- A chronic paradontitis
- An autoimmune disease (thyroiditis, MS, etc.)
- Any virus-related chronic infection.
- But most of all obesity with a BMI > 25 kg/m^2 is considered a major source of a so-called silent inflammation where fat-cells show an increased expression and release of active hormones such as leptin, resistin, adiponectin, interleukin-6 (IL-6), INF-γ and tumor necrosis factor alpha (TNF-alpha). These active hormones are indicators of an ongoing silent inflammation, which is reflected by a sedimentation rate higher than normal (3–13 mm/h) and an increase in CRP (C-reactive protein; 10–40 mg/L norm), sensitive clinical parameters for inflammation. Especially cells that comprise visceral, abdominal fat (Fig. 8.13) release those inflammatory cytokines and amounts of hormones than cells that make up subcutaneous fat.

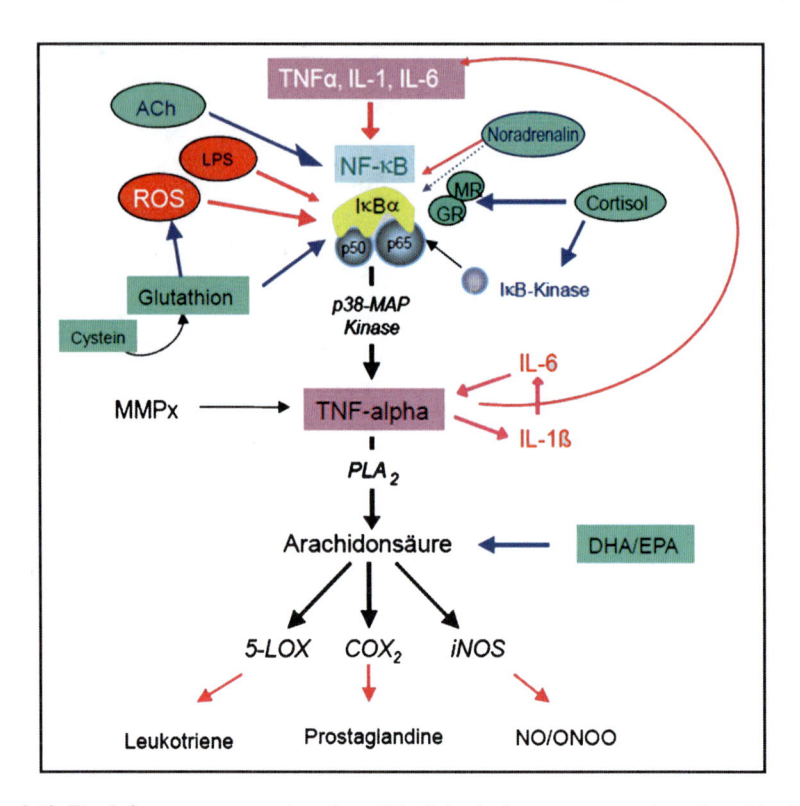

Fig. 8.12 The inflammatory cascade, where NF-κB in the immune system is activated by ROS resulting in the release of inflammatory components (TNF-alpha, IL-6, IL-1ß) which by themselves induce a vicious cycle. Green = sites within the cascade where therapy with micronutrients is effective. Adapted from [24]

8.4.2 Cryptopyrroluria, Nitrosative Stress and Mitochondrial Disease

Cryptopyrroluria is a genetically determined, metabolic disorder, often occurring as a biochemical-enzymatic familiarly during the chemical reaction of formation of the red blood pigment. A typical sign is the abnormally increased excretion of pyrrole (Malvene factor) in the urine. This can be given the highly probable diagnosis of cryptopyrroluria, and although this substance is detected in everybody's excretions, if the amounts are higher than ten micrograms per deciliter in the urine however, then there are corresponding subsequent damages, which follow cryptopyrroluria. In addition, cryptopyrrolics are an indirect indicator for a high NO (nitric oxide) synthesis, and is measured by detecting pathological high citrulline levels in the urine or by an increase of NO concentrations in the expired air. Being indicative for mitochondropathy, the pyrrole together with vitamin B_6 and zinc are excreted in excess in the urine. Evidence, whether sufficient vitamin B_6 and zinc

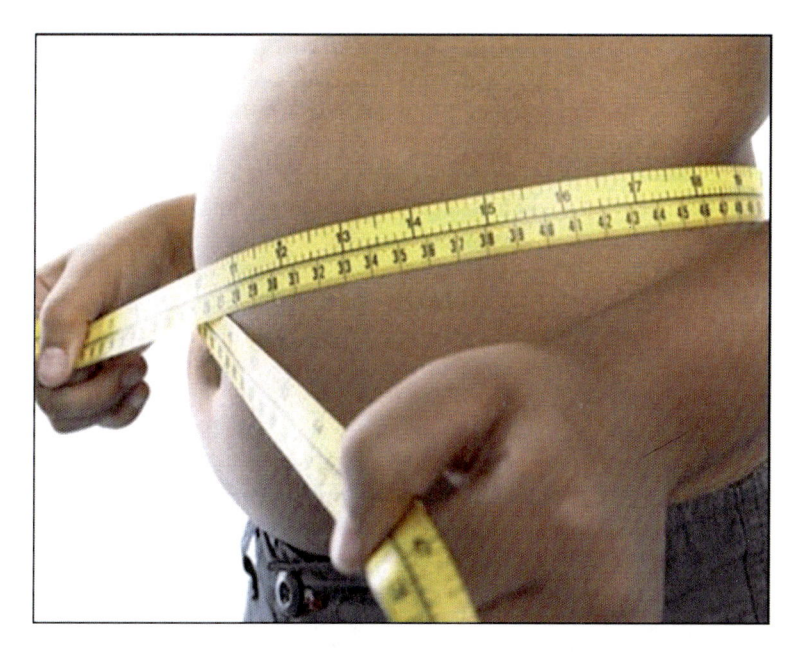

8.4 Avoiding Inflammation, Mediator of Nitrosative Stress

Fig. 8.13 Obesity as estimated by the waist-to-hip ration. For men, a ratio of .90 or less is considered safe, for women, a ratio of .80 or less is considered safe. In the US at least > 30% are considered obese with Blacks having a 51% higher prevalence of obesity, Hispanics a 21% higher obesity prevalence compared with whites. In Germany there is a 50% prevalence of obesity, a reason why Germans are considered the fattest people in Europe shortly followed by the British population

are available for cell metabolism can be derived only from intracellular blood analysis. Serum analyses in such case do not reflect the true deficit. Thus, *Cryptopyrroluria* is only a sign of a concomitant excess formation of NO, and once NO affects blood formation as well as the function of other blood components, cryptopyrroluria becomes evident. Over the day there is a marked fluctuation, which can be interpreted as a stress factor because of the nervous system leading to a deficit in vitamin B_6 and zinc. Such deficits are not detectable by serological analysis. More significant is the analysis of cystathionine in the urine. This parameter is increased, when the vitamin B_6 demand in the organism cannot be met, although serum values may still lie in the standard range or even in higher range. Together with a zinc deficit this can be indicative and is only diagnosed by intracellular analysis. Within the context of a mitochondrial disease, cryptopyrroluria is necessary for the survival of the body and an important regulator. Any zinc deficit is of importance, since numerous kinases become inactive because and phosphorylation of vitamin B_6 and vitamin B_1 is no longer fully possible. As a result secondary deficits, due to a vitamin B_6 deficit, with insufficient neurotransmitter formation, protein synthesis and transamination occur. The low zinc levels are probably desirable by the organism, because during low energy yield situations the intake of zinc would additionally inhibit the aconitase in the citric acid cycle and apart from that, zinc further activates the glutamate receptor (NMDA receptor). In such a situation the energy deficits and the state of irritability would increase. The brain has priority for the maintenance of its performance ability and for the securing of minimal energy availability. Disadvantages are vitamin B_6 utilization disturbances, but also the increased risk for further disorders/ diseases. Particularly rich in zinc is the retina, the prostate gland and the hippocampus, an organ responsible for the conversion of the short-term memory to long-term memory in the brain. By a zinc deficit, vitamin A cannot be transported either, since the protein binding retinol is dependent on zinc. This situation explains the fact that cryptopyrrolics often react with massive side effects to any zinc intake. Light forms instead, show rapid clinical, cerebral improvements by administration of vitamin B_6 and zinc. In light cases the affected persons indicate a strong appetite for meat dishes, in severe cases there is an aversion against meat, since the conversion of muscle protein to body own protein is dependent on vitamin B_6 and if there is a corresponding strong deficit in zinc and/or Vitamin B_6, the organism cannot break down normally foreign proteins. It is noteworthy that mitochondrial genetic damages are inherited from the mother, not from the father. This explains among others, that cryptopyrroluria can also be inherited. A chronic energy deficit in the nervous system often leads to an additional activation of C nerve fibers. These surround mast cells in a network form, so that by irritations increasingly histamine is released. The chronic histaminosis (histadelia) has a stimulatory action to the brain and peripherally it is a question of time, when allergies against external factors (pollen, dust etc.) would occur. The affected persons are not diseased because of the pollen, the mites etc. to allergic reactions, but rather to an increased release of histamine and a disturbed break down of histamine. The diamine oxidase needs copper and vitamin B_6 for the breakdown of histamine. Although in the brain there is no DAO available, many methylation reactions, which need S-adenosyl methionine (SAM), take place. SAM however, in order to maintain its reactivity, needs an adequate methionine supply obtained by meat as well as an adequate supply of vitamin B_6, folic acid and vitamin B_{12}. If not enough meat is consumed, affected persons have an additional deficit in vitamin B_6, Vitamin B_{12} and methionine. Because of this the brain and the peripheral organs cannot break down histamine adequately. Besides, SAM is necessary for the synthesis of adrenaline, the formation of the sleep hormone melatonin from serotonin and for the synthesis of polyamines such as spermidine and spermine, which play an important role during cell division and cell maturity. Any zinc deficit leads to functional weakness of the Zn/Cu superoxide dismutase. Apart from that, by zinc deficit glutathion-S-transferases become inert. Affected persons show a very sensitive reaction to foreign and harmful substances. In addition, NO inhibits the cytochrome P450 enzymes, which are very important for the phase-I-detoxification.

Mothers with KPU and mitochondrial diseases portray an increased risk for complications during pregnancy. NO activates the uterine contractions and opens the cervix. Consequences of this effect are premature births. Therefore, metabolic deficits and the chronic surplus of NO lead to disturbances in the maturity of the infant brain. Visual, acoustic and motoric centres, the interconnection of the right and left brain hemispheres, the dominance of the left temporal lobe are underdeveloped, exactly as the interconnection of neurons with each other. These children grow up with obvious neurological and psychological deficits. Children with ADHS have a brain volume smaller by 3% than without ADHS. Further symptoms of mitochondrial diseases are a higher risk for neurodermatitis, nasal polyps, and inflammations of the middle ear, bronchitis, allergies and many more diseases. The physician speaks of comorbidities, but they are not. The result of the above is that the treatment for pyrroluria only for light forms is the administration of B-vitamins, zinc and other micronutrients, because in the end cryptopyrroluria must be treated as a mitochondrial disease. This treatment is only possible by using micronutrients, which have an influence upon the electron transport chain of mitochondria and secure an improvement of the citric acid cycle. As one can see, the numerous diseases appearing in childhood such as ADS, ADHS, neurodermatitis, fatigue, pain and allergy of children, are not isolated diseases, but always the expression of a mitochondrial disease. This assumption is underlined by the fact that all such diseases are seen with an increased rate of NO of formation and of citrulline synthesis. Therefore, a consequent and cause-oriented therapy is only possible when mitochondria are treated accordingly, not however when using symptomatic treatment such as Ritalin®, corticoids or other medications, all of which further overactivate the immune system. Thus, the development of allergies against certain proteins in early childhood per se is not an allergic reaction, but rather the expression of an increased release of histamine, or an inhibition in the breakdown of histamine due to nitrosative stress. Numerous dipeptidases of the small intestine (enzymes breaking down proteins) can only act in the presence of zinc (thy are zinc- dependent). Some allergies such as protein allergies or cow-milk allergies are caused by such an effect. Since by these mitochondrial diseases, disturbances in the breakdown of lactose, fructose and gluten are unavoidable, such symptoms can develop. It would be wrong to place children and youths on a diet, since this is not a therapy treating the primary cause. High NO and histamine open and consequently damage the blood brain barrier. Also, one should consider the release of increased levels of NO^- in cases of instable cervical spine, which results in episodes of malperfusion with stress followed by the release of NO. This for instance is seen after whiplash syndrome, automobile collision, trauma of the head, forceps use during vaginal delivery etc. This way, disturbances of nerve cells in the long run are unavoidable. Specific markers for such pathology in persons are a pathological increase in the blood brain barrier protein S-100, or in neuron specific enolase. And finally, cryptopyrroluria is a symptom of a mitochondrial disease with the consequence, that cryptopyrrolurea means for the patient: a change of nutritional habits, intake of nutrition supplements, resulting in higher resistance to toxic substances and detoxification. Sequelae of such "Redox-Dysbalance" are the acquired mitochondrial disease. In this context it is important to consider how the $NO/ONOO^-$ activation explains the many unexplained properties of chronic multisystem diseases.

In summary, one can observe and explain the signs of cryptopyrroluria only by the concomitant symptoms of a significantly more serious disturbance in the mitochondrial level, which triggers performance deficits in several organs and if not attended properly, leads to massive health disorders in a multi organic level. Yet, question still remain, and for many it is questioned how a single cause of an underlying disease may end up in such a diversity of ailments. Thus one may ask:

1. How can a variety of stressors induce different cases of multisystem illnesses?
   All stressors either increase nitric oxide/peroxinitrate levels in the body or increase other elements within that cycle. This is because they all have the potential to stimulate the NO/ONOO⁻ cycle in specific regions of the body.
2. Why do the symptoms in chron. multisystem diseases differ from one patient to another?
   This is because the NO/ONOO⁻ cycle is basically local, being localized to different tissues in different individuals.
3. How do these different multisystem illnesses differ from one another?
   Each patient must have an overactive NO/ONOO⁻ cycle in one or more selective organ in order to meet the diagnosis of mitochondropathy in different tissues (Refs. [73, 128] in Chap. 7) [26].
4. How are the symptoms and signs of illness generated?
   Through the occurrence of one or more elements of the NO/ONOO⁻ cycle (Ref. [4] in Chap. 5, Refs. [73, 128] in Chap. 7) [26].
5. How should these illnesses (mitochondropathies) be treated?

By using agents, which regulate different aspects in the NO/ONOO⁻ cycle. Are these illnesses true ailments? Yes, albeit somewhat unusual ones. Because of the fundamentally local nature of the NO/ONOO⁻ cycle, there is an almost unlimited number of mitochondrial diseases or a broad variation of NO/ONOO⁻ cycle caused ailments, a reason why there is a wide spectrum of illnesses. For example, fibromyalgia and another overlapping section called chronic fatigue syndrome are just different clinical symptoms of a similar underlying entity. By using different definitions for CFS, this will define a somewhat different section of this huge spectrum. Medicine commonly views a specific disease as being qualitatively different from each other. For example, the tuberculosis bacillus always causes different cases of tuberculosis, whereas different infectious agents cause other infectious diseases. However, within the NO/ONOO⁻ cycle, the concept of differential diagnosis on which modern medicine is based, strongly has to be questioned.

8.4.3 Specific Treatment Options in Chronic Fatigue Syndrome

Dr Paul Cheney recognizes that there are three stages of chronic fatigue syndrome, which are characterized by different immunological and biochemical issues. Chronologically these stages are not completely distinct, one runs into others. However, if one can identify the stage that you are in, this will have implications for treatment.

- **The acute alarm stage** – this follows the original trigger for chronic fatigue, which may be sudden onset following viral exposure or chemical exposure, or maybe of gradual onset over some months or years. This stage is characterized

8.4 Avoiding Inflammation, Mediator of Nitrosative Stress   261

by lots of symptoms, which can be anything from acute fever, lymphadenopathy and malaise to headache, irritable bowel, muscle aching, anxiety, sleeplessness or whatever. This stage probably reflects immune over-activity, which might be appropriate in the case of the virus or bacterial infection, inappropriate in the case of allergy and autoimmunity and is accompanied by a heightened stress response with high levels of stress hormones to allow the body to gear up and cope with these increased energetic demands. Chronic work or emotional stress also results in such an alarm reaction.

- **The exhausted phase** is characterized by biochemical and hormonal failures. Nearly always there is foggy brain, can't think clearly, inability to multitask, inability to learn. We see more exhaustion with poor stamina and delayed fatigue, but the shopping list of symptoms starts to lessen. They may be replaced by chronic background pain and malaise. This is the stage where there are obvious biochemical failures due to exhaustion of micronutrients, toxic stress and poor healing and repair. Typically we see poor mitochondrial function, poor antioxidant status, poor quality of sleep, poor digestion of foods with hypoglycemia, depression of the hypothalamic pituitary adrenal axis, etc.
- **The stage of maladaption**. This seems to arise after some years of the above problems. This stage is characterized by relative freedom from the above symptoms so long as one stays strictly within limits. However, this stage is characterized by a very marked push and crash. What this means is that should the sufferer go very slightly above their permitted limits (which may be bed bound!), then this has the potential to make them ill for days. What Dr Cheney hypothesizes here is that the normal stress response which involves cortisol, thyroid hormones, growth hormones, insulin and so on, has flipped. Normally if the body is stressed, this hormonal response allows us to move up a gear, cope with that stress and then drop back to baseline functioning. In this third maladapted stage the stress hormones seem to have the complete opposite effect. Instead of allowing the sufferer to gear up, they make him crash instead. This is where Dr Cheney is focusing his efforts for therapy, but in order to understand this third stage we need to look a little more closely at the second stage of biochemical failures.

Cheney believes that all these stages result from how the body deals with free radical stress. He calls this the "redox" state. In the first stage to cope with increased demands we greatly increase our output of energy, but this brings free radical stress. It is our ability to deal with, which determines who recovers and who goes on to a fatigue syndrome.

– *The ALARM STAGE*
This is most often triggered by viral infection Obviously one cannot avoid all viruses, but do your best. There tends to be a mini-epidemic of colds at the start of every school term as every virus acquired is shared around! Do not travel to exotic locations and expose yourself to the risk of picking up some horrible bug. Vaccinations can certainly trigger flares of CFS – some are probably essential and less likely to cause CFS such as tetanus and oral polio. Most other

vaccinations have the potential to flare CFS. The view of many alternative therapists is that the nutritional approach is so effective at protecting against viral infection that vaccination against seasonal flu is not necessary. Vaccination has the potential to trigger other problems such as CFS as they have mercury as an enhancer. The best defense against viral infections is a healthy body and healthy immune system. Indeed it is believed that if you have a perfect immune system you should never get a cold. Getting a cold is a symptom of a poorly functioning immune system. Ensure a good micronutrient status by taking vitamin D. This is underlined in a creditable hypothesis that explains the seasonal nature of flu is that influenza is a vitamin D deficiency disease. Cannell and colleagues offer this hypothesis [27], where the author asked the question why we tend to see more coughs colds and flu in the Winter compared to the Summer. The answer is vitamin D – the only significant source is sunshine and vitamin D is highly protective against infections of all sorts. The advice is to take at least 2000 IU daily on days if one gets less than 20 min of good sunshine directly on the skin. Aside, any causes, such as chemical poisoning and/or psychological stress, should be avoided – the problem is that often they are not recognized at that time, which presents a major problem. Good quality sleep is essential since this is when healing and repair takes place, as many of the above problems are accompanied by insomnia.

- *EXHAUSTED PHASE* – the second stage of chronic fatigue syndrome
This is the stage of biochemical and hormonal failures characterized by poor mitochondrial function. Poor mitochondrial function has been clearly demonstrated in a published recently {Myhill, 2009 #5816}, which clearly demonstrates that the degree of disability is directly proportionate to mitochondrial function, which is the energy available to cells. There is considerable success treating this mitochondrial dysfunction (Fig. 8.14) by addressing all the issues that can result in poor energy supply resulting in CFS and the therapy with respect to diet, sleep, supplements support, mitochondria, correcting antioxidant status, etc.

Mitochondrial function is closely linked with free radical stress. Mitochondria are a major source of free radicals, and if not handled efficiently, are easily damaged by them. This is why antioxidant status is so important. The balance between producing free radicals and our ability to deal with them is called our "redox" state. This is important because redox state determines how well a person deals with acute and chronic infections. Cheney and coworkers have developed a biochemical tool looking at the energy available to cells. The cellular free energy can be measured in the heart using echocardiography. The heart is the most energetic organ in the body and indeed more than 50% of the heart is made up by mitochondria. Mitochondria make energy in the form of ATP and this ATP is used to pump calcium into the sarcoplasmic reticulum where the muscle fibers are. This it does rather slowly, but when the concentration becomes critical, calcium rushes back and the energy generated by this allows the muscle fibers to contract. If mitochondria go rather slowly it takes longer for this charging up to take place. So the time between the start of atria charging and mitral valve closing (which is called the isometric volume relaxation time) is a direct reflection of mitochondrial function.

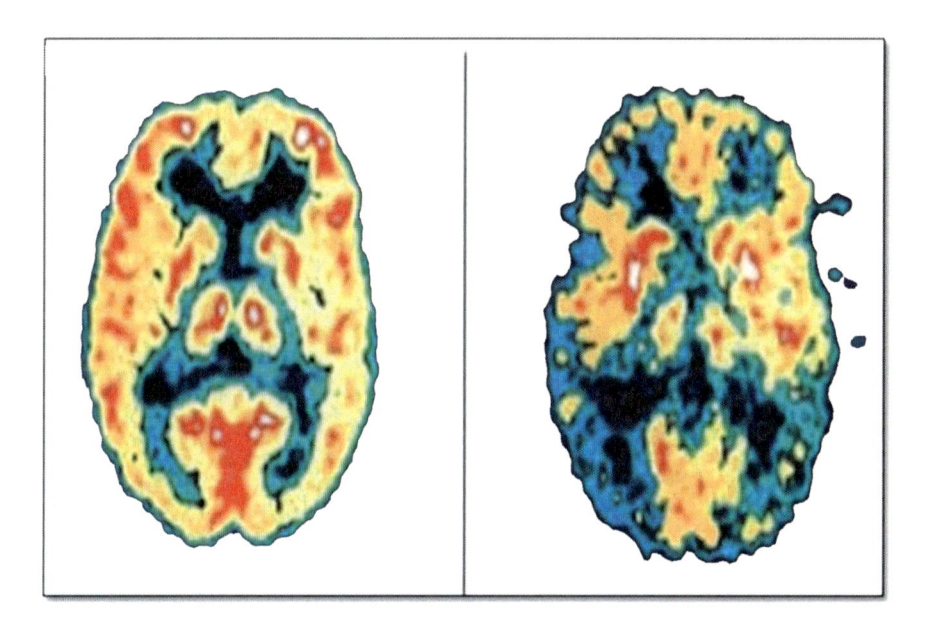

8.4 Avoiding Inflammation, Mediator of Nitrosative Stress 263

Fig. 8.14 Positron Emission Tomography (PET)-scan of a person with CFS (*right*) compared to a normal subject (*left*). The blue and black area indicate reduced activity resulting from damaged mitochondria

The more efficient the mitochondria, the more available cellular free energy and the shorter the IVRT. A normal result could be 75 milliseconds while heart failure occurs at 150 milliseconds. This test is highly accurate and reproducible within 2%. It allows Dr Cheney to assess within minutes the effect of various stressors on the availability of energy to cells in the heart. It is an instant measure of mitochondrial function. It is now known that people with chronic fatigue syndrome are in a low output cardiac state secondary to poor mitochondrial function and Dr Cheney has demonstrated this elegantly with his echocardiogram studies. In the process of doing these studies, he has identified two further problems which worsen the low cardiac output and can lead to a downward spiral in symptoms – one is a patent foramen ovale, the other is the problem of oxygen stress and free radical activity. Indeed he believes that how we deal with free radicals (our redox state) is what defines and controls CFS.

– *The maladapted stage*
  Cheney believes that this third and maladapted stage of chronic fatigue syndrome arises because of abnormal control mechanisms. Normally in a stress situation there is a release of stress hormones from the thyroid gland, the adrenal gland, and the pituitary gland in the brain. Dr Cheney has been looking at the heart response when these hormones are applied to controls and to patients with fatigue syndromes. He has developed extracts of these tissues and applies them transdermally. He sees a response in the heart within 30 s. This alone is quite

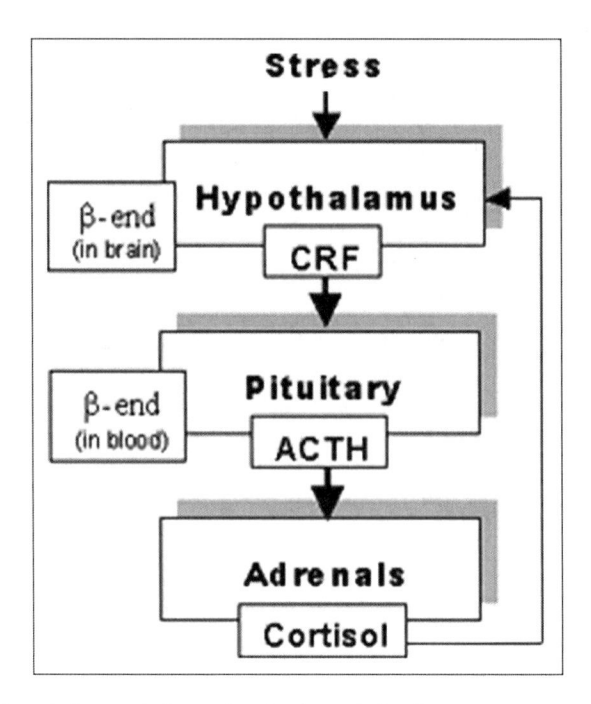

Fig. 8.15 The hypothalamus-pituitary-adrenal axis regulating the response to stressful situations in life. CRF = corticotropin releasing factor; ACTH = adrenocorticotropin hormone; ß-end = ß-endorphin, a natural opioid-like substance

amazing! What he finds is that when he applies these extracts to normal people is that the heart improves its function. However, when he applies these extracts to patients with chronic fatigue syndrome the heart response invariably gets worse. Cheney ran a trial whereby he took patients with CFS and for 6 months gave them transdermal liver extracts. Liver is rich in SODase and catalase. Although they improved with the heart tests, clinically they did not show any significant improvement. It was only when he added a heart extract, rich in glutathione peroxidase, that he saw some clinical recoveries. Cheney hypothesizes that he was using a "cell signaling factors" and believes that this may be a way of reversing this "flip" in the HPA (hypothalamus-pituitary-adrenal) axis (Fig. 8.15).

From an evolutionary perspective this may be the right way of thinking, since the tissues that have always been prized by primitive societies and wild animals are the meats of internal organs – heart, liver, kidney etc – over and above the muscle meat. The brain is wired to learn, and it does so by association and experience. Interestingly the NMDA-receptor is central to the learning process and we know this is activated in inflammation and pro-oxidant stress. It is therefore important to realize that CFS is a protective adaptive state. If one did not switch into a CFS, then the uncontrolled free radical stress would further impair the person ! Just complete lack of sleep for 2 weeks is sufficient! If we force the system against its will we risk creating more free radicals and making things much worse ! All the therapies that are being used have a logical basis:

- Reduce the generation of free radicals
- Rest and pacing for regeneration of mitochondria
- Lower/avoid the intake of toxins in the diet
- Get rid of toxins such as pesticides, heavy metals, VOCs and drug medication all of which generate free radicals
- Improve the ability to clear the body from free radicals
- Improve the micronutrient status by taking supplements
- Improve the length and the quality of sleep

The Redox state within mitochondria is greatly affected by pH resulting in massive changes of oxidation/reduction balance by small changes in pH. This is particularly true when major components of the redox system are heavily protein-buffered. In CFS patients, the degree to which they switch to anaerobic metabolism is one of the major factors. Those who 'struggle on regardless' soon begin to depend on anaerobic mechanisms and build tolerance to increased lactate formation being a major stress factor on general and the cell-specific redox balance.

In the leukocyte respiration studies, there is another major factor resulting in an uncoupling of the electron transport/oxidative phosphorylation chain. The one thing to keep in mind is that the catalase 'reduces' organic forms of metals including some detoxification conjugates back to simple organic forms of metals. This is a real problem in some cases because the inorganic non-fat soluble form cannot escape the cell. And because there is much catalase in cells, this process is inevitable, resulting in an accumulation at least of some toxic metals. With increased concentration of intracellular calcium, this promotes apoptosis with a normal cell replacement cycle. However, in some CFS patients it is now known that calcium-binding proteins are induced with Ca^{++}-Actin binding increased. If this happens with heavy metals then even more damage is possible and normal cell-replacement is not promoted because apoptosis is not triggered. The high affinity action-binding site of actin is capable of binding a wide range of metals, which for example has been demonstrated in terms of cadmium toxicity. Because of the reciprocal causality of Ca^{2+} and the actin cytoskeleton through actin-binding proteins this may be of importance not in terms of calcium but for the toxic metals, where catalase is worsening the situation to a great degree [28].

8.5 Therapeutics for Normal Mitochondrial Function

There are two additional important types of mechanisms that are important parts of the cycle and only recently have obvious from studies. Firstly, there are several known mechanisms by which peroxynitrite, superoxide and nitric oxide can lower energy metabolism in mitochondria. This lowering of energy metabolism is important both because it has a role in generating the symptoms of these diseases but also because it is part of the cycle itself. Specifically, it has important roles in producing an increase in NMDA activity resulting in pain, and it very likely, it also has an important role in producing elevated levels of intracellular calcium, being a part of the cycle. Aside from the substantial evidence for mitochondrial dysfunction in CFS and fibromyalgia, there is evidence from clinical trial studies that agents that improve mitochondrial function are helpful in the treatment of this group of illnesses, and where the NO/ONOO⁻ cycle presents a plausible mechanism for a number of multisystem illnesses. Within this scope, it must be possible to explain the symptoms and signs of a specific illness as being generated by one or more elements of that cycle (Fig. 6.1). Such explanations are necessary not only to explain both the specific symptoms and signs, but at the same time propose rational therapeutic interventions. Within this scope the hypothesis was corroborated in FMS patients who when taking an reduced glutathione (GSH) together with L-cysteine and N-acetyl-cysteine (NAC) for at least 2 months, demonstrated a marked reduction of malondialdehyde a surrogate marker in the urine reflecting corrosion (i.e. lipidoxidation) of monounsaturated fatty acids of cellular membrane by means of oxygen radicals (hydroperoxides) to chemical instable aldehydes and its major portion malondialdehyde (MDA; Fig. 8.16).

The reduction in MDA coincided with a significant increase of tolerance to pressure-induced pain at typical tender points being an indirect marker of the previously unregulated NMDA- and vanilloid excitatory receptors system within the pain transmission system (Fig. 8.17). These mechanism as outlined are established in other illnesses where there is an excess of NO and/or ONOO⁻ formation and present a plausible explanation which is based on such scientific data. Since the generation of such oxygen radicals occurs in a number of other multisystem ailments, this is consistent with the striking variation of symptoms and signs that are characteristic for such a diversity of illnesses. It may be argued that the symptoms and signs of CFS, MCS, FMS and PTSD but also of other chronic ailments should be found in all sufferers, in order to give the exact diagnosis.

In this respect, the arrows in the Fig. 6.1 outline the different interconnections within the NO/ONOO⁻ cycle represent different mechanisms by which one element can increase the level of another element of the cycle. Of these, 19 elements are scientifically documented and accepted by traditional medicine while there are three others that are less well recognized. Overall, the bulk of evidence underline the evidence for such mechanisms as outlined all of which have a common ending, i.e. leading into mitochondropathy. What one needs to do for each disease/ailment is to decide whether it is a possible candidate with the causative factor of an upregulated NO/ONOO⁻ cycle, and determine how good the fit is for that specific ailment. In the following, three types of evidence are discussed, that support the existence of the NO/ONOO⁻ cycle (Ref. [4] in Chap. 6, Ref. [128] in Chap. 7):

1. There is evidence from studies using two agents that release nitric oxide (i.e. nitroglycerine and Na-nitroprusside) that causes mammalian tissues to synthesize increased amounts of nitric oxide via the three nitric oxide synthases. Such data

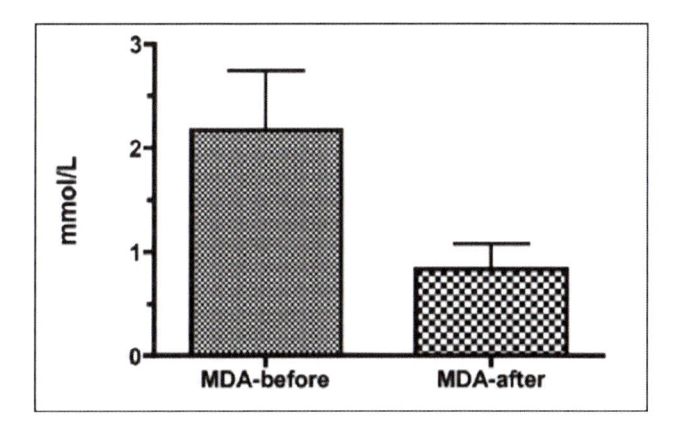

8.5 Therapeutics for Normal Mitochondrial Function

Fig. 8.16 Concentration of malondialdehyde (MDA) in urine of FMS patients before and after intake of the superoxide scavenger GSH plus NAC for 2 months (n = 12, mean ± SD). Adapted from [29]

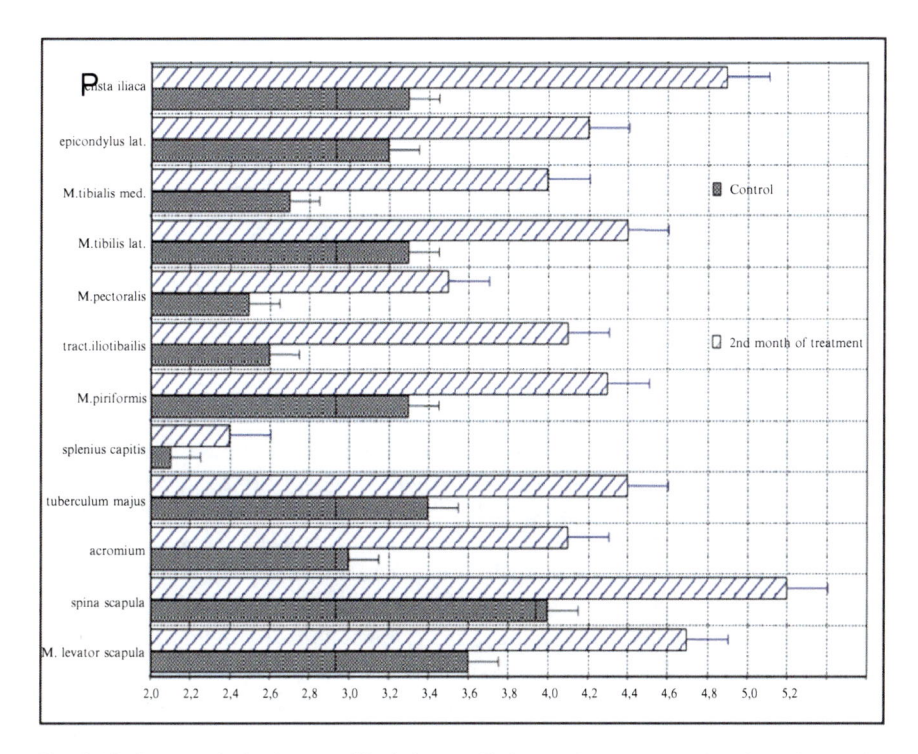

Fig. 8.17 Pressure in kg (mean ± SD) being applied to typical tender points in patients with fibromyalgia. Note, the significant rise of tolerance to pressure after 2 months of ingestion of oxygen radical scavengers. Adapted from [29]

support the existence of a vicious cycle involving all three nitric oxide synthases, as predicted by the NO/ONOO⁻ cycle, but do not say anything about their impacts of that cycle.
2. Research data from animal studies demonstrate, that any increase in NMDA activity essentially increases all other aspects of the NO/ONOO⁻ cycle as outlined in Fig. 6.1. Thus, any increase in NMDA-receptor activity directly increases intracellular calcium levels leading to an increase in nitric oxide levels [30, 31]. These studies demonstrate that most of the elements of the cycle can be increased simply by elevating intracellular calcium and nitric oxide, providing proof for a cycle that is similar or identical to the NO/ONOO⁻ cycle [32].
3. Animal pain models that involve all of the elements outlined in Fig. 6.1 result in excessive neuropathic pain or hyperalgesia. And while there is no rational to explain the development of neuropathic pain it all matches in when using the NO/ONOO⁻ cycle as an explanation (Ref. [123] in Chap. 7).

8.5.1 Antioxidants for Mitochondrial Protection

What allows us to live and our bodies to function are billions of chemical reactions in the body which occur every second. These are essential for the production of energy, which drives all the processes of life such as nervous function, movement, heart function, digestion and so on. If all these enzyme reactions invariably occurred perfectly, there would be no need for an antioxidant system. However, even our own enzyme systems make mistakes and the process of producing energy in mitochondria is highly active. When mistakes occur, free radicals are produced. Essentially, a free radical is a molecule with an unpaired electron, it is highly reactive and to stabilize its own structure, it will literally stick on to anything. That "anything" could be a cell membrane, a protein, a fat, a piece of DNA, or whatever. In sticking on to something, it denatures that something so that it has to be replaced. So having free radicals is extremely damaging to the body and therefore the body has evolved his own antioxidant system to eliminate these free radicals before they have a chance to do such damage. Free radicals come from inside the body (i.e. ROS during mitochondrial energy production, the P450 detoxification system in the liver and immune activity from inflammation) and from outside the body (i.e. poisoning from pesticides, volatiles organic compounds, heavy metals, radiation etc). In recent years more stress has been placed on our antioxidant system because we are increasingly exposed to internal toxins (modern diets) and external toxins (pollution), which often exert their malign influence by producing free radicals. Therefore, it is even more important than ever to ensure a good antioxidant status. Free radicals effectively accelerate the normal ageing process whilst antioxidants slow the normal ageing process. The best example that we have all seen is the effects of smoking – cigarette smoking produces large amounts of free radicals and people who have smoked for many years have prematurely aged skin. Smokers also die younger from cancer or arterial disease – problems one expects to see in the elderly. Conversely, people who live and eat in a healthy way age more slowly.

8.5.2 The Normal Antioxidant System

There are many substances in the body which act as antioxidants, but the three most important frontline antioxidants are

- **Co-enzyme Q10**. This is the most important antioxidant inside mitochondria and also a vital molecule in oxidative phosphorylation. Co-Q10 deficiency may also cause oxidative phosphorylation to go slow because it is the most important receiver and donator of electrons in oxidative phosphorylation. People with low levels of Co Q10 have low levels of energy. In general this level not only is low in CFS sufferers but almost always down in other chronic ailments that they can be corrected by taking Co-enzyme Q10 100 mg daily for 3 months, after which it should be continued with a maintenance dose of 50–80 mg/die.
- **Superoxide dismutase (SODase)** is the most important super oxide scavenger in muscles, containing zinc and copper SODase inside cells, manganese SODase inside the mitochondria and the zinc and copper extracellular SODase outside cells. Deficiency can explain muscle pain and easy fatigability in some patients. Since SODase is dependent on copper, manganese and zinc this should be maintained in people taking a physiological mix of minerals. However, when there is a deficiency, these minerals are taken separately. Experience shows that the best results are achieved by copper 1 mg in the morning, manganese 3 mg midday and zinc 30 mg at night. Low dose SODase may also result from gene blockage and this is also looked at when the SODase test is done. SOD blockages most often is caused by toxic stress, such as heavy metals and pesticides.
- **Glutathione peroxidase (GSH-Px)**. This enzyme is dependent on selenium and glutathione, a three amino acid polypeptide, and a vital free radical scavenger in the blood stream. There is a particular demand in the body for glutathione. Not only is it required for GSH-Px, which is an important frontline antioxidant, but it is also required for the process of detoxification. Glutathione conjugation is a major route for excreting xenobiotics. This means that if there are demands in one department, then there may be depletions in another, so if there is excessive free radical stress, glutathione will be used up and therefore less will be available for detoxification and vice versa. Of course, in patients with chemical poisoning or other xenobiotic stress, there will be problems in both, and it is very common to find deficiencies in glutathione: If there is a deficiency of GSH-Px, it is recommend that patients eat a high protein diet (which contains amino acids for endogenous synthesis of glutathione), take a glutathione supplement 250 mg daily, together with selenium 200 μg daily or take the precursor of glutathione, i.e. N-Acetyl-Cysteine (NAC) in a dosage of at least 2 g/die.

These molecules are present in parts of a million and are in the frontline process of absorbing free radicals. When they absorb an electron from a free radical both the free radical and the antioxidant are effectively neutralized, but the anti-oxidants re-activate themselves by passing that electron back to **second line antioxidants** such as vitamins A and beta carotene, some of the B vitamins, vitamin D, vitamin E, vitamin K and probably many others. These are present in parts per thousand.

Again, accepting an electron neutralizes these, but that is then passed back to the ultimate repository of electrons, namely vitamin C, which is present in higher concentrations. Most mammals can make their own vitamin C, but humans, fruit bats and guinea pigs are unable to do so. They have to get theirs from the diet and Linus Pauling, the world authority on vitamin C, advised that a person needs vitamin C in several gram doses per day. Presently, a minimum of 2 g of vitamin C daily and for some patients up to 6 g is recommended. Pauling himself advocated larger doses. Government recommend only 30 mg per day which is just sufficient to prevent scurvy, but insufficient for optimal biochemical function especially in a patient with increased radical oxygen stress.

- Paraoxonase is an antioxidant that sits on good cholesterol (HDL) and protects this and the bad cholesterol (LDL) from oxidation. Levels of paraoxonase are determined genetically. This enzyme detoxifies organophosphate pesticides and if deficient, it makes the organophosphate much more toxic. Since there is a deficiency in about one third of the population this explains why about one third of those farmers exposed to organophosphates become ill.
- There are many other antioxidants present in vegetables, nuts, seeds and fruits which the body takes advantage of when they are present in the diet. Other substances such as melatonin also have profound antioxidant properties.
- Vitamin B_{12} is an excellent antioxidant and if I have a patient with particularly poor antioxidant status then I often recommend B_{12} by injection. Effectively this provides instant antioxidant cover and protects the patient from further damage whilst they take the necessary micronutrients to heal and repair their own antioxidant system. All the above antioxidants can be measured and routinely the frontline antioxidants, i.e. Coenzyme Q10, superoxide dismutase (SODase) and glutathione peroxidase should be measured.

8.5.3 Principles/Practice of the "Cave Man Diet"

Human beings evolved over millions of years eating a particular diet. Neanderthal man was a carnivore and rarely ever ate meat, fish and shellfish. More recently Paleolithic man expanded the diet to include root vegetables, fruits, nuts and seeds, which he could scavenge from the outdoors. It is only in the last few thousand years since the Persians, Egyptians and Romans that we began farming, and grains and dairy products were introduced into the human diet. A few thousand years from an evolutionary point of view is almost negligible. Many people have simply failed to adapt to cope with grains and dairy products and it is very likely that these foods cause a range of health problems in susceptible people. Modern studies on ancient tribes who continue to eat a "Cave Mans (Paleolithic) Diet" show that these people did not suffer from diabetes, obesity, heart disease or cancer. If they had survived the ravages of infectious diseases, childbirth and war wounds, then these people

8.5 Therapeutics for Normal Mitochondrial Function

lived a healthy life and reached a high age. Thus the current evolving notion is, that whatever our medical problem may be, or even if one simply wants to stay well, one should avoid all prepacked and processed diet from mass production and move towards a care man diet based on organic vegetables, nuts, seeds, with meat, fish and eggs from a farmer not spraying pesticides, herbicides or fungicides. Recent Western diets get 70% of their calories from wheat, dairy products, sugar and potato and it is no surprise that these are the major causes of modern health problems resulting in as cancer, heart disease, diabetes, obesity and other degenerative disorders. It is noteworthy, that traditional Chinese diets have no dairy products, no gluten grains, no alcohol and no fruit. There are five aspects of modern Western diet and gut function, which commonly cause symptoms from irritable bowel syndrome to fatigue. These are:

1. High carbohydrate intake–this is probably the largest single cause of modern diseases such as hypertension, obesity, syndrome X, heart disease and cancer
2. Food allergy
3. Toxins in the diet. Lectins naturally are present in foods. However, artificial additives, colorings, flavorings; artificial sweeteners (i.e. aspartame, sodium saccharin, sucralose) pesticide residues, plasticizer residues, etc, and social chemicals (i.e. alcohol, caffeine, tobacco etc.)
4. Fermentation of food instead of digestion – see Fermentation in the gut and CFS.
5. Poor digestion of food due to low stomach acid – see hypochlorhydria and poor pancreatic enzyme production.

The Stone Age diet tries to address the top three problems at the same time, since they often co-exist in the same patient. This is the diet patients with IBS should eat for a long term. This is because it is the evolutionarily correct diet and by changing the diet one can avoid long-term health problems and postpone degenerative conditions.

As a general principle it is important to remember that:

- **Carbohydrates** (CHO) tend to cause fatigue, even in "normal" people. One should be eating protein and fat (especially MCFA) in the day and saving carbohydrates (CHO) until the evening, because they help to sleep. The present Western diets are completely upside down because people eat cereals and a toast at breakfast, sandwiches at lunch and meat in the evening. Naturally such diet makes one feel tired in the day and wakes one up at night!
- **Food allergy** is a common cause of many symptoms such as irritable bowel, asthma, mood swings, headache, arthritis, allergic muscles and of course fatigue. The commonest offenders are grains, dairy, yeast and toxins in the diet.
- **Chemicals** in the diet inhibit enzyme systems and slow up metabolism – this applies to drugs as well as food additives and pesticide residues, hormone residues, antibiotic residues etc. Inshore sea fish can be expected to have a

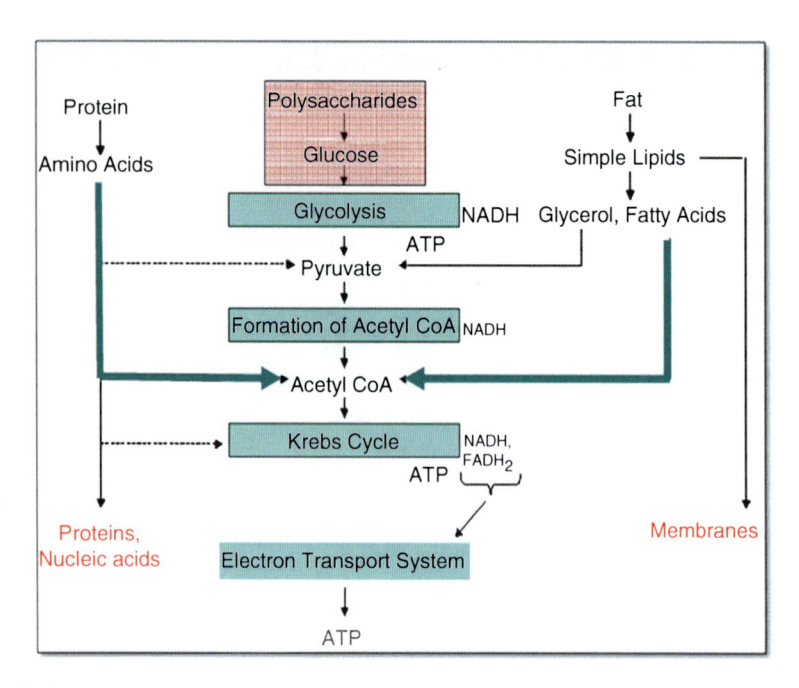

Fig. 8.18 By-passing the pyruvate stack in carbohydrate metabolism in the Krebs-cycle due to insufficiency of pyruvate-dehydrogenase activity resulting in an increase utilization of amino- and fatty acids

mercury load. Avoid additives, colorings, flavorings etc. and especially avoid plastic wrappings (especially if heated!) on food and try to switch to organic foods wherever possible.
- **Gut dysbiosis** and **poor digestion** of foods, whereby foods are fermented instead of being digested, can also cause fatigue symptoms.

This *"Cave Man Diet"*, therefore, contains foods of a low glycemic index (GI) in the day and moderate GI index in the evening; it avoids the common allergens, avoids moldy foods and foods of high fermentable substrate and is as free as possible from chemicals. In the long term this is a diet for life mimicking Stone Age principles and being an evolutionarily correct diet. Once the diet is established, one does not have to follow it slavishly, but it should make up the main diet and qualitatively all forbidden foods should become treat foods and not the main food supply. The rational for such a diet is derived from the fact that energy from carbohydrates cannot be used sufficiently due to the blockade of pyruvate dehydrogenase by $NO/OOHNO^-$ in the Krebs-cycle resulting in a pile-up of pyruvate. Contrary, amino and fatty acids are still being utilized as they are shifted into the Krebs-cycle via acetyl coenzyme A (Fig. 8.18) providing sufficient energy substrates for mitochondrial cellular respiration and the synthesis of ATP.

8.5.4 Unstable Blood Sugar Results in Reactive Oxygen/Nitrosative Stress

It is critically important for the body to maintain blood sugar levels within a narrow range. If the blood sugar level falls too low, energy supply to all tissues, particularly the brain, is impaired. However, if blood sugar levels rise too high, then this is very damaging to arteries and the long-term effect of arterial disease is heart disease and strokes. This is caused by sugar sticking to proteins and fats to make AGEs (Advanced Glycation End-products), which accelerate the ageing process. Normally, the liver controls blood sugar levels. It can create the sugar from glycogen stores inside the liver and releases sugar into the blood stream minute by minute in a carefully regulated way to cope with body demands, which may fluctuate from minute to minute. Muscles can take up excess sugar flooding into the system after a meal, but only so long as there is space there to act as a sponge. This only occurs when we exercise. This system of control works perfectly well until it is distressed by eating the wrong food or not exercising. For instance, eating excessive sugar at one meal, or excessive refined carbohydrates, which are rapidly digested into glucose, can overwhelm the muscle and the liver's normal uptake of glucose resulting in high blood glucose levels, which by themselves induce a release of ROS and even NOS in mitochondria [8]. Human mankind has evolved over millions of years eating a diet that was very low in sugar and had no refined carbohydrate. Control of blood sugar therefore largely occurred as a result of eating this Stone Age diet and the fact that the hunter at that times had to exercise vigorously while going after his pray, any excessive sugar in the blood was quickly burned off. Nowadays the situation is different: we eat large amounts of sugar and refined carbohydrates and do not exercise enough in order to burn off this excessive glucose. The body therefore has to cope with such excessive sugar loads by other mechanisms, setting free insulin from the pancreas, which deposits the excess glucose into the fat cells. For instance, when food is digested, the sugars and other digestive products go straight from the gut in the portal veins to the liver, where they should all be mopped up by the liver and processed accordingly. If excessive sugar or refined carbohydrate overwhelms the liver, the sugar spills over into the systemic circulation. If not absorbed by muscle glycogen stores, high blood sugar results, which is extremely damaging to arteries. If one would exercise hard, this would be quickly burned off. However, if one is not, then other mechanisms of control are brought into play. The key player here is insulin, a hormone excreted by the pancreas. This is very good at bringing blood sugar levels down and it does so by converting the sugar into fat. Indeed, this includes the "bad" cholesterol LDL. There is then a rebound effect and blood sugars may well go too low. Low blood sugar is also dangerous to the body because the energy supplied to all tissues is impaired. When the blood sugar is low, this is called "hypoglycemia". Subconsciously, people quickly work out that eating more sugar alleviates these symptoms, but of course they invariably overdo things; the blood sugar level then goes high and one ends up on a rollercoaster ride of blood sugar level going up and down throughout the day. Ultimately, this leads to metabolic syndrome or syndrome X – a major cause of disability and death in Western societies, since it is the forerunner of diabetes, obesity, cardiovascular disease, degenerative conditions and even cancer.

8.5.5 Recommendations for Diet in Mitochondropathy

8.5.5.1 Sugar and Fast Carbohydrates, Contraproductive in Mitochondropathy

The problem is that people feel boosted by a high level of blood glucose and it has been demonstrated elsewhere that sugar has addictive properties (Ref. [291] in Chap. 7) [4, 33]. This is because they have a good energy supply to their muscles and brain – albeit short-term. The problem arises when blood sugar levels dive as a result of insulin being released and energy supply to the brain and the body is suddenly impaired resulting in hypoglycemia. This, however, is followed by a whole host of symptoms: the neuronal cells of the brain can not function sufficiently, resulting in symptoms which include difficulty in thinking clearly, feeling spaced out and dizzy, poor word finding ability, foggy brain and sometimes even blurred vision or tinnitus. The body-symptoms include a sudden feeling of weakness and lethargy, feeling of faint and slightly shaky, a rumbling tummy and a craving for sweet things. Sufferers may look as if they are about to faint (and indeed often they do) and have to sit down and rest. Eating something sweet can quickly alleviate the symptoms, i.e. containing the fast reabsorbable glucose or fructose and the sufferer gradually recovers. These symptoms of hypoglycemia can be brought upon by missing a meal or the usual sweet snack such as a sweet drink, by vigorous exercise or even by alcohol. This is similar to diabetics who may become hypoglycemic if they use too much of their medication. So the brain likes sugar. Running on a high blood sugar allows the brain to function efficiently and also releases the neurotransmitters GABA and serotonin, which have a calming effect. We all know this because comfort-eating foods are carbohydrates. The second problem is that we have a "thermostat" for the right blood sugar, i.e. a measure against a set value of a blood sugar being consistently compared and controlled. Once the blood sugar level consistently runs high, there is a new set at higher levels. This for instance can be observed in people with diabetes who consistently run on a high blood sugar levels but feel hypoglycemic once their blood sugars drops below a value of 7 or 8 mmol/L. So, whatever interventions one makes to control high blood sugars must be done slowly so that this "thermostat" can be gradually reset.

8.5.5.2 The Five White Devils in the Daily Food Chain

1. **Sugar** has no nutritional value at all it, however, is highly addictive. It is an extract of the sugar beet and is chemically bleached. New studies point out that the rate of consumption of sweetened drinks, such as soda, juice, Kool-Aid and sports drinks,

has been on the rise in past decades showing a close correlation with obesity in children accompanied by metabolic syndrome issues, like diabetes and cardiovascular disease, high blood pressure, elevated levels of the blood triglycerides (saturated fats) and low levels of the artery-protecting HDL cholesterol. Still, a new research has found that drinking more than one soda daily, even the light variant, is linked to a higher incidence of metabolic syndrome, obesity and even heart disease [34].

2. **All bakery with refined and bleached (whitened) flours,** since in white flour the outer layer which contains most of the nutrients (and vitamins) and fiber that has been stripped of, is the result of a cultural aspect. The western society has become used to immaculate white bread, and the demand for white bread has progressively overthrown the old traditional grayish, yellowish bread from former times. Once the phenomenon of the color changes in air-aged flour was understood, millers started to use bleaching agents. As a result, consumptions habits changed and the increase in demand of white in opposition the old traditional bread has become the norm. Bleaching of flours traditionally is done in several ways all of which results in an oxidation process since these chemicals are radicals a fact why bleaching of flour is being banned in the EU and Australia.

- Azodicarbonamide, a hydrogen acceptor.
- Organic peroxides, also named benzoyl peroxide.
- Calcium peroxide (CaO_2) a solid peroxide.
- Nitrogen dioxide, the gaseous chemical compound NO_2
- Chlorine dioxide: a chemical compound with the formula ClO_2.

3. **Salt** (NaCl) in modern prepared foods is far too high, because a high proportion of one kind of the mineral sodium can cause deficiencies in others and also results in hypertension [35]. Use of sea salt for which the sodium content has been reduced and so is much closer to our physiological requirements than table salt (sodium chloride).

4. *Ultra*-**pasteurized milk,** ultra-heat treated (UHT), or condensed milk should be avoided. This is because it has been demonstrated in feeding studies with different types of refined milk, that the first generation of offspring's from cats showed cranio-facial dysplasia and other deformations, which was worst when the so called condensed milk was fed, indicating that heat-labile elements in the milk are destroyed by heat being only available in raw milk [36].

5. **Preservatives of chemical origin** regularly added in ready-to-make foods. As soon as something is refined the decay process begins and the food and loses it's vitality. Especially one should avoid "dead" food in tins and packets, and instead buy fresh, "alive" food. However, high quality frozen products probably are a second choice. Help maintain what's there by cooking foods lightly. Vegetables, which are boiled for too long, loose most of their trace elements and vitamins in the water. If however, one is used to like vegetables done this way then one should drink the cooking water or use it in the gravy. Eat foods as unprepared as possible, because raw food is an excellent vitamin donors.

Therefore one should choose food that is alive and eat the meat, which has "had a life". The fatty acid content of factory farmed fish, pork and poultry reflects that the food the animals eat is of poor quality. Eating free range -lamb and beef offers the best value in the country. There is a myth that chicken is healthy meat – if one could see the conditions under which chickens are kept, the amount of antibiotics added to their chow, and the quality of food they eat, than one would understand why chicken meat is of very low quality. As a general rule one should make breakfast a substantial meal. In principle, proteins and fats are more sustaining, carbohydrates are stupefying. It is recommend eating protein and fats at breakfast and carbohydrates in the evening. Breakfast like an emperor, lunch like a king, supper like a pauper! This is especially important for people suffering from fatigue.

8.5.5.3 Problems That May Arise When Consuming Wheat

- Refined wheat (i.e. white flour used to make white bread, biscuits, cake, pasta, etc) is the third most-produced cereal after maize (784 million tons) and rice, is quickly digested and has a **high glycemic index**. This means that blood sugar levels will run high after eating wheat, which in the long term causes obesity, diabetes, syndrome X and heart disease. Running a high blood sugar stimulates release of insulin that is a growth promoter and is undesirable if you wish to avoid cancer.
- Wheat contains gluten. This small tough molecule, especially in a person with inefficient digestion, gets into the bloodstream, where it acts as an endogenous opiate. This means that it has morphine like qualities and so in susceptible individuals will be **addictive**. This may cause problems such as:
- Gluten endogenous opiate like activity may be a cause of symptoms in autism.
- Wheat may be addictive for some people and cause psychiatric problems.
- Endogenous opiates switch off natural killer cell activity, which may reduce resistance to viral infections and cancer.
- Wheat bran for some people is directly irritating to the gut and can be a cause of **diarrhea**.
- Wheat contains toxic substances called lectins, which in susceptible individuals can cause **hemolysis** of the blood and other damage such as **celiac disease**, which may be due to lectin in the gut.
- Wheat **allergy** is extremely common. It is estimated that 40% of the population are allergic and wheat allergy may cause headaches, migraine, irritable bowel syndrome, anxiety, depression and fatigue. Wheat allergy in the gut can also present as celiac disease. Undiagnosed celiac disease is a major risk factor for **stomach lymphoma** (a cancer).
- Wheat has high levels of phytic acid. This chelates minerals, thereby preventing their absorption and puts one at risk of **mineral deficiency** syndromes such as anemia and osteoporosis.
- Wheat protein, gluten, is a tough molecule and readily passes from the gut into the bloodstream without having been properly digested. In some individuals this

will elicit an **antibody response** and if it just so happens that those antibodies fit one's own antigens, then this is the basis of autoimmunity. Indeed, eating wheat has been associated with **autoimmune disorders** such as an underactive thyroid, pernicious anemia, Addison's disease and so on.

8.5.5.4 Problems When Consuming Dairy Products

Dairy products are foods meant for growing calves. Nature did not mean them to be consumed by humans, or even worse adult humans. Whilst dairy products are promoted for their nutritional content, such as B vitamins and calcium, actually there are great problems associated with eating a lot of dairy products:

- **Fresh milk contains lactose**, which requires an enzyme lactase for its digestion. Only 10% of the world's population carries this enzyme and so the vast majority of people will be **unable to digest lactose**. Lactose intolerance can present with irritable bowel syndrome, wind, gas, pain and diarrhea. Following gastroenteritis, even those people who can digest lactose will develop temporary lactose intolerance. With any gut upset, all sufferers should avoid eating dairy products for at least one week.
- **Milk proteins** are very **allergenic** and allergy to milk is common. The common symptoms are catarrhal problems, headache, irritable bowel syndrome, depression, anxiety and chronic fatigue, but allergy is the great mimic and sensitivity to dairy products can cause almost any symptom. Milk passes readily from the gut into breast milk, where it may cause colic and projectile vomiting in the baby.
- **Dairy allergy** is the commonest allergy in children, often presenting with hyperactivity, catarrhal conditions, asthma, recurrent tonsillitis and sore throats, sinusitis, migraine and later on, premenstrual tension.
- **Antibody formation**, as 30% of the population makes antibodies to whey protein, which cross-react with platelets to make them stickier. Eating dairy products, therefore, causes sticky blood, which is a major risk factor for arterial disease.
- **The ratio of calcium to magnesium** in dairy products is 10:1, whereas our physiological requirements are for two parts calcium to one part magnesium. Since calcium and magnesium compete for absorption, taking dairy products induces a relative **magnesium deficiency**. Magnesium deficiency is common in heart disease, arthritis, chronic fatigue and osteoporosis. This magnesium supplementation is favored especially in patients with the risk of coronary heart disease or ventricular arrhythmia. There Mg intake has shown to elicit a preventive nature in the outbreak of the ailment [37, 38].
- **Milk** is generally believed to be a good source of vitamin D – this is not the case. The worst cases of Rickett's occur in Asian inner city children, who avoid sunshine but have a high consumption of dairy products. Dairy products are a poor source of vitamin D and all should be taking 2000 IU vitamin D daily.
- **Dairy products** are meant for young growing calves. From an evolutionary point of view calves had to grow very quickly in order to avoid being eaten by a saber toothed tiger and therefore **dairy products contain growth promoters**.

This is undesirable in anybody who wishes to avoid, or indeed has cancer. In China, where there is no dairy products consumed because everybody is allergic to them, there is a very low incidence of breast cancer and prostate cancer. Anyone who is in doubt about this, I recommend that they purchase "Your Life in Your Hands" by Professor Jane Plant, who cured herself from breast cancer simply by avoiding dairy products [39]. There she claims that today's dairy practices have changed so that pregnant cows (which produce high levels of estrogen) continue to be milked. Significant levels of conjugated estrogen are therefore present in cows' milk. The extent to which these are activated in the human gut and how much estrogen the consumer is exposed to is largely unknown.

- **Correct minerals in soil**. In addition to choosing the right foods to eat, one can further improve the nourishing value. The first is that ideally these foods should be grown from a soil in which the mineral content has been corrected. Modern farming simply applies three elements, namely NPK (nitrogen, phosphorus and potassium), resulting in soils, which are grossly out of balance. There is a net loss of minerals such as selenium, zinc, magnesium, cobalt, copper, manganese (just to name but a few) from the soil into plants, animals and humans, and out of this natural cycle. Especially selenium, which is needed for thyroxin synthesis should be considered as a supplement.
- **Free from toxins**. Foods must also be as free from toxins as possible, such as pesticide residues, hormone residues, antibiotic residues etc. To achieve this, buy organic as much as possible. One of the criticisms people have against organic farming, however, is they do not routinely treat the soil and correct mineral levels.

> **Note: Inshore sea fish can be expected to have a mercury load.**

- **Fresh foods, unprocessed**. Food that tastes good is likely to be of good nature. Taste is a sense that is trace element sensitive – i.e. foods, which are deficient in trace elements, don't taste so palatable. Furthermore zinc is necessary for taste buds to work and so food for the zinc deficient is tasteless. These people tend to go for salty, sweet or spicy foods to compensate and end up eating junk food, thereby worsening the zinc deficiency. Most people can tell the difference between home grown fresh vegetables and 3-day-old shop vegetables. The true free-range chicken is a rare fowl but quite different in taste from the factory bird. This "food vitality" may be difficult to quantify but it makes it no less real!
- **Locally grown foods in season**. As soon as food is stored, it starts to rot and loses quality. Eat fresh, while in season. Varying the diet with the seasons protects against developing allergies and a varied diet makes it less likely to become deficient in any one micronutrient.

- **Healthy fats**. People of western nations have been brainwashed into believing that fat is bad for you. This suits the food manufacturers well because fats are expensive and it is difficult to profit from them except cooking oil and margarine, which they have erroneously convinced us have health benefits. For example, margarine is bad for you. It is artificially prepared by heating oils to high temperatures. This causes formation of trans fatty acids, which are poorly metabolized in the body. Use best quality "cold pressed, virgin" olive oil for cooking and salads. Other oils have often been heated and therefore denatured. The mono-unsaturated fats are thought to be best in protecting against heart disease.
- **Western diet deficient in Ω-3 and Ω-6 fatty acids** – eat oily fish twice weekly and use a variety of cold pressed organic nut, seed and vegetable oils. Carbohydrates, however, can be bought cheap and sold expensive -potatoes can be bought for €100 per ton and sold for €10,000 per ton as crisps. Fat is good for you – it is a case of eating the right sort of fats.
- **Meat**. Vegetarianism is not necessarily healthier. In fact, I encourage CFS patients not to be vegetarians – vegetarian diets are artificially restricted and so people risk picking up food allergies. It is also quite hard work to prepare and eat the right foods for optimum balance of protein, fat, carbohydrates and essential micronutrients. Proteins are essential in a diet, especially for someone who is ill or stressed. The problem with meat appears to be how we cook it (primitive man would have eaten his meat raw!). Burned fats are oxidized fats, which are full of free radicals, which are damaging to arteries and possibly carcinogenic. If you do have highly cooked meats (e.g. roasts and barbeques), make sure you have something with it to "neutralize" the free radicals, for example, lots of vegetables. Meat that is boiled (stews, soups) does not contain oxidized fats. However, one should only eat meat that is sold from known organic farming, not using antibiotics (as regularly done in mass chicken breeding), being free of hormones (e.g. human growth hormone as being used in hog farming) or meat from genetically engineered cows.
- **Water**. Drink good quality water. Spring water (direct or bottled) is undoubtedly the best. Some people with severe allergies can only tolerate water in glass bottles. Second best is filtered water (water filters should be changed regularly), with tap water a poor third. Many drinks (tea, coffee, cacao, alcohol, pop) contain substances, which are diuretic, and make you pee out minerals. Tea is the main cause of iron deficiency anemia in the country. Tea is high in tannin, which binds minerals into insoluble tannates that cannot be absorbed. It is better to drink these beverages between meals and not with food. Instead, fruit juice (not fruit "drink") will enhance absorption of trace elements because of its vitamin C content, so drink this at mealtimes, but it must be diluted otherwise this will present a high sugar load.
- **Variety in diet**. Have as varied a diet as possible – everything in moderation is the key. Eat at least 8 oz of green (organic) vegetables daily bought from a reliable grocery store. Don't forget nuts and seeds – these are one of the richest sources of trace elements and vitamins.

8.5.6 Treatment Package for Failing Mitochondria

The biological basis of treatment in CFS therefore consists of the following:

- **Reduce the pace** – do not use up energy faster than your mitochondria can supply it.
- **Feed the mitochondria** – supply the raw material necessary for the mitochondria to heal themselves and work efficiently. This means feeding the mitochondria correctly so they can heal and repair.
- **Address the underlying causes** as to why mitochondria have been damaged. This must also be put in place to prevent any ongoing damage to mitochondria. In order of importance this involves:
- **Reduce all activities** to avoid undue stress to mitochondria
- **Getting sufficient sleep** so mitochondria can repair
- **Supply mitochondria with excellent nutrition** with respect to:
  - Taking a good range of micronutrient supplements
  - Stabilizing blood sugar levels
  - Identifying allergies to foods

- **Detoxify to unload** heavy metals, pesticides, drugs, social poisons (alcohol, tobacco, etc) and volatile organic compounds (VOCs), all of which poison mitochondria.
- **Address** the common problem of **hyperventilation**
- **Address** the secondary damage caused by mitochondrial failure such as **immune disturbances** resulting in allergies and autoimmunity, poor digestive function, hormone gland failure, and slow liver detoxification.

AMP can be recycled, but slowly. Interestingly, the enzyme which does this is cyclic AMP, which van be activated by caffeine! So the perfect pick-me-up for CFS sufferers could be a real black organic coffee with a teaspoon of D-ribose!

8.5.6.1 CFS is Low Cardiac Output Secondary to Mitochondrial Malfunction

Two papers have came out recently, which make great sense of both clinical observations and the idea that CFS is a symptom of mitochondrial failure. The two symptoms to look for in CFS in order to make the diagnosis is firstly very poor stamina and secondly delayed fatigue. These symptoms can now be explained in terms of what is going on inside cells and the effects on major organs of the body (primarily the heart). More importantly, there are major implications for a test for CFS and of course management and recovery. If mitochondria (the little engines found inside every cell in the body) do not work properly, then the energy supply to every cell in the body will be impaired. This includes the heart. Many of the symptoms of CFS could be explained by heart failure because the heart muscle cannot work properly. Cardiologists and other doctors are used to dealing with heart failure due to poor blood supply to the heart itself. In CFS the heart failure is caused by poor muscle function and therefore strictly speaking is a cardiomyopathy. This means the function of the heart will be very abnormal, but traditional tests of heart failure, such as ECG, ECHOs, angiograms etc, will be normal. Thanks to work by Dr Arnold Peckerman (www.cfids-cab.org/cfs-inform/Coicfs/peckerman.etal.03.pdf) it is now know that cardiac output in CFS patients is impaired. Furthermore the level of impairment correlates very closely with the level of disability in patients. Dr Peckerman was asked by the US National Institutes of Health to develop a test for CFS in order to help them to judge the level of disability in patients claiming Social Security benefits. Peckerman is a cardiologist and on the basis that CFS patients suffer low blood pressure, low blood volume and perfusion defects, he surmised CFS patients were in heart failure To test this he came up with Q scores. "Q" stands for cardiac output in liters per minute and this can be measured using a totally non-invasive method called Impedance Cardiography. This allows one to accurately measure cardiac output by measuring the electrical impedance across the chest wall. The greater the blood flow the less the impedance. This can be adjusted according to chest and body size to produce a reliable measurement (this is done by using a standard algorithm). It is important to do this test when supine and again in the upright position. This is because cardiac output in healthy people will vary from 7 L per min when lying down to 5 L per min when standing. In healthy people this drop is not enough to affect function. But in CFS sufferers the drop may be from 5 L lying down to 3.5 L standing up. At this level the sufferer has a cardiac output, which causes borderline organ failure. This explains why CFS patients feel much better lying down. They have acceptable cardiac output lying down, but standing up they are in borderline heart and organ failure. CFS is therefore the symptom, which prevents the patient developing complete heart failure. Actually, everyone feels more rested when they are sitting down with their feet up! The subconscious mind has worked out that the heart has to work less hard when you are sitting down with your feet up – so we do so because we feel more comfortable!

8.5.6.2 Low Cardiac Output Explains the Symptoms of CFS

The job of the heart is to maintain blood pressure. If the blood pressure falls, organs start to fail. If the heart is working inadequately as a pump then the only way blood pressure can be sustained is by shutting down blood supply to organs. Organs are shut down in terms of priority, i.e. the skin first, then muscles, followed by liver, gut, brain and finally the heart, lung and kidney. As these organ systems shut down, this creates further problems for the body in terms of toxic overload, susceptibility to viruses, which damage mitochondria further, thus exacerbating all the problems of the CFS sufferer. This is called POTS or *postural orthostatic tachycardia*

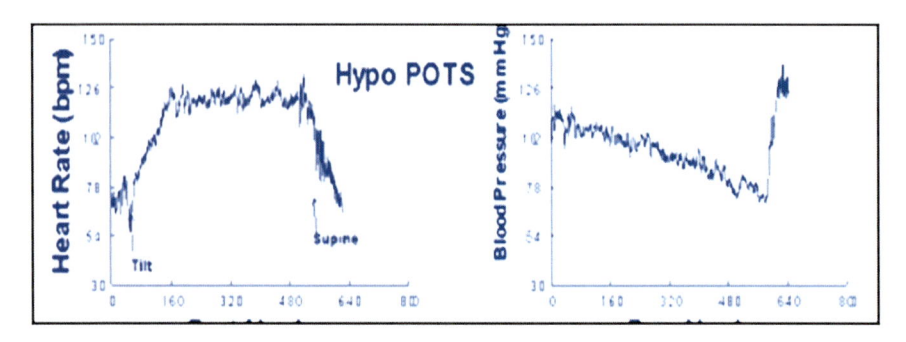

Fig. 8.19 Representative example of a patient with abnormal POTS patterns of heart rate (*left panels*) and blood pressure (*right panels*) abnormally in an adolescent

syndrome (Fig. 8.19). POTS has been proposed as a mechanism for symptoms of the Chronic Fatigue Syndrome in a series of adult patients. POTS and CFS may share a common pathophysiology particularly in the young. Recently, a review of patients with delayed orthostatic hypotension (delayed POTS) demonstrated a high degree of association with chronic fatigue.

Chest Pain in CSF

This is a common symptom in CFS. Chest pain results when energy delivery to the muscles is impaired. There is a switch to anaerobic metabolism, lactic acid is produced and this results in the symptom of angina. Doctors recognize one cause of poor blood supply, i.e. the supply of fuel and oxygen is impeded. However this fuel and oxygen has to be converted to ATP by mitochondria, so if this is slow, the same symptom of angina will result. One molecule of sugar, when burnt aerobically by mitochondria, will produce 36 molecules of ATP. In anaerobic metabolism, only 2 molecules of ATP are produced. This is very inefficient and lactic acid builds up quickly. The problem is that to convert lactic acid back to sugar (pyruvate) 6 molecules of ATP are needed, being done in the Cori cycle. So in CFS the chest pain is longer lasting because this conversion back to pyruvate is very slow. Clinically this does not look like typical angina. Many patients are told they have non-typical chest pain with the implication that nothing is wrong ! Actually they have a mitochondrial failure of the heart muscle.

Effects on the Skin

If one shuts down the blood supply to the skin, this has two main effects. The first is that the skin is responsible for controlling the temperature of the body. This means that CFS patients become intolerant to heat. If the body gets too hot then it cannot lose heat through the skin (because it has no blood supply) and the core temperature

8.5 Therapeutics for Normal Mitochondrial Function

increases. The only way the body can compensate for this is by switching off the thyroid gland (which is responsible for the level of metabolic activity in the body and hence the generation of heat). As a results so one gets a compensatory underactive thyroid. This, however, worsens the problems of fatigue.

The second problem is that if microcirculation in the skin is shut down, the body cannot sweat. Because this is a major way through which toxins, particularly heavy metals, pesticides and volatile organic compounds are excreted; the CFS patient's body is accumulating toxins, which of course further damages the mitochondria.

Symptoms in Musculature

If the blood supply to muscles is impaired, then muscles quickly run out of oxygen when one starts to exercise. With no oxygen in the muscles the cells switch to anaerobic metabolism, which produces lactic acid, which by itself results in muscles aches. Aside from the above problem, muscles in the CFS patient have very poor stamina because the mitochondria, which supply them with energy, are malfunctioning.

Symptoms in Liver and Gut

Poor blood supply to the gut results in inefficient digestion, poor production of digestive juices and a leaky gut syndrome. Leaky gut syndrome causes many other problems such as allergies, autoimmunity, malabsorption, etc., which further multiplies the problems of CFS. Also, when liver circulation gets inadequate, this will result in poor detoxification, not just of heavy metals, pesticides and volatile organic compounds, but also of toxins produced as a result of fermentation in the gut again further multiplying the poisoning of mitochondria.

Effects on the Brain

Functional scans of the brains of CFS patients demonstrate a reduced blood supply to some area of the brain, which may even look as if they had a stroke. The default is temporary and with rest, the blood supply recovers. However, this explains the multiplicity of brain symptoms these patients suffer from, such as poor short-term memory, difficulty in performing multi-tasks, a slow mental processing and so on. Furthermore, brain cells are not particularly well stocked with mitochondria and therefore they run out of energy very quickly.

Effects on the Heart

There are two effects on the heart. The first effect of poor microcirculation to the heart is disturbance of the electrical conductivity, which causes dysrhythmias.

Many patients with chronic fatigue syndrome complain of palpitations, missed heartbeats, bigeminus, tachyarrhythmia etc. This is particularly the case in patients, which are poisoned by chemicals since these chemicals also directly intoxicate the conductive system of the heart. In a pilot study (Freye et al, unpublished results) it was demonstrated that in patients with FMS who developed missed heart beats or bigeminus, a tea spoon full of D-ribose (a necessary constituent for energy generation within mitochondria) was able restore normal heart rhythm within 30 min. The second obvious result is poor exercise tolerance. The heart muscle fatigues in just the same way that other muscles fatigue. Symptomatically this causes chest pain and fatigue. In the long term it can cause heart valve defects because the muscles, which normally hold the mitral valve open also fatigue.

The difference between this type of heart failure and a clinically recognized congestive cardiac failure is that patients with CFS protect themselves from organ failure because of their fatigue symptoms. Patients with congestive cardiac failure initially do not get fatigue and often show organ failures of the kidney or overt heart failure. At present it cannot be explained why there is such difference.

The approach in treating heart failure in CFS is exactly the same regardless of the underlying cause.

So patients with angina, high blood pressure, heart failure, cardiomyopathy, and a valve defect as well as patients with cardiac dysrhythmias also have mitochondrial problems and will respond in the same way to nutritional therapies and replenish therapies as patients with CFS.

Effects on Lung and Kidney

The lung and the kidney are relatively well protected against poor microcirculation because they have the *Renin-Angiotensin*-Aldosteron-system, which keeps the blood pressure up in these vital organs. Therefore clinically one does not see patients with CSF developing kidney failure or pulmonary hypoperfusion.

8.5.7 Why Is There Fatigue in CFS and FMS Patients

Energy to the body is supplied by mitochondria, which firstly use NAD (nicotinamide adenosine diphosphate) from the Krebs's-citric acid cycle to power oxidative phosphorylation, to further generating ATP (adenosine triphosphate). These molecules are the of energy donators within the body. Almost all energy requiring processes in

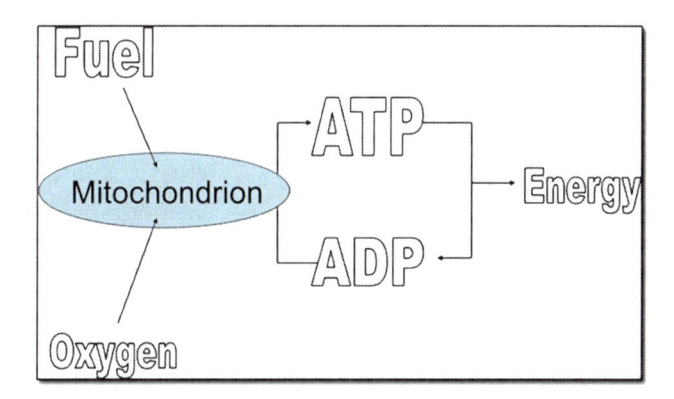

8.5 Therapeutics for Normal Mitochondrial Function

Fig. 8.20 Outline of mitochondrial function converting ADP into the energetically higher molecule ATP which is used for maintaining bodily function resulting in the formation of ADP which thereafter is recycled for the generation of fresh ATP

the body has to be "paid for" with NAD^+ and ATP, but largely ATP. The reserves of ATP in cells are very small, and the production lasts only for a short moment resulting in heart muscle cells to contract for only ten beats. Thus the mitochondria have to be extremely efficient in recycling ATP to keep the cell constantly supplied with energy. If, however, the cells are not too efficient at recycling ATP, then the cell quickly runs out of energy causing symptoms of weakness and poor stamina. The cells literally have to "hibernate" and wait until more ATP has been produced. Within the production process of energy, ATP (three phosphates) is used up leading into ADP (two phosphates) after which ADP is recycled back through mitochondria to produce ATP (for further information see translocator activity of mitochondria). However, if cardiac cells are pushed (i.e. stressed) with not sufficient ATP being available, then they will start to use ADP instead. The body uses the energy from ADP resulting in AMP (one phosphate) formation. However, the problem is that AMP cannot be recycled via the mitochondrial membrane to generate fresh new ADP or ATP. The only way that ADP can be regenerated is by making it from fresh molecules (e.g. D-ribose), which will take days for regeneration (Fig. 8.20). This biochemical process explains delayed fatigue symptoms seen in chronic fatigue syndrome but also in fibromyalgia.

To summarize, the basic pathology in CFS (but also in FMS and in burn-out syndrome) is an insufficient production of ATP together with a slow recycling of ADP back to ATP again. If patients push themselves with a high demand of energy within mitochondria, then ADP is converted to AMP which cannot be recycled resulting in a loss of recyclable ADP with insufficient ATP production and a delayed fatigue. This is because it takes the body several days to make fresh ATP from new molecules (Fig. 8.20). When patients overdo things and get exhausted, this is because they have no ATP or ADP within the cellular organelles to function at all.

8.5.8 Implications for Treatment of CFS and FMS

Basically, patients with symptoms of CSF or FMS get the basic treatment schedule with respect to vitamins and minerals, diet, pacing and sleep. However, many need a specific package of supplements, to further support mitochondrial activity, which includes D-ribose, CoQ10, acetyl-l-carnitine, NADH, magnesium and B_{12} injections. All these things must be put in place to repair and prevent ongoing damage to mitochondria, giving them the chance for recovery. For mitochondria to recover they need all the essential vitamins, minerals, essential fatty acids and amino acids to reactivate the cellular machinery and restore normal function. In this respect, the mitochondrial function tests allow us to identify the site of the lesions, which can be corrected by putting all emphasis on nutritional supplements, improvement of the antioxidant status, detoxification, avoiding hyperventilation, getting sufficient sleep, repairing mitochondrial cellular lining with specific fatty acids (e.g. MCFA). Because CFS sufferers have limited reserves of physical, mental and emotional energy the mitochondrial function test allows us to direct treatment regiments into a most fruitful line of approach.

The onset of fatigue is pre-dated by PIMs (Psychological, Irritable bowel syndrome, Migraine and headaches).

1. Psychological: mood swings, depression, PMT, anxiety
2. IBS: wind, gas, bloating, abdominal pain, alternating constipation and diarrhea,
3. Migraines or headaches, if undiagnosed and not avoided often go on to produce fatigue.

Clinically the above issues indicate, that there are two clear stages of fatigue:

1. Mild chronic fatigue syndrome – in mild fatigue there is mild failure of mitochondria. If mitochondria go slow then cells go slow. If cells go slow then organs go slow. The body will become generally less efficient. So for example somebody mildly affected would not be able to increase their fitness – if they try to exercise they would quickly switch into lactic acid metabolism and would be forced to stop. Indeed we now know that mitochondria are responsible for controlling the normal ageing process. Therefore many of the symptoms and diseases associated with ageing are actually the result of mitochondrial function declining. Indeed many of these ageing diseases have now been attributed to mitochondrial failure such as loss of tissues (loss of muscle bulk), organ failures, neurodegenerative conditions, heart disease and cancer. Many symptoms, which are attributed to ageing, are due to mitochondria. It is not that we can stop the mitochondria from ageing, but we can certainly slow it down by using a sufficient supplementation, a diet free from additives, freedom from environmental toxic stress, a healthy lifestyle and so on.
2. Severe chronic fatigue syndrome – in severe chronic fatigue all the above factors apply. However, there is an additional problem. The most metabolically demanding organ in the body is the heart and if mitochondria cannot supply the heart with

sufficient energy then the heart will go into a low output state. This compounds the problem of all mitochondria. If the heart is in a low output state then blood supply is poor and therefore the fuel and oxygen necessary for the engine to work are also impaired. So this aggravates all the above problems and makes them proceed even quicker with patients ending up in a greater disability.

It is suspected that a combination of the underlying poor mitochondrial function, which when it comes to cardiac output at one point reaches a critical peak, precipitating a more severe illness in someone who is already compromised.

8.5.8.1 Causes of Fatigue in Patients with CSF and FMS

There are four aspects of what we eat, which commonly may present a cause for fatigue:

- High carbohydrate diet
- Food allergy
- Gut dysbiosis (wrong bacteria in the gut)
- Chemical overload (including prescription drugs)

> **High carbohydrate diet addiction is the problem in chronique fatigue.**

- The main addiction is to carbohydrates because the brain likes to run on a high blood sugar. The body does not like a high sugar, so insulin is released which brings the blood sugars down. When blood sugar runs high (hyperglycemia), this is potentially dangerous to tissues resulting in a reduced blood and oxygen supply to mitochondria with mitochondrial function being switched off.
- When, however the blood sugar is low (hypoglycemia), there is insufficient nutrition and mitochondria are being switched off.
- Fluctuating blood sugar levels therefore switch mitochondria on and off (Fig. 8.21).

Measuring blood sugar levels is not a terribly useful test for hypoglycemia, partly because they fluctuate too much and partly because by the time one shows symptoms of a low blood sugar, the blood sugar level is already being corrected. A much better test is to measure short chain fatty acids in blood collected in the morning before breakfast. When blood sugar levels are low, the body switches to short chain fatty acids (SCFAs) metabolization. In addition, there is a complication with hyperglycemia, which triggers insulin release. When being secreted the hormone acts like a stress hormone and a proinflammatory mediator resulting in a rebound hypoglycemia which both can be a cause of stress or the result of stress, leading into one of those vicious cycles that are often seen in a mitochondrial disease state.

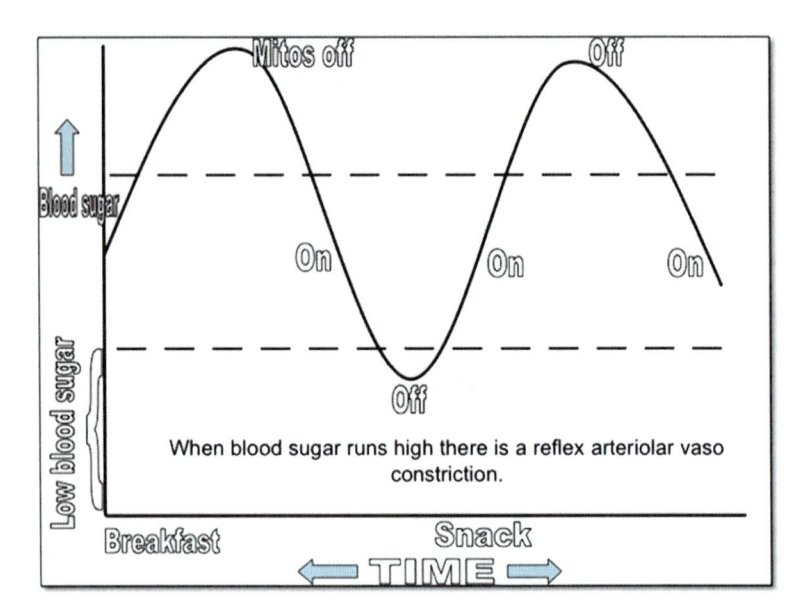

Fig. 8.21 The main cause of mitochondrial malfunction is the addiction to carbohydrates because the brain likes to run a high blood sugar. The body, however, does not like a high sugar, so insulin is released which brings blood sugars down. When blood sugar runs high, this is potentially dangerous to tissues so the blood supply is narrowed and oxygen supply to mitochondria reduced, and mitochondria are switched off. When blood sugar runs low, there is insufficient nutrition and mitochondria are switched off. Fluctuating blood sugar levels therefore switch mitochondria on and off while a high blood sugar level acts as a proinflammatory mediator

8.5.8.2 How to Avoid Phases of Hypoglycemia

- Use carbohydrates with low glycemic index diet
- Correct gut flora, since yeast ferment sugars to alcohol, which destabilizes blood sugar levels.
- Use probiotics, which ferment sugars to short chain fatty acids
- Check for allergies as they may cause hypoglycemia.
- Stress, including sleep loss, causes hypoglycemia
- Take micronutrients for protection, especially high dose niacinamide 500 mg tds (if nauseated, it could indicate an abnormal but rare liver function) plus chromium, 2 mg at night for 2 months.
- Check for thyroid and adrenal abnormalities with low plasma levels of adrenalin, epinephrine, and thyroxine
- Check for possible exposure to xenobiotics, such as environmental pollutants and heavy metal toxins since they tend to destabilize blood sugar levels.
- Consider an allergy to foods, which is common and can cause any symptom. The commonest allergens are grains, dairy, yeast and artificial food additives (for further information on food additives see page 94 ff)

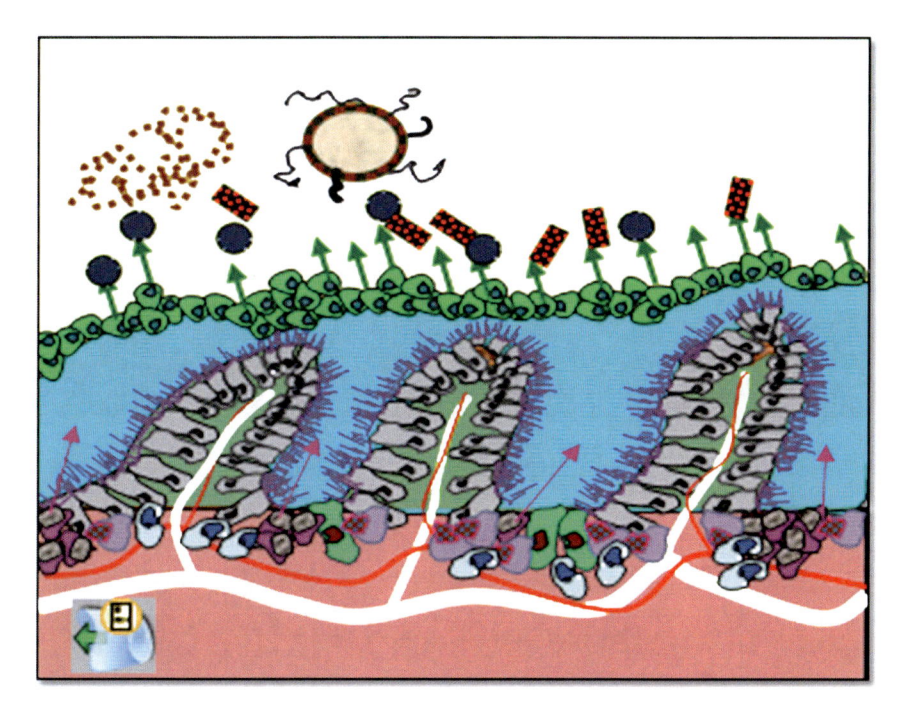

8.5 Therapeutics for Normal Mitochondrial Function

Fig. 8.22 The intestinal inner lining with a "carpet-like" layer of mucus and bacteria, necessary constituents for well being

A very common problem is gut dysbiosis (=the wrong bacteria in the gut), which arises from early childhood due to poor inoculation with mother's gut bacteria at birth. However, often the use of antibiotics, diets consisting of highly refined carbohydrates, and use of female sex hormones (e.g. anticontraceptives) are the cause for such intestinal maldigestion and often are comorbidities in mitochondropathy. This is because of an insufficiency of the barrier at the inner lining of the intestine brought about by the wrong bacteria (Fig. 8.22), with a disruption in cellular tight junctions resulting in an opening and a direct contact and activation of the immune system (Fig. 8.23) with an ensuing chronic intestinal bowel syndrome (IBS).

A. **Physiological microflora.** Intact inner lining with stable microecological compositions.
B. **Gaps in colonization** with bacteria support the growth, increase and intrusion with Candida. Immune reaction is possible because of the tight contact of antigens with Payers plaques.
C. **Pathological mycoflora.** Candida has managed to establish itself; via enzyme activity invasive growth with infection is likely resulting in a dissection of tight-junctions with antigen material getting into contact with the immune system (Peyer plaques)

Following traverse of antigen material with the aid of specific M-cells, and presented to T-cells, they in return activate B-cells with the antigen information

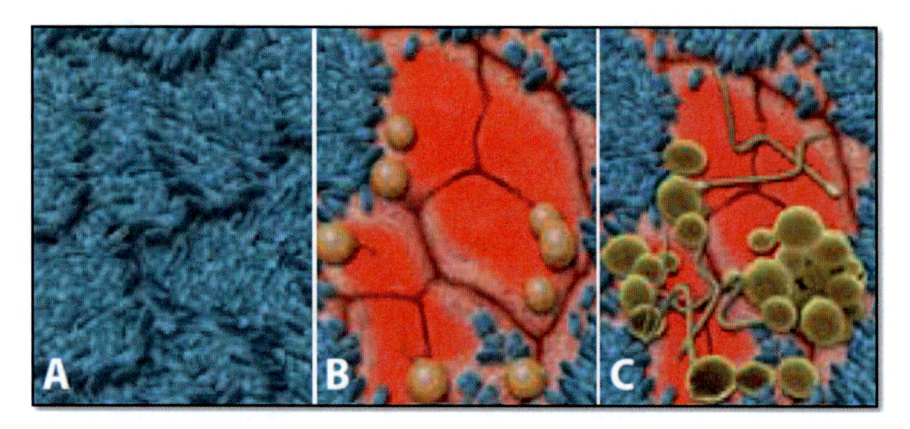

Fig. 8.23 Changes in colonization of the gut with different pathology

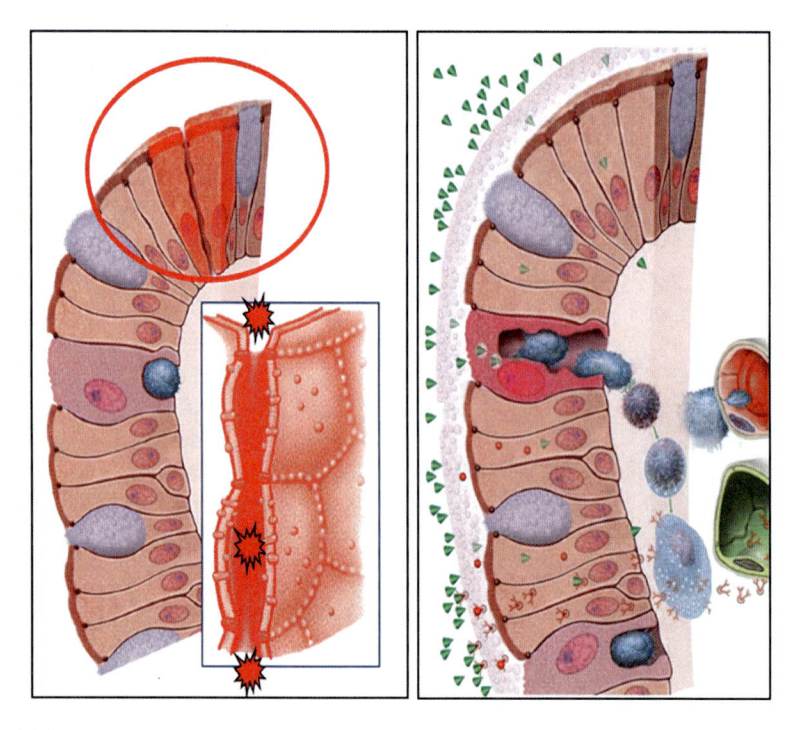

Fig. 8.24 In case of disruption of tight junctions within the cellular lining of the gut (*left*), antigens are permanently taken up by M-cells (*green at the right*) and presented to T-cells at the gut-associated lymphatic system, inducing secretion of cytokines of the TNF-alpha and the IL-1 type

thus changing into antibody producing plasma cells setting free cytokines, which in return induce a sensibilization of gut nerve endings resulting in colic-like pain, a typical symptom in IBS (Fig. 8.24).

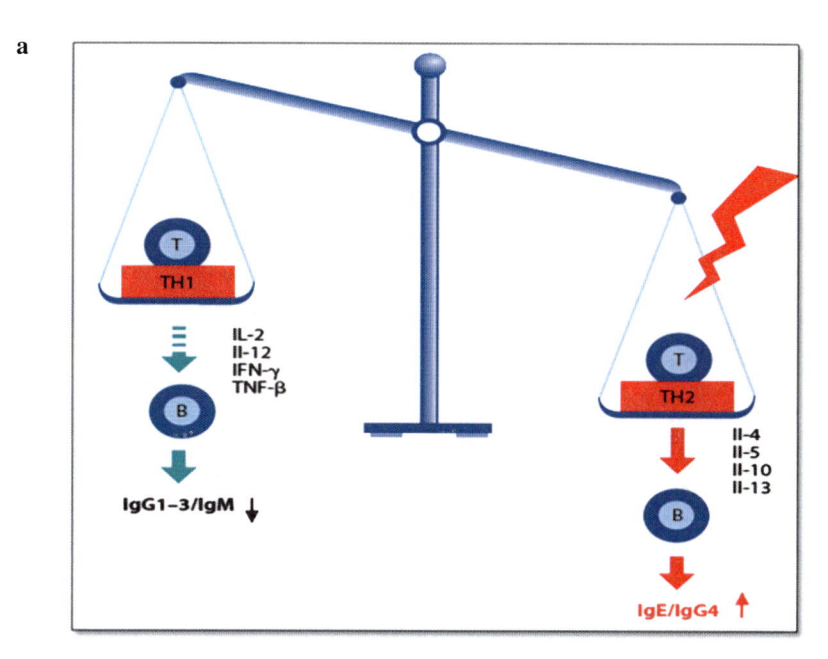

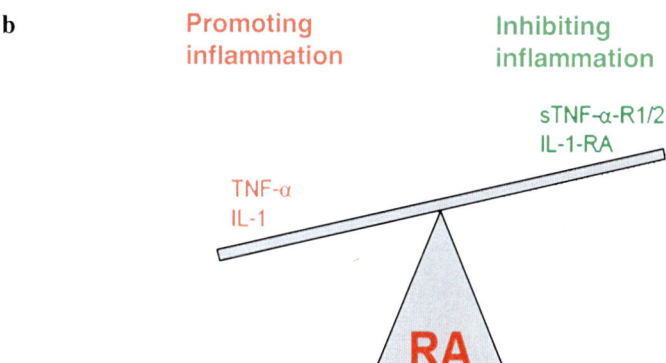

8.5 Therapeutics for Normal Mitochondrial Function

Fig. 8.25 (a) Shift in the Th1/Th2 balance in favor of TH2 resulting in allergic reaction with even distant symptoms within mucous membranes of the respiratory tract, while dominance in Th1 cells results in inflammatory reactions, typically observed in rheumatoid arthritis. Optimal is a balance between Th1/Th2 cells where none of the factors outweighs the other. (b) In inflammation there is a shift in dominance of Th1 cells with a secretion of TNF-alpha and interleukine-1, those cytokines, which keep the inflammatory reaction going for instance in rheumatoid arthritis (RA)

Once an allergen gets in contact with the immune system of the gut it induces an immune reaction (a Th1/Th2 shift) with different cytokines being released into the blood stream (Fig. 8.25a, b) followed by secondary and even distant pathological symptoms (e.g. skin, mucous membranes; for further information on the T1/Th2-balance see page 19).

On the contrary, gut dysbiosis can induce a shift in the TH1/TH2 balance in favor of TH2, which in the long run effectively can only be cured by adjusting colonization in the gut taking into account that certain diets induce the growth of pathogens within the intestine with local inflammation. Therefore, by obeying the following basic rules, there is a good chance for recovery:

(a) Avoiding all glucose containing beverages or diets
(b) Avoiding all fructose containing beverages or diets
(c) Avoiding monosodium glutamate, regularly found in Chinese, but also in other dishes.
(d) Avoiding all kinds of sugar replacements, especially aspartame and saccharin.
(e) Use probiotics to induce growth of bifidus-, and lactobacillus (Lactobacillus acidophilus, L. bifidus), those bacteria found in natural yoghurt or "kefir".
(f) Avoiding preprocessed food or food heated in the microwave oven.
(g) Use prebiotics, found in whole grain, seedflax, artichoque, asparagus, chicoree, onions, garlic and pectins in apples.
(h) Correct the nutritional status by using an "antiallergic" composition of amino acids plus vitamins, which is necessary for normal growth and function of intestinal mucosa cells living in symbiosis with benign bacteria fighting off pathogens.
(i) Correct the previous malabsorption-induced vitamin B_{12} and iron deficiency.
(j) Correct the underlying candida dysbiosis with a specific agent.

In the beginning, getting worse on the diet is almost to be expected. The reasons for worsening are as follows:

- Hypoglycemia – this is the commonest reason for worsening and may take weeks to settle. There are some nutritional interventions which help greatly (see treatment of hypoglycemia)
- Caffeine withdrawal – again common. Usually results in headache, which clears in 4 days.
- Food allergy withdrawal – this may cause many different symptoms. Some people report "flu-like" feeling. Typically this last 4 days, but with symptoms like eczema, arthritis, allergy, muscle pain and fatigue it can take weeks to clear. One patient with prostatism took 4 months to clear!

If the withdrawal symptoms are too severe, relax the diet a little, wait a few days, then tighten things up again. Most notably, specific cytokines, such as interleukin-1 (IL-1), have been implicated to be involved closely in acute neurodegeneration, following stroke and traumatic head injury. In spite of their diverse symptoms, a common inflammatory mechanism may contribute to many neurodegenerative

Table 8.5 Traditional therapeutic concepts in inflammatory diseases

Group of Agents	Examples	Clinical symptoms	Radiologic regression
NSAIDs	Ibuprofen, Naproxen	Relief	None
Glucocorticoids	Prednisolone	Relief	Questionable
Disease Modifying Anti-Rheumatic Drugs (DMARS)	Methotrexate, Hydrochloroquine, Sulfasalazine, Lefunamid	Relief	Positive effect
Biologics, cave: they depress the immune system resulting in TBC, cancer etc.	TNF-α antagonists: Infliximab, Etanercept, Adalimumab	Relief	Positive effect

disorders and in some (e.g. multiple sclerosis) inflammatory modulators are used clinically. While in conventional rheumatoid arthritis (RA) or osteoarthritis (OA), different therapeutic regimens are being used, all of which however are known to produce some kind of side effect most notably via an intoxication/depression of mitochondria (Table 8.5).

A viable prognostic marker for rheumatoid arthritis is the cyclic citrullinated peptide (CCP) with a sensitivity of 68% and a specificity of 97%, using the principle of an anti-CCP-ELISA test. In comparison to rheumatoid arthritis factors, the test shows a higher specificity and sensitivity to identify those patients which are bound to develop a severe form of the disease [40, 41].

8.6 Basics in Therapy of Mitochondrial-Related Diseases

There are two main approaches that clearly underline the kind of therapy in mitochondropathy:

- **Avoid any stressors that up-regulate the ROS or NOS activity.**
- **Use specific agents that lower the NO/ONOO⁻ related symptoms.**

ALL the different stressors that up-regulate the ROS or NOS formation may vary from one disease to another and even from one patient to another. However, some basic considerations should be observed.

- Clearly, in MCS patients it is essential to avoid chemical exposure.
- Since (silent) infections can up-regulate NO/ONOO⁻ and poses the major problem especially in CFS patients, it should be treated accordingly.
- Also, since excessive exercise can up-regulate NO/ONOO⁻ formation in CFS patients, it should be prevented.
- Any exposure to allergens, especially food allergens should be evaded in many of such patients.

- High and especially continuous psychological stress prompts NMDA activity, which in return triggers off the NO/ONOO⁻ production, presumably the most important detail in patients with PTSD syndrome.
- In patients with other chronic ailments such as hypertension, diabetes type 2, Parkinson, and/or another neurodegenerative disease, the energy production within neuronal tissue has reached the lowest possible level where up to 60% of mitochondria are damaged and/or present a deficient function (for more details see Sect. 3 on page 13).

Above all, therapy of the different ailments should aim to downregulate the formation of ROS and NOS together with protective measurements of mitochondria from stressors in conjunction with elements, which support mitochondrial activity and the formation of ATP (Table 8.6). In combination with any type of mitochondrial therapy, numerous other symptoms/ diseases of an overactive NO/ONOO⁻ cycle such as

Table 8.6 Agents used to down-regulate the formation of NO/ONOO⁻ and which have established efficacy in clinical trials

Agent or product	Mode of action
NMDA antagonists	Depresses excessive NMDA activity in FMS, CFS, Parkinson, Alzheimer
Magnesium	Acts as a blocker at the NMDA receptor site, which is upregulated in hypertension, FMS, CFS.
Long chain Ω-3-fatty acids	Substantial anti-inflammatory activity; act on neuronal tissue, in plasma membranes and inner mitochondrial membrane
Coenzyme Q10	Improves mitochondrial function; protects mitochondria from peroxynitrite mediated damage. Used in FMS, CFS, Alzheimer, Parkinson
Ecklonia cava extract	Polyphenolic chain breaking antioxidant; scavenges peroxynitrite and superoxide in FMS
D-Ribose	Helps to restore ATP formation from ADP at complex V within mitochondria. Significant effect in FMS and CFS
Vitamin B_6 including pyridoxal phosphate	Stimulates glutamate decarboxylase activity, limits, excitotoxicity, necessary fort normal mitochondrial function
Niacin (B_3) or Nicotinic acid and its amide, nicotinamide	Restores NAD/NADH pools after peroxynitrite-mediated poly-ADP ribosylation resulting in pool depletion; important to spark energy production and restoration
Riboflavin including 5'-phosphate	Exhaustion limits glutathione reductase activity, essential to restore reduced glutathione pools
Carotenoids including natural ß-carotene, lutein, lycopene	Helps to scavenge peroxynitrite in biological membranes
Natural vitamin E, including α, ß, γ -tocopherol, and tocotrienols	γ-tocopherol has a special role in scavenging NO_2 radicals, formed from peroxynitrite; has anti-inflammatory properties; tocotrienols protect from excitotoxicity and mitochondrial oxidation
Taurine	Lowers excitotoxicity by raising GABAergic activity

(continued)

Table 8.6 (continued)

Agent or product	Mode of action
Zinc, manganese and copper	Used in modest doses; putatively rate-limiting in the synthesis of the three forms of superoxide dismutase
α-lipoic acid	Antioxidant effect of the reduced form of α-lipoic acid; restores reduced glutathione pools
N-acetyl-cysteine (NAC)	Restores reduced glutathione pools by serving as a precursor (de novo pathway); high doses are used to minimize possible excitotoxicity
Selenium as seleno-L-methionine	Precursor of selenoproteins with antioxidant functions; a variety of seleno compounds are peroxynitrite scavengers including selenomethionine; selenium levels are low in multisystem diseases
Betaine (trimethylglycine)	By raising purine nucleotide pools it produces increased degradation leading to accumulation of the scavenger of peroxynitrite products, uric acid; also increases methylation of CNS transmitters serotonin, dopamine
High-dose (7.5 g) ascorbate i.v.	According to clinical data, act by scavenging peroxynitrite, while reducing BH_3 back to BH_4
Vit B_{12} s.c. used as methyl- or hydroxycobalamin	Very effective, a specific nitric oxide scavenger in doses > 2,000 µg/die
Sauna therapy	A cost saving approach to increase BH_4 levels; works like a preconditioning stress response followed by adaptation
High-dose folates	Raises levels of 5-methyltetrahydro-folate, a potent peroxynitrite scavenger
Tetrahydrobiopterin (BH_4) or its precursors, sepiapterin or biopterin	Restores the tetrahydrobiopterin (BH_4) level, which in clinical studies, plays a pivotal key role in the down-regulation of iNOS and more important, mitochondrial nitric oxide synthase (mtNOS) activity
Inosine, RNA or D-ribose	By raising purine nucleotide pools there is an increased degradation leading to the accumulation of the scavenger of peroxynitrite products, uric acid.
Vasoactive intestinal peptide (VIP)	By inducing GTP cyclohydrolase I, it leads into an increased de novo synthesis of BH_4
Reduced glutathione, as nasal spray, IV, inhaled or as a liposomal formulation	Reduces BH_2 back to BH_4 and the natural body's own antioxidant
Ellagic acid, some flavonoids and other polyphenolic antioxidants	Scavenge the breakdown products of peroxynitrite; may also lower oxidation of BH_4

neurodermatitis, fatigue syndrome, chronic exhaustion, pain within the cervical spine and the joints and all types of allergy can be treated successfully at the same time.

While the aim of the above mentioned agents is to down-regulate the NO/ONOO⁻ production within the cells, there is emerging data demonstrating that an important part of that NO/ONOO⁻ production, is also an important issue in terms of therapy, i.e. down-regulating the generation of nitric oxide in the mitochondrion via the mitochondrial nitric oxide synthase (designated as mtNOS). However, such approach has yet to be found. And since this overactive mtNOS seems to be a key-regulator in the production of NO/ONOO⁻ (Fig. 8.26) additional forms of therapy have emerged, all of which are targeted to down-regulate the hyperactive mtNOS (Table 8.7).

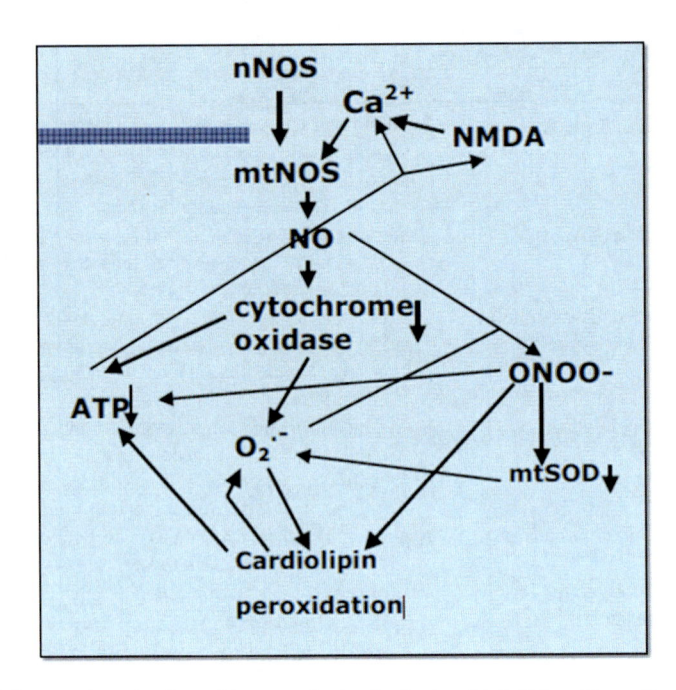

Fig. 8.26 Proposed mechanism of the central role of mitochondrial nitric oxide synthase (mtNOS) in the production of nitric oxide (NO) and peroxynitrite (ONOO-). mtSOD = mitochondroal superoxide dismutase mtNOS = mitochondrial nitric oxide synthase NMDA = N-methyl-D-aspartate; excitatory receptor nNOS = neuronal nitric oxides synthase

Table 8.7 Agents and new therapeutic approaches for down-regulation of mtNOS responses

Agents/therapeutic approach	Mode of action
Thyroid hormone (T3)	T3 lowers both nNOS transcription and its transport into mitochondria to form mtNOS. It raises thyroid hormone levels in patients from low or low normal into mid or upper normal ranges
Intermittent hypoxic oxygen breathing	Damaged mitochondria are prompted to perform apoptosis while signaling healthy mitochondria to replicate; only to be used under medical supervision. A second line of action is boosting the body's store of the enzyme superoxide dismutase (SOD)
Sauna therapy	Cold raises mtNOS activity while warmth lowers it; this is probably mediated through the heat-shock Hsp90 protein lowering mtNOS activity
Phospholipids	The basis of NT factor treatment, prepared and unoxidized phospholipids are used as a precursor for regeneration of unoxidized cardiolipin
Nicotinamid-Adenin-Dinucleotid-Hydrid (NADH) or Coenzyme 1	Being the initial spark to catalyse energy production in every mitochondrion, it also repairs damaged mtDNA while at the same time being a powerful antioxydant

Once clinically available, there may be other agents that could inherit attractive therapeutic possibilities, such as superoxide dismutase (SOD) mimitics, and also nitric oxide synthase inhibitors that lower mtNOS activity.

8.7 Reasoning Supplemental Therapy in Mitochondropathy

The basic objective is, that the NO/ONOO-cycle gives an explanation for chronic human ailments and a therapy that focuses on the down-regulation of the activated NO/ONOO$^-$ cycle and the ensuing mitochondropathy, rather than on the sole treatment of symptoms. There are up to 30 distinct currently available agents or classes of agents that have shown to down-regulate this excess in activity. While some studies suggest that D-ribose is effective in FMS, there are findings provided support that FMS is associated with biochemical abnormalities in glycolysis, which require appropriate metabolic therapy for cells (Ref. [273] in Chap. 7) [42], Ecklonia cava extract, was also found to show a similar beneficial benefit [43]. Some of these suggestions have been found in clinical trial studies to provide significant improvement in one or more of the multisystem illnesses (Ref. [4] in Chap. 6). Efficacy of others is supported by clinical observations and/or anecdotal evidence whereas in other there is no such evidence available. Generally, when efficacy of a single agent is reported, it effect is modest. This is because there is no "magic bullet" treatment, and owing to the complexity of the NO/ONOO$^-$ cycle several agents act additively or synergistically with each other, so that a complex treatment protocols is much more effective than just a single treatment. The mechanisms predicting that all these agents to one or a lesser degree down-regulate the upregulated NO/ONOO$^-$ cycle biochemistry is related to one or more of these multisystem illnesses, was demonstrated in clinical trials, clinical observation and anecdotal report, resulted in various complex treatment protocols from Dr. Teitelbaum (Table 8.8), or Dr. Niolson (Table 8.9). By containing up 14–18 different agents or classes of agents predicted to down-regulate the overexpressed NO/ONOO$^-$ cycle, each of them appears to be effective in the treatment of the multisystem disease based either on clinical observations or, in two cases (Ref. [265] in Chap. 7) [44], in clinical trial studies. In the two other from Dr. Paul Cheney (Table 8.10) and Dr. Nash Petrovic (Table 8.11) protocol and specifically the Drs Pall/Ziem protocol (Table 8.12), treatment plans were used to relieve symptoms in patients with multiple chemical sensitivity (MCS), while others have been used to treat CFS patients FMS patients (Ref. [4] in Chap. 6) [26]. Some of the agents in these protocols presently were eliminated because they obviously did not lower the activated NO/ONOO$^-$ cycle. However, above all it is important to note for any agents or classes of agents to become effective, that the individual patient avoids the intake of stressor(s) that up-regulate the NO/ONOO$^-$ cycle. Among these include:

1. All chemical exposure in MCS patients
2. Food antigens in patients who have developed a food allergy with IBS

Table 8.8 Agents from Dr. Teitelbaums protocol predicted to down-regulate the overexpressed NO/ONOO⁻ cycle

Vitamin B – complex – high dose B_6 (pyridodoxine), B_2 (riboflavin), B_1 (thiamine), B_3 (niacin) and folic acid, are necessary for sufficient function of the respiratory cycle in mitochondria.

Betaine hydrochloride (HCl) – lowers excessive anxiety; those should only take the hydrochloride form, who are deficient in stomach acid.

Magnesium – as magnesium glycinate and magnesium malate, lowers NMDA activity, applied orally or by injections.

Alpha-Lipoic acid – antioxidant, which helps to regenerate reduced glutathione.

Vitamin B_{12} – by intramuscular injection of 3 mg, using either cyano- or hydroxocobalamin, which act as nitric oxide scavenger.

Fish oil – source of long chain Ω-3-fatty acids. Lowers iNOS activity and has an anti-inflammatory effect.

Vitamin C – potent radical oxygen scavenger

Grape seed extract – source for flavonoids, e.g. oligomer procyanidine (OPC) a powerful antioxidant.

Vitamin E – natural antioxidant consisting of γ-tocopherol or tocotrienols.

Protein formula – precursors for glutathione and/or serotonine/dopamine synthesis.

Zinc – antioxidant properties and a precursor for copper/zinc superoxide dismutase (SOD).

Acetyl-L-carnitine – a necessary adjunct for restoration of mitochondrial function.

Coenzyme Q10 – both an important antioxidant and a necessary coenzyme in mitochondrial function.

D-Ribose – increases regeneration of adenine nucleotides, and ATP formation when energy synthesis in mitochondria is impaired; also help to restore reduced glutathione pools.

Table 8.9 Agents from Dr. Nicolson protocol predicted to down-regulate the overexpressed NO/ONOO⁻ cycle

Phosphatidyl polyunsaturated lipids – these lipids and the phosphatidyl choline are predicted to help restore the oxidative damaged mitochondrial inner membrane

Magnesium – lowers NMDA activity, and aids in the formation of energy within mitochondria

Taurine – antioxidant, which lowers excitoxicity reducing increased NMDA activity

Artichoke extract – a source for flavonoids

Spirulina – blue-green alga, a source of high antioxidant concentrations

Vitamin E – including γ-tocopherol and tocotrienols as potent antioxidants

Calcium ascorbate – source of vitamin C as an antioxidnat

Alpha-lipoic acid – antioxidant and key role in the regeneration of reduced glutathione

Vitamin B_6 – balances glutamate and GABA levels at the excitatory NMDA receptor

Vitamin B_3 (Niacin) – restoring function of respiratory cycle within mitochondria necessary for ATP production

Vitamin B_2 (Riboflavin) – reduces oxidized glutathione back to the active reduced glutathione; also restores mitochondrial function for ATP synthesis

Vitamin B_1 (thiamin) – restoration of mitochondrial function for ATP synthesis

Vitamin B_{12} – effective as nitric oxide scavenger

Folic acid – lowers nitric oxide synthase uncoupling which is a source of increased NO production

3. Excitotoxins such as monosodiumglutamate (MSG), saccharin and aspartame which are metabolized to formaldehyde, a neurotoxin that may be expected to increase NMDA activity with resultant pain
4. Excessive exercise resulting in post- training malaise in CFS patients
5. Persistent psychological stress, especially in PTSD patients

8.7 Reasoning Supplemental Therapy in Mitochondropathy

Table 8.10 Agents from Dr. Cheney protocol predicted to down-regulate the overexpressed NO/ONOO⁻ cycle

Vitamin B_{12} – high dose hydroxocobalamin (1,000 μg) injections, a potent nitric oxide scavenger
Protein formulation – precursor for glutathione synthesis
Guanifenesin – used as a vanilloid (TPR) receptor antagonist for reduction of pain transmission
NMDA antagonists – block overactive excitatory transmission, e.g. memantine, amantadine
Magnesium – lowers increased NMDA activity
Taurine – antioxidant, also lowers excitotoxicity and NMDA activity
GABA agonists – GABA acts as an inhibitory neurotransmitter to lower NMDA activity, include the agent neurontin (Gabapentin®) and pregablin (Lyrica®)
Histamine antagonist – blocks release of histamine from activated mast cells, which both activate nitric oxide synthesis and vanilloid release at the receptor site
Betaine hydrochloride (trimethylglycine) – lowers excessive stress, activates the methyl cycle, hydrochloride form should only be used in low stomach acid.
Flavonoids – found in olive leaf extract, organic botanicals, hawthorn extract, with antioxidant properties
Vitamin E – potent antioxidant
Coenzyme Q10 – acts both as an antioxidant and as a necessary coenzyme for restoration of mitochondrial function
Alpha-lipoic acid – antioxidant and key role in the regeneration of reduced glutathione
Selenium – functional role in the antioxydant enzyme systems and reduces risk of cancer
Ω-3- and Ω-6-fatty acids – lower iNOS activity with an anti-inflammatory effect
Melatonin – as an antioxidant and an inducer for regulation of the sleep cycle
Pyridoxal phosphate – active form of vitamin B_6, which comprises three natural organic compounds, pyridoxal, pyridoxamine and pyridoxine, improves glutamate/GABA ratio and necessary for sufficient function the respiratory cycle in mitochondria
Folic acid – lowers nitric oxide synthase uncoupling which is a source of increased NO production

Table 8.11 Agents from Dr. Petrovic protocol predicted to down-regulate the overexpressed NO/ONOO⁻ cycle

Valine and isoleucine – branched chain amino acids involved in energy metabolism of mitochondria, to stimulate ATP production; modest levels may also lower excitotoxicity
Vitamin B_6 (Pyridoxine) – improves balance between glutamate and GABA, lowers excitotoxicity
Vitamin B_{12} – as cyanocobalamin, which is converted to hydroxocobalamin in the body, a nitric oxide scavenger, where the latter does not require this step for conversion
Vitamin B2 (Riboflavin) – reduces oxidized glutathione back to the active reduced glutathione; also restores mitochondrial function for ATP synthesis
Carotenoids (i.e. alpha-carotene, bixin, zeaxanthin and lutein), lipid soluble peroxonitrite scavengers
Flavonoids (i.e. flavones, rutin, hesperetin) – radical oxygen scavengers
Ascorbic acid (vitamin C) – potent antioxidant
Tocotrienols – types of vitamin E, assumed to lower excitotoxicity
Vitamin B_1 (thiamine, aneurin) – involved in energy production of mitochondria
Magnesium – lowers NMDA activity and aids in energy production of mitochondria
Zinc – precursor of superoxide dismutase (SOD) the inborn natural antioxidant
Betaine hydrochloride (HCl) – lowers excessive stress and a methyl donor, hydrochloride form should only be used when deficient stomach acid
Essential fatty acids – including long chain Ω-3-fatty acids; predicted to help restore the oxidative damaged mitochondrial inner membrane
Phosphatidyl serine – lowers iNOS induction

Table 8.12 Agents used in the Dr Pall/Ziem protocol predicted to down-regulate the overexpressed NO/ONOO⁻ cycle

---

Inhaled, reduced glutathione- a powerful radical oxygen scavenger system

Inhaled/sublingual hydroxocobalamin- a precursor of vitamin B_{12}, a powerful scavenger of NO

Mixture of natural tocopherols – including γ-tocopherol, radical oxygen scavenger system

Vitamin C, buffered- an antioxidant

*Magnesiummalate-*Mg-salt for fast absorption, lowers NMDA activity and necessary for the energy production in mitochondria

Flavonoids found in ginkgo biloba extract, cranberry extract, silymarin, and bilberry extract, potent radical oxygen scavengers

Selenium – from grown yeast, part of enzyme that binds oxygen radical formation in lipid layers

Coenzyme Q10 – acts both as an antioxidant and as a necessary coenzyme for restoration of mitochondrial function

Folic acid- lowers nitric oxide synthase (iNOS) uncoupling, which is a source of increased NO production

Carotenoids- i.e. lycopene, lutein and beta-carotene,

Alpha-lipoic acid- antioxidant and key role in the regeneration of reduced glutathione

Trace elements- i.e. zinc (modest dose), manganese (low dose) and copper (low dose), necessary for enzyme synthesis

Pyridoxal phosphate – active form of vitamin B_6, which comprises three natural organic compounds, pyridoxal, pyridoxamine and pyridoxine, improves glutamate/GABA ratio and necessary for sufficient function the respiratory cycle in mitochondria

Riboflavin 5'-phosphate (FMN) –activated form of

Betaine (trimethylglycine)- lowers excessive stress, used as a methyl donator for melatonin, serotonin, and adrenalin. Also used in peripheral polyneuropathy to induce lecithin synthesis of nerve sheaths; it also increases metabolization of histamine reducing allergic effects on CNS. The hydrochloride should only be used when deficient stomach acid is obvious

Green tea extract- a herbal derivative from green tea leaves (*Camellia sinensis*), containing antioxidant ingredients, mainly epigallocatechins (EGTC[a])

Hawthorn extract- Active ingredients are tannins, flavonoids (such as vitexin, rutin, quercetin, and hyperoside), oligomeric proanthocyanidins (OPCs, such as epicatechin, procyanidin, and particularly procyanidin B-2 as potent antioxidants), flavone-C, triterpene acids (such ursolic acid, oleanolic acid, and crataegolic acid), and phenolic acids, such as caffeic acid, chlorogenic acid, and related phenolcarboxylic acids. Used in conventional medicine for treatment of cardiac insufficiency

l-carnitine- required for the transport of fatty acids from the cytosol into the mitochondria during the breakdown of lipids (fats) for the generation of metabolic energy

---

[a]Epigallocatechin gallate (EGCG), also known as epigallocatechin 3-gallate, is the ester of epigallocatechin and gallic acid, and is a type of catechin. EGCG is the most abundant catechin in most notably tea, among other plants, and is also a potent antioxidant [45] that may have therapeutic properties for many disorders including cancer [46, 47]. It is found in green tea, but not black tea, as EGCG is converted into thearubigins in black teas. In a high temperature environment, an epimerization change is likely to occur, because heating results in the conversion from EGCG to GCG. It therefore is considered inappropriate to infuse green tea or its extracts with overheated water

The fact that so many agents are necessary to down-regulate the overactive NO/ONOO⁻ cycle in treatment protocols is not coincidental, since the NO/ONOO⁻ cycle attacks a number of biochemical pathways, thus being useful predictors in terms of therapy. Conversely, the apparent efficacy of both individual and combinations of such agents provides support for a NO/ONOO⁻ cycle etiology of these illnesses. Since science is continuously developing, new aspect in the treatment of mitochondropathy have emerged, which gave their way into new avenues of treatment. For instance, the importance of adding nutraceuticals, a term combining the words "nutrition" and "pharmaceutical", *is coined to a food or food product that provides health and medical benefits, including the prevention and treatment of a disease.* Such products may range from isolated nutrients, dietary supplements and specific diets to genetically engineered foods, herbal products and processed foods such as cereals, soups, and beverages. With recent breakthroughs in cellular-level nutraceuticals agents, researchers, and medical practitioners are developing templates for integrating assessing and assessing information from clinical studies on complimentary and alternative therapies into responsible medical practice. Dr. Stephen L. DeFelice, founder and chairman of the Foundation of Innovation Medicine (FIM), Crawford, New Jersey, originally defined the term nutraceutical. Since the term was coined by Dr. DeFelice, its meaning has been modified by Health Canada, which defines nutraceutical *as a product isolated or purified from foods, and generally sold in medicinal forms not usually associated with food and demonstrated to have a physiological benefit or provide protection against chronic disease.* Typical examples are beta-carotene and lycopene. The definition of nutraceutical that appears in the latest edition of the *Merriam-Webster Dictionary* is as follows: *A food stuff (as a fortified food or a dietary supplement) that provides health benefits.* Nutraceutical foods are not subject to the same testing and regulations as pharmaceutical drugs. The American Nutraceutical Association works with the Food & Drug Administration in consumer education, developing industry and scientific standards for products and manufacturers, and other related consumer protection roles. The FDA provides a list of dietary supplement companies receiving warning letters about their products.

8.8 Food Used as Medicine in Mitochondropathy

The Indians, Egyptians, Chinese, and Sumerians are just a few civilizations that have provided evidence suggesting that foods can be effectively used as medicine to treat and prevent disease. Ayurveda, the 5,000 year old ancient Indian health science has mentioned benefits of food for therapeutic purpose. Documents hint that the medicinal benefits of food have been explored for thousands of years [48].

Hippocrates, considered by some to be the father of Western medicine, recommended to people (Fig. 8.27).

"Let food be thy medicine"

Fig. 8.27 Considered as father of Western medicine, Hippocrates advocated the healing effects of food

The modern nutraceutical market began to develop in Japan during the 1980s. In contrast to the natural herbs and spices used as folk medicine for centuries throughout Asia, the nutraceutical industry has grown alongside the expansion and exploration of modern technology [49]. New research conducted among food scientists show that there is more to food science than what was understood just a couple decades ago [49]. Until just recently, analysis of food was limited to the flavor of food (sensory taste and texture) and its nutritional value (composition of carbohydrates, fats, proteins, water, vitamins and minerals). However, there is growing evidence that other components of food may play an integral role in the link between food and health. These chemical components are derived from plant, food, and microbial sources, and provide medicinal benefits valuable to long-term health. Examples of these nutraceutical chemicals include probiotics, antioxidants, and phytochemicals (Fig. 8.28) . Nutraceutical products were considered alternative medicine for many years. Nutraceuticals have become a more mainstream supplement to the diet, now that research has begun to show evidence that these chemicals found in food are often effective when processed effectively and marketed correctly.

8.8 Food Used as Medicine in Mitochondropathy

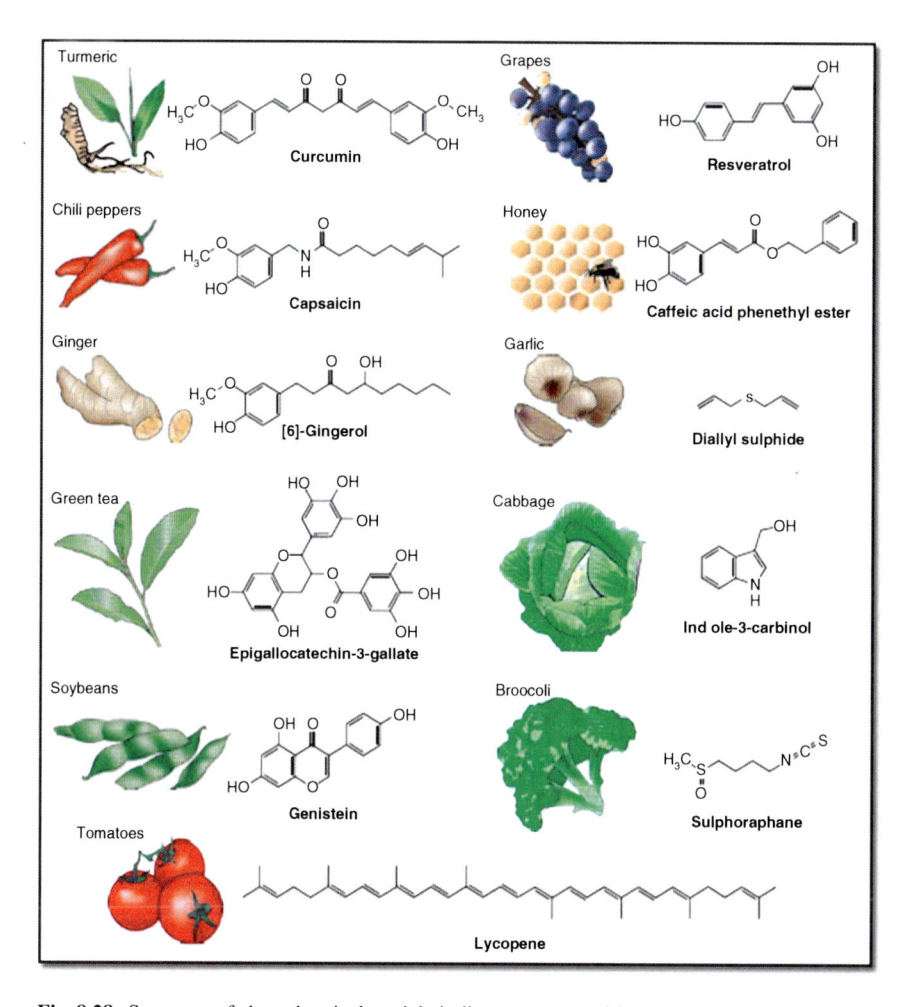

Fig. 8.28 Summary of phytochemicals and their dietary sources with antioxidative (and anticancer) properties

8.8.1 Classification of Nutraceuticals

Nutraceuticals is a broad umbrella term used to describe any product derived from food sources that provides extra health benefits in addition to the basic nutritional value found in foods. Products typically claim to prevent chronic diseases, improve health, delay the aging process, and increase life expectancy [50]. There is minimal regulation over which products are allowed to display the nutraceutical term on their labels. Because of this, the term is often used to market products with varying uses and effectiveness. The definition of nutraceuticals and related products often depend on the source. Members of the medical community desire that the nutraceutical term be more clearly established in order to distinguish between the wide varieties of products out there. There are multiple different types of products that fall under the category of nutraceuticals.

Dietary supplements, such as the vitamin B supplements are typically sold in pill form. A dietary supplement is a product that contains nutrients derived from food products that are concentrated in liquid or capsule form. The Dietary Supplement Health and Education Act (DSHEA) of 1994 defined generally what constitutes a dietary supplement. "A dietary supplement is a product taken by mouth that contains a "dietary ingredient" intended to supplement the diet. The "dietary ingredients" in these products may include: vitamins, minerals, herbs or other botanicals, amino acids, and substances such as enzymes, organ tissues, glandulars, and metabolites. Dietary supplements can also be extracts or concentrates, and may be found in many forms such as tablets, capsules, soft gels, gel caps, liquids, or powders [50]. Dietary supplements do not have to be approved by the U.S. Food and Drug Administration (FDA) before marketing. Although supplements claim to provide health benefits, products usually include a label that says: "These statements have not been evaluated by the Food and Drug Administration. This product is not intended to diagnose, treat, cure, or prevent any disease."

Functional foods are designed to allow consumers to eat enriched foods close to their natural state, rather than by taking dietary supplements manufactured in liquid or capsule form. Functional foods have been either enriched or fortified, a process called nutrification. This practice restores the nutrient content in a food back to similar levels from before the food was processed. Sometimes, additional complementary nutrients are added, such as vitamin D to milk. Health Canada defines functional foods as "ordinary food that has components or ingredients added to give it a specific medical or physiological benefit, other than a purely nutritional effect " [51]. In Japan, all functional foods must meet three established requirements: foods should be

1. Present in their naturally-occurring form, rather than a capsule, tablet, or powder.
2. Consumed in the diet as often as daily; and
3. Should regulate a biological process in hopes of preventing or controlling a disease [52]

Medical foods on the other hand are not available as an over-the-counter product to consumers. The FDA considers medical foods to be "formulated to be consumed or administered internally under the supervision of a physician, and which is intended for the specific dietary management of a disease or condition for which distinctive nutritional requirements, on the basis of recognized scientific principles, are established by medical evaluation." Nutraceuticals and supplements do not meet these requirements and are not classified as Medical Foods. Medical foods can be ingested through the mouth or through tube feeding. Medical foods are always designed to meet certain nutritional requirements for people diagnosed with specific illnesses.

Medical foods are regulated by the FDA and will be prescribed/monitored by medical supervision.

Farmaceutials – according to a report written for the United States Congress entitled "Agriculture: A Glossary of Terms, Programs, and Laws", the word Farmaceuticals is a melding of the words farm and pharmaceuticals. It refers to medically valuable compounds produced from modified agricultural crops or animals (usually through biotechnology). Proponents believe that using crops and possibly even animals as pharmaceutical factories could be much more cost effective than conventional methods (i.e., in enclosed manufacturing facilities) and also provide agricultural producers with higher earnings, "At issue in the United States has been whether the current system for regulating biotechnology is adequate for ensuring the safety (to humans, animals and crops, and the environment) of newly emerging applications, such as farmaceuticals. The term farmaceuticals is more frequently associated, in agricultural circles, with medical applications of genetically engineered crops or animals."

8.8.2 Oxygen Radical Absorbance Capacity of Food

Oxygen Radical Absorbance Capacity (ORAC) is a method of measuring antioxidant capacities in biological samples in vitro [53]. A wide variety of foods has been tested using this methodology, with certain spices, berries and legumes rated highly [54]. Correlation between the high antioxidant capacity of fruits and vegetables, and the positive impact of diets high in fruits and vegetables, is believed to play a role in the free-radical theory of aging. However, there exists no physiological proof in vivo that this theory is valid. Consequently, the ORAC method, derived only in test tube experiments, cannot currently be applied to human biology. The assay measures the oxidative degradation of the fluorescent molecule (either beta-phycoerythrin or fluorescein) after being mixed with free radical generators such as azo-initiator compounds. Azo-initiators are considered to produce the peroxyl radical by heating, which damages the fluorescent molecule, resulting in the loss of fluorescence. Antioxidants are considered to protect the fluorescent molecule from the oxidative degeneration. The degree of protection is quantified using a fluorometer. Fluorescein is currently used most as a fluorescent probe. Equipment that can automatically measure and calculate the capacity is commercially available (Biotek, Roche Diagnostics).

Scientists with the United States Department of Agriculture have published lists of ORAC values for plant foods commonly consumed by the U.S. population (fruits, vegetables, nuts, seeds, spices, grains, etc.). Values are expressed as the sum of the lipid soluble (e.g. carotenoid) and water-soluble (e.g. phenolic) antioxidant fractions (i.e., "total ORAC") reported as in micromoles Trolox equivalents (TE) per 100 g sample, and are compared to assessments of total polyphenol content in the samples (Table 8.13). Drawbacks of this method are: (1) only antioxidant activity against particular (probably mainly peroxyl) radicals is measured; however,

Table 8.13 Lists of ORAC values for plant foods commonly consumed by the U.S. population (fruits, vegetables, nuts, seeds, spices, grains, etc.). Values are expressed as the sum of the lipid soluble (e.g. carotenoid) and water-soluble (e.g. phenolic) antioxidant fractions (i.e., "total ORAC") reported as in micromoles Trolox equivalents (TE) per 100 g sample, and are compared to assessments of total polyphenol content in the samples (Source: [58])

Food	Serving size	Total antioxidant capacity per serving size. in units of mmol of trolox equivalents
Raw unprocessed cacao	100 g	28,000
Small red bean	½ cup, dried	13,727
Wild blueberry	1 cup	13,427
Red kidney bean	½ cup, dried	13,259
Pinto bean	½ cup	11,864
Blueberry	1 cup	9,019
Cranberry	1 cup	8,983
Artichoke hearts	1 cup, cooked	7,904
Blackberry	1 cup	7,701
Prune	½ cup	7,291
Raspberry	1 cup	6,058
Strawberry	1 cup	5,938
Red delicious apple	1 apple	5,900
Granny Smith apple	1 apple	5,381
Pecan	1 oz	5,095
Sweet cherry	1 cup	4,873
Black plum	1 plum	4,844
Russet potato	1 cooked	4,649
Black bean	½ cup, dried	4,181
Plum	1 plum	4,118
Gala apple	1 apple	3,903

peroxyl radical formation has never been proven; (2) the nature of the damaging reaction is not characterized; (3) there is no evidence that free radicals are involved in this reaction; and (4) there is no evidence that ORAC values have any biological significance following consumption of any food. Moreover, the relationship between ORAC values and a health benefit has not been established. Although research in vitro indicates that polyphenols are good antioxidants and probably influence the ORAC value, antioxidant effects in vivo are probably negligible or absent [55]. By *non*-antioxidant mechanisms still undefined, flavonoids and other polyphenols may reduce the risk of cardiovascular disease and cancer [56]. As interpreted by the Linus Pauling Institute and EFSA, dietary polyphenols have little or no direct antioxidant food value following digestion [55, 57]. Not like controlled test tube conditions, the fate of polyphenols in vivo shows they are poorly conserved (less than 5%), with most of what is absorbed existing as chemically-modified metabolites destined for rapid excretion. The increase in antioxidant capacity of blood seen after the consumption of polyphenol-rich (ORAC-rich) foods is not caused directly by the polyphenols, but most likely results from increased uric acid levels derived from metabolism of flavonoids [57].

8.8.3 Rational for Specific Supplements in Mitochondropathy

Before going into therapy of mitochondropathy after all it is relevant to know if this upregulation in the production of NO/ONOO⁻ has any specific physiological meaning in normal life.

Cobalamine (Vitamin B_{12}) is a nitric oxide scavenger and deficient in the majority of chemically ill patients. The cyano form is not recommended, since patients don't need cyanide and the hydroxy and methyl forms work much better in the brain and nerve cells. Superoxide dismutase is deficient in a significant portion of chemically ill patients and its cofactors, copper, zinc, and manganese must be adequate. These are often reduced in chemically injured patients and should be tested and replaced in well-absorbed and transported forms, for example, picolinates. Antioxidant function is usually inadequate in chemically ill patients, and increased lipid peroxides and other free radicals are common. Intervention to help reduce this vicious biochemical cycle includes: methyl or hydroxycobalamine sublingually or intramuscular (not oral due to poor absorption), general antioxidants (vitamin C, E, selenium), glutathione by nebulizer due to poor oral absorption, and ample alpha-lipoic acid to reactivate the glutathione in the many damaged lipid tissues (cell membranes, mitochondria, lymph, brain, etc,). Trimethylglycine (betaine) is recommended as a methyl donor to reduce the effects of peroxynitrite. Magnesium should be ample because deficiency is very common with toxic injury and adequate magnesium decreases NMDA activation. Peroxynitrite scavengers such as a mixture of carotinoids are also recommended, while carotinoids tend to be more organ-specific. An inclusion of gingko (brain), silimarin (liver), bilberry (collagen stabilizing, capillary permeability, vision), cranberry (urinary) and other mixed carotenoids is advised. Mineral levels should be measured and followed by intracellular (e.g. RBC) or lipid functional (e.g. lymphocyte mitogenesis, a SpectraCell technology). Functional lymphocyte evaluation and follow-up of glutathione, alpha-lipoic acid, total antioxidant function, C, E and zinc is also suggested. At this time this technology is only available through Spectra cell laboratory located in the US.

First and most important of all, since NF-κB elevation and consequent elevation of the inflammatory cytokines with an induction of iNOS and other inflammatory responses are key elements in activating the immune system, other defensive responses are necessary to manage and overcome acute infections. Aside, blood pressure control is substantially regulated by angiotensin II, which increases activity of the NADPH oxidase consequently resulting in superoxide production. This in turn directly activates the central element iNOS resulting in BH_4 depletion, which presents a key role in the production of an increased blood pressure.

Second, mitochondrial activity is regulated by the mitochondrial nitric oxide synthase (mtNOS), whose activity in return is ruled both by both hypoxia and the thyroid hormone T3. Hypoxia and low T3 levels both yield an increased activity of mtNOS with lowering in mitochondrial function by competitively inhibiting cytochrome oxidase activity [59]. Such increase in activity of mtNOS consecutively results

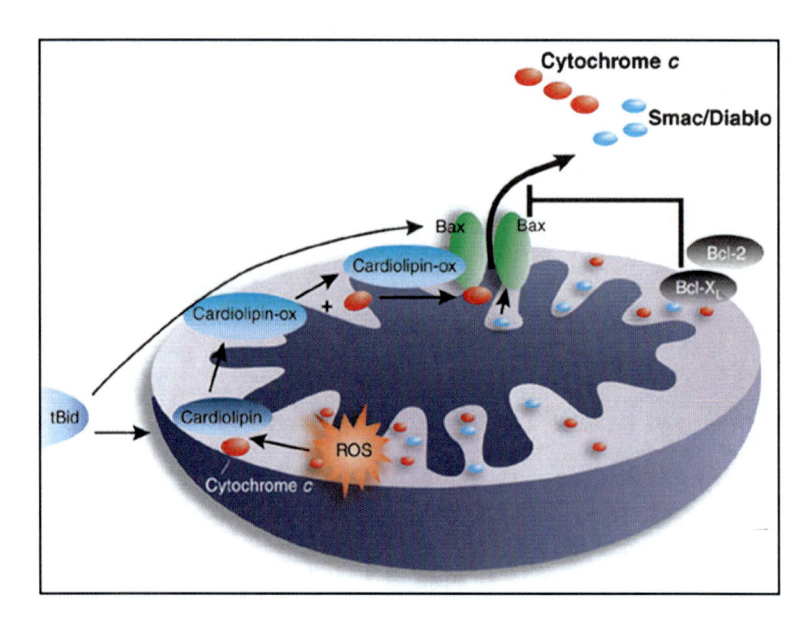

Fig. 8.29 Cytochrome *c* is anchored to the outer surface of the mitochondrial inner membrane by electrostatic and hydrophobic interactions with cardiolipin. During the early phase of apoptosis, mitochondrial ROS production is stimulated, and cardiolipin is oxidized by a peroxidase function of the cardiolipin–cytochrome *c* complex. The hemoprotein is then detached from the mitochondrial inner membrane and can be extruded into the soluble cytoplasm through pores in the outer membrane. Cardiolipin also serves as a mitochondrial target to the C-terminal cleavage product of the Bcl-2 protein, Bid, which promotes pore formation in the outer membrane by Bax or Bak, a process that is inhibited by Bcl-2 or Bcl-X_L. Finally, permeabilization of the outer membrane is further enhanced by cardiolipin hydroperoxides, which stimulate the release into the cytoplasm of cytochrome *c* and Smac/Diablo. Cardiolipin-ox, i.e. oxidized cardiolipin. Adapted from [61]

in the release of large amounts of superoxide from the electron transport chain. The lowering of cytochrome oxidase activity in response to hypoxia may seem counterproductive, but it has been argued that it allows downstream tissues from the site of hypoxia to obtain at least some oxygen, thus avoiding in dying of complete anoxia. It follows, that if substantial amounts of tissues died in response to hypoxic conditions, this would rapidly lead into death of the organism. Therefore, this reaction may be an important short-term regulatory response for survival. However, if activated over a longer period of time, such compensatory mechanisms lead to an inactivation of protein function and oxidative physphorylation of mitochondrial cardiolipin. Since the inner membrane is a lipid bilayer containing a high proportion of cardiolipin (diphosphatidylglycerol) induction of apoptosis results in cytochrome c release and cardiolipin oxidation, both major changes resulting from ROS and NO/ONOO⁻ production (Fig. 8.29). **Third**-Cardiolipin sufficiently reduces the membrane's permeability to protons making the membrane especially impermeable to ions and establishing the conditions that allow a proton-motive force to be established across it. Thus when being oxidized, cardiolopin-ox looses its function which can be tested by determining cardiolipin in white blood cell mitochondria where the percentage of inner membrane lipids that fluorescence-stain as cardiolipin can be measured using fluorescence microscopy. This is of importance, since cardiolipin (CL) is essential for the functionality of several mitochondrial proteins. Its distribution between the inner and outer leaflet of the mitochondrial internal membrane is crucial for ATP synthesis. In the heart, functional cardiolipin is linoleic acid rich, and the loss of linoleic acid content is associated with cardiac disorders including ischemia and reperfusion, heart failure, and diabetes. Dietary interventions (such as R-lipoic acid, the R-enantiomer of alpha-Lipoic acid; with higher bioavailability) may restore cardiolipin to its linoleic acid-rich form and improve cardiac function by redirecting the remodeling process [60]. Therefore blocked translocator sites (TL) may be due to the following:

- *Elevated Hydrogen Sulphide levels*: Elevated H_2S levels caused by the fermentation of sugar by bad bacteria and fungi in the digestive tract – H_2S attaches to the mitochondrial enzyme cytochrome c oxidase and weakens oxidative phosphorylation and ATP production.
- *A too low pH at the membrane (e.g. too acidic)*: The translocator (TL) protein site on the inner mitochondrial membrane is extremely pH sensitive and can be affected by local or general acidosis. Also, the organic acid accumulations from over-dependence on anaerobic metabolism (the dominant form of metabolism where oxygen levels are too low) affect the ability of the TL sites.
- *Elevated intracellular Calcium and reduced intracellular Magnesium*: The efficiency of TL site can also be compromised by increased intracellular Calcium or reduced intracellular Magnesium. Enzymatic reactions are required to transport minerals in and out of the cells of the body, which require ATP. ATP shortages can have knock on effects as above which further exacerbate the problem.

As a result, mitochondrial dysfunction may in turn affect hypothalamic/hormonal dysfunction, poor liver and kidney functioning, cardiac capability and digestive efficiency. Mitochondrial function will impact all the cells of the body and their normal function to some degree, impacting all the organs and glands (some more than others), and their ability also to produce enzymes and hormones as they should. Symptoms of mitochondrial dysfunction may include a lack of physical energy, lack of mental energy and ability to concentrate ('brain fog'), tendency to crash and burn, muscle and joint weakness, cardiac weakness/insufficiency, digestive inefficiency, and perhaps even muscular control. The exact effects vary according to the individual.

In order to repair the mitochondrial membranes, one must ingest or produce enough of the requisite Phospholipids and ingest sufficient Essential Fatty Acids. The problem that often occurs in CFS patients is that the body is unable to produce high volumes of phospholipids on account of a blockage in the methylation pathway (for further information see page 189 ff), which is dependent on the bioavailability of the correct amino acids as well as active forms of folate and B_{12}. Thus, it is very important for those with damaged or leaking mitochondrial membranes to supplement sufficient essential fatty acids (Ω-3 and Ω-6) and also phospholipids, especially phosphatidyl choline.

Fourth, and most important, there are several factors in modern life style that makes people much more susceptible to NO/ONOO⁻ upregulation and its resultant mitochondrial dysfunction. Most of all, western diet results in a marginal deficiency of vitamin D_3 which not only is involved in bone density, but lately has been shown but to be engaged in immune function, outbreak of cardiovascular disease and also tumor growth [62]. For instance, studies have shown conclusively that people with low vitamin D levels are more prone to develop myocardial infarction and outbreak of cancer, both clinical findings which have led to the recommendation supplementation of least 1,000 μg of vitamin D to the daily diet in patients with such ailments [63]. Because of the consequence of such modern lifestyle, vitamin D levels are less than half those found in people who are living outdoors and taking a so called "cave man" diet [64]. Lastly, the administration of vitamin D, as was pointed out by other researchers has a major impact in therapy as it greatly lowers NF-κB responses [65].

Also, many of us are marginally deficient of magnesium, which is due to the highly processed foods in western diets, leading to an increased NMDA-activity and possibly leading to lowered energy metabolism within mitochondria [66].

There are at least four factors in our modern life which actively upregulate the NO/ONOO⁻ activity leading into mitochondrial dysfunction:

- Monosodium glutamate (MSG) is an agonist of the NMDA receptors, and while it has been argued that the blood brain barrier (BBB) will prohibit excess glutamate from entering the brain, this is at best an oversimplification.
- Most important, chemical toxins in the today's diet all affect mitochondrial function to an extensive degree. Among them the most common ones are:
  - Lectins naturally present in foods
  - Artificial additives, colorings, flavorings; artificial sweeteners (for instance aspartame in diet soft drinks)
  - Pesticide residues, plasticizer residues in food and drinks.
  - Social chemicals (alcohol, caffeine, tobacco etc), and
  - Dioxin, which according to recent news, inadvertently has entered the food chain during the processing of pellets that were fed to chicken, cows, pigs and turkeys. The term **dioxin** is used in chemistry to describe a heterocyclic six-membered ring where two carbon atoms have been substituted by oxygen atoms. While this moiety can appear in a wide range of compounds, the most important and widely known group are the polychlorinated dibenzodioxins, a group of highly toxic environmental toxins commonly referred to as simply "dioxins".
- Fructose as well as glucose levels in many of our diets, produce highly fluctuating blood sugar levels, which in return results in large amounts of methylglyoxal **being** involved in the formation of advanced glycation end products (AGEs). In fact, methylglyoxal is proven to be the most important glycation agent (forming AGEs), down-regulate normal mitochondrial function with a significant increase in mitochondrial superoxide production [67]. This has been demonstrated in studies where high levels of glucose result in increased mitochondrial superoxide generation and ROS [68]. Glucose could, therefore, be viewed as an inflammatory mediator!

- Not only is the exposure to various chemicals of relevance for impairment of mitochondrial function, but also a number of Rx medication (for further information see page 91 ff) directly inhibit mitochondrial function, ATP formation **and** indirectly produce an increase in NMDA activity.
- Electromagnetic field exposures produce increased activity of the plasma membrane NADH oxidase, leading to increased superoxide production [69].
- And lastly, depletion of tetrahydroxybiopterin (BH_4), which is a core regulator in the formation of NO^- and $ONOO^-$ (see Fig. 6.3) and has been found to down-regulate iNOS and more important, mitochondrial nitric oxide synthase (mtNOS) activity (Ref. [13] in Chap. 7), although only few studies have been published in this area.

In summary, while, there are several reasons why a temporary increase in $NO/ONOO^-$ production is of advantage for survival, since it has important, or even essential regulatory significance in dealing with acute stressors, long exposures are deleterious. Over and over again, one finds specific elements in various chronic diseases are being elevated: Notably more or less, there is an elevated nitric oxide level, a high peroxynitrite concentration, signs of oxidative and nitrosative stress with the production of reactive oxygen and nitrogen species (ROS & RNS), an elevation in inflammatory cytokines, and a raise in NF-κB activity. But most of all, if one specifically looks for it, patients demonstrate a significant mitochondrial dysfunction with elevated excitotoxicity including an excessive NMDA activity. This pattern of elevation of the specific elements is quite consistent, although not all of them have been looked for at in every disease.

8.9 Newer Therapeutic Options in Mitochondropathy

8.9.1 *Rational for Supplementation with Vitamin D_3*

Calcitriol, or 1,25-dihydroxyvitamin D_3 ($1,25(OH)_2D_3$) is a well-known endocrine regulator of calcium homeostasis. More recently, local calcitriol production by immune cells was shown to exert autocrine or paracrine immunomodulating effects. Immune cells that produce calcitriol also express the vitamin D receptor (VDR) and the enzymes needed to metabolize vitamin D_3 (1α-, 25-, and 24-hydroxylases). Studies of animal models and cell cultures showed both direct and indirect immunomodulating effects involving the T cells, B cells, and antigen-presenting cells (dendritic cells and macrophages) affecting both innate and adaptive immune responses. The overall effect of Vitamin D_3 is a switch from the Th1/Th17 response to the Th2/Treg profile. These immunomodulating effects of vitaminD may explain the reported epidemiological associations between a low vitamin D status and a large number of autoimmune and inflammatory diseases (Fig. 8.30). Such associations have been suggested by observational studies not only in rheumatoid arthritis, lupus, inflammatory bowel disease, and type 1 diabetes; but also in infections, cancer [70], transplant rejection, and cardiovascular disease . In animal models for

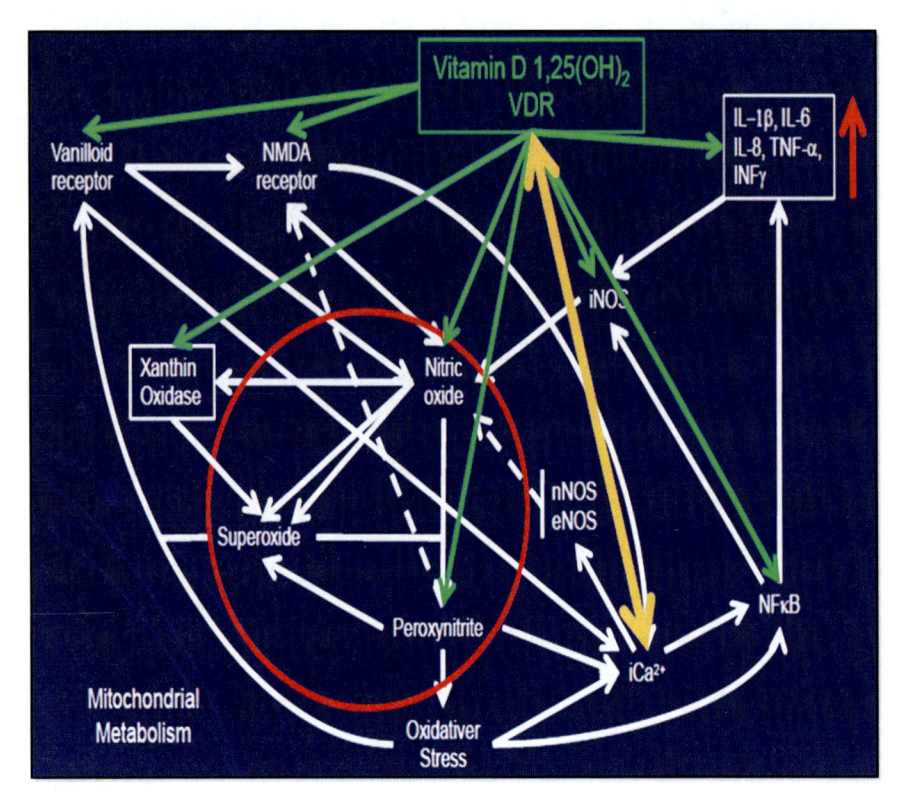

Fig. 8.30 Site of action and rational for the use of vitamin D_3 in nitrosative stress and the accompanying mitchondropathy with down-regulation of NF-κB activity, NO-, superoxide-, and peroxynitrite-production, reduced activity at the NMDA- and vanilloid receptor sites, all of which are highly overexpressed with NOS (nitrosative oxygen substrates) formation. VDR = Vitamin D Receptor

these diseases, vitamin D supplementation has been found to produce therapeutic effects. Thus, vitamin D_3 deficiency seems epidemic being a key focus for public health efforts and holds promise for the treatment of dysimmune diseases [71]. When substituting vitamin D_3 and in contrast to the RDA, at least a dose of 1,000–2,000 IU (= 25–50 μg) or even 5,000 IU per day is advocated, depending on clinical reaction. Such high dosages are well tolerated as presently no known side effects (even kidney stones!) have been reported in patients taking excessive amounts of that vitamin that no adverse events (not even potential ones) for anyone taking less than 40,000 IU/day (NOT a recommended amount, however!) [70]. Especially it was demonstrated that it takes more vitamin D to keep rising at higher levels. Due to the wide inter- and even intraindividual blood plasma levels in patients when taking the same high dose of vitamin D_3 over a long period of time, one has to reassure sufficient bioavailability. This is because this vitamin, besides Q10 and vitamin E, has lipophilic properties, which makes it difficult to enter the hydrophilic

8.9 Newer Therapeutic Options in Mitochondropathy

layer of the mucous membranes of the intestine. Therefore, a formulation with a high micellation factor is advocated.

In Conclusion

- D_3 is the natural form in human and animals; it is what we make in our skins on exposure to UV-B light
- D_2, once thought equivalent to D_3, is only ~50–60% as potent as D_3
- There is now a gradually growing acceptance of 75–80 nmol/L (30–32 ng/mL) as the lower end of the "normal" range.
- Serum 25(OH)D levels below 80 nmol/L are not adequate for any body system
- Levels of as high as 125 nmol/L may be closer to optimal
- Inputs from all sources combined are in the range of:
- ~4,000 IU/d to sustain 80 nmol/L, and
- ~5,000 IU/d to sustain 100 nmol/L

8.9.2 Supplementation with Neutraceuticals

Within this scope, neutraceuticals such as curcumin, N-acetyl-cysteine and Q10, together with NADH (Table 8.14), have led into a basic and individually tailored plan of therapy when mitochondropathy results in diverse pathology of different organs and organ failure, all of which aim to downregulate the overexpressed NO/ONOO⁻ cycle, block NF-κB activity with a reduction of inflammation and restore mitochondrial function.

8.9.3 NADH Essential Within the Electron Transport Chain

NADH or nicotinamide-adenine-dinucleotide, like all dinucleotides, consists of two nucleotides joined by a pair of bridging phosphate groups. The nucleotides consist of ribose rings, one with adenine attached to the first carbon atom (the 1′ position) and the other with nicotinamide at this position. NADH also termed as Coenzyme 1, is a dinucleotide, since it consists of two nucleotides joined through their phosphate groups, with one nucleotide containing an adenine base and the other containing nicotinamide (Fig. 8.31). It is involved in redox reactions, carrying electrons from one reaction to another. The coenzyme is, therefore, found in two forms in all cells: NAD+ is an oxidizing agent, i.e. it accepts electrons from other molecules and becomes reduced. This reaction forms NADH, which can then be

Table 8.14 Summary of neutraceuticals being used in patients with mitochondrial related diseases

Neutraceutical	Mode of action
Propolis	Resinous mixture of honey bees collected from tree buds, sap flows, or other botanical sources are used as a sealant for open spaces in the hive; due to its composition in sinapic acid, isoferulic acid, caffeic acid and chrysin, with anti- fungal-, anti-bacterial and antiviral properties [72]. Also, isoflavonoids 3-Hydroxy-8,9-dimethoxypterocarpan and medicarpin [73] are beneficial in flu, sore throat, allergy [74, 75], inheriting an antitumor growth factor [76] and an immunmodulatory action [75, 77]
Curcumin from Curcuma longa	Effective in inflammation as a COX2 -, and a NF-κB- inhibitor. Protects liver against ROS. A co-medication in cancer therapy. Proposed mechanism of action in mitochondropathy since it absorbs light quantum within the 420 nanolamda range, which is similar to mitochondrial cytochrome c, bridging the defect at complex IV-V induced by excess NO/OONO- formation
Siberian ginseng	Active ingredients are ginsenoides, used as an adaptogen in stress, fatigue, demonstrate antiinflammatory effect; increases alertness, and has an antidiabetic effect in type 2 diabetes
NADH, "nicotinamide adenine dinucleotide (NAD)+hydrogen (H)."	Necessary constituent in energy production within mitochondria (complex I). Rational for its use is that it Coenzyme Q10 needs NADH to be transformed into its reduced form. This is because only the reduced form of CoQ10 is pharmacologically active and can transport electrons across the two layers within mitochondria. Therefore NADH is the first necessary component resulting into the production of ATP improving mental clarity, alertness, concentration and memory. Also plays a role in the immune system, activates the biosynthesis of neurotransmitters adrenaline, dopamine and serotonin. Advocated for treating Parkinson & Alzheimer's disease via increased synthesis of BH_4, since NADH is the essential cofactor for this reaction of the enzyme quinoid dihydrobiopteridin reductase (DHPR) resulting in increased formation of BH4. Energy production improves athletic endurance and treatment of chronic fatigue syndrome (CFS) Considered in high blood pressure, high cholesterol, jet lag; opposes alcohol's effects on the liver and the hormone testosterone; reduces signs of aging; and protects against the side effects of the AIDS drug zidovudine (AZT)
N-Acetyl-Cysteine up to 2 g	A necessary precursor of the radical oxygen scavenger glutathion (GSH) serving as a SH-donator in synthesis of glutathione. Antioxidant for liver protection. Immune stimulator in AIDS, effective in influenza and protects against glucose toxicity
Anthocyanes, a group of the flavinoides, extracts from blueberries	Natural plant coloring agent, powerful antioxidant and radical oxygen scavenger, more potent than vitamin C & E, considered potent inhibitor of cancer growth; for increases in bioavailability natural raisins are added (i.e. propolis)

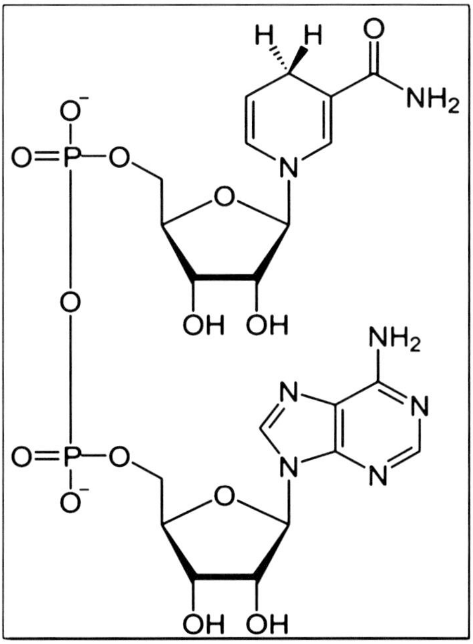

8.9 Newer Therapeutic Options in Mitochondropathy

Fig. 8.31 Chemical structure of NADH, a Nicotinamide-adenine- dinucleotide

used as a reducing agent to donate electrons. These electron transfer reactions are the main function of NAD+. However, it is also used in other cellular processes, the most notable one being a substrate of enzymes that add or remove chemical groups from proteins, in posttranslational modifications. Because of the importance of these functions, the enzymes involved in NAD+ metabolism are targets for drug discovery.

In metabolism, the compound accepts or donates electrons in redox reactions. Such reactions (summarized in formula below) involve the removal of two hydrogen atoms from the reactant (R), in the form of a hydride ion (H Template:−), and a proton (H$^+$). The proton is released into solution, while the reductant RH$_2$ is oxidized and NAD$^+$ reduced to NADH by transfer of the hydride to the nicotinamide ring.

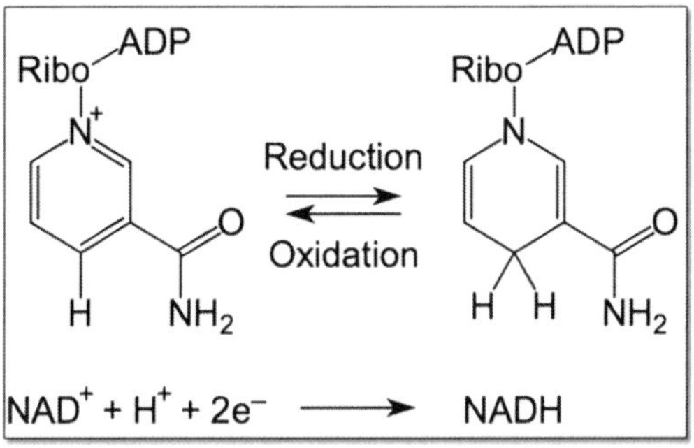

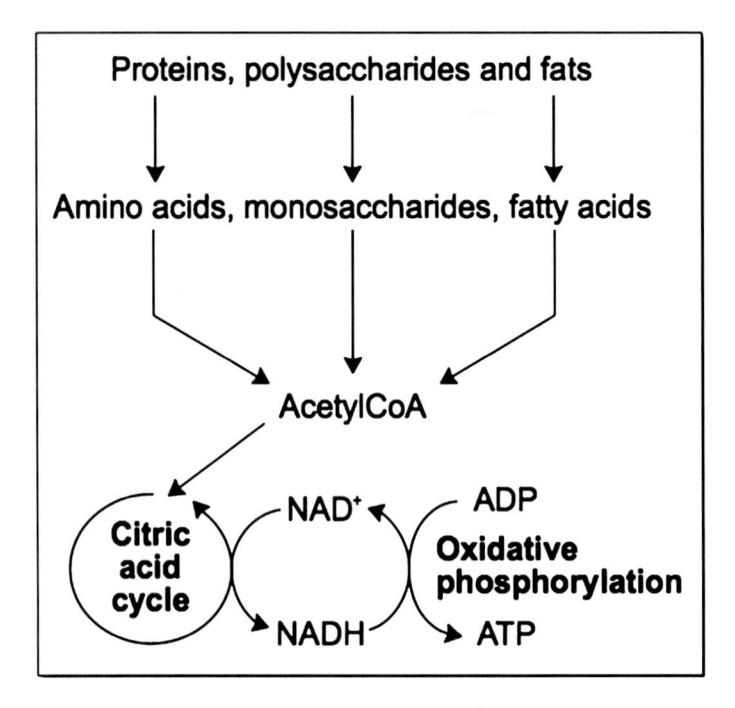

Fig. 8.32 A simplified outline of redox metabolism, showing how NAD⁺ and NADH link the citric acid cycle and oxidative phosphorylation

The balance between the oxidized and reduced forms of nicotinamide adenine dinucleotide is called the NAD⁺/NADH ratio. This ratio is an important component of what is called the *redox state* of a cell, a measurement that reflects both the metabolic activities and the health of cells. There is a high requirement for NAD⁺ results from the constant consumption of the coenzyme in reactions such as posttranslational modifications, since the cycling of NAD⁺ between oxidized and reduced forms in redox reactions does not change the overall levels of the coenzyme. These redox reactions catalyzed by oxidoreductases are vital in all parts of metabolism, but one particularly important area where these reactions occur is in the release of energy from nutrients. Here, reduced compounds such as glucose are oxidized, thereby releasing energy. This energy is transferred to NAD⁺ by reduction to NADH, as part of glycolysis and the citric acid cycle. In eukaryotes the electrons carried by the NADH that is produced in the cytoplasm by glycolysis are transferred into the mitochondrion (to reduce mitochondrial NAD⁺) by mitochondrial shuttles, such as the malate-aspartate shuttle. The mitochondrial NADH is then oxidized in turn by the electron transport chain, which pumps protons across the membrane to generate ATP through oxidative phosphorylation (Fig. 8.32).

NADH is potentially useful in the therapy of neurodegenerative diseases such as Alzheimer's and Parkinson disease [78]. Evidence on the use of NAD⁺ in neurodegeneration is mixed; studies in mice are promising [79], with one open clinical trial

8.9 Newer Therapeutic Options in Mitochondropathy

	median	minimum	maximum
Age	68	33	84
Mini mental score before	16	Acetylcholinesterase inhibitors	
after	24		
improvement	8	MMSE 2-3 points improvement	
Global deterioration before	4	2	6
after	2	1	4
improvement	2	1	2
Weeks of therapy	10	8	12

Fig. 8.33 Coenzyme Nicotinamide-Adenine-Dinucleotide improves dementia of the Alzheimer type using the Mini Mental Score (MMSE) before and after treatment with NADH 10 mg/die when compared to a classical acetylcholine esterase inhibitor. Adapted from [82]

showing beneficial effects [80], while a placebo-controlled clinical trial failed to show any effect [81] (Fig. 8.33).

The main role of NAD^+ in metabolism is the transfer of electrons from one molecule to another. Reactions of this type are catalyzed by a large group of enzymes called oxidoreductases. The correct names for these enzymes contain the names of both their substrates: for example NADH-ubiquinone oxidoreductase catalyzes the oxidation of NADH by coenzyme Q10. However, these enzymes are also referred to as dehydrogenases or reductases, with NADH-ubiquinone oxidoreductase commonly being called NADH dehydrogenase or sometimes coenzyme Q10 reductase (Fig. 8.34).

Pyridine nucleotide (a coenzyme) plays an essential role in the metabolic burst and the activity of macrophages. Neutrophils contain NADH-oxidase, which generates superoxide in the phagosomes, which destroy foreign organisms. The "*killing mechanism*" associated with phagocytosis is fueled by NADH and NADPH which are internally derived from an increased activity of the hexosemonophosphate shunt. NADH is therefore directly involved in the cellular immune defensive system. NADH may also increase the phagocyte capacity of leukocytes during the *metabolic burst*. In neuron cell cultures, adding NADH to the culture medium can increase dopamine production. In a dose dependent manner, NADH yields a six-fold increase production of dopamine. Furthermore, NADH stimulates *tyrosine* hydroxylase (TH), the key enzyme for the production of dopamine in a dosage dependent manner of up to 70%. The neurotransmitter dopamine undergoes what is called auto-oxidation. This process forms cytotoxic agents that may damage very sensitive areas in the basal ganglia. Research has found auto-oxidation occurs significantly more often in older

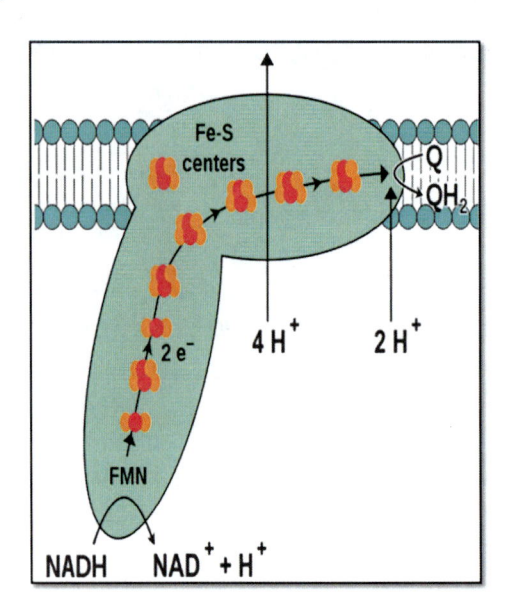

Fig. 8.34 Basic functional steps of NADH serving as an H⁺-donator at complex I of the electron transport chain, which eventually is used to reduce Q10 into its reduced and active form Q10H (uniquinol). It is only this form, which exhibits a potent antioxidative property

individuals. A dopamine auto-oxidation inhibiting substance is useful in reducing damage to certain areas within the brain, by retarding cell death and tissue degeneration. Also, a University of Paris research study proved that in brain tissue (neuron cell) cultures, the production of the brain neurotransmitters, such as dopamine, can be increased by adding NADH to the culture medium. The study's results showed that adding NADH yields a sixfold increase in the production of the neurotransmitter dopamine. It was also found coenzyme NADH stimulates the production of many different brain neurotransmitters, including dopamine, norepinephrine or noradrenaline, and serotonin.

NADH can stimulate the endogenous synthesis of dopamine by increasing activity of quinonoid dihydropteridine reductase, which recycles the inactive dihydrobiopterin to the active tetrahydrobiopterin, thereby providing reduction equivalents to tyrosine hydroxylase the rate limiting factor of dopamine biosynthesis (Fig. 8.35). However diminished dopamine synthesis, probably responsible for some aspects of depression in Alzheimer, is not known to play a significant role in its typical cognitive deficits.

NADH also stimulates dopamine production. Medical science has proven that increased dopamine production has a positive effect on the following brain functions: thinking, cognitive functions (like memory & decision making), sex drive, mood, drive, strength, coordination, movement, mobility and much more. Medical science has proven dopamine has a positive impact on growth hormone secretion. Growth hormone secretion is regarded as the key factor for the regeneration of cells and tissue. Increased dopamine production enhances the body's ability to repair or replace damage and wounded cells.

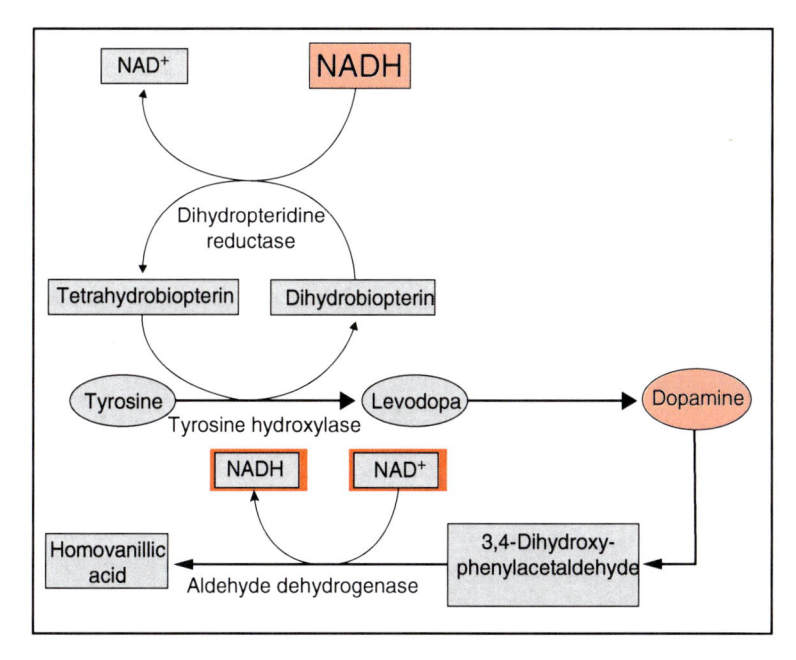

8.9 Newer Therapeutic Options in Mitochondropathy

Fig. 8.35 Proposed efficacy of NADH in Parkinson disease

Dopamine has been proven to reduce prolactin secretion, which is the cause of appetite. The higher the dopamine levels in the blood, the lower the appetite. It is believed that higher or normalized dopamine blood levels reduce a so-called "binge" eating.

8.9.4 NADH in Parkinson-Clinical Studies Demonstrate Efficacy

In a *double blind placebo* controlled study performed at a German university hospital, Parkinson patients were treated with NADH or a *placebo*. The patients receiving NADH showed elevated levels of L-dopa and dopamine in the blood. Medical science has proven that Parkinson's disease occurs when the brain cells that produce dopamine die. All Parkinson patients in this German study who were taking NADH improved in their condition. The likely explanation for such beneficial effect is that an elevated NADH level results in a significant elevation of pyridine dinucleotide levels in the brain. The brain has an active uptake mechanism for the accumulation of nicotinamide through the choroid plexus, indicating this is an important functional role for brain metabolism and oxidative defense. Taken all these factors into account one can conclude that NADH has a protective effect on neuronal structures being exposed to excitotoxicity.

8.9.5 Supplementation of NADH as a Powerful Antioxidant

This can be measured objectively by the so-called redox potential. The more negative this potential is, the greater the antioxidative capacity. Conversely, co-enzyme Q10 has a positive redox potential, and therefore is not an antioxidant at all. It can become an antioxidant in the body if it is reduced. This reduction is achieved in the cell and **only** by NADH. This simple fact implies two consequences:

1. The intake of commercially available CoQ10 is not very meaningful unless the organism has sufficient amounts of cellular NADH available to reduce CoQ10 and make it an antioxidant.
2. Even highly conditioned athletes have a measurable NADH deficit. If you take commercially available CoQ10 without an equivalent does of NADH, you may deplete the cell from NADH and thereby make the cell energy deficient and more prone to degeneration [83] (Fig. 8.36).

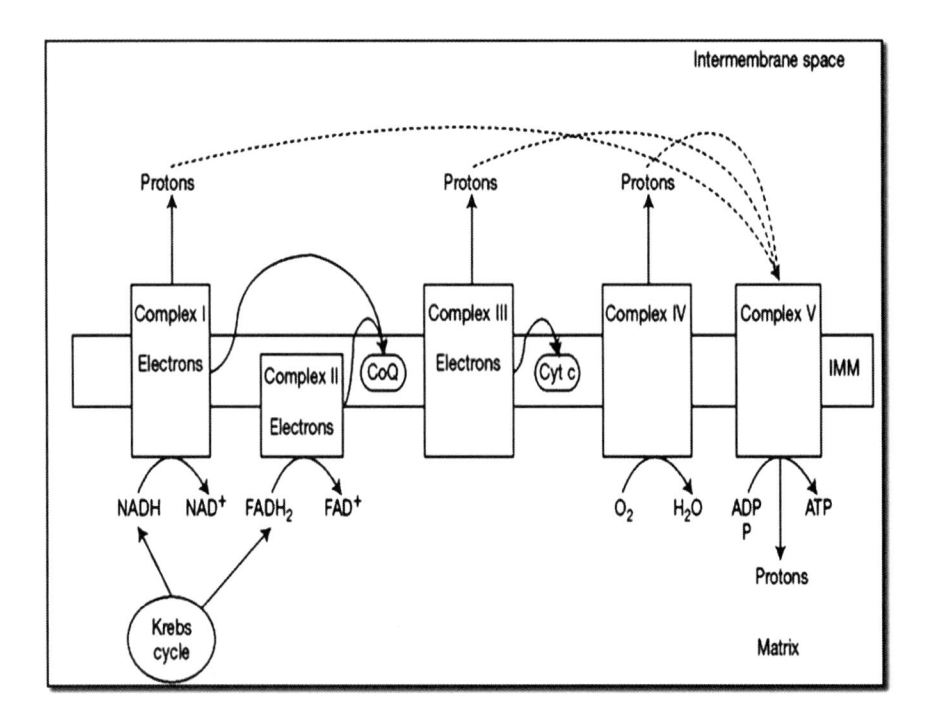

Fig. 8.36 Process of adenosine 5′-triphosphate (ATP) production through the electron transport chain (ETC) in mitochondria, also known as the respiratory chain, and its oxidative phosphorylation. 1–4 NAD$^+$ =1,4-nicotinamide-adenine-dinucleotide; NADH=reduced form of NAD$^+$; FAD$^+$=flavin-adenine-dinucleotide; FADH=reduced form of FAD$^+$; CoQ=coenzyme Q10; Cyt c=cytochrome c; O$_2$=oxygen; H$_2$O=water; ADP=adenosine 5′-diphosphate; P=phosphate; IMM=inner mitochondrial membrane

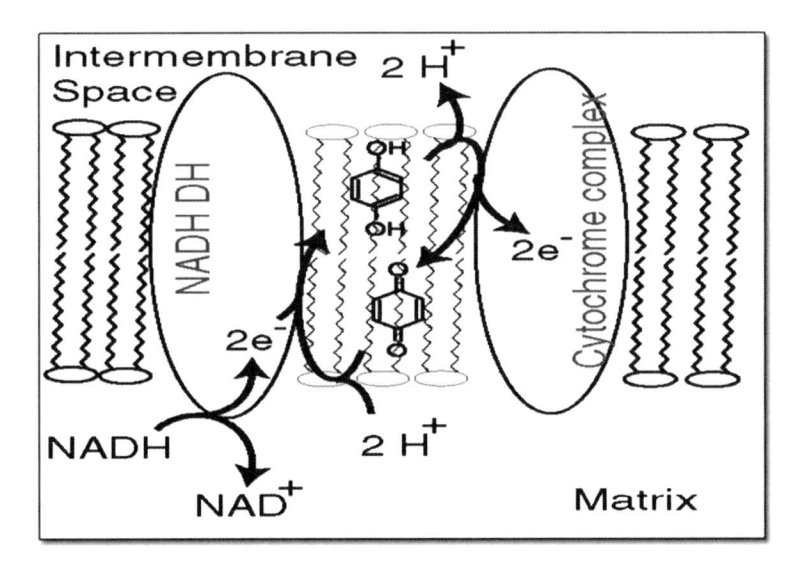

8.9 Newer Therapeutic Options in Mitochondropathy

Fig. 8.37 Conversation of NADH to NAD⁺ at complex I giving way to two negatively charged electrons into the matrix

Since NADH is the first step in the aerobic formation of energy it is an essential one. Because without this first step in electron transfer there is no progression to the next step (s) resulting in a loss in the final ATP production at complex V (Fig. 8.37).

In addition to the reported benefits of NADH in Parkinson, there is also a proposed efficacy of NADH in Morbus Alzheimer and one of the very few promising treatments for Alzheimer's is the coenzyme, NADH (nicotinamide adenine dinucleotide) [84] In a placebo-controlled study, Alzheimer's subjects given NADH for 6 months exhibited significantly better performances on verbal fluency, visual constructional ability and abstract verbal reasoning than the control subjects given a placebo. Why would NADH be effective in Alzheimer? In the process of converting pyruvate (which accumulates due the inability of neuronal or glia cells to use for energy production) to lactate, lactate dehydrogenase consumes oxygen by oxidizing NADH to NAD⁺ (Fig. 8.38).

The biosynthesis of NAD⁺ occurs through both salvage and de novo pathways [85]. The salvage pathways begin with either nicotinamide or nicotinic acid collectively referred to as niacin or vitamin B3. One salvage pathway leading from nicotinic acid (Na) to NAD⁺, known as the Preiss-Handler pathway, goes through two intermediates, nicotinic acid mononucleotide (NAMN) and nicotinic acid adenine dinucleotide (NAD). A parallel salvage pathway leading from nicotinamide (NAM) to NAD⁺ goes through one intermediate, nicotinamide mononucleotide (NMN). The "de novo" pathway leads from tryptophan to quinolinate, which connects to the Preiss-Handler salvage pathway through NAMN. Recently, nicotinamide riboside (NR) was also shown to be a precursor for NAD⁺ synthesis, connecting to the NAM salvage pathway through NMN [86]. The enzymatic actions of PARP-1 and SIRT1 release the NAM moiety from NAD⁺ to produce ADP-ribose-protein and

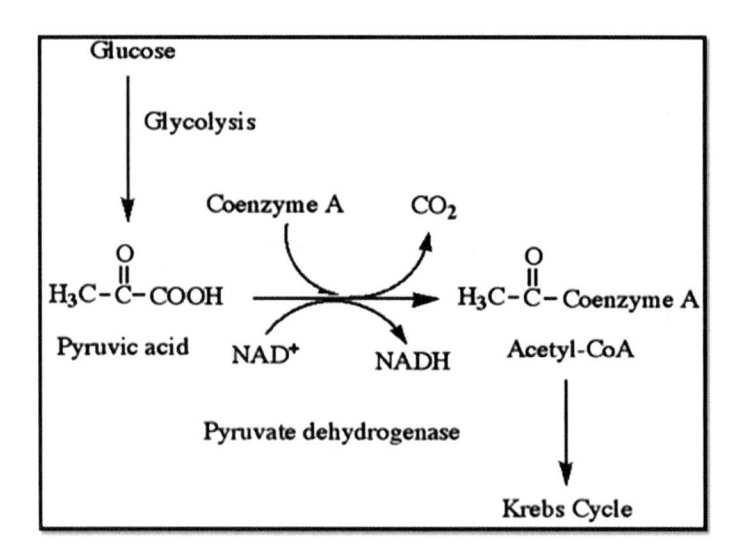

```
┌─────────────────────────────────────────────────────┐
│  Glucose                                            │
│     │                                               │
│     │ Glycolysis                                    │
│     │                                               │
│     ▼       Coenzyme A      CO₂                     │
│              O                    O                 │
│              ‖                    ‖                 │
│     H₃C-C-COOH   ─────▶    H₃C-C- Coenzyme A       │
│     Pyruvic acid  NAD⁺   NADH    Acetyl-CoA         │
│                                       │             │
│              Pyruvate dehydrogenase   │             │
│                                       ▼             │
│                                    Krebs Cycle      │
└─────────────────────────────────────────────────────┘
```

Fig. 8.38 Once the bioavailability of NADH is increased, it stands to reason that the astrocyte would have an enhanced ability to convert pyruvate to lactate, the critical step in the anaerobic metabolic pathway that is enhanced by amyloid-beta. The process, by absorbing the toxic oxygen, would reduce the damage to the lipids due to oxygen exposure, and would also provide lactate as a source of energy for the neurons

O-acetyl-ADP-ribose, respectively, understanding of the cellular consequences of complex I deficiency are absolutely necessary for the development of new treatment strategies to combat devastating disorders as being observed in mitochondrial related encephalopathy [87]. For the time being, the latter should focus on further elaboration of potential promising strategies including the use of selectively targeted bioactive molecules, affecting the mitochondrial activation of complex.

In this respect, NADH substitution seems to have a selective preferential effect in Alzheimer patients, because it increases the activity of

- NADH oxidoreductase, coupled to mitochondrial respiratory chain is increased in neocortex and hippocampus [88].
- NAD/NADH dependent 3-hydroxyacyl-coenzyme A dehydrogenase in human brain [89], which is identical with ß-amyloid peptide binding protein (ERAB) that might mediate neurotoxic effects of amyloid plaques in Alzheimer patients.

8.9.6 Benefits of Curcumin in Mitochondrial Disease

Curcumin is considered a multiple signal inhibitor especially that of the NF-κB cellular pathway (nuclear factor 'kappa-light-chain-enhancer' of activated B-cells) an important specific transcription factor, present in all cells and tissues. By binding at selective parts of genes it is able to affect the transcription of several genes with different properties, thus propagating in a number of symptoms involved in any

8.10 Mechanisms of Action of Agents for Therapy in CFS-FMS-Parkinson-Alzheimer   323

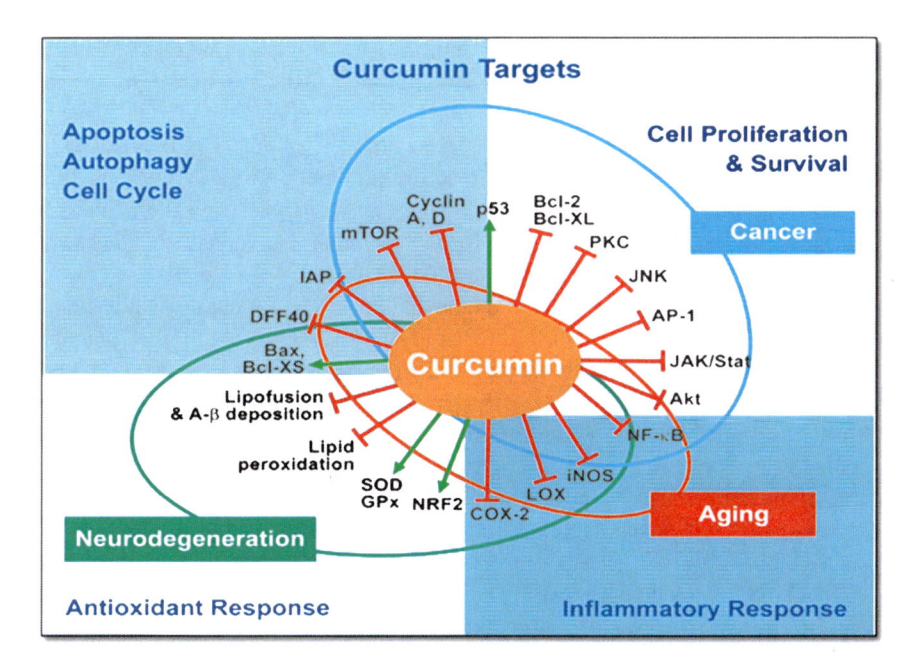

Fig. 8.39 Overview of the different modes of action of curcumin involved in reversing oxidant, inflammatory response, activation of apoptosis and autophagy with down-regulation of cell proliferation and survival of cancer cells

inflammatory process (Fig. 8.39). In addition, it blocks important pathways involved in cell proliferation of cancer cells and in neurodegeneration of the CNS with the disposition of lipofusion, lipid peroxidation while inflammatory cyclooxygenase is inactivated. At the same time the inborn antioxidant superoxydismutase (SOD) and apoptosis is activated through the action of the Bcl-2 family of proteins, which includes anti- and pro-apoptotic members such as *Bcl-xS* and Bax, which plays an important role in cancer growth (Fig. 8.39). Preliminary phase I studies in patients with cancer outside the gastroesophageal tract have demonstrated efficacy with very low side effects with a daily oral dose of 3.6 g [90].

8.10 Mechanisms of Action of Agents for Therapy in CFS-FMS-Parkinson-Alzheimer

Based on the biochemical pathways within mitochondria, the basic vitamins/trace elements and amino acids necessary for the electronic transport chain (ETC) to function properly can be summarized as follows

- **Magnesium** – is particularly important in regulating and stabilizing ATP, and is often displaced by the presence of heavy metals. With a deficiency of Magnesium, the ATP becomes over-regulated or inhibited, resulting in low energy levels. Magnesium helps to fire up various essential enzymatic reactions. Nutritional elements such as Magnesium, Zinc and Selenium have an important role in protecting the body from the effects of heavy metals, and those individuals with elevated heavy metal levels have a greater requirement for these nutrients (and a corresponding great need to detoxify the heavy metals from their bodies). Magnesium is best taken in chelated form, e.g. citrate or glycinate, together with the amino acid taurine to help carry it into the cells.

- **Active B_1** – Thiamine Pyro Phosphate (TPP), a.k.a. Thiamine Diphosphate (TDP) or Cocarboxylase, is the biologically active form of Vitamin B_1. Thiamine is used efficient in burning carbohydrate and removing excess lactic acid (a cause of muscle ache). Minimum dose is 100 mg/die.
- **Active B_2** – Riboflavin, Vitamin B_2, is also involved in the Krebs Cycle, as a cofactor in Complex I and II. There are two active, coenzyme forms of B_2, flavin mononucleotide (FMN) and flavin dinucleotide (FAD). FMN is the coenzyme form of B_2 found in supplement form. Minimum dose is 10 mg/die.
- **Active B_3** – A variant of Vitamin B_3 (Niacin), known as Nicotinamide Adenine Dinucleotide plus high-energy Hydrogen, or NADH for short, is also involved in the Krebs cycle. Niacin can be obtained from protein but it is dependent on efficient protein digestion and amino acid conversion in the body. NADH and NAD are an essential part of the ATP to ADP conversion. Active B_3 probably has most immediate noticeable effect on energy levels, but all B-vitamins are important to some degree in energy production. B_3 levels can drop significantly if there is Peroxynitrite-related oxidative damage occurring in the body (to poly (ADP-ribose) polymerase enzyme DNA), having a sparking effect on the mitochondrial electron transport chain. Minimum dose of NADH is 4–5 mg/die and of niacin is 100 mg/die.
- **Vitamin B_5** – Pantothene (a biologically active form of Vitamin B_5) is involved in the synthesis of Coenzyme-A (CoA). CoA is important in energy metabolism for pyruvate to enter the tricarboxylic acid cycle (TCA cycle) as acetyl-CoA, and for alpha-ketoglutarate to be transformed to succinyl-CoA in the cycle. Minimum dose is 100 mg/die.
- **Active B_6** – Pyridoxal-5-Phosphate the active form of Vitamin B_6 can catalyze transamination reactions that are essential for providing amino acids as a substrate for gluconeogenesis P-5-P is also a required coenzyme of the glycogen phosphorylase enzyme, that allows glycogenolysis to occur. Minimum dose is 100 mg/die.
- **Active B_{12}** – The enyzme Methylmalonyl Coenzyme A mutase (MUT) requires vitamin B_{12} in the form Adenosylcobalamin (Ado-B_{12}). MUT is involved in carbohydrate metabolism, converting Methylmalonyl-CoA (MMI-CoA), the coenzyme A link form of methylmalonic acid (MMA), into Succinic-CoA (Su-CoA), the coenzyme-A link form of succinic acid. It forms part of the Krebs cycle for the production of energy. Minimum dose of hydroxycobolamin is 1.5–2 mg s.c. If available methylcobalamin is the best option!

- **Acetyl-L-Carnitine** (ALC or ALCAR) – is an amino acid that and allows the transport and burning of fat and prevents mitochondria from shutting down. It is the acetylated ester of the amino acid L-Carnitine. Carnitine helps to transport ATP and ADP across the mitochondrial membranes, and to transport activated ATP to where it is actually used. It is manufactured intracellularly in the body from the Essential Amino Acids L-lysine and L-methionine, by a process of methylation i.e. does not float around in the blood stream (and thus urine) – and hence is usually not included as a parameter in amino acid analyses. Minimum dose is 1–5 g/die orally.

In addition other substitutes all aim to support and repair mitochondrial function, and their apparent mechanism of action is outlined in the following table. These distinct classes of agents present meaningful evidence for the underlying disease mechanism, predicted to down-regulate the overexpressed NO/ONOO⁻ cycle (Table 8.15):

The question is raised whether the action of these agents is compatible with the NO/ONOO⁻ cycle and specifically what parts of the cycle may be influenced from the mechanisms of action of certain micronutrients. This is important, because the efficacy of these agents provide evidence within specific parts of the NO/ONOO⁻ cycle being involved in the underlying disease. For example, the agents carnitine/acetyl carnitine and coenzyme Q10 are reported to be helpful by improvement of mitochondrial function, suggesting that mitochondrial dysfunction is an important underlying factor in the disease mechanism. In addition, medium chain fatty acids (MCF) are essential for restoration of mitochondrial function, especially since they act on the inner mitochondrial membrane thus being able to restore the function and lower the NO/ONOO⁻ cycle.

In addition, vitamin B_{12} seems essential as it lowers the toxic elevated nitric level (NO scavenger) at the site of mitochondria while agents like coenzyme Q10 and magnesium are essential for restoration of normal function of the electrical transport chain (ETC) within mitochondria at complex II and IV respectively. Next to this, the increase of the endogenous formation of a potent oxygen radical scavenger glutathione (GSH) is necessary in order to neutralize any ROS production of damaged mitochondria, giving them a chance for restoration. Because oral glutathione is rapidly oxidized within the stomach, an intravenous application is mandatory in order to achieve sufficient high plasma levels. Alternatively, N-acetyl-cysteine (NAC) can be given orally, as it incorporates the important sulfhydryl component, which is necessary for de novo endogenous formation of GSH. It should be noted that doses of up 2 g/day (!) are necessary as they have been demonstrated to result in a highly significant prolongation of survival in patients with HIV with marked mitochondrial recovery when compared to the sole use of HAART (highly active antiretroviral therapy) therapy [91]. The antiepileptic agent Lyrica® (pregablin) indirectly acts by lowering excitotoxicity including NMDA activity. This has been demonstrated in clinical trials of FMS patients with significant improvements and the agent is currently approved for therapy in the treatment of FMS patients [92]. Magnesium, which acts directly at the NMDA receptor site by lowering its activity, is probably the most widely reported agent for producing improvements in the

Table 8.15 Mode of action and indications of agents aimed to down-regulate the overexpressed NO/ONOO- cycle

Agent or class	Mechanism of action	Comments on use
Vitamin C (ascorbic acid)	Chain breaking antioxidant; lowers NF-kappa B activity; reported to scavenge peroxynitrite and also helps to restore tetrahydrobiopterin (BH_4) levels	May require high doses to be effective with the latter two mechanisms, the basis of so called 7.5 g iv "megadose therapy" with vitamin C. Otherwise use Ca-ascorbate of up to 2,000 mg/day
Vitamin E including tocopherols and tocotrienols	Lipid soluble antioxidant; gamma-tocopherol is particularly useful in scavenging breakdown products of peroxynitrite and in lowering chronic inflammatory responses; tocotrienols are important in protecting from excitotoxicity and protecting mitochondria; lowers NF-κB activity	High dose alpha-tocopherol, most commonly used form of vitamin E induces an enzyme that degrades other forms of vitamin E; thus high dose alpha-tocopherol should be avoided; preferentially use a mixture of alpha-, beta-, delta- and gamma-tocopherol totaling up to a minimum of 200 IU
Magnesium	Lowers NMDA activity associated with pain, useful in improving energy metabolism and ATP formation at the mitochondrial level	Magnesium is useful in the treatment of any multisystem illnesses, especially with pain. Minimum necessary dose is 500 mg/die
N-acetyl-cysteine (NAC)	Precursor, and activator in synthesis of reduced glutathione. Binds heavy metals	In case of sensitivity to NAC, reduce dose; doses up to 2 g should be taken with meals
Fish oil (long chain Ω-3-fatty acids)*	Lowers iNOS induction; important for brain function; lowers production of inflammatory protaglandins	Highly susceptible to lipid peroxidation and may, therefore be depleted
Flavonoids	Chain breaking antioxidants; some scavenge peroxynitrite, some scavenge superoxide; some reported to induce SOD; All three types are found in FlaviNox; some also help to restore BH_4 levels; lowers NF-κB activity	Flavonoids rapidly increase in plasma after consumption but also drop rapidly; four times/day for maintenance of high blood levels over the day

8.10 Mechanisms of Action of Agents for Therapy in CFS-FMS-Parkinson-Alzheimer

Carotenoids including beta-carotene, lycopene, and lutein	Scavenges peroxynitrite in lipids, such as biological membranes	Only natural forms are used; the natural form of beta-carotene has substantial amounts of cis double bonds, whereas synthetic beta-carotene is predominantly all trans and largely inactive as a scavenger; other carotenoids are very active, and may be more active than even natural beta-carotene
Selenium in the form of seleno-methionine	Serves as a precursor for selenoproteins including three forms of the antioxidant enzyme glutathione peroxidase and also a selenoprotein reported to be a peroxynitrite scavenger	Since peroxynitrite reacts with many selenium compounds this will lead to lowered selenium levels which have been found in multisystem illnesses
Acetyl L-carnitine	Transports fatty acids into mitochondria; important for energy metabolism and restoration of oxidized fatty acid residues, produced by superoxide in the cardiolipin of the inner mitochondrial membrane	Lowers ROS and NOS stress
Ecklonia cava extract	Polyphenolic chain breaking antioxidant; scavenges both peroxynitrite and superoxide; may also help to restore BH_4 levels	Stays in the body longer than flavonoids, helpful in a clinical trial study of fibromyalgia
Vitamin B_6 including pyridoxal phosphate	Activates enzyme glutamate decarboxylase which converts excitatory NMDA into inhibitory GABA-receptor	Restores balance between glutamate and GABA; lowers excitotoxicity and excessive NMDA activity. Long term use up to 100 mg/day
Methyl cobalamin active form of vitamin B_{12}	Potent nitric oxide scavenger, lowers nitric oxide and peroxynitrite levels	Take orally four times 2,000 μg/day for constant levels; higher levels can be obtained by injection, inhalation or nasal spray
Folic acid	Lowers partial uncoupling of the nitric oxide synthases, helps to restore tetrahydrobiopterin	Reacts with oxidants and therefore may be depleted due to the overexpressed NO/ONOO- cycle. Dose of 0.4-5 mg/day
Vitamin B_3 or Niacin	Helps restore NAD/NADH pools that can be depleted by peroxynitrite mediated poly ADP-ribosylation	Important cofactor in mitochondrial energy dysfunction

(continued)

Table 8.15 (continued)

Agent or class	Mechanism of action	Comments on use
Vitamin B_2 or Riboflavin including 5'-phosphate	Stimulates glutathione reductase, a key enzyme for maintenance of reduced glutathione. Activates flavoprotein-monoxygenase, which dissolves xenobiotics & pesticides from cellular binding	Maintaining depleted glutathione (GSH) levels, which reacts with ROS in the overexpressed NO/ONOO- cycle. Important in detoxification from xenobiotics. Use 2–3 tablets of 100 mg/day
Vitamin B_1 or thiamine	Important cofactor in energy production of mitochondria, coenzyme for transketolase in the pentose phosphate shunt to enter Krebs cycle, required for oxidative decarboxylation of pyruvate to acetyl-coenzyme A, which generates NADPH	Critical for NADPH which can act to regenerate reduced glutathione; reacts with oxidants and therefore may be depleted due to the overexpressed NO/ONOO- cycle
R-α-lipoic acid	Important antioxidant; helps restore reduced glutathione levels; lowers NF-kappa B activity, also scavenges peroxynitrite, superoxide and toxic heavy metals	Rapidly converted in the body to reduced lipoic acid and lipoamide, the most active forms; possibly one of the most important agents but not tested in clinical trials for multisystem illnesses
Biotin, and Pantothenic acid	Biotin depleted with alpha-lipoic acid supplementation; pantothenic acid important for energy metabolism	Coenzyme A (produced from pantothenic acid), a thiol compoundv is depleted with ROS, NOS; another mechanism producing energy
Betain (trimethyl-glycine)	Lowers ROS, NOS stress; helps to generate S-adenosyl-methionine (SAM)	SAM generation may be of concern, since enzyme methionine synthase is inhibited by ROS & NOS, leading to lowered SAM and lowered methylation. Minimum dose is 50 mg/day
Q10 ubiquinonel/ubiquinol	Essential in mitochondrial function for generation of ATP at complex II; important antioxidant; scavenges peroxynitrite; a must in every patient taking statins, an alternative in heart insufficiency	Dosage 2–3 × 200 mg/day, may vary considerably among individuals; take in the morning as it can keep you up at night. The reduced form ubiquinol has higher bioavailability and faster onset of action. Take together with NADH

8.10 Mechanisms of Action of Agents for Therapy in CFS-FMS-Parkinson-Alzheimer

Zinc, copper, manganese, and selen	All precursors of the antioxidant enzyme superoxide dismutase (SOD); can be rate limiting for its synthesis	A low dose is important as too high dosages can cause problems; thus take Zi, Cu, Mn, Se in doses of 20–50 mg, 1–2 mg, 5–10 mg, and 100–200 µg/day respectively
Ribonucleic acid (RNA)	Has two important functions: Provides adenosine for restoring adenine nucleotide pools after energy metabolism dysfunction; when catabolized, the purine bases generate uric acid, a peroxynitrite scavenger	D-ribose and inosine act similarly. Possible disadvantages: D-ribose is a glycating agent. Inosine stimulates mast cells. Commercial source of RNA is yeast, which may present a problem in yeast allergy.
Taurine	Lowers excitotoxicity including NMDA activity; helps restore balance between glutamate and GABA activity	Reported to be depleted in the overexpressed NO/OONO− cycle
Glutathione- reduced form	The body's own antioxidant, for detox of xenobiotics, blocks the NF-κB activity	Preferentially used as iv Tationil® (Roche/Italy) because oral formulation has low bioavailability, 3 × 600–1,200 mg/week
Bioflavonoides	Antioxidants that neutralize nitric oxide, peroxynitrite superoxide, they reduce NF-κB activity, potentiate vitamin C	Used as epigallocatechin gallate (EGCG), a major constituent in green tea, use 2–3 L/day
NADH or coenzyme 1	Nicotinamid-adenin-dinucleotid, important energy supplier sparking off oxidative phosphorylation with ATP formation in mitochondria	5–10 mg (Fairvital or ENADA) relieves parkinson symptoms, also claimed affective in Alzheimer, foggy brain, helps attain higher level of vigilance
Melatonin	Reactivates antioxidants GSH, SOD, and catalase which catalyses hydrogen peroxide, a powerful and potentially harmful oxidizing agent, to water and oxygen. Neutralizes aggressive NO and peroxynitrite.	Dosages above the sleep-induction (3–5 mg/70 kg) should be in the range of 5 mg/kg body weight to be effective

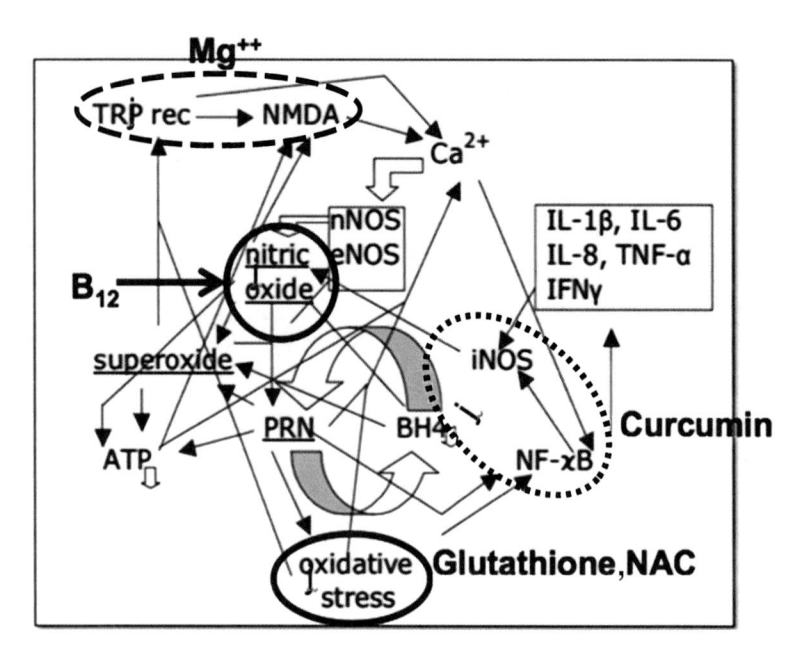

Fig. 8.40 Site of action of different agents for reduction of an activated NO/ONOO⁻ cycle. *Black circle* = GSH; *broken circle* = NMDA-antagonists & magnesium; *dotted circle* = curcumin

multisystem illnesses (Fig. 8.40). The agent also acts at the mitochondrial level by improving energy metabolism resulting in an increase in ATP formation.

8.10.1 Rational for Taking High dose B_{12} in Mitochondropathy

The polyphenolic antioxidants, flavonoids and Ecklonia cava extract, have been reported to produce improvements in clinical trials, suggesting a therapeutic role in oxidative stress. Algae supplements may also act primarily by providing substantial doses of antioxidants and have been reported to produce improvements in clinical trials. High-dose vitamin C therapy, in dosages upwards of 1,000 times the U.S. Recommended Dietary Allowance (RDA) or Daily Reference Intake (DRI), has been shown in legitimate clinical studies to cure all sorts of illnesses [93]. The primary role may be a different one than simply acting as a chain breaking antioxidant. What, however, has been demonstrated conclusively, is the potential benefit of high dose vitamin B_{12} in depressing excess formation of NO/ONOO⁻, downregulation of iNOS, and immune medulation of NF-κB in mitochondropathy [94]. High dose hydroxycobalamin, a form of vitamin B_{12} in doses of > 2,000 μg/day, is a potent nitric oxide scavenger and in a placebo-controlled trial was reported to produce statistically significant improvements in a group of CFS patients [95]. Historically, vitamin B_{12} has been used for over 60 years to treat CFS patients and has also been used to treat FMS and MCS patients [96]. It appears that cyanocobalamin is less effective than is hydroxocobalamin, presumably because the conversion of cyanocobalamin to hydroxocobalamin in the body is only partial. Hydroxycobalamin has been used therapeutically in migrineurs for reversal nitric oxide [97] and its potent role as a nitric oxide scavenger is extensively recognized in the scientific literature. There are several types of evidence that suggest that in nitrosatrive stress hydroxycobalamin is acting primarily as a nitric oxide scavenger rather than by alleviating B_{12} deficiency. There was no correlation between the initial B_{12} blood levels and the effectiveness of the treatment regimen in a placebo-controlled trial with diabetics taking metformin [98]. Secondly, it appears that a much higher dose is needed to get a good clinical response than is needed to alleviate a low level. Generally, the hydroxycobalamin is given by IM or subcutaneous injection, by nasal spray or occasionally, as a nebulized inhalant. Each of these are expected to give much higher blood levels than one can get from an oral supplement, because of the limited absorption of oral B_{12} due to limited availability of intrinsic factor, a glycoprotein that has an important role in the absorption of B_{12}.

Fish oil supplements with Ω-3-fatty acids have been shown in clinical trials to be helpful with these multisystem illnesses and fish oil supplements are known to have antiinflammatory effects. This is the most probable mode of action in nitrosative stress, providing some support for an important inflammatory aspect to these multisystem illnesses. Thus from clinical trial studies, there is ample evidence for the roles of oxidative stress, excess NMDA activity, mitochondrial dysfunction, nitric oxide/peroxynitrite activity, increase in inflammation and a BH_4 depletion in multisystem illnesses. There is also evidence of such roles in other types of ailments, including genetic correlates in the chronic phase of the ailment, and paradigms in animal model studies and gene expression studies for CFS (Ref. [4] in Chap. 6, Ref. [128] in Chap. 7) [26, 99]. All these data corroborate the underlying activity of the NO/ONOO$^-$ cycle being the central causal mechanism for these illnesses.

Instead of trying to take all neutraceuticals at once, start with the first for 3 days to see if it is well tolerated, adding a second for 3 days, and so forth. By doing this one finds out if there is any products that is not well tolerated, that can be eliminated for the time being and perhaps be taken later with either the same or possibly lower dosage. By doing so, it will take 21 days to get to the end of the period, where all products are being taken. It can be seen from the above-described combination of nutritional supplements supply nutrients that help to down-regulate various aspects of the NO/ONOO$^-$ cycle. Many act as antioxidants, lowering oxidative stress, blocking oxidative chain reactions, and in some cases scavenging such oxidants as peroxynitrite and superoxide or acting to increase superoxide dismutase activity.

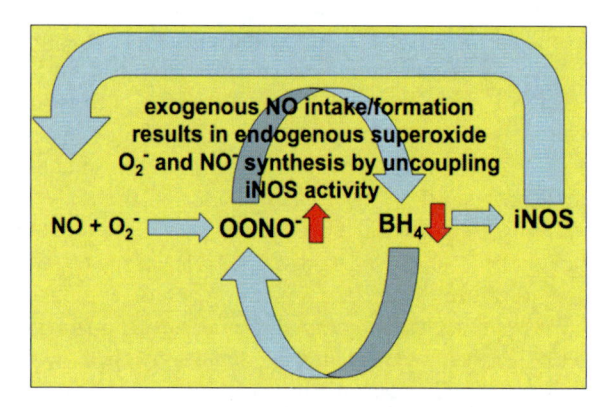

Fig. 8.41 The vicious BH_4/$ONOO^-$ cycle where interaction of high peroxynitrite level induce a reduction in tetrahydroxybiopterin formation (BH_4) a necessary agent, where the depletion results in an uncoupling of all nitric oxide synthases with excess formation of NO/ONOO⁻

8.10.2 Tetrahydroxybiopterin (BH_4) – Key Candidate in Mitochondropathy

The basic problem in not obtaining total cures may be that the central couplet of the NO/ONOO⁻ cycle is not adequately lowered. That what is called the central couplet is the reciprocal relation between peroxynitrite and tetrahydrobiopterin (BH_4). Peroxynitrite oxidizes BH_4 leading to partial uncoupling of the nitric oxide synthases (iNOS), which in return now produces more peroxynitrite (Fig. 8.41).

There are several agents that, in part, do lower either peroxynitrite levels or help to restore BH_4 levels, but it is not clear that they are effective at the levels that can be easily obtained in the body from oral supplements. It may well be that the main reason for not getting a substantial number of total cures from the above outlined therapeutic interventions, is the inability to sufficiently lower this central couplet. Folic acid supplements have been reported to be helpful in clinical trial studies as well and these may be expected to help restore BH_4 levels [100]. Similar high dose IV vitamin C has been reported to produce substantial improvements in CFS studies and in MCS such high dose IV ascorbate also is to help in restoring BH_4 levels. All these studies provide some evidence that BH_4 depletion has a significant role in multisystem illnesses where the beneficial effect of vitamin C on endothelial function is best explained by an increased intracellular BH_4 content and a subsequent enhancement of eNOS activity, which appears to be independent of the ability of vitamin C to scavenge superoxide anions [101, 102]. While there are some arguments that for instance sauna therapy may act by primarily by increasing BH_4 availability in the body, the best agent to lower this central couplet is high dose intravenous ascorbate (up to 7.5 g/day 3 times/week), an agent which, as mentioned above has been shown to produce substantial improvements in CFS and was reported to produce similar improvements in MCS. High dose ascorbate is predicted to act in three ways to lower this central couplet:

1. Ascorbate is a peroxynitrite scavenger, although all of the evidence available suggests that it does not work very well at the normal levels typically found in the body or easily obtainable by using oral ascorbate. However, IV ascorbate can generate 30 times or higher levels, which should be much more effective in scavenging peroxynitrite.
2. When BH_4 is oxidized by peroxynitrite, the initial oxidation product designated BH_3 (the one electron oxidation product) can be reduced back to BH_4 by ascorbate. However, the BH_3 by itself is unstable, so that an efficient reduction can only be expected with relatively high levels of ascorbate (vitamin C).
3. High dose ascorbate gets oxidized in the body, generating substantial amounts of hydrogen peroxide, and hydrogen peroxide is known to activate an enzyme called GTP cyclohydrolase I. This enzyme is the first and rate-limiting enzyme in the de novo pathway for the production of BH_4. What follows is that high dose ascorbate is predicted to produce more BH_4 in the body via this pathway.

> Before giving high dose ascorbate it is important to perform a test for iron binding saturation, because high free iron reacts with ascorbate generating high levels of Fenton chemistry, resulting in the formation of higher oxidation iron state intermediates.

8.10.3 NADH in Parkinson–de Novo Pathway for the Production of BH_4

While NADH also seems to be effective in CFS patients [103] other studies confirmed and extended previous reports on the clinical benefit of NADH in Parkinson patients [82, 104, 105]. A new finding in another study was that the oral form of NADH shows a beneficial clinical effect comparable to that of the intravenously applied NADH (Ref. [119] in Chap. 7). The galenic formulation of the oral form of NADH is a critical factor with regard to clinical efficacy, since NADH when ingested orally gets into contact with the acid condition of the stomach where part of it is downgraded. This is because NADH is rapidly oxidized below pH 7.6 the conditions in the stomach will inactivate NADH by converting it to NAD^+. The investigations of this report were therefore performed with NADH capsules coated with an acid-stable film and a release time of 2–3 h. With this galenic formulation of NADH an improvement in availability is achieved. When first used, NADH in gelatin capsules a beneficial effect was not convincing. This was most likely due to the rapid dissolution of the capsules (approximately 10–15 min) leading to a rapid release in the stomach. Once coated capsules were used release of NADH was delayed yielding comparable plasma values comparable to that of intravenously applied NADH. It should be pointed out that most of the patients in this study, in addition to NADH received the classical medication for Parkinson's disease such as Madopar® or Sinemet® with or without addition of Selegiline®, bromocryptine or amantadine. In many of these patients the daily dose of L-DOPA could be reduced considerably, while in some patients it could be omitted totally.

The question is whether or not the well-established L-DOPA therapy should be replaced by NADH treatment? Arguments in favor of the new NADH supplementation become apparent when the biochemical and pharmacological differences behind these two therapeutic concepts are confronted. L-DOPA therapy follows the principle of substitution, i.e. a restoration of dopamine deficit by means of substitution with its immediate precursor L-DOPA. However, exogenous substitution of biological constituents results in a depression of the organism's own biosynthesis. This not only holds true for cortisol, thyroxin, aldosterone, but also for many other hormones and metabolic compounds, and it is certainly valid for L-DOPA substitution. In other words the exogenous administration of L-DOPA will inhibit its endogenous biosynthesis. As demonstrated elsewhere, the L-DOPA producing enzyme tyrosine hydroxylase (TH) is markedly reduced in Parkinson patients [106, 107]. In addition, it has been shown that tyrosine hydroxylase is inhibited by its end product L-DOPA [108, 109]. Such results underline the assumption, that TH activity, already being insufficient in Parkinson patients, will further be inhibited by the exogenous administration of L-DOPA. Therefore L-Dopa therapy results in an additional down-regulation of enzyme activity. Whether or not this is the cause for a frequent observed "off"-effect in Parkinson patients, in particular during long-term treatment with L-DOPA, remains to be elucidated. An NADH therapy on the other hand pursues an opposite strategy, namely stimulation of endogenous L-DOPA biosynthesis by activation of the key enzyme tyrosine hydroxylase (TH). There are a number of arguments in favor of NADH treatment one of which is that patients who do not respond to the classical L-DOPA therapy even after higher dosages, show some improvement after NADH treatment, and the stimulation of L-DOPA biosynthesis may occur via enhanced production of the tyrosine hydroxylase coenzyme BH_4. As demonstrated by Nagatsu and coworkers, levels of BH_4 in the brain and the cerebrospinal fluid of Parkinson patients are reduced up to 50% in comparison to age matched healthy individuals [110]. The cause of this BH_4 deficiency is still obscure. If the deficit in BH_4 is due to a decreased biosynthesis, the biochemical mode of action of NADH can easily be explained. This is, because BH_4 is formed from dihydrobiopterin (BH_2) via the enzyme quinoid dihydrobiopteridin reductase (DHPR), NADH is the essential cofactor for this reaction [111]. Since enzyme is responsible for a necessary step in the synthesis pathway, recycling the molecule tetrahydrobiopterin (Fig. 8.42).

Tetrahydrobiopterin or BH_4 modulates several enzymes, including neuronal nitric oxide synthase (nNOS), which generates nitric oxide (NO), and aromatic amino acid hydrolase. It is also involved as a cofactor in the synthesis of neurotransmitters of the brain, i.e. serotonin, melatonin, dopamine, norepinephrine & NO, which transmit signals between nerve cells (Fig. 8.43). nNOS is also under the control of

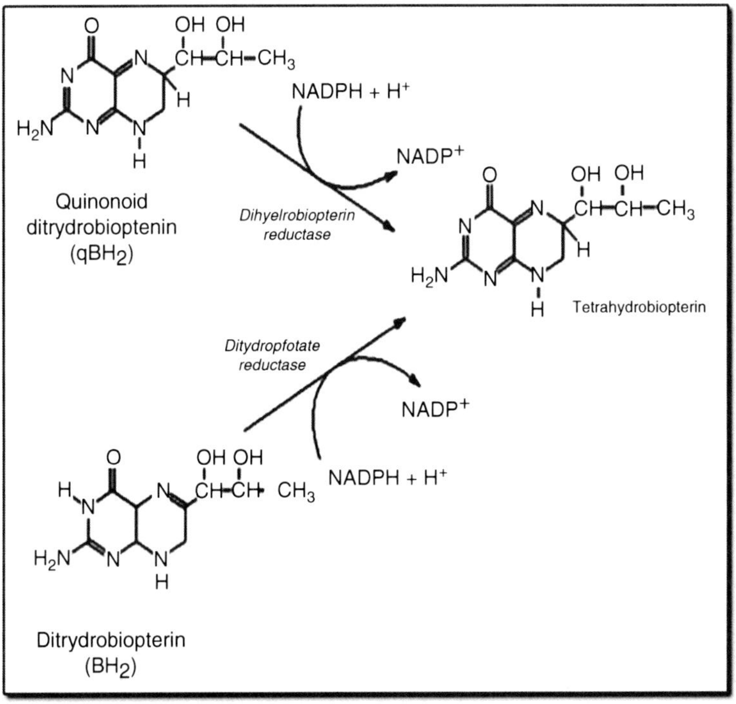

8.10 Mechanisms of Action of Agents for Therapy in CFS-FMS-Parkinson-Alzheimer

Fig. 8.42 Tetrahydrobiopterin (BH_4) is a cofactor that carries electrons for REDOX reactions, as in the oxidation of phenylalanine to tyrosine. It is formed from dihydrobiopterin (BH_2) through the action of the enzyme dihydrofolate reductase, or from the quinonoid form of dihydrobiopterin through the action of dihydropteridine reductase, using NADH as an essential cofactor

NMDA (*N*-methyl-D-aspartate) receptors through their regulation of intracellular calcium ions. Because it helps enzymes carry out chemical reactions, tetrahydrobiopterin is known as an important cofactor. When tetrahydrobiopterin interacts with enzymes during chemical reactions, the cofactor is oxidized and must be recycled in order to yield a reduced, active form. Quinoid dihydropteridine reductase is one of two enzymes that recycles tetrahydrobiopterin (BH_4) in the body where NADH is essential (Fig. 8.43). In addition, tetrahydrobiopterin (BH_4) also plays a critical role in the processing pathway of several protein amino acids in the body. For example, using the enzyme phenylalanine hydroxylase the amino acid phenylalanine is converted to the other amino acid tyrosine (Fig. 8.43).

Also, there is indirect evidence that the enzyme DHPR affects tyrosine hydroxylase activity via BH_4, because agents which competitively inhibit DHPR such as 1-methyl-4-phenyl-1,2,3,6 tetrahydropyridine (MPTP) induce Parkinson symptoms in addicts using this toxic by-product when trying to synthesize meperidine [112–114]. A valid proof for the clinical efficacy of NADH in stimulation of L-DOPA biosynthesis would be the measurement of an increase in L-DOPA formation in the brain, in particular the substantia nigra. For obvious reasons it is impossible to

```
Guanosine triphosphate (GTP)
            │ GTP cyclohydrolase 1 (GTPCH)
Dihydroneopterin triphosphate(NH₂P₃)
            │ 6-pyruvoyl tetrahydropterin synthase(PTPS)
6-Pyruvoyl-tetrahydropterin(6-PPH₄)
            │ Sepiapterin reductase (SR)
Tetrahydrobiopterin (BH4)      Tyrosine        Tryptophan        Phenylalanine
Dihydropteridin                Tyrosine        Tryptophan        Phenylalanine
Reductase                      hydroxylase     hydroxylase       hydroxylase
(DHPR)                         (TH)            (TPH)             (PAH)
Quinonoid    ← Pterin 4a-      L-Dopa          5-OH-Tryptophan   Tyrosine
Dihydrobiopterin  carbinolamine
(qBH2)
            Pterin             Dopamine        Serotonin
         carbinolamine
         dehydratase (PCD)
                               Homovanillic    5-Hydroxyindolacetic
                               acid (HVA)      acide (5-HIAA)
```

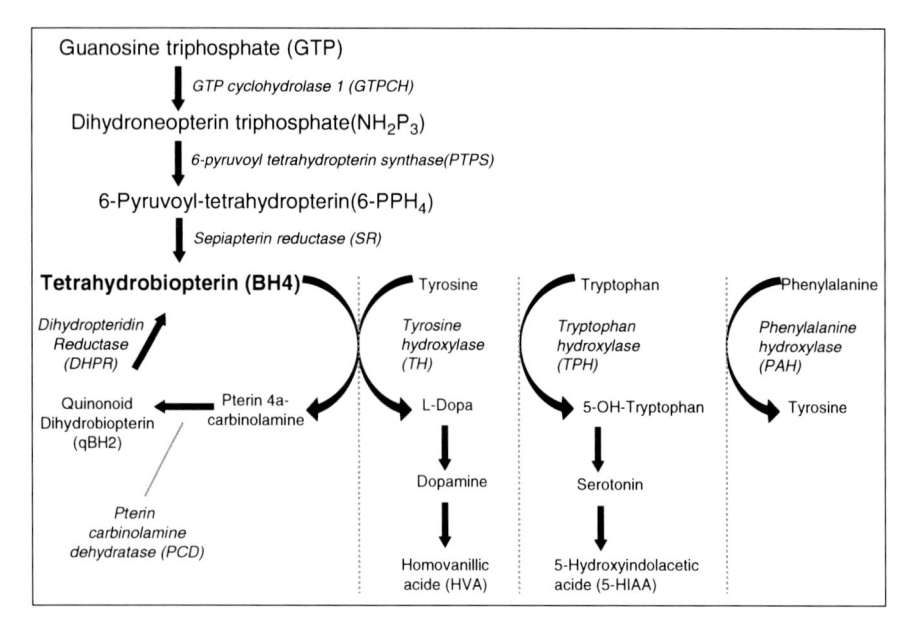

Fig. 8.43 Synthesis of tetrahydrobiopterin being also involved in the biosynthesis of the neurotransmitters dopamine and serotonin. Because of toxic interference the rate of formation and of recovery may be insufficient to meet individual demands

gain such data because, for the time being, it is not possible to measure L-DOPA concentration in the substantia nigra before and after NADH treatment. Therefore one has to rely on indirect evidence by determining the metabolic product of dopamine, homovanillic acid (HVA). There, it was demonstrated that the concentration of this metabolite increased significantly after NADH treatment, which was in parallel with an improvement in disability. Furthermore, experiments in tissue cultures using dopamine producing neuroblastoma cells, showed an increase in the production of dopamine when adding NADH to the medium. Furthermore, NADH directly stimulated enzymatic activity of TH when added to the culture medium. These findings indicate that NADH acts directly on tyrosine hydroxylase (TH) activity resulting in a stimulation of dopamine biosynthesis, corroborating the clinical concept for stimulating endogenous dopamine production by means of NADH supplementation. It may be argued that the observed beneficial clinical effects after NADH medication are not of central nature but due to a peripheral effect. If this holds true, an increase in L-DOPA in the blood would be the consequence from which a certain percentage will reach the brain using the same transport mechanism as exogenously supplied L-DOPA. Indirect evidence for this assumption is derived from the observation that the DOPA decarboxylase inhibitor carbidopa, when given in combination with NADH in a number of patients, yielded a better and longer lasting clinical improvement than the sole use of NADH. In regard to the central and important part in NO/ONOO⁻ upregulation, depletion of the compound called tetrahydrobiopterin (BH$_4$), which is oxidized by peroxynitrite (PRN) is the essential part

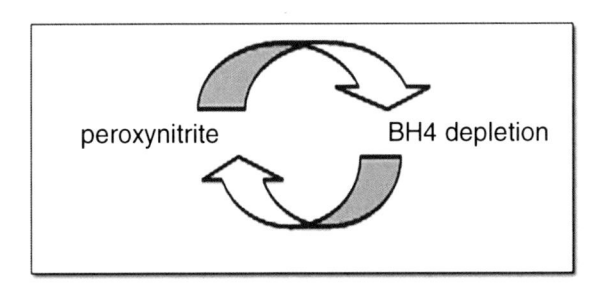

8.10 Mechanisms of Action of Agents for Therapy in CFS-FMS-Parkinson-Alzheimer

Fig. 8.44 Interaction of high peroxynitrite level inducing reduction of tertahydroxybiopterin (BH$_4$) a necessary agent, where depletion results in an uncoupling of the nitric oxide synthases (NOS) with excess formation of NO/ONOO$^-$

in the pathological process, which directs to the potential benefit of a therapeutic approach with this agent.

BH$_4$ is what is known as a cofactor in the nitric oxide synthases (NOSs), and tetrahydrobiopterin depletion produces what has been called partial uncoupling of the NOSs. When a NOS enzyme is missing BH$_4$, it produces superoxide in place of nitric oxide. The consequence of this is that in cells and tissues that have high NOS activity and partial uncoupling, one has many adjacent enzyme molecules, some producing nitric oxide and others producing superoxide and these will react rapidly with each other to form more peroxynitrite. This will, in turn oxidize more BH$_4$, producing more partial uncoupling. This reciprocal relationship between peroxynitrite and BH$_4$-depletion is a potential vicious cycle within the larger NO/ONOO$^-$ cycle and may constitute the essential core of the cycle. Lowering of this central couplet will be expected to produce clinical improvement in these diseases, but will produce an increase in nitric oxide. So while one might think that the net effect of nitric oxide in these diseases is a negative one, agents that increase nitric oxide by lowering this central couplet should be helpful. The new therapeutic principle for treating Parkinson's disease, the stimulation of the endogenous L-DOPA biosynthesis, could overcome the drawback of the L-DOPA treatment, reducing further destruction of residual active nigra cells, which is caused by oxygen radicals formed in considerable quantities by auto-oxidation of L-DOPA [115].

While some agents act to help restore tetrahydrobiopterin (BH$_4$) levels and thus lowers partial uncoupling of the nitric oxide synthases (iNOS), other agents lower nitric oxide levels directly, certain agents act in the mitochondria to restore energy (ATP) levels, while others act to lower excitotoxicity including excessive NMDA activity. Selected agents act by lowering NF-κB activity or by reducing inflammatory prostaglandin synthesis. All together, the overexpressed NO/ONOO$^-$ cycle should be lowered to at least to a certain extent in order to show clinical efficacy and approximately 85% of the patients who use these micronutients for therapy get substantial improvement. Finally, the central couplet in the reciprocal relation between peroxynitrite and tetrahydrobiopterin (BH$_4$) should be kept in focus. This is because peroxynitrite oxidizes BH$_4$ leading to partial uncoupling of the nitric oxide synthases, which in return produces more peroxynitrite (Fig. 8.44).

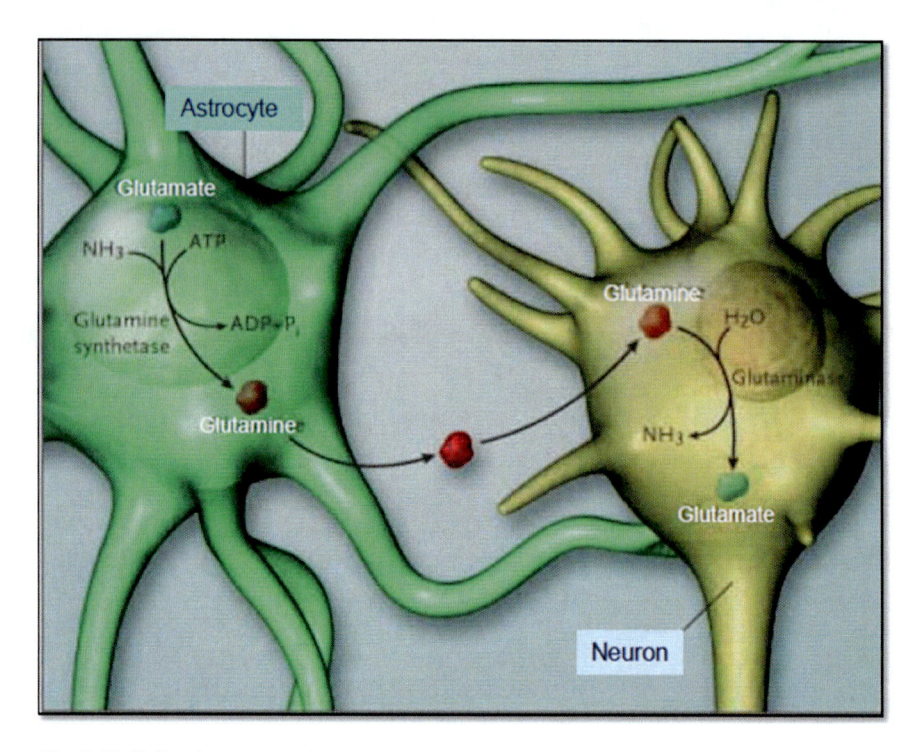

Fig. 8.45 Following uptake of glutamate, glutamnine synthesis in astrocytes via glutamine synthetase is activated though betain which gives way to NH$_3$-goups. Glutamine is then released and taken up by glutamatergic neurons. Within the neuron glutamin is hydrolyzed to form glutamate. Only the astrocytes take up glutamate, whereas the neurons take up glutamine

The nutritional support as outlined above, does not contain any reduced glutathione, although it covers agents that may serve as precursors for reduced glutathione in the body (i.e. N-acetyl-cysteine). Taking reduced glutathione in a liposomal oral supplement, a nasal spray or even as a nebulized inhalant may be helpful for s substantial number of individuals. However, people with asthma symptoms tend to be sensitive and should therefore, start on low doses. In addition, the one agents that directly lowers NF-κB activity is curcumin and it is possible that this NF-κB lowering agents may be helpful. Also magnesium lowers NMDA activity, which is another reason, why this agent should be added to the therapeutic protocol.

8.10.4 Rational for Betain in Mitochondropathy

Betain or **trimethylglycine** is used in regard to its effect on CNS function aiding in the synthesis of glutamine within astrocytes for taking effect on glutaminergic neurons (Fig. 8.45).

Fig. 8.46 The LOGI (= low glycemic and insulinemic) diet results in a lesser increase of blood glucose levels followed by a reduced insulin release

In addition, methylation aids in the inactivation of toxins within the liver, since betaine is a methyl-donator, where methylation via trimethylglycine is being completed.

8.10.5 Dietary Considerations in Mitochondropathy

Aside from nutritional supplementation a new insight in carbohydrate metabolism, advocates a strict change in nutrition by avoiding all sugar for the purpose of flavor. Instead of fructose or glucose one should use palatinose and change to the

- LOGI diet with a low glycemic index where carbohydrates resulting in a rapid increase in blood glucose levels are evaded. Also termed the *cave man diet* it consists in the reduction of carbohydrate intake to only 20% of total food in exchange to fat and oils, omitting all sugar, spaghetti, french fries, potatoes, rice, pizza, cake, chocolate, any kind of sweets, and ice cream (Fig. 8.46).

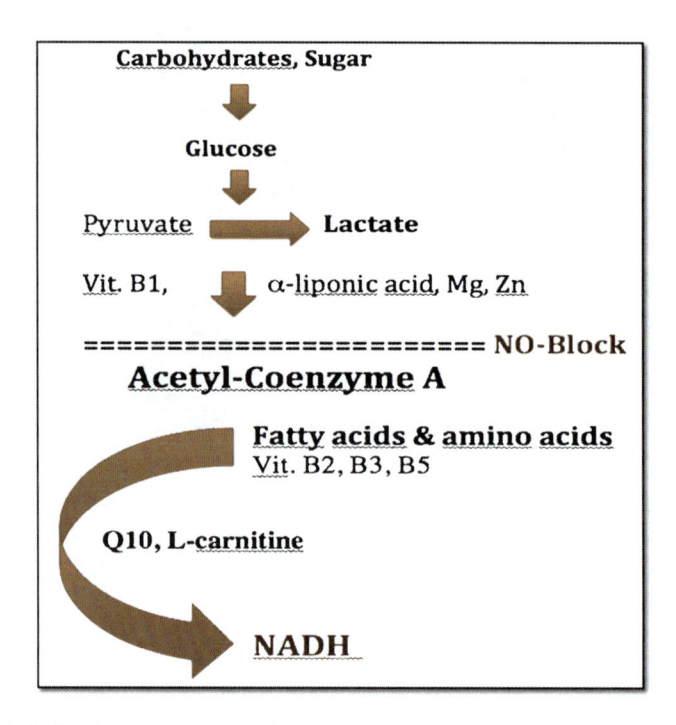

```
Carbohydrates, Sugar
        ⬇
     Glucose
        ⬇
Pyruvate  ➡  Lactate
Vit. B1,    α-liponic acid, Mg, Zn
        ⬇
========================= NO-Block
  Acetyl-Coenzyme A
         Fatty acids & amino acids
         Vit. B2, B3, B5

      Q10, L-carnitine

              NADH
```

Fig. 8.47 Principle in overcoming the NO-block and pile-up of pyruvate in the Krebs-cycle by strict reduction in carbohydrate intake to reduce pyruvate stock and substitute with B1, B6, Mg, Zn supplements for intracellular pyruvate breakdown plus B12 or folic acid for reduction in nitrosative stress

The scientific rational behind this reduction contains the fact that in cases of pyruvate dehydrogenase deficiency via high dose NO/ONOO⁻ levels, the next step in carbohydrate mobilization (i.e. the formation of Acetyl Coenzyme A) is blocked with an accumulation of pyruvate, which now can only be used metabolically to form lactate (Fig. 8.47). The result is an insufficient metabolization of carbohydrates to be shifted into the Krebs-cycle with a pile-up of pyruvate, resulting in insufficient utilization of energy carrier for the production of ATP within mitochondria. And since fatty acids as well as amino acids can by-pass this pyruvate block, they preferentially are used for energy production. As a result the diet should be changed putting all emphasis on the consumption of fats, oils (such as MCF) and proteins, a proposal which very much is in contradiction to the present recommendations, where a marked reduction in any kind of fat is endorsed since it allegedly should reduce cholesterin levels. In contrast to such sanctions data from overweight patients who markedly have cut down on highly utilizable carbohydrates and substituting their diet with medium chain fatty acids and proteins, showed a documented decline in cholesterol level accompanied by a reduction in their BMI.

In addition,

- Substitute with vitamin D_3 1000–2000 IU/die, which lowers inflammatory responce and greatly lowers NF-κB responses.
- Avoid histamine liberating diet to control silent inflammations.
- Determine thyroid gland activity with possible substitution of thyroxin.

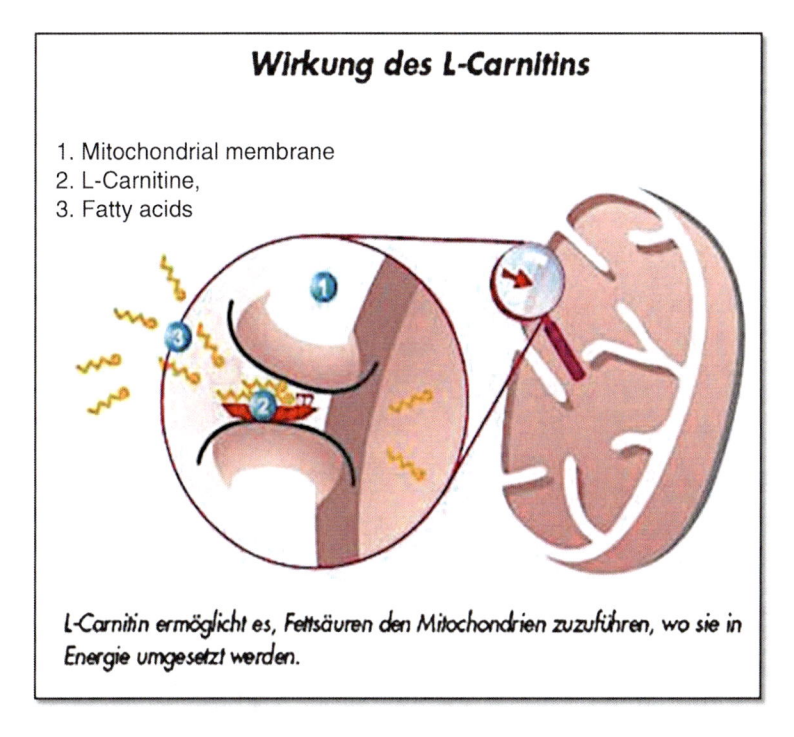

8.10 Mechanisms of Action of Agents for Therapy in CFS-FMS-Parkinson-Alzheimer

Wirkung des L-Carnitins

1. Mitochondrial membrane
2. L-Carnitine,
3. Fatty acids

L-Carnitin ermöglicht es, Fettsäuren den Mitochondrien zuzuführen, wo sie in Energie umgesetzt werden.

Fig. 8.48 Carnitine is the important vehicle for the transport of fatty acids into mitochondria

- Infusion of reduced, bioavailable and active form of glutathione (GSH; the mother of all antioxidants), e.g. Recancostat™.
- Substitute Q10 in combination with NADH to restock energy level in mitochondria, since this is the weakest link the oxidative phosphorylation chain. For correct plasma levels determine the lipid-corrected Q10 levels because in hyperlipidemia Q10 also binds to lipids.

Any deficit in the Q10 plasma level is always related to an underlying mitochondropathy.

- Substitute with L-Acetyl Carnitine, the transporter system of fatty acids for repair of mitochondrial membranes (Fig. 8.48).
- Substitution of vitamin B_{12}, preferentially methyl- or hydroxycobolamine, both of which act as a NO scavenger, reducing $ONOO^-$ formation as outlined below

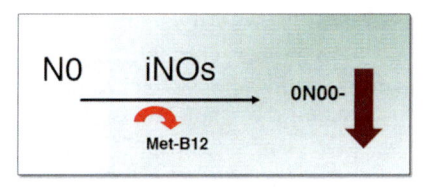

NO $\xrightarrow{\text{Met-B12}}$ iNOs $\longrightarrow$ ONOO-

- Take D-Ribose 2×5 g/die to increase ADP to ATP exchange in mitochondria at translocase complex V.

Fig. 8.49 The pentose phosphate pathway (PPP) and the point of entry for ribose into the pathway. 5-phosphoribosyl-1-pyrophosphate is represented as PRPP

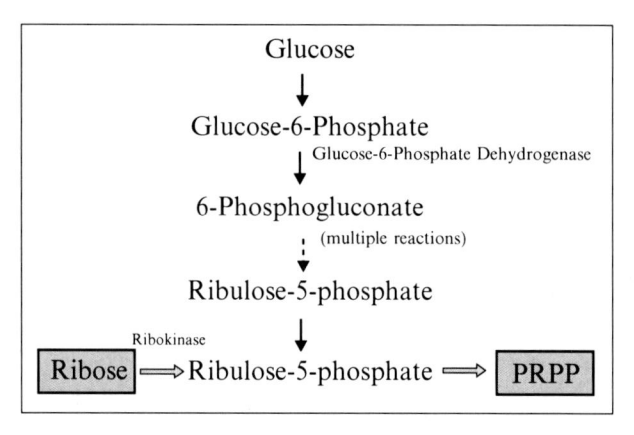

```
                    Glucose
                       ↓
              Glucose-6-Phosphate
                       │ Glucose-6-Phosphate Dehydrogenase
                       ↓
              6-Phosphogluconate
                       ┊ (multiple reactions)
                       ↓
              Ribulose-5-phosphate
                Ribokinase    ↓
        Ribose ⟹ Ribulose-5-phosphate ⟹ PRPP
```

Fig. 8.50 The role of ribose in de novo synthesis of ATP

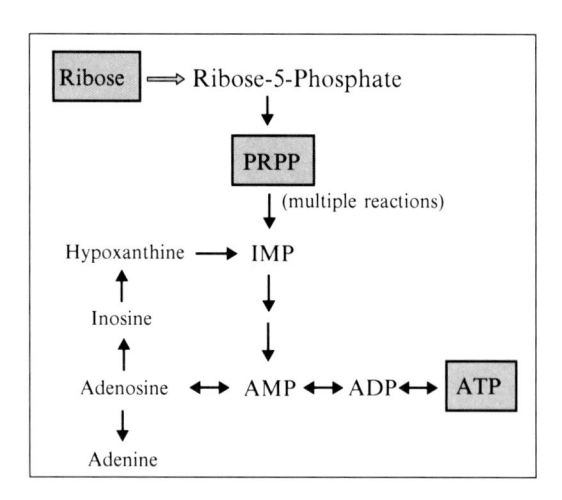

```
        Ribose ⟹ Ribose-5-Phosphate
                       ↓
                      PRPP
                       ↓ (multiple reactions)
    Hypoxanthine ⟶ IMP
        ↑              ↓
     Inosine
        ↑              ↓
    Adenosine ↔ AMP ↔ ADP ↔ ATP
        ↓
     Adenine
```

8.10.6 Ribose – Option for Restoring Energy Depletion

Ribose is a naturally occurring pentose monosaccharide. It is used by the body to synthesize nucleotides, nucleic acids, glycogen, and other important metabolic products. Ribose is formed in the body from conversion of glucose via the pentose phosphate pathway (PPP, also known as the hexosemono-phosphate shunt or the phosphogluconate pathway; Fig. 8.49).

Supplemental ribose enters the PPP by being phosphorylated to R-5-P by ribokinase. The R-5-P thus formed can be utilized to (a) generate glucose by reverse flux up the PPP [116, 117]; (b) form pyruvate through glycolysis [117, 118], or (c) synthesize nucleotides 23 which are needed for ATP production. In this way ribose is utilized in animals and man in many different tissues, including the heart and skeletal muscle. Ribose is the substrate for formation of 5-phosphoribosyl-1-pyrophosphate (PRPP). PRPP is, in turn, used in de novo synthesis of nucleotides such as ATP, adenosine, and inosine (Fig. 8.50) [119, 120]. PRPP is also an essential participant in the salvage pathways for ATP regeneration (Figs. 8.50 and 8.51) [119–121].

8.10 Mechanisms of Action of Agents for Therapy in CFS-FMS-Parkinson-Alzheimer

Fig. 8.51 The role of 5-phosphoribosyl-1-pyrophosphate (PRPP) in the ATP salvage pathway

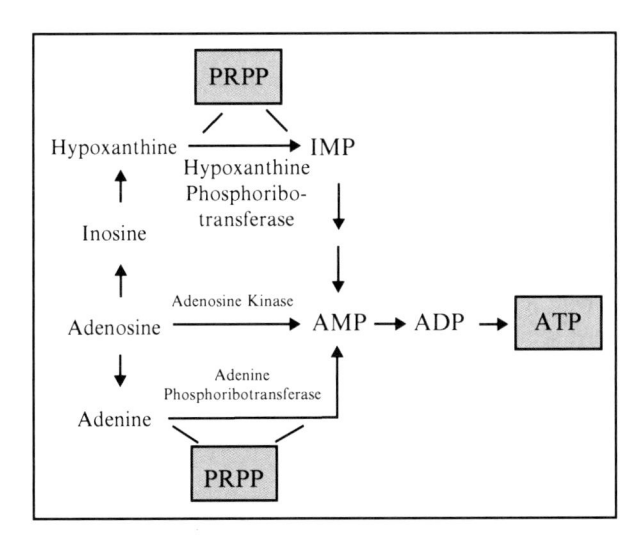

Nucleotides, including ATP, are essential energy sources for basic metabolic reactions and play important roles in protein, glycogen and nucleic acid synthesis (ribonucleotides and deoxyribonucleotides), cyclic nucleotide metabolism, and energy transfer reactions.

Ribose plays a pivotal role in both myocardial and skeletal muscle metabolism, largely through its participation (as a precursor to PRPP) in the synthesis of ATP, adenine nucleotides, and nucleic acids. In these tissues the PPP is inefficient due to low availability of glucose-6-phosphate dehydrogenase. Supplemental ribose administration allows the rate-limiting glucose-6-phosphate dehydrogenase step in the PPP to be bypassed, thereby directly elevating PRPP levels [120–123]. Elevated PRPP levels are then available for increased adenine nucleotide biosynthesis, which accelerates replenishment of depleted cardiac and skeletal muscle adenine nucleotide pools. This is the key to recovery of depleted ATP levels after ischemia or strenuous exercise. The metabolic basis for the effectiveness of D-ribose is apparently not species specific because glucose-6-phosphate dehydrogenase is the rate limiting enzyme in the heart and skeletal muscle PPP for rats, dogs, and swine, as well as humans. Since this enzymatic reaction is the rate-limiting step in the PPP that limits the available PRPP pool and thus the adenine nucleotide levels, the enzymatic basis for the effectiveness of ribose, i.e., the formation of PRPP by bypassing the G-6-PDH reaction step, is the same for these different species.

8.10.6.1 Effect of Ribose on the Heart

When the myocardium becomes oxygen depleted due to ischemia (restricted blood flow to the heart) resulting from occluded arteries, heart attack, heart surgery, organ transplantation or other surgery, myocardial levels of ATP will fall dramatically and can take up to 10 days to recover [124–127]. Under conditions of such energetic depletion myocardial function is compromised and there is an increased risk of permanent loss of myocardial tissue. And even in lower risk situations, such as healthy individuals who are pushing their physical limits by intense exercise, ATP reserves can become depleted and take several days to recover [128–132]. Slow replenishment of ATP in both myocardial and skeletal muscle tissues has been attributed to the low rate of de novo synthesis and slow recovery of ATP and its precursors via the salvage pathways [119, 131, 133, 134]. Since replenishment of ATP is likely to enhance the functional recovery of these tissues investigators have sought methods of improving the salvage rates and increasing de novo synthesis. Interestingly, in a wide range of studies several investigators have found that ATP recovery can be stimulated in both myocardial and skeletal muscle tissues by administering a simple sugar called ribose [119, 122, 123, 127, 131, 134–138]. Knowledge concerning the effect of ribose in the heart has been gathered from many laboratory and clinical studies of human and animal myocardial tissue and function. These studies have documented several positive effects of ribose including improved ventricular function and enhanced recovery of myocardial ATP and adenine nucleotide levels following ischemia, increased exercise tolerance in patients with stable coronary artery disease, and improved thallium-201 redistribution in cardiac imaging applications. If myocardial tissue becomes oxygen depleted when blood flow to the heart is restricted. A persistent consequence of this ischemia is a substantial lowering of tissue energy, as evidenced by decreased myocardial ATP levels. These lowered energy levels are in turn correlated with depressed cardiac function [119, 133]. The correlation between decreased ATP levels and depressed myocardial performance has spurred researchers to develop methods of metabolic intervention into adenine nucleotide degradation and/or biosynthesis in order to restore myocardial ATP levels. In a series of oxygen depletion studies in the myocardium using asphyxia recovery and ATP depletion models evidence was gathered that PRPP availability limits adenine nucleotide synthesis by both the de novo and salvage pathways [134, 135, 139, 140]. By providing ribose to the myocardium a pronounced stimulatory effect on PRPP synthesis occurs. The presence of ribose allows the rate-limiting step in the pentose phosphate pathway, the G-6-PDH enzymatic reaction, to be bypassed, leading to the production of PRPP. This increase in PRPP levels is noted to be accompanied by accelerated cardiac adenine nucleotide synthesis and improved global heart function. Thus, ribose restores cardiac energy reserves and positively affects myocardial function.

The effect of orally administered ribose on exercise tolerance in stable coronary artery disease patients has also been studied [141]. Two positive baseline treadmill studies were performed for eligibility into this study. The criterion for inclusion was development of moderate angina and/or ST-segment depression (an indicator of ischemia) on the electrocardiogram. Patients were randomized into two groups. Ten patients received placebo (glucose) for 3 days and another 10 patients received ribose dissolved in water for the same time period. A final treadmill evaluation was performed in all patients after taking the supplement. In the ribose-treated group, the mean walking time to ST-segment depression was significantly greater than in the placebo group ($p<0.002$). The time to both ST-segment depression and onset of moderate angina was also prolonged significantly in the ribose group compared to its pre-ribose baseline ($p<0.005$). These results show that patients who had been given ribose were able to exercise longer without chest pain or evidence of ischemia than patients who did not receive ribose.

Ribose also enhances the detection of hibernating myocardium during diagnostic procedures such as thallium imaging or dobutamine stress echocardiography. In two swine models, ribose infusion after transient ischemia modified thallium-201 ($^{Tl}201$) clearance in both ischemic and non-ischemic myocardial regions, resulting in faster $^{Tl}201$ redistribution [142, 143]. Furthermore, placebo-controlled clinical trials have also found that intravenous ribose infusion enhances thallium-201 redistribution in humans [144, 145]. One such trial addressed whether or not an intravenous infusion of ribose could facilitate 201TI redistribution after transient myocardial ischemia in patients with coronary artery disease and thus improve the ability to detect jeopardized but viable myocardium [144]. Seventeen patients with documented coronary artery disease and chronic, stable angina were enrolled. Each patient underwent two separate exercise tests, one with saline infusion and one with ribose, performed 1–2 weeks apart. In each test an injection of ^{201}TI was given and two subsequent imaging procedures were performed. Post-exercise and initial imaging, patients received the infusion of either ribose or saline. Imaging was performed again at 1 h, followed by a rest period of 4 h. Following the rest period imaging was performed one final time. The results revealed that at both 1 and 4 h post-exercise there were significantly more reversible defects identified when patients were given ribose versus saline. In another $^{Tl}201$ study with a similar protocol, but with imaging at 4 and 24 h, results showed that there were more defects detected at 4 h post-exercise when ribose infusion was given than at 4 and 24 h with saline infusion [145]. The conclusions from both of these studies imply that ribose substantially improves the identification of viable ischemic myocardium using $^{Tl}201$ imaging after exercise, suggesting improved post-ischemic myocardial function with ribose administration. Another research study points out that ribose infusion in conjunction with dobutamine stress echocardiography increased the contractile response in hibernating regions of the heart [146]. In a placebo-controlled double-blind study 25 patients with ischemic cardiomyopathy were infused with either D-ribose or dextrose placebo for the 4 h prior to dobutamine stress echocardiography. On day two the patients were crossed over to the alternate treatment. During dobutamine stress echocardiography more dysfunctional wall segments responded with improved wall motion when D-ribose was infused prior to the procedure as compared to placebo ($p=0.02$). In patients who then underwent coronary artery bypass surgery the predictive sensitivity for functional recovery of the segments identified during the D-ribose infusion was greater than those identified during placebo infusion.

A recent review provides the background and rationale for the use of ribose in metabolic support of the heart [147]. Evidence such as that discussed above is presented in support of the main hypothesis that ribose is the rate-limiting component in the pathways necessary for the heart to restore depleted adenine nucleotide levels. Thus D-ribose has been shown not only to be efficient in CFS/FMS, but can also be considered a basic form of treatment in other ailments, all of which are related to reduced formation of the energy substrate ATP in mitochondria [148, 149] (Ref. [273] in Chap. 7). For instance, decades of research have shown that ribose has a profound effect on heart function in patients with congestive heart failure, coronary artery disease, and cardiomyopathy (a weakened heart muscle) [119, 136]. Like the muscles in patients with fibromyalgia, sick hearts are energy starved. This energy deprivation keeps the heart from relaxing between heartbeats, making it impossible for the heart to completely fill with blood [138]. Using ribose to restore the energy level in the heart allows it to fully relax, fill, and empty completely to circulate blood to the outer reaches of the body [138, 150]. Circulating more blood means muscles in the arms and legs, and the tissues of the brain, get the oxygen they need to function normally. This result was made evident in several important studies in patients with congestive heart failure and angina. In one study conducted at the Cologne University in Germany, patients with congestive heart failure were treated with either 10-grams of ribose or a sugar placebo every day for 3 weeks [151]. They were then tested for heart function, exercise tolerance (a measure of fatigue), and quality of life using a questionnaire designed for this purpose. In this study, ribose therapy had a significant effect on all measures of diastolic heart function, showing that increased energy in the heart allowed the heart to relax, fill, and pump more normally. Patients in the study were also much more tolerant to exercise when they were on ribose, and, through their responses to the questionnaire, showed they had a higher quality of life as a result. Two additional studies went on to help explain how ribose therapy in congestive heart failure may affect fatigue and exercise tolerance [144, 145]. These studies showed that ribose treatment increased ventilatory and oxygen utilization efficiency, a medical way of saying that the patients were able to breathe better and use the oxygen they inhaled more efficiently. Improving the patient's ability to use oxygen means more oxygen is available to go into the blood and out to the tissues. Having more oxygen available allows the muscle to burn fuel more efficiently, helping it keep pace with its energy demand. The result is less fatigue, a greater ability to tolerate exercise, and a higher quality of life. An added benefit to improving ventilatory efficiency is that ventilatory efficiency is a dominant predictor of mortality in congestive heart failure. Increasing ventilatory efficiency with ribose therapy is, therefore, a direct correlate to prolonging life in this patient population. There are very few nutritional therapies that can legitimately boast of having this profound of an effect on the tissues they target. None, other than ribose, can claim such an effect in cell or tissue energy metabolism. Ribose is a unique and powerful addition to our complement of metabolic therapies in that it is completely safe, proven by strong, well designed clinical and scientific evidence, natural, and fundamental to a vital metabolic process in the body [152–154]. Ribose therefore can be considered a simple and cost saving alternative that regulates how much energy we have in our bodies, and for those suffering from fatigue, muscle soreness, stiffness, and a host of related medical complications, the relief found in energy restoration can be life changing [155]. This is why D-Ribose is recommended in all CFS/FMS patients starting with 5 g (1 scoop 3 × day) for 2–3 weeks then twice a day [148]. It is critical to take the 3 scoops a day for the first few weeks to see the optimal effects. Although many other treatment options with neutraceuticals in mitochondria-related diseases may take several weeks to show an optimal effect, most people feel the difference with D-Ribose by the end of at least a few days and in some instances within 1 h. This has been demonstrated in patients with heart beat irregularities in FMS, where 5 g of D-Ribose orally resulted in normal heat beat within 1 h (unpublished observation in 30 female patients by the author).

8.10.7 *Mitochondria Remodeling with Intermittent Hypoxia Therapy (IHT)*

The binding of nuclear receptors by glucocorticoids, heat radiation, nutrient deprivation, viral infection and hypoxia results in an increase intracellular calcium concentration all result in Ca^{2+}-influx [156, 157]. For example, damage to the cell membrane, can all trigger the release of intracellular apoptotic signals within a number of other cellular components. Based on such apoptotic triggering there is an imperative need for exploring and implementing mitochondria-rejuvenative interventions that can bridge the current gap toward the step-by step realization of strategies for engineering negligible senescence (SENS). This is because recently discovered in mammals, natural mechanism mitoptosis – a selective "suicide" of **mutated** mitochondria – can facilitate continuous purification of mitochondrial pool in an organism from the most reactive oxygen species (ROS)-producing mitochondria. Mitoptosis, which is considered to be the first stage of ROS-induced apoptosis, underlies the concept of follicular atresia, which is a "quality control" mechanism in female germline cells that eliminates most germinal follicles in female embryos leading to a remodeling and rearrangement of mitochondrial membranes, and leading to mitochondria rupture and disruption of degenerated, less active mitochondria [158]. Mitoptosis can be also activated in adult postmitotic somatic cells by evolutionary conserved phenotypic adaptations to intermittent oxygen restriction (IOR) and synergistically acting intermittent caloric restriction (ICR). IOR and ICR are common in mammals and seem to underlie extraordinary longevity and augmented cancer resistance in bowhead whales (*Balena mysticetus*) and naked mole rats (*Heterocephalus glaber*). Furthermore, in mammals IOR can facilitate continuous stromal stem cells-dependent tissue repair. A comparative analysis of IOR and ICR mechanisms in both mammals, in conjunction with the experience of decades of biomedical and clinical research on emerging preventative, therapeutic, and rehabilitative modality – the intermittent hypoxic training/therapy (IHT) – indicates that the notable clinical efficiency of IHT is based on the universal adaptational mechanisms that are common in mammals. Further exploration of natural mitochondria-preserving and -rejuvenating strategies can help refinement of IOR- and ICR-based synergistic protocols, having value in clinical human rejuvenation of mitochondria [159]. Based on such assumptions, another option that has evolved in the treatment of acquired mitochondropathy, which is termed as intermittent hypoxia therapy or training. Interval hypoxic training (IHT) has been used with success in autoimmune disease, chronic infections, Diabetes type 2, obesity, burnout-syndrome, depression, migraine, chronic farigue syndrome, fatigue associated with cancer, allergy, asthma, Parkinson, and Alzheimer. There are case reports of Alzheimer, where nearly all symptoms dissipated with IHT treatment, inspite the poor prognosis of this ailment. This is hard to understand, since Alzheimer patients have no chance for any recovery. However, when considering this disease as a mitochondrial-related disorder with all its symptoms, than this disease can only be treated with a therapeutical approach that directly affects mitochondria such as intermittent hypoxia therapy. When using the described technique for regaining healthy mitochondria and exterminate those with already mutated mtDNA arrays, the patient via a respiratory mask, first inhales excessive followed by lowered ambient oxygen concentration. In the latter, diseased mitochondria will be activated to initiate an apoptosis program (isolated extinction). But before that program of self-destruction is activated they send a signal to the more vibrant and healthy mitochondria for replication. This yields a higher net concentration of energetic, viable and active mitochondria resulting in an increase in productivity and restoration with an increase in energy levels of all cellular organelles within the body. Interval oxygen deprived therapy so far has been used in a number of ailments, all of which are linked to impaired mitochondrial productivity, which has been documented in a number of scientifically based papers and also in cancer studies [160, 161].

Positive effects if breathing oxygen-deprived air can be derived from nature, where bowhead whales living north of the polar circle, accumulate oxygen when being at the water surface. Thereafter they dive for 15 min into the deep ocean sea, during which oxygen concentration within the cells drops to extremely low values. This processes is repeated 100 times over the whole day. They feed themselves from plankton and small fish, which are only available for half a year. Female bowhead whales give birth at the average of every third year until they reach 90 years of age and in some cases animals reach 200 years of age without showing any signs of cancer. In addition, people living continuously at high altitude for a period of months or years is not an option for most people. Russian doctors, in trying to find a solution to acclimatizing pilots, athletes, mountaineers and cosmonauts to low-oxygen environments, discovered that adaptation to low-oxygen environments could occur rapidly in a clinical setting and without the side effects of altitude sickness. The oxygen levels found at high altitude were administered in a controlled manner and given in measured intermittent doses. The technique became known as Intermittent Hypoxic Training (IHT), whereby a machine known as the Hypoxicator™, which separates air through a semi-permeable membrane, administers oxygen levels of between 10% and 15% (equivalent to an altitude range of 2,500–6,000 m). A **hypoxicator** is a medical device intended to provide a stimulus for the adaptation of an individual's cardiovascular system by means of breathing reduced oxygen hypoxic air and triggering mechanisms of compensation (Fig. 8.52).

The aim of intermittent hypoxic training or hypoxic therapy conducted with such a device is to obtain benefits in physical performance and wellbeing through

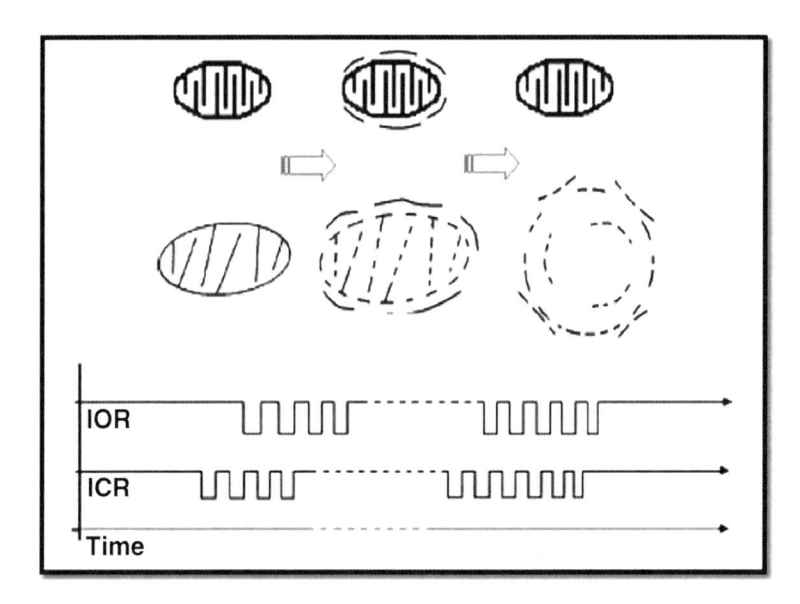

8.10 Mechanisms of Action of Agents for Therapy in CFS-FMS-Parkinson-Alzheimer 349

Fig. 8.52 Principle of mitoptiosis as a physiological response of hypoxia-reoxygenation and calorie restriction. Oscillations of oxygen and cellular nutritient (*IOR* intermittent oxygen restriction and *ICR* intermittent caloric restriction) stimulate impulse ROS production in mitochondria, consequently overloading mitochondrial antioxidative defense. Wild type mitochondria (*above*) respond by increased production of antioxidative enzymes and survive. Mutated or defective mitochondria (*below*) are much more sensitive to oxidative stress going into apoptosis, thus being selectively eliminated by mitoptosis

improved oxygen metabolism. Advanced hypoxicators have a built-in pulse oximeter used to monitor and in some cases control the temporary reduction of arterial oxygen saturation that results in physiological responses evident at both systemic and cellular levels even after only a few minutes of hypoxia [162], the Hypoxic Training Index (HTI) can be used to measure the delivered therapeutic dosage over the training session. The underlying mechanisms of adaptation to mild, non-damaging, short-term (minutes) hypoxic stress (also called – intermittent hypoxic training) are complex and diverse [163], but are part of normal physiology and are opposite to pathophysiological effects of severe sleep apnea hypoxia. There are a number of types of hypoxicators that can be distinguished by the method of producing hypoxic air and its delivery to the user's respiratory system. Commonly used are air separation systems employing semi-permeable membrane technology or pressure swing adsorption or (PSAS). There are also non-powered hand-held devices – rebreathers-hypoxicators. The term hypoxicator was suggested by Russian scientists in 1985 to describe a new class of devices for Intermittent Hypoxic Training (or Therapy) (IHT) – an emerging drug-free treatment for a wide range of degenerative disorders and for simulated altitude training used to achieve greater endurance performance [164] as well as offering pre-acclimatization for mountaineers – minimizing the risk of succumbing to acute mountain sickness on a subsequent ascent.

8.10.8 Practice in Applying Intermittent Hypoxic Training

The hypoxia challenge of IHT is normally delivered in an intermittent manner: 3–7 min of hypoxic air breathing alternated with 1–5 min of normoxic or hyperoxic air. The **hypoxicator** allows automated and pre-programmed delivery of the required hypoxic and hyperoxic or normoxic air and safety monitoring. The therapeutic range of arterial oxygen desaturation for IHT is $SpO_2 = 75-88\%$ and must be selected based upon the recommendation of a medical specialist. The person is asked to breathe the high-altitude air for just a few minutes at a time while their blood oxygen levels are continuously monitored. They then breathe ambient or normal air for a few minutes, giving their body time to adjust back to normal conditions. The time spent alternating between low oxygen air and normal air is 60–90 min at a time. The procedure is generally carried out once or twice a day for a total of 16–30 sessions. Researchers found that this technique allowed adaptation to high altitudes with less stress to the body than continuous exposure to low oxygen. Because the dose and the blood oxygen levels are totally controlled, there is no danger of altitude sickness. The intermittent nature of the hypoxic exposure means the adaptation is low and is not lost, as it is the case in normal acclimatization to high altitude. Adaptation to intermittent hypoxia has the unique attribute of activating the body's own internal production of antioxidants in brain, liver and heart as a result of the frequently repeated reoxygenation that occurs on breathing room of hyperoxic air. Intermittent hypoxic training (or therapy) (IHT) is a non-invasive, drug-free technique aiming to improve human performance and well being using the phenomena of adaptation to reduced oxygen (Fig. 8.53).

An IHT session constitutes a few minutes interval of breathing low oxygen (hypoxic) air alternated with an ambient or hyperoxic air over a 45- to 90-minute session per day. A full treatment course is 3–4 weeks. As a rule, the patient remains stationary and breathes hypoxic air via a hand-held mask. Therapy is delivered using a hypoxicator during that course, while the dosage is monitored. Biofeedback can be controlled using a pulse oximeter (Fig. 8.54). The idea of IHT is that it delivers a non-damaging training stimulus that naturally triggers a cascade of beneficial adaptive responses without adverse effects. The response is almost instant [165] and is evident at various levels, from systemic down to cellular level [166, 167].

This technique differs from continuous hypoxia, which actually reduces oxygen over a long period of time resulting in adaptation.

Since the beginning of time, living things, from simple viruses and bacteria to humans, have shown the seemingly miraculous ability to adapt to changes in their environment. In humans the process of adaptation is more efficient and more fascinating than in any other species. Our genetic characteristics and potentials are not fully expressed until something challenges us to adapt. In this way, challenges and stresses help us to develop strengths and abilities we might not otherwise have developed. As we adapt to one level of any environmental or even emotional stress, we become capable of handling even larger doses in the future.

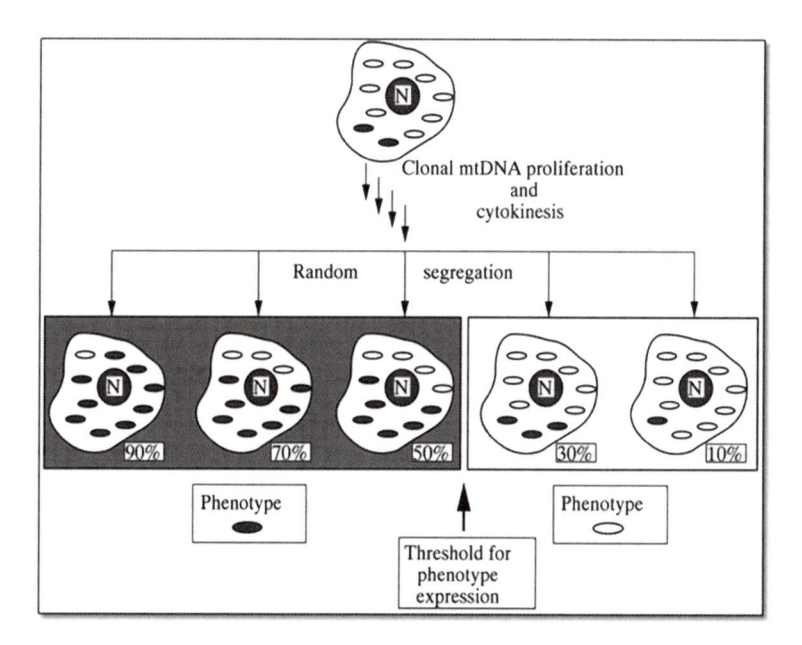

Fig. 8.53 Oscillatory inhalation from normal to low and to increased oxygen tension eliminates mutated mitochondria, which due to false oxphos reactions act by themselves like oxygen radical cannons. Intermittent hypoxic episodes induce programmed cell death of mutated mitochondria while at the same time activate replication of wild type healthy mitochondria

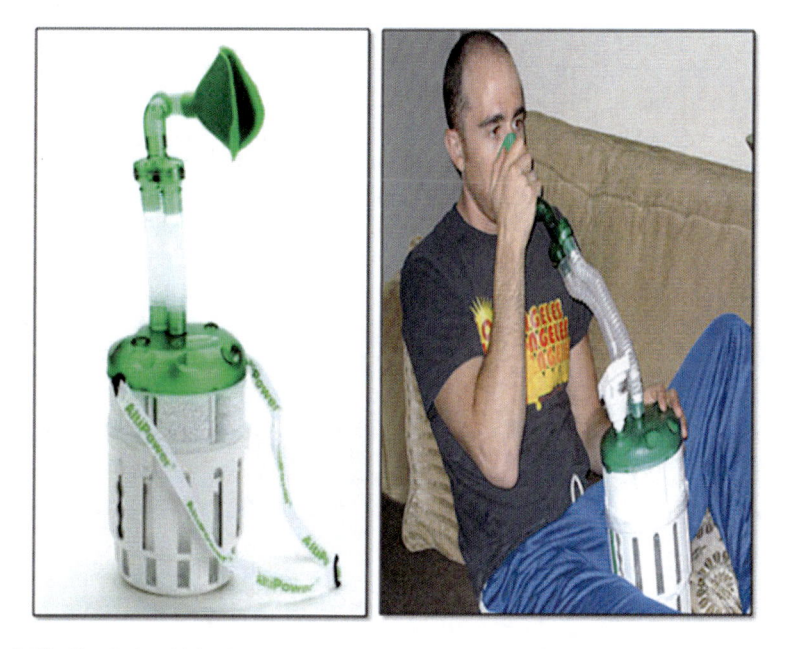

Fig. 8.54 Simulating high altitudes with a is a small, portable Hypoxicator™ (GO2Altitude®) including a breathing unit, an oxygen analyzer and 2 disposable cartridges, with the pro model including a pulse oximeter which is useful for fine tuning altitude training. It works by extracting the carbon dioxide from the user's exhaled air

This ability is what enables, for example, weight lifters to lift unimaginable loads well beyond their own bodyweights. This is why our species (and all other currently living species) has survived and even thrived despite environment changes and times of extreme difficulty and social upheaval. IHT forms an important branch of an emerging new discipline called Adaptive Medicine. The Russian Academy of Medical Sciences has inaugurated a Department of Adaptive Medicine, with one of its aims being to look at the therapeutic potential of IHT. An International Academy of Adaptive Medicine was formed in 1990 by an interdisciplinary group of scientists and clinicians from many countries including Japan, Australia, Germany and the USA. The academy aims to develop further understanding and share information about the ways in which the adaptive process enables the body to respond to different stress stimuli, with a view to treating and preventing different diseases. The scientific definition of stress as given by Dr. Hans Selye, one of the first researchers to really study stress in the laboratory, is: "A stressor is anything that challenges an organism to adapt." In this context, heat, cold, physical exercise, electrical stress, lack of food, hypoxia of altitude and even emotional or psychological turmoil are stress factors that can be used to strengthen us if we experience the minute amounts we can tolerate and to which we scan adapt. Almost all of these factors have been used as means to help people restore health. On the other hand, too much of any of these stress factors, carried on for too long without sufficient recovery time, can exhaust the adaptive mechanism and contribute to disease rather than health. If people fasts, use saunas or cold baths, they begin with an exercise program to get fit or undertake IHT, engaging the principles of adaptation for the restoration of health. One of the amazing things about adaptation is what's known as "cross adaptation". Adaptation to one type of stress or load will, to some extent, increase the body's ability to cope with stresses of another type. It is well know that a regular exercise program is associated with an increased tolerance to stress. Stress-related diseases such as hypertension, heart disease, ulceration of the stomach or duodenum, diabetes, dermatological diseases and disordered immunity have all been shown to have improved outcome with both exercise and IHT. Protection slowly develops because the body now becomes more tolerant and resistant to stress with

- A fade away of the reactions to stress
- An increased activity of the central and peripheral system limiting stress
- A desensitization of target organs.

8.10.9 Improving Antioxidant Status with Bicarbonate

From the theoretical background and the fundamental data demonstrating the important role of tetrahydrobiopterin for the synthesis of NO and for the regulation of the NO-producing enzyme NO-synthase, clinical studies on the manipulation of

8.10 Mechanisms of Action of Agents for Therapy in CFS-FMS-Parkinson-Alzheimer

the metabolism of tetrahydrobiopterin are still pending; however such therapy may be promising in the future with regard to the treatment of endothelial dysfunction and mitochondropathy. After having mentioned new and most of all a safe approach in the treatment of acquired mitochondropathy with nutraceuticals, there are many good reasons to believe that there is nothing in mainstream medicine that addresses deacidification, detoxification, correcting nutritional deficiencies, modulating and boosting the immune system, and increasing body function. Medical science has failed in its attempts to cure degenerative, metabolic, or autoimmune diseases. Without removing toxins and acids from all organs, cells and tissues, and without providing the essential nutritional components like magnesium, the body will not be able to heal himself. Unless a treatment regimen actually removes acid toxins from the body and increases oxygen, any nutritional therapy will shortfall like most medical interventions. Most of the allopathic medicines by themselves are mitochondrial poisons. What happens is that they often change the symptomatology after which they always drive the underlying disorders into a deeper chronic state. When the body's tissues and cells in conjunction with mitochondrial deficiency become too acidic the stage is set for tissue inflammation and degeneration and become breeding ground for anaerobic pathogens. It is time for allopathic medicine to realize that viruses, bacteria and fungi all thrive in acid conditions. Tissues and cells are like factories with mitochondria everywhere with acid wastes that have to be cleared away every millisecond we live. There is no way to avoid acid waste that can accumulate quite rapidly during the metabolic process. There is an increase in oxidative stress, which correlates exceptionally well with any pH changes lading into acidity, especially affecting mitochondria, which results in oxidative and nitrosative stress. Many physicians working in alternative medicine believe that there is not drug on the market that effectively reduces the acidity of the body or addresses any kind of nutritional deficiency. This however is not true! There happens to be two exceptional agents that are excellent in addressing most of the above-mentioned issues. Both magnesium chloride and sodium bicarbonate can be considered as medicines in their own as they are available in an injectable forms and both almost immediate provide a relief from physiological problems. Patients receiving sodium bicarbonate achieve an urine pH of 6.5 as opposed to a pH of 5.6 in patients receiving sodium chloride. Such an alkalization is hypnotized to have a protective effect against the formation free radicals that may cause mitochondropathy. One of the fundamental approaches in medicine is the alkalization of the body so it can easily dispose acids from cells tissues and organs. As a caregiver, we can support this in many ways but often doctors has to do this in an emergency situation. In other times, when we have a cooperative patient we can use food as medicine and accomplish healing gradually with time. The most powerful alkalizing foods on earth are those, which have the highest content in chlorophyll. In chlorophyll magnesium has a central position in the molecule; without magnesium, there would be no chlorophyll, and life simply would not exist. One cannot beat the cell-restoring potential of green foods such as wheat, barley, kamut, alfalfa, and oat grasses that are in the dame line as spirulina and chlorella. All these foods have the highest content in magnesium and act as food medicines. They are thousands of times more powerful than ordinary green vegetables, because there is a concentration in chlorophyll, alkaline minerals, rare trace minerals, vitamins, phytonutrients, and enzymes. Today there is always the quest for exceptionally powerful medical agents that in reality are not really medicines in the way one would normally think of medicines. Both, sodium bicarbonate and magnesium chloride are common items that when not injected are considered safe for consumption. However, these two substances are affective for chronic and acute disorders when used orally and transdermally; we do not need to go to the emergency room for injections. Magnesium bicarbonate is a complex hydrated salt that under specific conditions exists only in water. The magnesium ion is Mg^{2+}, and the bicarbonate ion is HCO_3^-. So, magnesium bicarbonate must have two bicarbonate ions: $Mg(HCO_3)_2$. Magnesium chloride and sodium bicarbonate taken with water separately present an ideal way to supply magnesium ions and bicarbonate ions to body cells. In this respect, magnesium and bicarbonate enriched mineral waters are rapidly absorbed and may have a health benefit. Likewise, in small doses, the two together make up an ideal treatment system for distilled and reverse osmosis. Adding these two substances to taste not only will remineralize highly processed water but also will provide the body with a constant supply of the ultimate mitochondrial cocktail. Good drinking water would contain approximately 125 mg of magnesium and 650 mg of bicarbonate per liter. When consumed together, magnesium chloride and sodium bicarbonate work very well together to reverse medical problems. This is when tissues become too acidic and lack sufficient magnesium for ATP production, the cellular metabolism declines leading into insufficient ATP production, which eventually may even end up in obesity and diabetes.

Few clinicians are aware how these two substances work by enhancing each other – they are mutual reinforces because magnesium functions as a bicarbonate co-transporter into the cell, and bicarbonate acts as a transporter of magnesium into mitochondria. According to the Dietary Reference Intake guide from the Institute of Medicine magnesium influx is linked with bicarbonate transport, while any magnesium transport in or out of cells requires the presence of carrier-mediated transport system [168]. The ATPase reaction has a broad pH optimum centering with neutral pH, with a significantly reduced activity when the pH is above 9.0 or below 5.0. Therefore, any change away from the overall acidity toward an alkaline pH into the neutral zone results in an enhancement of cell metabolism by means of optimizing mitochondrial function. Alkalosis also enhances magnesium reabsorption in the juxtamedullary proximal nephron. It was the dedicated work of Dr Russell Beckett that paved the way to understand the significance of bicarbonate acting in conjunction with magnesium. Dr. Beckett's theoretical expertise and experimental research has resulted in the understanding how important, both bicarbonate and magnesium ions are in human physiology and how they work together to optimize human health and the ability to recover from a disease. Bicarbonate ions work alongside with magnesium creating a condition for an increased glucose transport across cell membranes. Bicarbonate ions without any doubt create an alkaline environment for optimal activity of pancreatic and intestinal enzymes. Bicarbonate neutralizes acid conditions required for inflammatory reactions, which is why sodium bicarbonate is of benefit in the treatment in a number of chronic inflammatory and autoimmune diseases, and research has found that bicarbonate stimulates ATPase by acting directly on it.

Magnesium does not easily reach the mitochondrion; only if plenty of bicarbonate is available in which case it acts as a carrier system for magnesium into mitochondria. The problem is that few magnesium bicarbonate products are available; they are expensive and difficult to shop. So using magnesium chloride and sodium bicarbonate individually for later combination is the idea. A person gets more control over both bicarbonate and magnesium physiology when magnesium is taken in its chloride form and when bicarbonate is taken as sodium bicarbonate. When using magnesium chloride for oral consumption you must use the highest quality sources. Seawater evaporation magnesium oils are not appropriate for this type of application.

8.11 New Avenues in Treatment of Mitochondropathy

Improve antioxidant status is considered a main therapeutic avenue in mitochondropathy. For instance, as soon as the muscle becomes painful and releases toxic metabolites, there is secondary muscle damage by free radicals. Having a good antioxidant status helps to protect against such secondary damage. The most important antioxidants to measure on a regular basis are Co-enzyme Q10, glutathione peroxidase, and superoxide dismutase. There is another antioxidant, which has been trialed in few studies with good results called astaxanthine, and the daily dose would be 4 mg/die.

It is now common knowledge from the work done by Otto Warburg in 1930 that the difference between cancer cells and normal cells is that cancer cells can only function on glycolysis as their energy source, whilst normal cells function on mitochondria. It has been thought up until recently that this was a secondary effect of cells becoming cancerous. However, it now looks very much as though this is the primary effect that is to say this is a cause of cancer. The problem with cells switching to glycolysis is that it is mitochondria that control survival of the whole cell because they activate apoptosis, which is the mechanism by which normal cells self-destruct. When cells switch mitochondria off, they become immortal and this is, of course, the basis of cancer. Therefore, this is one possible basis for treating cancer and indeed has already been shown to be an effective treatment. By using a drug, namely dichloroacetate (DCA), cells can be switched from glycolysis and back to normal mitochondrial function (this of course pre-supposes the existence of healthy mitochondria!). In cancer, some sort of genetic damage probably causes the switch from mitochondria to glycolysis. By feeding tumor cells DCA, glycolysis is switched off, mitochondrial function is switched on, mitochondria recognize that the cell is damaged and causes it to self-destruct. Therefore this is an extremely cheap and effective way of treating all cancers, whatever that cancer may be. However, it is conceivable that this could also be a useful way of switching people back from glycolysis to normal aerobic respiration. It may well be that part of the problems in fibromyalgia and chronic fatigue syndrome is that metabolism gets in a rut.

It literally gets used to glycolysis because that is the route that it has been switched into for reasons above. Once the mitochondrial lesion has been repaired, we then need a switch to kick people back into oxidative phosphorylation instead of a tendency to lapse into glycolysis. Remember that from an evolutionary point of view, glycolysis is a more primitive and therefore possibly more desirable root because it has been longer established. Therefore, this presents a possible therapeutic avenue if all the above interventions do not have an impact in fibromyalgia. The only problem is that this drug is only available on a named patient basis and the only compounding pharmacy that can supply may be a pharmacy, which deals with alternative medicine.

8.11.1 Pyrroloquinoline Quinone, Another CoQ10 for Neuroprotection

Pyrroloquinoline quinone (PQQ) is a B vitamin-like cofactor that occurs naturally in plant foods and in most organs and tissues. The most important effects of PQQ is the growth of new mitochondria otherwise known as mitochondrial biogenesis [169]. Over 15 years of research has underscored the unique properties of PQQ in maintaining neuronal integrity. The diverse cellular mechanisms of PQQ distinguish this powerful quinone antioxidant from other neuroprotective dietary supplements. The effects of PQQ on neuronal function partially stem from its influence on mitochondrial homeostasis. While many nutritional supplements support mitochondrial bioenergetics by acting as metabolic cofactors, where PQQ generates *de novo* synthesis of mitochondria by the activation of the transcription factor CREB (cAMP response element binding protein) and the genomic coactivator PGC1α (PPARγ coactivator 1α) [170–172]PGC1α promotes the expression of a network of genes involved in the synthesis of mitochondria [169]. Thus, the mitochondrial effects of PQQ complement are similar to the well-known properties of CoQ10. While CoQ10 plays a bioenergetic role within the mitochondrial membrane, PQQ supports the coordinated process through which new mitochondria are generated (Fig. 8.55).

PQQ also supports neuronal health through intracellular targets that are distinct from CoQ10 and are largely unprecedented in the realm of nutritional supplementation. For example, PQQ maintains healthy levels of DJ-1, a protein that plays a role in antioxidant defenses in dopaminergic neurons. PQQ also supports the biosynthesis of nerve growth factor (NGF), an important signaling molecule in the brain and sensory nerves [170–174]. Perhaps the most salient discovery over the past two decades of research on PQQ and neuronal function is its ability to balance activation of ionotropic glutamate receptors. The N-methyl D-aspartate (NMDA) receptor is the primary conduit for excitatory glutamate neurotransmission and is regarded as an important target for maintaining neuronal health. PQQ supports synaptic and intracellular responses by maintaining healthy NMDA receptor activity [175]. Recent studies indicate that PQQ supports cognitive function. In one study involving rats, PQQ supplementation supported performance in a water maze test [176]. In a randomized, double blind study of 71 middle aged

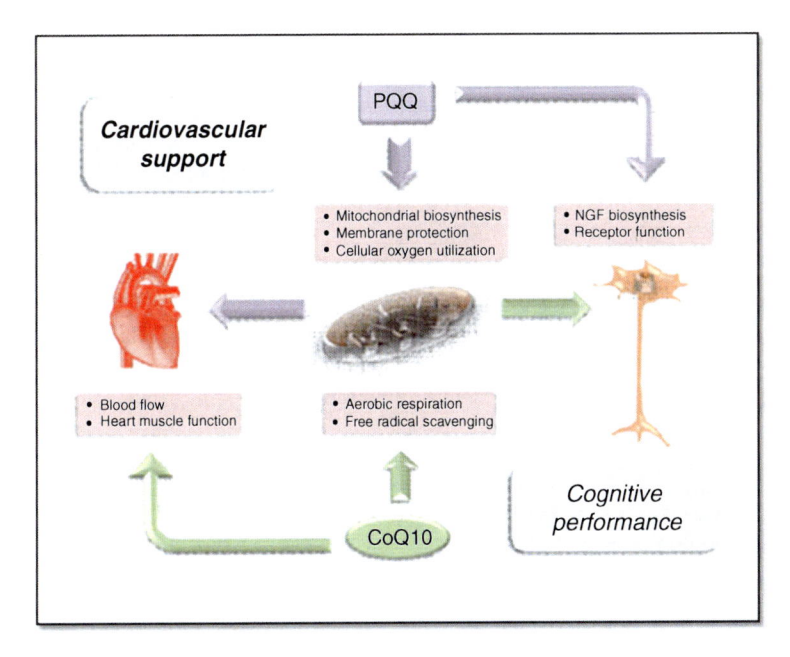

8.11 New Avenues in Treatment of Mitochondropathy

Fig. 8.55 Sites of action of PQQ within the neuronal network and the generation of fresh mitochondria

individuals, PQQ supplementation over 12 weeks promoted mental processing. In the same study, a combination of PQQ and CoQ10 was more effective than either agent alone [177]. PQQ offers wide range neural and cognitive support, including antioxidant activity, mitochondrial synthesis and nerve cell receptor function. While additional clinical investigations are necessary, existing evidence suggests that PQQ provides powerful multifaceted neuroprotection.

8.11.2 The Positive Effects of Saturated Fats

The much-maligned saturated fats are not the cause of modern diseases. They play in the chemical processes in the human body, many important roles: Saturated fatty acids constitute at least 50% of the cell membranes. They give our cells necessary stiffness and strength; thus,

1. They play a vital role in the health of our bones. For calcium to be effectively integrated into the formation of the skeleton must be at least 50% of dietary fats, saturated fats.
2. They lower lipoprotein alpha, a substance in the blood, which indicates a tendency to heart disease. They protect the liver against the effects of alcohol and other toxic substances such as paracetamol (Tylenol®).
3. They support the immune system.

4. They are necessary for the proper utilization of essential fatty acids in the body. Elongated Ω-3 fatty acids are better retained in the tissues when the diet of saturated fats.
5. Saturated 18 – carbon – and 16-carbon stearic acid – palmitic acid are the preferred foods for the heart. Therefore, the fat around the heart muscle is highly saturated. The heart draws in stressful situations on these reserves.
6. Short and medium-chain saturated fatty acids have important antimicrobial properties. They protect us against harmful microorganisms in the digestive tract.
7. Honestly evaluated study data revealed no confirmation of the claim that saturated fats lead to clogging of arteries and heart disease. In fact, investigations revealed the fat in arterial plaques that are in these, only 26% of fat, saturated fatty acids. The rest is unsaturated, of which more times than half.

8.11.3 Cholesterol, the Culprit for Arteriosclероris?

Our blood vessels can become damaged in various ways – for example through the action of free radicals or because they are structurally weak – and when that happens, then the body begins to repair, till the damage resolves by using a substance called cholesterol. Cholesterol is a high molecular weight, heavy alcohol that is produced in the body in the liver and most body cells. Like saturated fats, including cholesterol, which is produced in the body or absorbed from food plays a series of vital roles:

1. Along with saturated fats, cholesterol makes the membranes of cells necessary stiffness and strength. When the diet contains an excess of polyunsaturated fatty acids, these replace saturated fatty acids in the cell membrane, making the cell walls flaccid.
2. When this happens, cholesterol from the blood into the tissues to give them structural integrity. That is the reason that the serum cholesterol levels may go down temporarily when we replace saturated fats in the diet with polyunsaturated oils.
3. Cholesterol is a precursor to vital corticosteroids, hormones that help to protect against with stress and protect against heart disease and cancer. They are also the sex hormones like androgen, testosterone, estrogen and progesterone.
4. Cholesterol is a precursor of vitamin D, a fat-soluble vitamin, which is essential for healthy bones and nervous system, proper growth, mineral metabolism, muscle tone, insulin production, reproduction and immune function.
5. The bile salts are made from cholesterol. They are irreplaceable in the digestion and assimilation of fats in the diet.
6. New research shows that cholesterol acts as an antioxidant. This is a plausible explanation for the fact that cholesterol levels increase with age. As an antioxidant, cholesterol protects us against free radical damage that leads to heart disease and cancer.

8.11 New Avenues in Treatment of Mitochondropathy

7. Cholesterol is necessary for the proper function of serotonin receptors in the brain. Serotonin is the natural "feel-good" chemical in the body. Low cholesterol levels are associated with a bad temper and aggressive behavior, depression and tendency to commit suicide together.
8. Breast milk contains extremely high amounts of cholesterol and a special enzyme that helps the baby to absorb this nutrient. Babies and children need cholesterol-rich foods throughout their growing years to ensure proper development of the brain and nervous system.
9. Dietary cholesterol plays an important role in maintaining the health of the stomach and intestinal walls. That is the reason why low-cholesterol vegetarian diets can lead to leaky gut syndrome and other intestinal problems.
10. Cholesterol is not the cause of heart disease but rather a potent antioxidant against free radicals in blood, and a repair material for repairing damage in the arteries (though), which include arterial plaques, as such, very little cholesterol.
11. However, cholesterol can – like the fat – are damaged by heat and the access of oxygen. This damaged or oxidized cholesterol seems to both injury in the arteries feeding do than to promote the pathological buildup of plaque in the vessel walls. Damaged cholesterol is found in powdered eggs, powdered milk (skim milk to the substance of the increase) was added and in meat and fat, which has been heated to high temperatures in frying and other high-temperature processes.
12. High cholesterol levels are often an indication that the body needs cholesterol to protect against high levels of altered, free-radical-containing fats. Just as in a place with a high crime rate, an increased need for police there, so in an elevated amount of cholesterol is in a malnourished or improperly nourished body is needed to protect them from the risk of cardiovascular disease and cancer.
13. A thyroid – hypothyroidism is often expressed in an elevated cholesterol level. When thyroid function is low – generally because of high sugar consumption and a lack of usable iodine, fat-soluble vitamins and other nutrients – your body shakes as an adaptive and protective measure cholesterol in the blood and thus provides a sufficient amount of materials needed to heal tissues and to produce protective steroids.

The cause of heart disease is not the consumption of animal fat and cholesterol into consideration, but a series of factors arising from the so-called "modern" diet, including excess consumption of vegetable oils and hydrogenated fats, excess consumption carbohydrate in the form of sugar and white flour, mineral deficiencies (especially magnesium and iodine), a vitamin deficiency (especially vitamin C which is necessary for the maintenance of the vessel walls), and a lack of antioxidants like selenium and vitamin E to protect against free radicals, and finally, the disappearance of saturated fats in the diet (in the form of animal fats and tropical oils).

The modern process of oil extraction is purely insane.

It is important to realize that of all human food, especially polyunsaturated oils are the ones who most easily are at danger by food processing, especially the chemical unstable Ω-3-linolenic acid. One has to keep in mind that the naturally occurring fatty acids have go through an industrial processes before they appear on the consumer's table:

1. **Extraction:** Oils in the naturally occurring fruits, grains and nuts, must first be extracted. In ancient times this was done by slow-moving stone presses. But modern, large mills do this by crushing and heating the oil-bearing plant parts to about 110°C. Then the oil is squeezed out at pressures of up to 1 ton per square inch, resulting in additional heating of the material. During this process the oils are exposed to damaging light and oxygen. In order to extract the remaining crushed the last (approximately) 10% of the remaining oil content, the pulp is mixed with various solvents. In the usually hexane. The solvent is then boiled off, although up to 100 ppm (parts per million may remain) in the oil. Such solvents that are toxic in themselves retain traces of toxic pesticides in the oil, which were contained before processing begins in the seeds or fruits.
2. **Heating.** Due to the high processing temperatures to break apart the fragile bonds between the carbon atoms in the unsaturated fatty acids, especially triple incurred as unsaturated linolenic acid, thereby creating dangerous free radicals. In addition, antioxidants such as fat-soluble vitamin E, which protect the body from the ravages of free radicals, are neutralized and destroyed by high temperatures or pressures. To this vitamin E and others, through the manufacturing process to replace destroyed natural preservative is often added to the oil BHT and BHA, two artificial substances, which in turn are suspected of causing cancer and brain damage.
3. **Storage.** There is a safe modern method of extraction, in which the grains 'drilled' and its oil content, with its precious cargo of antioxidants under low temperature and minimal exposure to light and oxygen. This expeller – pressed, unrefined oil is cooled and bottled in dark bottles for a long time preserved. Crushing olives between stone or steel rollers produces extra virgin olive oil. The process runs smoothly and keeps the fat and the various conservative elements of the oil intact. If olive oil is stored in opaque containers, it is its freshness and precious antioxidants for many years.
4. **Hydrogenation:** This is the process that is liquid at room temperature from the polyunsaturated oils makes a solid fat at room temperature, ie margarine and cooking fats. Producers should use the cheapest oils-soy, corn, sunflower, cotton seeds, canola, are already rancid from the extraction. These oils and mix them with tiny metal particles-usually nickel oxide. The oil with its nickel catalyst is then exposed to a high-pressure and high temperature reactor in a hydrogen atmosphere. Next, soap-emulsifiers and starch are squeezed into the mixture to increase the consistency. In the subsequent steam cleaning the oil a second time will be exposed to a high temperature. This cleaning removes the oil and the bad smell. The resulting margarine has an unsightly gray color will bleach. However, still dyes are needed and strong flavors are added to make it

resemble butter. Finally, it is pressed into blocks or pressed into container to be sold as an alleged "healthy" food. Partially hydrogenated margarines and shortenings are even worse for you than the highly refined vegetable oils from which they are produced, because during the hydrogenation process chemical changes that occur. A shortening is a semisolid fat in food preparations especially baked goods, and is so called because it promotes a short or crumbly texture. The nickel catalyst effected under the influence of high temperatures (250–350°C) changes in the position of hydrogen atoms in the fatty acid chains.

5. **Transformation.** Before hydrogenation, the hydrogen atoms occur in pairs in the fatty acid chain, causing the chain to bend slightly, creating a concentration of electrons at the double bond. This is called the cis-formation, the majority occurring in nature in this configuration. In the hydrogenation one of the hydrogen atoms is shifted to the other side of the chain, which stretches the molecule. This is called the Transform, which is a very rare configuration in nature. Most of these man-made trans fats (trans-form of the molecules) are toxic to the body, but unfortunately our digestive system does not recognize them as such. Instead of being eliminated, the intestinal system absorbs them into the cells as if they were the natural cis-fatty acids. However, when taken to the interior of the cell, they cannot be processed because of the misplaced hydrogen atom, and because the cell mechanisms are not prepared for such an unnatural atomic arrangement. The blocked cells now cannot perform any other tasks no more. They must be dismantled and disposed off by the immune system.

6. **Homogenization:** This is the process where the product with its fat particles of cream is pressed at high pressure through fine pores. The resulting fat particles are so tiny that they float in the milk, rather than sell at the surface. This makes fat and cholesterol more oxidizable and they become easily rancid; some studies even suggest that homogenized fats contribute to the rise in heart disease. The constant attack of the media against saturated fats is extremely suspect. The claim that butter has increased cholesterol levels, however, this can not be substantiated by research, although some studies showed a slight, temporary increase through butter, while other studies have cholesterol lowering by stearic acid, the main component of beef fat, actually. Margarine results in chronically high cholesterol (!) and is also significantly associated with heart disease and cancers. Even the new "soft" margarine developments do not change anything, because they are made from rancid, oxidized vegetable oils and also include more artificial additives

7. In the 1940s, researchers found a strong correlation between cancer and the consumption of fat. The fats used were hydrogenated fats although the results were presented as if saturated fats were the culprit. And until recently in the United States, saturated fats were lumped together with trans fats when it came to understand the link between dietary habits and disease. Thus, natural saturated fats have been drawn into the vortex of the negative brush of unnatural hydrogenated vegetable oils (=trans fats).

8. Altered partially **hydrogenated vegetable oils** in the body cells block the intake of essential fatty acids and are therefore the cause of many health problems

including sexual dysfunction, elevated blood cholesterol and paralysis of the immune system. The intake of trans fats is associated with a host of other serious diseases, not only cancer but also atherosclerosis, diabetes, obesity, immune system dysfunction, underweight babies, birth defects, visual acuity, sterility, decreased ability to breastfeed, problems with muscles and tendons. Nevertheless, hydrogenated (hardened) fats continue to be healthy for food advocates. The popularity of such partially (hydrogenated) margarine over butter reflects a triumph of advertising over common sense. The best defense against such nightmare is to take these products, like a disease, of the road.

9. Contrary, **butter** contains the following elements: Fat-soluble vitamins are: That includes natural vitamin A or retinol, vitamin D, vitamin C and vitamin E as well as their naturally occurring cofactors needed to obtain maximum effect. Vitamin A is added actually easier from the butter from the body and as recovered from other sources. Fortunately, these vitamins are relatively stable and survive the pasteurization process. When Dr. Weston Price visited isolated traditional communities around the world, he discovered that butter was a staple in their diets. He did not find any ethnic group, which consumed polyunsaturated oils. The groups, whose way of life he studied particularly valued the deep yellow butter from grazing milk cows fed with fast-growing grass. Their natural intuition led them to the positive qualities being especially beneficial for expectant mothers and infants. When Dr. Price analyzed this deep yellow butter he found it an unusually high doses of fat-soluble vitamins, particularly vitamin A. He called these vitamins "catalysts" or "activators," without which the body according to his conviction is not in a position to absorb minerals no matter how many we are feeding him with food. He was even the view that fat-soluble vitamins necessary for the absorption of water-soluble vitamins are.

10. **Vitamins A and D** are essential for growth, for healthy bones, for the function of the brain, the nervous system and the function of the reproductive organs. Many studies have shown the importance of butterfat for reproduction. Its absence results in "nutritional castration," the inability to train male and female sexual characteristics. To the extent to which decreased consumption of butter in the U.S., increased rates of infertility and sexual problems. If you replace the butter fat in calves, they are no longer are able to grow and to reproduction is no longer maintained. The only good source of fat-soluble vitamins in the diet is butter. Butter added to vegetables or spread on bread, and cream added to soups and sauces, ensures proper digestion of vitamins, minerals and water from grain products, vegetables and meat.

11. The **Wulzen Factor**: The "anti-stiffness factor" a phenomenon, which has been known to occur in raw animal fat. Researcher Rosalind Wulzen discovered that this substance protects humans and animals from calcification of the joints – degenerative arthritis. It also protects against hardening of the arteries, cataracts and calcification of the pineal gland. Calves that were fed on pasteurized milk or skim milk develop joint stiffness and do not

thrive. These symptoms disappear when raw butterfat was added to their feed. The pasteurization process destroys this factor, which is only available in raw milk, butter and cream (cream).

12. The **Price Factor or Activator X:** The Discovered by Dr. Price, Activator X is a powerful catalyst which contributes such as vitamins A and D in the body to absorb minerals and exploit. Butter can be an especially rich source of Activator X when it comes from the milk from grazing cows that feed in spring and autumn of fast-growing grasses. It is not contained in the butter from the milk of cows with cottonseed cake, or are fed high with protein Sojafutter. The X-activator is fortunately not destroyed by pasteurization.

13. One frequently voiced objection to the consumption of butter and animal fats is that they tend to accumulate environmental poisons. Fat-soluble toxins such as DDT, accumulate in fat tissue actually, but water-soluble toxins such as antibiotics and growth hormones, accumulate in the watery part of milk and meat.

Vegetables and grains accumulate poisons. Crops in the growing season, from planting to harvesting are treated with chemical poisons, while cows generally graze on land which was not sprayed.

The assumption is correct that all of our foods, whether they are plants or of animal origin, may be contaminated. The solution to this problem does not consist in the exclusion of animal fats since overall health is so important, but in the search for organic meat from animals raised accordingly and the use of butter from free-grazing cows, followed by a non-chemically grown vegetables, fruits and cereals. You can get these products in specialty shops for organic farming, in special sections in supermarkets or sold directly from the farmers. In short, the choice of fats, which are included in our diet is of extremely important. Most people benefit more health with more than less fat, and this is especially true for young children and adolescents. But that what we consider as fat, must be carefully selected. We all should avoid polyunsaturated oils and hydrogenated (hardened) vegetable oils (vegetable oils) each containing such industrially prefabricated products, oils or fats.

On the other hand, use traditional products such as cold pressed (or virgin) olive oil and small amounts of unrefined flax seed oil. Get used to the benefits of coconut oil for baking and animal fats for occasional frying. Eat egg yolks (only valuable in the raw state). And finally, do not eat so much of good butter as you like, with the lucky feeling that it is a positive holistic food for you and your entire family. Butter organic origin, cold pressed (extra virgin) olive oil and flax oil in opaque containers are available in health food stores and gourmet shops. Edible Coconut Oil is available in stores that sell Indian or Caribbean cuisine as well as in health food stores.

References

1. Kane E, Kane P (2005) Phosphatidcholine-Life's designer molecule. In: BodyBio Bulletin. Milville/New York, pp 1–6
2. DesMaisons K (2000) The sugar addict's total recovery program. Ballantine Books, New York
3. Blass E, Fitzgerald E, Kehoe P (1987) Interactions between sucrose, pain and isolation distress. Pharmacol Biochem Behav 26:483–489
4. Avena NM, Rada P, Hoebel BG (2008) Evidence for sugar addiction: behavioral and neurochemical effects of intermittent, excessive sugar intake. Neurosci Biobehav Rev 32:20–39
5. Abbas S et al (2008) Serum 25-hydroxyvitamin D and risk of post-menopausal breast cancer–results of a large case-control study. Carcinogenesis 29:93–99
6. Kofke WA et al (2004) The effect of apolipoprotein E genotype on neuron specific enolase and S-100β levels after cardiac surgery. Anesth Analg 99:1323–1325
7. Levi B, Werman MJ (1998) Long-term fructose consumption accelerates glycation and several age-related variables in male rats. J Nutr 128:1442–1449
8. Kohli R et al (2011) High-fructose medium-chain-trans-fat diet induces liver fibrosis & elevates plasma coenzyme Q9 in a novel murine model of obesity and nonalcohol steatohepatitis. Hepatology 52:934–944
9. Seneff A et al (2011) Nutrition and Alzheimer's disease: the detrimental role of a high carbohydrate diet. Eur J Int Med 22:134–140
10. Noh HS, Kim YS, Choi WS (2008) Neuroprotective effects of the ketogenic diet. Epilepsia 49:120–123
11. Henderson ST (2008) Ketone bodies as a therapeutic for Alzheimer's disease. Neurotherapeutics 5:470–480
12. Henderson ST et al (2010) Ketone bodies as a therapeutic for Alzheimer's disease. In: 25th international conference of Alzheimer's Disease International (ADI), Athens
13. Newport MT (2010) Caregivers reports following dietary intervention with medium chain fatty acids in 47 persons with dementia. In: 25th international conference of Alzheimer's Disease International (ADI), Athens
14. Newport MT (2011) Alzheimers: What if there was a cure the story of ketones. Basic Health Publications. Laguna Beach/CA (in print)
15. Reger MA et al (2004) Effects of beta-hydroxybutyrate on cognition in memory-impaired adults. Neurobiol Aging 25:311–314
16. Slater P, Longman DA (1979) Effects of diazepam and muscimol on GABA-mediated neurotransmission: interactions with inosine and nicotinamide. Life Sci 25:1963–1967
17. Piper J et al (1998) Mechanisms of anticarcinogenic properties of curcumin: the effect of curcumin on glutathione linked detoxification enzymes in rat liver. Int J Biochem Cell Biol 39:445–456
18. Weissenberger J et al (2010) Dietary curcumin attenuates glioma growth in a syngeneic mouse model by inhibition of the JAK1,2/STAT3 signaling pathway. Clin Cancer Res 16:5781–5795
19. Lev-Ari S et al (2005) Celecoxib and curcumin synergistically inhibit the growth of colorectal cancer cells. Clin Cancer Res 11:6738–6744
20. Nihal M et al (2005) Antiproliferative and proapoptotic effects of (−)-epigallocatechin-3-gallate on human melanoma: possible implications for the chemoprevention of melanoma. Int J Cancer 114:513–521
21. Goodson AG et al (2009) Use of oral N-acetylcysteine for protection of melanocytic nevi against UV-induced oxidative stress: towards a novel paradigm for melanoma chemoprevention. Clin Cancer Res 15:OF1–OF7
22. Ghafourifar P et al (1999) Mitochondrial nitric oxide synthase stimulation causes cytochrome c release from isolated mitochondria. Evidence for intramitochondrial peroxynitrite formation. J Biol Chem 274:1185–1188

23. Gröber U (2010) Vitamin D–an old vitamin in a new perspective [Article in German]. Med Monatsschr Pharm 33:376–383
24. Allan SM, Rothwell NJ (2003) Inflammation in central nervous system injury. Philos Trans R Soc Lond B Biol Sci 358:1669–1677
25. Nickerson M et al (1979) Isobutyl nitrite and related compounds. Pharmex, Ltd., San Francisco
26. Pall ML (2009) Multiple chemical sensitivity: toxicological questions and mechanisms. In: Ballantyne B, Marrs TC, Syversen T (eds) General and applied toxicology. Wiley, London
27. Cannell JJ et al (2006) Epidemic influenza and vitamin D. Epidemiol Infect 134:1129–1140
28. Chun JT, Santella L (2009) Roles of the actin-binding proteins in intracellular Ca2+ signalling. Acta Physiol Oxford 195:61–70
29. Freye E, Strobel HP, Levy JV (2004) New therapeutic approach in the management of fibromyalgia syndrome (FMS). Avicenna 3(2):2–10
30. Eisenberg E, Pud D (1998) Can patients with chronic neuropathic pain be cured by acute administration of the NMDA receptor antagonist amantadine? Pain 74:37–39
31. Parsons CG (2001) NMDA receptors as targets for drug action in neuropathic pain. Eur J Pharmacol 429:71–78
32. Woolf CJ, Mannion RJ (1999) Neuropathic pain: aetiology, symptoms, mechanisms, and management. Lancet 353:1959–1964
33. Minovitz O, Driol M (1989) The sexual abuse, eating disorder and addiction (SEA) triad: syndrome or coincidence ? Med Law 8:59–61
34. Nettleton JA et al (2009) Diet soda intake and risk of incident metabolic syndrome and type 2 diabetes in the Multi-Ethnic Study of Atherosclerosis (MESA). Diabetes Care 32:688–694
35. O'Shaughnessy KM, Karet FE (2004) Salt handling and hypertension. J Clin Invest 113:1075–1081
36. Pottenger FMJ (1995) In: Pottenger EE, Pottenger RTJ (eds) Pottenger's cats: a study in nutrition. Price-Pottenger Nutrition Foundation, Lemon Grove/CA
37. Al-Delaimy WK et al (2004) Magnesium intake and risk of coronary heart disease among men. Am Coll Nutr 23:63–70
38. Klevay LM, Milne D (2002) Low dietary magnesium increases supraventricular ectopy. Am J Clin Nutr 75:550–554
39. Plant J (2007) Your life in your hands – understanding, preventing and overcoming breast cancer. Virgin Books, London
40. Zendman AJW, Vossenaar ER, van Venrooij WJW (2004) Autoantibodies to citrullinated (poly)peptides: a key diagnostic and prognostic marker for rheumatoid arthritis. Autoimmunity 37:295–299
41. Mimori T (2005) Clinical significance of anti-CCP antibodies in rheumatoid arthritis. Int Med 44:1122–1126
42. Eisinger J et al (1994) Glycolysis abnormalities in fibromyalgia. J Am Coll Nutr 13:144–148
43. Kang KA et al (2005) Eckol isolated from Ecklonia cava attenuates oxidative stress induced cell damage in lung fibroblast cells. FEBS Lett 579:6295–6304
44. Nicolson GL, Ellithorpe R (2006) Lipid replacement and antioxidant nutritional therapy for restoring mitochondrial function and reducing fatigue in chronic fatigue syndrome and other fatiguing illnesses. J Chron Fatigue Syndr 13:57–68
45. Katiyar S, Elmets CA, Katiya SK (2007) Green tea and skin cancer: photoimmunology, angiogenesis and DNA repair. J Nutr Biochem 18:287–296
46. Pyrko P et al (2007) The unfolded protein response regulator GRP78/BiP as a novel target for increasing chemosensitivity in malignant gliomas. Cancer Res 67:9809–9816
47. Aktas O, Waiczies S, Zipp F (2007) Neurodegeneration in autoimmune demyelination: recent mechanistic insights reveal novel therapeutic targets. J Neuroimmunol 184:17–26
48. Wildman REC (2001) Handbook of nutraceuticals and functional foods, 1st edn, Series in modern nutrition. CRS Press, Boca Raton/FL

49. Takayuki S et al (2008) In: A.C. Society (ed) Functional food and health, vol 993, ACS symposium series. ACS Publications, Washington, DC
50. Government U (2009) Overview of dietary supplements. Food and Drug Administration, Washington, DC
51. Health Canada (2010) http://www.hc-sc.gc.ca/sr-sr/biotech/about-apropos/gloss-eng.php
52. Hardy G (2000) Nutraceuticals and functional foods: introduction and meaning. Nutrition 16:688–689
53. Ou B, Hampsch-Woodill M, Prior R (2001) Development and validation of an improved oxygen radical absorbance capacity assay using fluorescein as the fluorescent probe. J Agric Food Chem 49:4619–4626
54. Laboratory ND et al (2007) Oxygen Radical Absorbance Capacity (ORAC) of selected foods – 2007. U.S. Department of Agriculture (USDA), Beltsville
55. Williams RJ, Spencer JP, Rice-Evans C (2004) Flavonoids: antioxidants or signalling molecules? Free Radic Biol Med 36:838–849
56. Arts IC, Hollman PC (2005) Polyphenols and disease risk in epidemiologic studies. Am J Clin Nutr 81:317S–325S
57. Lotito SB, Frei B (2006) Consumption of flavonoid-rich foods and increased plasma antioxidant capacity in humans: cause, consequence, or epiphenomenon? Free Radic Biol Med 41:1727–1746
58. Nutrition Data Laboratory (eds) (2010) USDA Database for the oxygen radical absorbance capacity (ORAC) of selected foods, Release 2. U.S. Department of Agriculture. Agriculture Research Service Beltsville Human Nutrition Research Center Nutritient Data Laboratory, Beltsville/MD
59. Lacza Z et al (2001) Mitochondrial nitric oxide synthase is constitutively active and is functionally upregulated in hypoxia. Free Radic Biol Med 31:1609–1615
60. Sparagna GC, Lesnefsky EJ (2009) Cardiolipin remodeling in the heart. J Cardiovasc Pharmacol 53:290–301
61. Orrenius S, Zhivotovsky B (2005) Cardiolipin oxidation sets cytochrome c free. Nat Chem Biol 1:188–189
62. Milliken EL et al (2005) EB1089, a vitamin D receptor agonist, reduces proliferation and decreases tumor growth rate in a mouse model of hormone-induced mammary cancer. Cancer Lett 229:205–215
63. Wang TJ et al (2008) Vitamin D deficiency and risk of cardiovascular disease. Circulation 117:453–455
64. Cordain L (2002) The paleo diet: lose weight and get healthy by eating the food you were designed to eat. Wiley, New York
65. Hoeck AD, Pall ML (2011) Will vitamin D supplementation ameliorate diseases characterized by chronic inflammation and fatigue? Med Hypotheses 76:208–213
66. Rivat C et al (2008) Polyamine deficient diet to relieve pain hypersensitivity. Pain 137:125–137
67. Shinohara M et al (1998) Overexpression of glyoxalase-I in bovine endothelial cells inhibits intracellular advanced glycation endproduct formation and prevents hyperglycemia-induced increases in macromolecular endocytosis. J Clin Invest 101:1142–1147
68. Du X et al (2003) Inhibition of GAPDH activity by poly(ADP-ribose) polymerase activates three major pathways of hyperglycemic damage in endothelial cells. J Clin Invest 112:1049–1057
69. Johann S et al (1993) Influence of electromagnetic fields on morphology and mitochondrial activity of breast cancer cell line MCF7. Bioelectrochem Bioenerg 30:127–132
70. Garland CF et al (2011) Vitamin D supplement doses and serum 25-hydroxyvitamin D in the range associated with cancer prevention. Anticancer Res 31:617–622
71. Guillot X et al (2010) Vitamin D and inflammation. Joint Bone Spine 77:552–557
72. Qiao Z, Chen R (1991) Isolation and identification of antibiotic constituents of propolis from Henan (in Chinese). Zhongguo Zhong Yao Za Zhi 16:481–482
73. Salomão K et al (2009) Brazilian green propolis: effects in vitro and in vivo on trypanosoma cruz. Evid Based Complement Altern Med 2011:1–11

74. da Silva FB, Almeida JM, Sousa SM (2004) Natural medicaments in endodontics - a comparative study of the anti-inflammatory action. Braz Oral Res 18:174–179
75. Park YK et al (1998) Antimicrobial activity of propolis on oral microorganisms. Curr Microbiol 36:24–28
76. Kunimasa K et al (2009) Brazilian propolis suppresses angiogenesis by inducing apoptosis in tube-forming endothelial cells through inactivation of survival signal ERK1/2. Evid Based Complement Altern Med 2011:12–19
77. Brätter C et al (1999) Prophylactic effectiveness of propolis for immunostimulation: a clinical pilot study. Forsch Komplementärmed 6:256–260
78. Belenky P, Bogan KL, Brenner C (2007) NAD⁺ metabolism in health and disease. Trends Biochem Sci 32:12–19
79. Kaneko S et al (2006) Protecting axonal degeneration by increasing nicotinamide adenine dinucleotide levels in experimental autoimmune encephalomyelitis models. J Neurosci Res 26:9794–9804
80. Birkmayer GD (2004) Nicotinamide adenine dinucleotide (NADH)–a new therapeutic approach to Parkinson's disease. Drugs Exp Clin Res 30:27–33
81. Swerdlow RH (1998) Is NADH effective in the treatment of Parkinson's disease? Drugs Aging 13:263–268
82. Birkmayer W, Birkmayer JGD (1989) Nicotinamidadenindinucleotide (NADH): the new approach in the therapy of Parkinson's disease. Ann Clin Lab Sci 18:38–43
83. Birkmayer JGD, Vank P (1996) Reduced coenzyme 1 (NADH) improves pyschomotoric and physical performance in athletes. In: White paper report. Menuco Corp, New York
84. Demarin V et al (2004) Treatment of Alzheimer's disease with stabilized oral nicotinamide adenine dinucleotide: a randomized, double-blind study. Drugs Exp Clin Res 30:27–33
85. Rongvaux A et al (2003) Reconstructing eukaryotic NAD metabolism. Bioessays 25:683–690
86. Bieganowski P, Brenner C (2004) Discoveries of nicotinamide riboside as a nutrient and conserved NRK genes establish a Preiss-Handler independent route to NAD⁺ in fungi and humans. Cell 117:495–502
87. Smeitink JAM et al (2004) Cell biological consequences of mitochondrial NADH: ubiquinone oxidoreductase deficiency. Curr Neurovasc Res 1:29–40
88. Nakamaru-Ogiso E et al (2010) The ND2 subunit is labeled by a photoaffinity analogue of asimicin, a potent complex I inhibitor. FEBS Lett 584:883–888
89. Eaton S et al (2002) Control of mitochondrial β-oxidation at the levels of [NAD⁺]/[NADH] and CoA acylation. Adv Expert Med Biol 466:145–154
90. Sharma RA et al (2004) Phase I clinical trial of oral curcumin: biomarkers of systemic activity and compliance. Clin Cancer Res 10:6847–6854
91. Herzenberg LA et al (1997) Glutathione deficiency is associated with impaired survival in HIV disease. Proc Natl Acad Sci USA 94:1967–1972
92. Crofford LJ et al (2005) Pregabalin for the treatment of fibromyalgia syndrome: results of a randomized, double-blind, placebo-controlled trial. Arthritis Rheum 52:1264–1273
93. Cabanillas F (2010) Vitamin C and cancer: what can we conclude–1,609 patients and 33 years later? Puerto Rico Health Sci J 29:215–217
94. Pall ML (2001) Cobalamin used in chronic fatigue syndrome therapy is a nitric oxide scavenger. J Chron Fatigue Syndr 8:39–45
95. See reference [94]
96. Ellis FR, Nasser S (1973) A pilot study of vitamin B12 in the treatment of tiredness. Br J Nutr 30:277–283
97. van der Kuy PH et al (2002) Hydroxocobalamin, a nitric oxide scavenger, in the prophylaxis of migraine: an open, pilot study. Cephalalgia 22:513–519
98. Bauman WA et al (2000) Increased intake of calcium reverses vitamin B12 malabsorption induced by metformin. Diabetes Care 23:1227–1231
99. Pall ML (2003) Elevated nitric oxide/peroxynitrite theory of multiple chemical sensitivity: central role of N-methyl-D-aspartate receptors in the sensitivity mechanism. Environ Health Perspect 111:1461–1464

100. Moat SJ et al (2006) Folic acid reverses endothelial dysfunction induced by inhibition of tetrahydrobiopterin biosynthesis. Eur J Pharmacol 530:250–258
101. Baker TA, Milstien S, Katusic ZS (2001) Effect of vitamin C on the availability of tetrahydrobiopterin in human endothelial cells. J Cardiovasc Pharmacol 37:333–338
102. d'Uscio LV et al (2003) Long-term vitamin C treatment increases vascular tetrahydrobiopterin levels and nitric oxide synthase activity. Circ Res 92:88–95
103. Forsyth LM et al (2000) Therapeutic effects of oral NADH on the symptoms of patients with chronic fatigue syndrome. Ann Allergy Asthma Immunol 84:639–640
104. Birkmayer W, Birkmayer JGD (1989) Nicotinamidadenindinucleotide (NADH) and Nicotinamidadenindinucleotidephosphate (NADPH). Acta Neurol Scand 126:183–187
105. Birkmayer W et al (1990) Nicotinamide adenine dinucleotide (NADH) as medication for Parkinson's disease. Experience with 415 patients. New Trends Clin Neuropharmacol 4:7–24
106. Lloyd KG, Davidson L, Hornykiewicz O (1975) The neurochemistry of Parkinson's disease: effect of L-DOPA therapy. J Pharmacol Exp Ther 195:453–464
107. Riederer P et al (1978) CNS modulation and adrenal tyrosine hydroxylase in Parkinson's disease and metabolic encephalopathies. Neural J Transm 14:121–131
108. Ames M, Lerner P, Lovenberg W (1978) Tyrosine hydroxylase activation by protein phosphorylation and end product inhibition. J Biol Chem 253:27–31
109. Nagatsu T, Levitt M, Udenfriend S (1964) Tyrosine hydroxylase: the initial step in norepinephrine synthesis. J Biol Chem 239:2910–2917
110. Nagatsu T et al (1982) Biopterine in human blood and urine from controls and Parkinsonian patients: application of a new radioimmunoassay. Clin Chim Acta 109:305–311
111. Nichol CA, Smith GK, Duch DS (1985) Biosynthesis and metabolism of tetrahydrobiopterin and molybdopterin. Ann Rev Biochem 54:729–764
112. Blair J et al (1984) Etiology of Parkinson's disease. Lancet 1:167–170
113. Langston JW et al (1984) Pargyline prevents MPTP-induced parkinsonism in primates. Science 225:1480–1482
114. Langston JW et al (1983) Chronic parkinsonism in humans due to a product of meperidine-analog synthesis. Science 219:979–980
115. Serra PA et al (2000) Manganese increases L-DOPA auto-oxidation in the striatum of the freely moving rat: potential implications to L-DOPA long-term therapy of Parkinson's disease. Br J Pharmacol 130:937–945
116. Hiatt HH (1957) Glycogen formation via the pentose phosphate pathway in mice in vivo. J Biol Chem 224:851–859
117. Segal S, Foley JT (1958) The metabolism of D-ribose in man. J Clin Invest 37:719–735
118. Bloom B, Eisenberg FJ, Stetten DJ (1955) Glucose catabolism in liver slices via the phosphogluconate oxidation pathway. J Biol Chem 215:461–466
119. Pasque MK, Wechsler A (1984) Metabolic intervention to affect myocardial recovery following ischemia. Ann Surg 200:1–10
120. Coffey RG, Morse H, Newburgh RW (1965) The synthesis of nucleic acid constituents in the early chick embryo. Biochim Biophys Acta 114:547–558
121. Zimmer H-G, Schad J (1984) Ribose intervention in the cardiac pentose phosphate pathway is not species-specific. Science 223:712–713
122. Zimmer H-G (1996) Regulation of and intervention into the oxidative pentose phosphate pathway and adenine nucleotide metabolism in the heart. Mol Cell Biochem 160/161:101–109
123. St Cyr JA et al (1989) Enhanced high energy phosphate recovery with ribose infusion after global myocardial ischemia in a canine model. J Surg Res 46:157–162
124. Benson ES, Evans GT, Hallaway BE (1961) Myocardial creatine phosphate and nucleotides in anoxic cardiac arrest and recovery. Am J Physiol 201:687–693
125. Lee HT, LaFaro RJ, Reed GE (1995) Pretreatment of human myocardium with adenosine during open heart surgery. J Card Surg 10:665–676
126. Jennings RB, Stanbergen CJ (1985) Nucleotide metabolism and cellular damage in myocardial ischemia. Ann Rev Physiol 47:727–749
127. Ward HB et al (1984) Recovery of adenine nucleotide levels after global myocardial ischemia in dogs. Surgery 96:248–255

128. Stathis CG et al (1994) Influence of sprint training on human muscle purine nucleotide metabolism. J Appl Physiol 76:1802–1809
129. Hellsten-Westing Y et al (1993) The effect of high-intensity training on purine metabolism in man. Acta Physiol Scand 149:405–412
130. Hellsten-Westing Y et al (1993) Decreased resting levels of adenine nucleotides in human skeletal muscle after high-intensity training. J Appl Physiol 74:2523–2528
131. Tullson PC, Terjung RL (1991) Adenine nucleotide synthesis in exercising and endurance-trained skeletal muscle. Am J Physiol 261:C342–C347
132. Tullson PC et al (1995) IMP metabolism in human skeletal muscle after exhaustive exercise. J Appl Physiol 78:146–152
133. Foker JE, Einzig S, Wang T (1980) Adenosine metabolism and myocardial preservation. J Thorac Cardiovasc Surg 80:506–516
134. Zimmer H-G, Gerlach E (1978) Stimulation of myocardial adenine nucleotide biosynthesis by pentoses and pentitols. Pflugers Arch 376:223–227
135. Zimmer H-G (1980) Restitution of myocardial adenine nucleotides: acceleration be administration of ribose. J Physiol Paris 76:769–775
136. Pasque MK et al (1982) Ribose-enhanced myocardial recovery following ischemia in the isolated working rat heart. Thorac Cardiovasc Surg 83:390–398
137. St Cyr JA et al (1986) Long term model for evaluation of myocardial metabolic recovery following global ischemia. In: Bratbar N (ed) Myocardial and skeletal muscle bioenergetics. Plenum, New York, pp 401–414
138. Chatham JC et al (1985) Studies of the protective effect of ribose in myocardial ischaemia by using 31P-nuclear-magnetic-resonance spectroscopy. Biochem Soc Trans 13:885–886
139. Zimmer H-G, Ibel H, Steinkopff G (1980) Studies on the hexose monophosphate shunt in the myocardium during development of hypertrophy. In: Tajuddin M et al (eds) Advances in myocardiology. University Park Press, Baltimore, pp 487–492
140. Zimmer H-G, Ibel H (1983) Effects of ribose on cardiac metabolism and function in isoproterenol-treated rats. Am J Physiol 245:H880–H886
141. Pliml W et al (1992) Effects of ribose on exercise-induced ischaemia in stable coronary artery disease. Lancet 340:507–510
142. Angello DA et al (1989) Recovery of myocardial function and thallium-201 redistribution using ribose. Am J Card Imaging 3:256–265
143. Angello DA, Wilson RA, Gee D (1988) Effect of ribose on thallium-201 myocardial redistribution. J Nucl Med 29:1943–1950
144. Perlmutter NS et al (1991) Ribose facilitates thallium-201 redistribution in patients with coronary artery disease. J Nucl Med 32:193–200
145. Hegewald MG et al (1991) Ribose infusion accelerates thallium redistribution with early imaging compared with late 24-hour imaging without ribose. J Am Coll Cardiol 18:1671–1681
146. Gradus-Pizlo I et al (1999) Effect of D-ribose on the detection of the hibernating myocardium during the low dose dobutamine stress echocardiography. Circulation 100(Suppl):3394
147. Pauli DF, Pepine CJ (2000) D-ribose as a supplement for cardiac energy metabolism. J Cardiovasc Pharmacol Ther 5:249–258
148. Teitelbaum JE, Bird B (1995) Effective treatment of severe chronic fatigue: a report of a series of 64 patients. J Musculoskelet Pain 3:91–110
149. Gebhart B, Jorgensen JA (2004) Benefit of ribose in a patient with fibromyalgia. Pharmacotherapy 24:1646–1648
150. See reference [147]
151. Omran H, St Cyr J, Lüderitz B (2004) D-Ribose aids congestive heart failure patients. Exp Clin Cardiol 9:117–118
152. Brault JJ, Terjung RL (1999) Purine salvage rates differ among skeletal muscle fiber types and are limited by ribose supply. Med Sci Sports Exerc 31:1365
153. Zarzeczny R et al (2000) Purine salvage is not reduced during recovery following intense contractions. Med Sci Sports Exerc 32(Suppl 5):214
154. Brault JJ, Terjung RL (2000) Attempted expansion of resting muscle ATP content by a prolonged period of adenine salvage. Med Sci Sports Exerc 32(5):213

155. Lund E et al (2003) Muscle metabolism in fibromyalgia studied by P-31 magnetic resonance spectroscopy during aerobic and anaerobic exercise. Scand J Rheumatol 32:138–145
156. Cotraqn R, Kumar V, Robbins SR (1989) Pathologic basis of disease. W. B. Saunders Company, Philadelphia
157. Mattson MP, Chan SL (2003) Calcium orchestrates apoptosis. Nat Cell Biol 5:1041–1043
158. Tinari A et al (2007) Mitoptosis: different pathways for mitochondrial execution. Autophagy 3:282–284
159. Prokopov A (2007) Theoretical paper: exploring overlooked natural mitochondria-rejuvenative intervention: the puzzle of bowhead whales and naked mole rats. Rejuvenation Res 10:543–560
160. Zhang X et al (2004) Enhancement of hypoxia-induced tumor cell death in vitro and radiation therapy in vivo by use of small interfering RNA targeted to hypoxia-inducible factor-1α. Cancer Res 64:8139–8142
161. Belozerov VE, Van Meir EG (2005) Hypoxia inducible factor-1: a novel target for cancer therapy. Anticancer Drugs 16:901–909
162. Ciulla MM et al (2007) Effects of simulated altitude (normobaric hypoxia) on cardiorespiratory parameters and circulating endothelial precursors in healthy subjects. Respir Res 8:58–63
163. Manukhina EB, Downey HF, Mallet RT (2006) Role of nitric oxide in cardiovascular adaptation to intermittent hypoxia. Expt Biol Med 231:343–365
164. Serebrovskaya TV (2002) Intermittent hypoxia research in the former soviet union and the commonwealth of independent States: history and review of the concept and selected applications. High Alt Med Biol 3:2005–2221
165. Huang LE et al (1998) Regulation of hypoxia-inducible factor 1alpha is mediated by an O2-dependent degradation domain via the ubiquitin-proteasome pathway. Proc Natl Acad Sci USA 95:7987–7992
166. See reference [163]
167. Gore CJ, Clark SA, Saunders PU (2007) Nonhematological mechanisms of improved sea-level performance after hypoxic exposure. Med Sci Sports Exerc 39:1600–1609
168. Mason P (1997) VI. Magnesium. In: I.o. Medicine (ed) Dietary reference intakes – calcium, phosphorus, magnesium, vitamin D, and fluoride. National Academy of Science, Patterson, pp 6–11
169. Chowanadisai W et al (2010) Pyrroloquinoline quinone stimulates mitochondrial biogenesis through cAMP response element-binding protein phosphorylation and increased PGC-1 alpha expression. J Biol Chem 285:142–152
170. Stites T et al (2006) Tchaparian. Pyrroloquinoline quinone modulates mitochondrial quantity and function in mice. J Nutr 136:390–396
171. Bauerly KA et al (2006) Pyrroloquinoline quinone nutritional status alters lysine metabolism and modulates mitochondrial DNA content in the mouse and rat. Biochim Biophys Acta 1760:1741–1748
172. Zhu BQ et al (2006) Comparison of pyrroloquinoline quinone and/or metoprolol on myocardial infarct size and mitochondrial damage in a rat model of ischemia/reperfusion injury. J Cardiovasc Pharmacol Ther 11:119–128
173. Yamaguchi K, Anderson JM (1992) Biocompatibility studies of naltrexone sustained release fomulations. J Control Release 19:299–314
174. Urakami T et al (1997) Synthesis of monoesters of pyrroloquinoline quinone and imidazopyrroloquinoline, and radical scavenging activities using electron spin resonance in vitro and pharmacological activity in vivo. J Nutr Sci Vitaminol Tokyo 43:19–33
175. Aizenman E et al (1992) Interaction of the putative essential nutrient pyrroloquinoline quinone with the N-methyl-D-aspartate receptor redox modulatory site. J Neurosci 12:2362–2369
176. Ohwada K et al (2008) Pyrroloquinoline quinone (PQQ) prevents cognitive deficit caused by oxidative stress in rats. J Clin Biochem Nutr 42:29–34
177. Nakano M et al (2009) Effect of pyrroloquinoline quinone (PQQ) on mental status of middle-aged and elderly persons. FOOD Style 21:50–53

Chapter 9
Lab Tests for Mitochondropathy, Nitrosative Stress

In order to diagnose a potential nitrosative stress and mitochrondriopathy as well as the underlying biochemical upregulated NO/OOHNO⁻ cycle, and before starting therapy, it is imperative to do:

1. A detailed anamnestic exploration is mandatory, which often leads to the exact diagnosis. In contrast to inborn mitochondropathy with clinically is a heterogeneous multisystem disease characterized by defects of brain–mitochondrial encephalopathies and/or muscle–mitochondrial myopathies due to alterations in the protein complexes of the electron transport chain of oxidative phosphorylation, acquired mitochondrial disease can be considered the main cause in a number of modern western ailments. This is often observed following previous long-term intake of antibiotics, long term to the exposure to xenobiotics, regular intake of medication resulting in NO-stress (see page 95), a past history of coxsaccae virus infection, or a past accident with head trauma (i.e. whip-lash syndrome, concussion of the head, etc). Patient complain of repetitive flu-like infections, showing signs of early prostration, loss of stamina, brain fog, depression, lack in concentration, regular sleep problems in addition to abdominal cramps or heart rate disturbances and/or retardation to cope with daily chores are suspicions for a low synthesis of energy within mitochondria and of mitochondropathy.

2. If in doubt, basic laboratory tests will lead to the correct diagnosis of ROS/NOS⁻induced mitochondriopathy:

    (a) Increased concentration of citrulline in the urine (normal < 100 µmol/g creatinine; Fig. 9.1)
    (b) Increase in methylmalonic acid (MMA) in urine, an indicator of insufficient B_{12}, folate levels (norm <1.6mg/g creatinne)
    (c) Increase in nitrotyrosine concentration in plasma (normal < 10 nmol/l, Fig. 9.2)
    (d) Low Q10 concentration in plasma (norm 0.9 µg/l). The lipid-corrected Q10 value is a more reliable parameter, since some of the Q10 is bound to cholesterol (norm >2.5 mg/l)

$$\text{Arginin} + O^2 \xrightarrow[\text{Enzyme: i,e,n,mNOS}]{} NO^* + \text{Citrullin}$$

Fig. 9.1 The significance of citrulline formation, which is excreted via the kidneys being an indicator for excess nitric oxide formation in patients

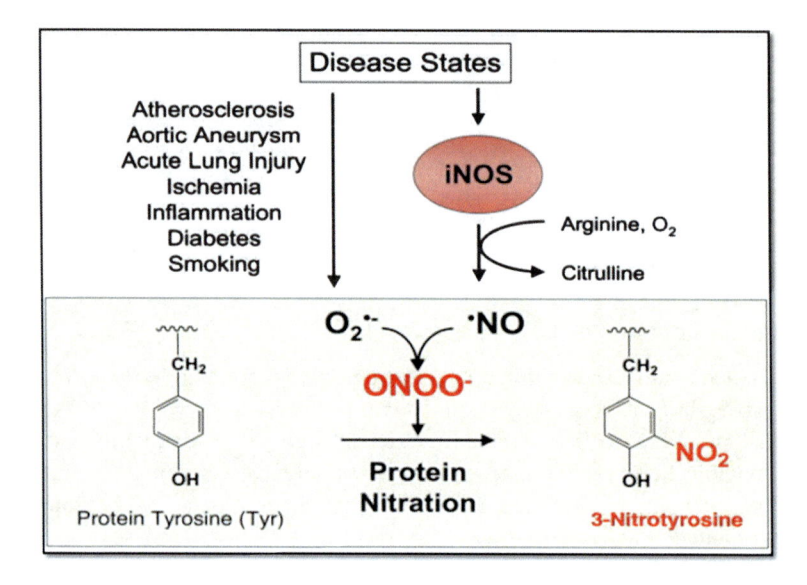

Fig. 9.2 Formation of peroxynitrite and protein nitration of tyrosine to nitrotyrosine a product, which is indicative for the formation of peroxynitrite in a patient

- (e) Lactate (norm 4.5–19.8 μmol/l) and pyruvate (norm 39–82 μmol/l) and the lactate/pyruvate ratio (norm <1:20) as an indicator of NO-induced pyruvate block within glucogenesis resulting in excess formation of lactate.
- (f) Increase in sedimentation rate as an indicator of chronic (silent) infections (norm 3 mm/h)
- (g) Eosinophily of lymphocytes (> 3%) as an indicator of high NO-levels
- (h) Vitamin D_3 levels (Serum 25-norm hydroxyvitamin D)>20 ng/ml
- (i) In suspected case of myopathy CK and CKMB levels (norm <170 IU/l)
- (j) Check for gut dysbiosis using a specific lab analysis for colonization
- (k) Check for possible food allergy.

9.1 The Extended Lab Tests

- (a) Serotonin level in plasma
- (b) Anti-CCP-Test for cyclic citrullinated peptides in plasma since excess levels of citrulline bind to peptides which act like antigens, resulting in an immune reaction with formation of antibodies; also considered as a test for rheumatic disease.

9.1 The Extended Lab Tests

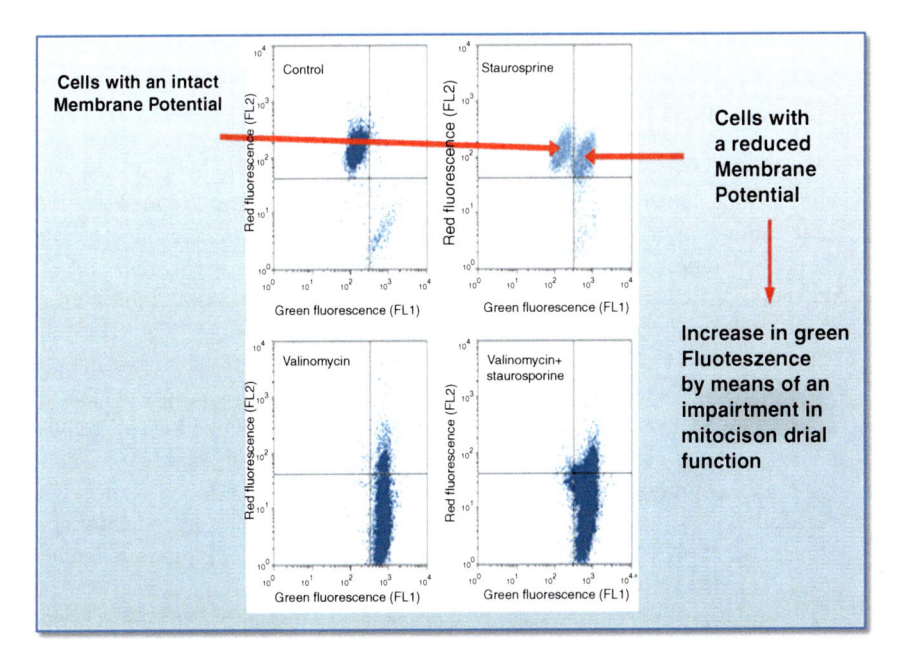

Fig. 9.3 Measurement of electrochemical (membrane) protone gradient using the flow cytometry technique in mitochondria of granulocytes

(c) Levels of intracellular (!) B_1, B_2, B_6, B_{12}, zinc, selenium, magnesium, level of serum 1,25-dihydroxy-vitamin D, intracellular superoxide dismutase, CRP, carnitine, holo-transcobalamine < 35 pmol/l for lowB_{12}-levels, homocysteine, as well as the proinflammatory cytokines such as IL-6, IL-8 and TNF-alpha.

(d) The electrochemical transmembrane potential ($\Delta\psi$) within mitochondria, a cytofluorimetric (FCM) technique by using the lipophilic cation JC-1 measured in % [1], which is lowered in patients with autoimmune diseases (MS, rheumatoid arthritis), neurodegenerative diseases (Parkinson, Alzheimer), asthma, inflammation of the intestinal tract (i.e. IBS, colitis, morbus Chrohn), and in cancer (Fig. 9.3).

(e) The S-100 protein as a surrogate for severe head trauma with excessive NO-formation, where the protein from glia cells leaks through a ruptured blood-brain-barrier into the blood stream (normal value < 0.07 µg/l).

(f) The ATP-test determining the efficacy of ATP synthesis in neutrophils, offered by laboratories in the UK, and in Germany by Biovis. For ATP-measurement, granulocytes are isolated from CPDA blood probes, which should be as fresh as possible. Use express postal service, since the probe shoul not be old > 30 hrs. For measurement, 10^8 nucleated cells are used and following isolation, they are broken up by adding the light-emitting biological pigments luceferin together with the enzyme luciferase. ATP concentration is determined by measuring the concentration dependent luminescence yielding normative values > 500 pmol/10^8 cells (basal ATP). Thus, determination of ATP synthesis within complex V of mitochondria is possible.

In addition to assessing basal ATP concentration, functional integrity of complex IV can be estimated by means of reversible inhibition with sodium acid. By adding this agent ATP-concentration is markedly reduced (ATP stress test), which after removal of the acid should result in a near to complete regeneration of ATP concentration (recovery ATP). Recovery should be at least 25% with healthy subjects yielding values above 50%. Once there is a recovery rate <25% mitochondrial dysfunction is diagnosed. Aside from determining basal ATP concentration, the stress test allows determination of function and the regeneration potential of mitochondria by means of a reversible inhibition of the electron transport chain (ETC).

(g) The "Mitochondrial Function Profile". The central problem of chronic fatigue syndrome is mitochondrial failure resulting in poor production of ATP. ATP is the currency of energy in the body and if its production is impaired then all cellular processes will go slow. It is not sufficient enough to measure absolute levels of ATP in the cells since this will simply reflect how well rested the sufferer is. The perfect test is to measure the rate at which ATP is recycled in cells. John McLaren Howard in the UK has developed this test and he calls it "ATP profiles" being a test for mitochondrial function. Not only does this test measure the rate at which ATP is made, it also looks at where the problem lies. Production of ATP is highly dependent on the magnesium status and the first part of the test studies this aspect. Another important co-factor in the production of energy in cells is D-ribose. It is used up so quickly by cells that measuring levels is unhelpful, however low levels of ATP imply low levels of D-ribose. The second aspect of the test measures the efficiency with which ATP is made from ADP. If this is abnormal then this could be as a result of magnesium deficiency, of low levels of Co-enzyme Q10, low levels of vitamin B_3 (NAD) or of acetyl L-carnitine. The third possibility is that the protein, which transports ATP and ADP across mitochondrial membrane (the translocator protein at complex V) is impaired and this is also measured. The advantage of the ATP profiles test is that we now have an objective test of the chronic fatigue syndrome (and other mitochondria related diseases), which clearly shows that this illness has a physical basis. This test definitively shows that cognitive behavior therapy (CBT), graded exercise and anti-depressants are irrelevant in addressing the root cause of this illness. To get the full picture it is recommend to combine this test with measuring levels of Co-enzyme Q10, SODase, Glutathione Peroxidase, L-carnitine, NAD and cell-free DNA. Cell free DNA is very useful because it reflects the severity of the illness. When cells are damaged and die, they release their contents into the blood stream, and the cell free DNA determines the extent of this damage. The levels in CFS patients are similar to those from patients recovering from major infections, trauma, surgery or chemotherapy – so this test puts CFS firmly in the realm of a major organic pathology. SODase is an important antioxidant, which clears up the free radicals produced of all the inefficient chemical reactions produced by damaged mitochondria. Dr John McLaren-Howard from Acumen laboratory in the UK has developed a serum L-carnitine test, which is also included in the "Mitochondrial Function Profile". All seven tests have now been combined in a "Mitochondrial Function Profile" and can be ordered from the practice of Dr, Myhill in the UK. To order

9.1 The Extended Lab Tests

the test, go to the online order form at the bottom of the test page and complete the Medical Questionnaire. The "mitochondrial function profile test" measures the ATP-content in granulocytes, the ADP transport to the inside of mitochondria, and the intramitochondrial ATP concentration, related to the efficacy of the respiratory chain yielding five independent numerical factors from 3 series of measurements on blood samples (neutrophils):

- ATP concentration in the neutrophils is measured in the presence of excess magnesium, which is needed for ATP reactions. This gives the factor ATP in units of nmol per million cells (or fmol/cell), the measure of how much ATP is present in the cell (see acumenlab@hotmail.co.uk).
- A second measurement is made with just endogenous magnesium present. The ratio of this to the one with excess magnesium is the ATP-ratio. This tells us what fraction of the ATP is available for energy supply.
- The efficiency of the oxidative phosphorylation process is measured by first inhibiting the ADP to ATP conversion in the laboratory with sodium azide. This chemical inhibits both the mitochondrial protein cytochrome c, the last step in the ETC and ATP synthase (Ref. [7] in Chap. 5). ATP should then be rapidly used up and have a low measured concentration. Next, washing and resuspending the cells in a buffer solution remove the inhibitor. The mitochondria should then rapidly replete the ATP from ADP and restore the ATP concentration. The overall result gives OxPhos, which is the ADP to ATP recycling efficiency that makes more energy available as needed.
- The translocase complex V switches a single binding site between two states. In the first state ADP is recovered from the cytosol for re-conversion to ATP, and in the second state ATP produced in the mitochondria is passed into the cytosol to release its energy. Trapping the mitochondria on an affinity chromatography medium makes measurements. First the mitochondrial ATP is measured. Next, an ADP-containing buffer is added at a pH that strongly biases the translocase towards scavenging ADP for conversion to ATP. After 10 min the ATP in the mitochondria is measured. This yields the number translocase OUT. This is a measure of the efficiency for transfer of ADP out of the cytosol for reconversion to ATP in the mitochondria.
- In the next measurement a buffer is added at a pH that strongly biases the TL in the direction to return ATP to the cytosol. After 10 min the mitochondria are washed free of the buffer and the ATP remaining in the mitochondria is measured and this gives the number translocase IN. This is a measure of the efficiency for the transfer of ATP from the mitochondria into the cytosol where it can release its energy as needed. Adapted from: The "ATP profile"

Mitochondrial dysfunction is inherited only from the mother to the child – check for reactive nitrogen species (RNS) in children with attention deficit, concentration-memory impairment, learning problems, combined with hyperactivity, physical weakness early prostration after exercise, a high incidence of infections, repetitive headache, and fluctuations in mood.

tests were developed and are carried out at the Biolab Medical Unit, London, UK (www.biolab.co.uk) or acumenlab@hotmail.co.uk

(h) Serology for Lyme disease (or borelliosis using the lymphocyte transformation, Melisa® test), coxsackie virus, human herpes virus (HHV) Epstein Barr virus (EBV), and
(i) The heavy metals (Melisa®) test for chromium, titanium, aluminum, cadmium, plumb, cobalt, titan, iridium, copper, molybdenum, platinum, mercury, titan oxide, vanadium, and zinc. If all fails consider gene-analysis in order to evaluate if mitochondrial dysfunction is of genetic nature.

9.2 Diagnosing a Possible Allergy

1. While the allergen is the same, the symptom changes through life starting with colic and projectile vomiting as a baby, followed by toddler diarrhea, catarrh and recurrent infections, growing pains, headaches, depression, irritable bowel syndrome, PMT, asthma, irritable bowel syndrome, arthritis etc. and ending eventually in chronique fatigue.
2. There are other obvious allergic disorders, which often are present in childhood such as asthma, eczema, urticaria, rhinitis and catarrh, colic etc. and are due to food allergies.
3. In case there a positive family history. Often this is not much of a problem, which runs in families but gives an answer to the problems. Often a patient who is dairy allergic who does not have a first-degree relative (parent, sib, child) who also has symptoms caused by dairy products !
4. Allergy tests however are not safe. In addition, they are not reliable, they are expensive and it is difficult to persuade a patient to avoid a selective food if the test says it is safe !
5. Avoid toxins in the diet, such as lectins, which are naturally present in foods, artificial additives, colorings, flavorings, and artificial sweeteners.

By all means, be on the safe side, and avoid pesticide residues, plasticizer residues, etc. in processed food and use of social chemicals such as alcohol, excess consumption of caffeine, and tobacco, and/or habit-forming drugs. All this should be evaded, since it directly damages mitochondrial function.

9.2.1 Testing for a Toxic Load in the Organism

Many of the functional biochemical tests regularly show evidence of toxic stress, which we can then be investigated by other means. For example, when measuring levels of superoxide dismutase, it is common to find the gene that codes for SODase is blocked. This can be investigated further by doing DNA adducts.

9.2 Diagnosing a Possible Allergy

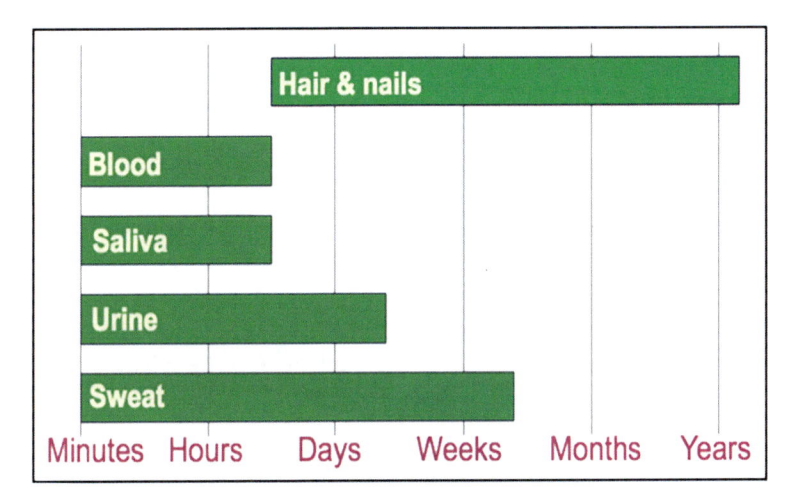

Fig. 9.4 The relative detection times of drugs in biologic specimens. Blood and saliva maintain detectable levels of drugs for hours, urine for days, and sweat for weeks with a cumulative device, and hair and nails for several years (Adapted from [3])

1. DNA adducts looks at a whole range of chemicals that may be stuck onto DNA; this is important to know because a chemical stuck onto DNA is potentially a pre-malignant condition.
2. Tests of mitochondrial function often show blockage to translocator protein. This can be investigated further by looking at a possible chemicals stuck on to the translocator protein.
3. Fat biopsy can be extremely useful way of looking at what may be stored one's fat and can pick up a whole range of pesticides and volatile organic compounds (VOCs). It is a very easy test to do just the fat contained in the bore of the sampling needle is sufficient for analysis.
4. The Kelmer test is a challenge test to look for heavy metals in urine. The problem with sweat tests and hair tests is that some people are poor detoxifiers, as they do not dump heavy metals efficiently in sweat and hair, or even urine and therefore these tests are misleading. A Kelmer test is a useful way of getting around this problem. The idea here is to take a substance such as the Kelmer agent (a chelating agent), or a therapeutic dose of selenium or zinc in order to displace heavy metals. This can give one an idea of metal toxicity. Anybody who has any amount of dental amalgam filling will test positive for mercury with a Kelmer tests.
5. The sweat tests can be a useful way of picking up heavy metals, but some people who are poor detoxifiers do not seem to get rid of heavy metals in sweat (Fig. 9.4), so this can be unreliable.
6. Hair analysis can be a useful way of picking up heavy metals (Fig. 9.4), but some people who are poor detoxifiers do not seem to get rid of heavy metals in hair. Again, this may be unreliable.

Some substances cannot be tested that easily because they get into the body, cause damage and then leave the body. Some examples are formaldehyde, fluoride, noxious gases, carbon monoxide, sulphur dioxide, nitrous oxide and radiation damage. Drugs of addiction such as heroin, cannabis, ecstasy, amphetamines can all be detected by urine drug screening tests (e.g. UDS) in a person who has recently consumed such a drug [2]. Many so called social toxins such as alcohol or nicotine but also prescription medications (e.g. such as in aberrant drug behavior) cause damage, but are not looked for or do not come up in routine tests.

Silicones cannot be detected because they are so closely related to glass, furthermore, silicone is universally used in sampling equipment such as needles. There is no one test for all chemical poisoning. However, companies having done many hundreds of these tests have a feeling for what comes up more often than others. Obviously it depends on somebodies occupation and the location where he previously had been exposed. For example organophosphates come up very commonly in farmers with sheep dip flu, but rarely in others. What one finds most commonly in subjects with mitochondropathy are as follows:

- Nickel and other metals such as mercury (dental amalgam), or cadmium (smoking).
- Polybrominated biphenyls, which are known carcinogens used as fire retardants in soft furnishings.
- Lindane and other organochlorines used as timber treatment in houses and gardens.
- Molecules indicating a poor antioxidant status such as malondialdehyde and other lipid peroxides.
- Nitrosamines resulting from smoked food or smoking.
- Hair dyes used especially by women and it is frightening how often diazole hair dyes are stuck on to DNA.
- Triclosan, a commonly used disinfectant.
- Toxic fats, which are trans fats or fats that result from cooking at high temperatures such as diolein.
- Many toxins cause an injury to mitochondria and then leave the body such as formaldehyde and other noxious gases resulting from combustion e.g. Sox, Nox, and Cox (i.e. generic terms for mono-nitrogen oxides NO and NO_2, monosulfur oxides, and monocarbon oxides).

However, the major causes of toxicity are in the following order of importance:

1. Dental amalgam from fillings of teeth (mercury release).
2. Air pollution from the heavy metal industry. This is a very major cause of asthma and respiratory disease, heart disease, cancer and birth defects.
3. Indoor air pollution like fire retardants, formaldehyde and other such volatile organic compounds (VOCs)
4. Cosmetics, especially hair dyes and aluminum containing deodorants, wash detergents and cleaning agents.
5. Cooking resulting in the release of nickel from stainless steel saucepans, trans fats from poor quality of food, or burnt food.
6. Pesticide residues in food.

9.2 Diagnosing a Possible Allergy

7. Smoking resulting in the inhalation of nitrosamine and cadmium.
8. Occupational exposure of farmers, Gulf War veterans, firemen with 9/11 syndrome, aeroplane industry (for further information see www.aerotoxic.org)
9. Silicone prostheses as they are being used in cosmetic surgery for breast implants.
10. Traffic pollution resulting in benzene accumulation. Other pollutants such as noxious gases are not picked up in these tests.

Since we all have vicious toxins on board which cause ongoing damage to the body, higher levels of antioxidants (vitamins ACE and selenium) help to protect, sufficient levels of B vitamins induce detoxification, while sufficient levels of protein, essential fatty acids and other minerals help to repair the damage induced at mitochondria. A good example of this assumption came to the open when doing research on the cause of thalidomide-related birth defects. Thalidomide was an agent prescribed to women in pregnancy as a safe hypnotic in pregnancy, which caused serious birth defects if the women took it between the 38th and 42nd day of gestation. However, not all babies were affected. This drug was tested in rats, but no offspring's were abnormal. This was a mystery to researchers, until someone had the bright idea of putting the rats into nutritionally depleted diets. The offspring's than got the congenital fetal disorder involving the limbs or phocomelia ("flipper limbs"). It was the results of a combination of toxic stress induced by the drug and the concurrent nutritional deficiency, both of which caused the problem to become apparent.

Non- contaminated nutrition is highly protective against toxic stress – this is a good reason to take nutritional supplements.

Look, I wear a label at my ear, which indicates that I am not contaminated with dioxin, pesticides, or a genetically engineered growth hormone leading to mitochondropathy or a high insulin-like growth factor (IGF-1) in your body!

References

1. Cossarizza A, Ceccarelli D, Masini A (1996) Functional heterogeneity of isolated mitochondrial population revealed by cytofluorimetric analysis at the single organelle level. Exp Cell Res 222:84–94
2. Freye E, Levy JV (2009) Pharmacology and abuse of cocaine, amphetamines, ecstasy and related designer drugs. Springer, Dordrecht, p 300
3. Caplan YH, Goldberger BA (2001) Alternative specimens for workplace drug testing. J Anal Toxicol 25:396–399

List of Certified Labs in Europe for Diagnosing Mitochondropathy (Not Claimed for Completeness)

1. **Biovis Diagnostik MVZ GmbH**
   Konrad Adenauer Straße 17
   55218 Ingelheim/Germany
   www.biovis-diagostik.de

The company has developed a recently released urine dry spot technology for measurement of adrenaline, noradrenalin, dopamine, GABA, glutamate, tryptophan, and citrulline for neurostress, CSF and mitochondropathy respectively, In addition, a blood dry-spot is available for measurement of nitrotyrosine and lactate-pyruvate ratio in mitochondropathy. The obvious advantage is that just a drop of capillary blood with no need for venous puncture, or a drop of midstream urine for citrulline levels (dry urine test – DUT) is necessary, being soaked up by blotting paper, stored for drying and then shipped to the laboratory for measurement. This triangle shaped dry-spot test (Fig. A.1) can be stored for several months with no degradation of the relevant information in plasma contents.

Biovis now also offers the highly sensitive and quantitative detection of ATP by luciferase driven bioluminescence in viable somatic cells where as the basic principle, the luciferase from Photinus pyralis (American firefly) catalyses the following reaction: $ATP + D\text{-luciferin} + O_2 \rightarrow \text{oxyluciferin} + PP_i + AMP + CO_2 + \text{light}$. The quantum yield for this reaction is about 90% with the resulting green light having an emission maximum at 562 nm.

E. Freye, *Acquired Mitochondropathy – A New Paradigm in Western Medicine Explaining Chronic Diseases: The Safety Guide for Prevention and Therapy of Chronic Ailments*, DOI 10.1007/978-94-007-2036-7,

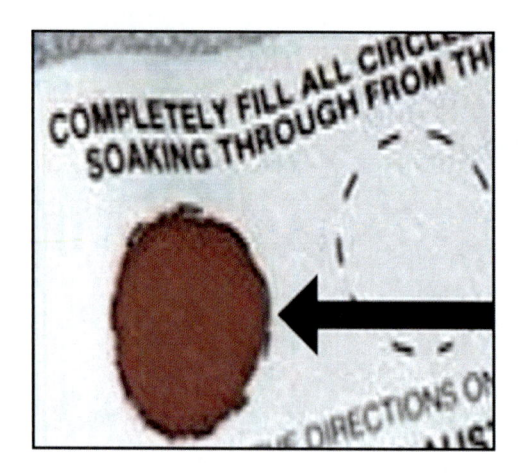

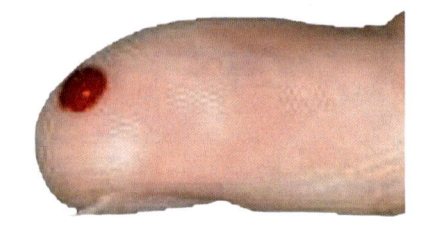

Fig. A.1 The dry blood spot (DBS) test for determination of mitochondropathy

List of Certified Labs in Europe for Diagnosing Mitochondropathy

2. Ganzimmun Diagnostics AG
   Hans-Böckler-Straße 109
   D-55128 Mainz/Germany
   www.ganzimmun.de

3. Bio Lab Medical Unit
   9 Weymouth St London W1W 6DB, UK
   www.biolab.co.uk

**4. Acumen lab, UK gives the ATP profile test together
   with an analysis of a medical questionnaire
   Email: Hania on hania@doctormyhill.co.uk,
   Enter TEST ORDER in the subject of your email**

   For payment methods, go to ordering tests. One can also post a note requesting the test with a paper copy of the questionnaire and a payment (a cheque for £295, i.e. £225 for the tests and £70 for a letter to the GP, made payable to Sarah Myhill Limited) to the office at Upper Weston, Llangunllo, Knighton, Powys LD7 1SL. On receipt of the questionnaire and payment a test kit will be sent. The price for the letter reflects the fact that in that 10–14 page letter Dr. Myhill interprets seven separate tests as well as giving advice about all the various health problems reported in the questionnaire.

5. Orthomedis Speziallabor AG
   **Fluhstrasse 30, 8640 Rapperswil/ Switzerland
   Tel. ++41 (0) 55 210 24 68, Fax: ++41 (0) 55 210 05 43**
   e-mail: orthomedis@bluewin.ch

6. Protea Biopharma in Belgium
This lab, specifically does the Neurotoxic metabolite test and the **Hydrogen sulphide test** in the urine for gut dysbiosis, both of which can be ordered online directly from **e-mail: www.redlabs.com**

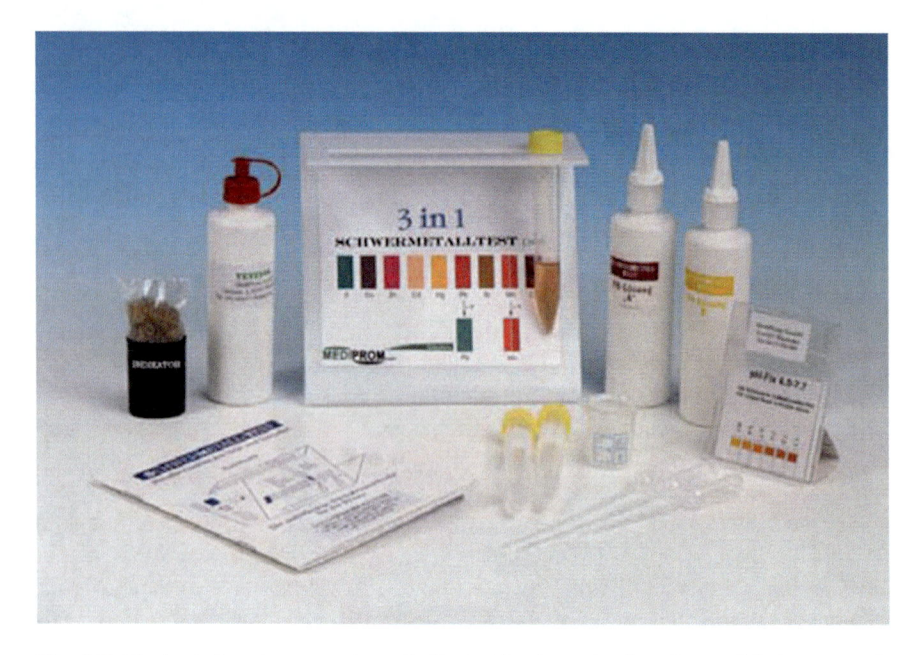

Fig. A.2 The basic three in one semiquantitative test for determination of up to eight heavy metals in aqueous solutions

7. Tarvalin® company has developed a sensitive diagnostic tool - the transketolase-like-1-Protein (TKTL1) - as a new marker to detect activity of cancer cells within the body (EDIM technology)

   TAVARLIN AG, Landwehrstr. 54
   64293 Darmstadt/Germany
   Phone: +49 6151-950 55 62
   Fax: +49 6151-950 55 51
   Mail: www.tavarlin.de   www.tavarlinshop.de.

   Since TKTL-1 is responsible for shifting energy production within tumor cells from oxidative to anaerobic metabolism in the presence of oxygen, this marker it can be also used to detect possible resistance against conventional chemo- and/or radiotherapy and the efficacy of any cancer-related therapy.

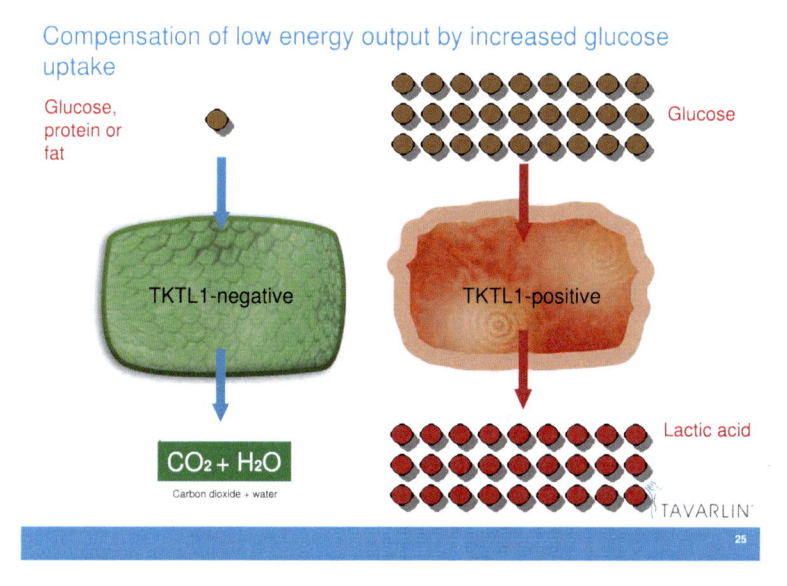

List of Certified Labs in Europe for Diagnosing Mitochondropathy

Compensatory energy production for cancer-related fermentation within cells by means of an increased demand for glucose using the transketolase-like-1-Protein (TKTL-1)

Index

A
Aconitase, 47, 50, 51, 81, 86, 87, 150, 257
Advanced glycation end-products (AGEs)
  cellular peroxynitrite, 234–237
  fructose, 233
  glucose, 233
  processed carbohydrates, 233, 234
Allergy, 23, 95, 126, 148, 158, 164–166, 169, 180, 183, 187, 194, 196, 198, 247, 259, 261, 271, 276, 277, 287, 288, 292, 295, 297, 314, 329, 348, 372, 376–379
Alpha-lipoic acid (Ω–3 fatty acid), 96, 139, 141, 279, 294, 298–300, 307, 309, 326, 328, 331, 358
Alzheimer, 19, 38, 57, 81, 86, 96, 101, 102, 104–107, 109–112, 131, 139, 141, 142, 148, 162, 164, 174, 190, 219, 230–234, 236–241, 294, 314, 316–318, 321–355, 374
Anthocyanes, 314
Antioxidants, 1, 18, 56, 65, 82, 84, 88, 98, 101, 104, 127, 143–146, 152, 158, 161, 171, 186–188, 196–199, 201, 237, 246, 253, 261, 262, 268–270, 286, 294, 295, 298–300, 302, 305–307, 314, 320–323, 326–331, 341, 350, 352–360, 374, 378, 379
Apoptosis, 16, 29–31, 66, 81, 84, 107–114, 232, 234–236, 244–246, 265, 296, 308, 323, 347–349, 355
Arteriosclerosis, 23, 38, 50, 80, 93, 147, 225
Ascorbic acid (Vit C), 195, 299, 326
ATP profile test, 158, 374, 375, 383
Attention deficit hyperactivity syndrome (ADHS), 80, 96, 104, 259

Autoimmune diseases, 14, 17, 19, 23, 31, 37, 50, 56, 57, 59, 86, 149, 249–251, 255, 311, 348, 353, 355, 373

B
Beta-carotene, 145, 269, 294, 300, 301, 327
Betaine hydrochloiride (trimethylglycine), 191, 295, 299, 300, 307
Bicarbonate, 178, 196, 352–355
Bioflavonoides, epigallocatechin gallate (EGCG), 246, 300, 329
Brain-derived neurotrophic factor (BDNF), 113, 114
Broken heart syndrome-Takotsubo-cardiomyopathy, 38

C
Calcitriol, 311
Cancer, 1, 14–16, 19, 23, 30, 37, 46, 61, 73, 81, 90, 92, 96, 97, 134, 139–143, 171, 174, 175, 180, 190, 234, 242, 244–251, 268, 270, 271, 274, 276, 278, 286, 293, 300, 306, 310, 311, 314, 323, 347, 348, 355, 358–362, 373, 378
Carotenoids, 156, 294, 299, 300, 305–307, 327
Cave Man's Diet, 170, 172, 178, 221, 270–272, 310, 339
Cell free DNA, 158, 159, 374
Cholesterol, 47, 48, 85, 89, 90, 96, 131, 145, 163, 171, 217, 227, 231–241, 248, 249, 254, 270, 273, 275, 314, 340, 358–363, 371

E. Freye, *Acquired Mitochondropathy – A New Paradigm in Western Medicine Explaining Chronic Diseases: The Safety Guide for Prevention and Therapy of Chronic Ailments*, DOI 10.1007/978-94-007-2036-7, © Springer Science+Business Media B.V. 2012

Chronic fatigue syndrome (CFS),
  alcohol intolerance, 147, 165–166, 202
  foggy brain, 125, 131, 147, 164–167
  gut fermentation, 133, 271, 283
  hypocortisolism (adrenal insufficiency), 135–146
  hypoglycemia, 222 ff
  Isovolumetric relaxation time (IVRT)
  low cardiac output, 125–127, 280–284
  muscle pain, 186
  poor stamina, 41, 122, 123, 125, 128, 132, 280, 283
  postural orthostatic tachycardia syndrome (POTS), 281, 282
  sleep depreviation, 148, 167
Chronic (neuropathic) pain, 73, 85
Citrullination, 48, 49, 293, 372
Citrulline, 29, 48, 63, 64, 74, 85, 92, 249, 256, 259, 371, 372, 381
Coconut oil, 130, 131, 147, 148, 162, 164–166, 220, 237–240, 363
Coenzyme Q10 (Ubiquinone), 6, 11, 56, 81, 82, 89, 95, 98, 144, 146, 154, 155, 158, 161, 180, 187, 199, 248, 269, 270, 294, 298–300, 314, 317, 320, 325, 355, 374
Coenzyme Q10H (Ubiquinol), 318
Cori cycle, 43, 123, 124, 128, 132, 282
Cryptopyrroluria (KPU), 256–260
Curcuma longa (curcumin), 245, 246, 313, 314, 322–323, 330, 338
Cyano-, Methyl-, cobalamin (B12), 189, 191, 192, 299, 331, 373
Cyclic citrullinated peptides (CCP) test, 372
Cytochrome c, 6, 8, 9, 11, 47, 66, 87, 91, 95, 105, 246, 308, 309, 314, 320, 675

D
Depression, 21, 23, 43, 80, 101, 125, 127, 146, 148, 149, 153, 185, 219, 223, 261, 276, 277, 286, 293, 318, 334, 344, 348, 359, 371, 376
Detoxification
  antioxidants, 146, 269, 286, 379
  Cave Man's Diet, 178
  chlorella protothecoides, 156
  far-infrared saunaing (FIRS), 155, 156
  liver P450 cytochrome, 181, 201–203
  superoxide dismutase (SODase), 49
DHA. *See* Docosahexanoic acid (DHA)

Docosahexanoic acid (DHA), 164, 165
D-ribose, 41, 42, 123, 128, 152, 154, 159–161, 165, 191, 280, 284–286, 294, 295, 297, 298, 329, 341, 343, 345–347, 374

E
Ecklonia cava extract, 294, 297, 327, 330
Electrochemical transmembrane potential ($\Delta\Psi$), 373
Endothelial NOS (eNOS), 29, 30, 32–34, 47, 60, 66, 73, 74, 76, 77, 105, 152, 236, 237, 332

F
Fibromyalgia syndrome (FMS)
  tender points, 150, 267
  TRPV-receptor (pain), 120, 151
Flavonoids, 295, 298–300, 306, 326, 327, 330
Folic acid, 63, 77, 92, 179, 188, 189, 258, 298–300, 327, 332, 340
Food as poison, 43, 95
Fructose and obesity, 172–174, 233

G
γ-amino butyric acid (GABA), 48, 49, 119, 121, 175, 232, 244, 274, 281, 294, 298–300, 327, 329
Gluconeogenesis, 48, 81, 324
Glutathion (GSH), 17, 19, 56, 58, 78, 79, 101, 104, 141, 144–146, 149, 153, 154, 158, 187, 188, 191, 192, 197, 199, 202, 221, 244, 245, 258, 264, 266, 269, 270, 294, 295, 298–300, 307, 314, 325–329, 338, 341, 355, 374
Glycolysis, 10, 11, 15, 49, 51, 57, 81, 94, 124, 159, 175, 244, 245, 249, 297, 316, 342, 355, 356

H
Hashimoto thyreoiditis, 23, 49, 57, 86, 249
Heart insufficiency, 19, 81, 90, 126, 328
Heavy metal intoxication, 22
Heavy metal test, 384
Hemp oil, 164, 165, 220
Hydrogenated vegetable oils, 361
Hydrogen peroxide (H_2O_2), 45, 56, 65, 66, 73, 106, 329, 333

Index

Hydrogen sulfide (H_2S), 180, 185, 197, 198, 309
Hypercholesterinemia, 85
Hypertonus, 71
Hypoxicator?, 348–351

I

Immune system
  allergy, 23
  $CD4^+$T-cells, 19
  cytokines, 19–22, 31, 60, 105
  INF-gamma, 255
  inflammation, 1, 22, 23, 30, 60, 66, 227, 251
  interleukin–3 (IL–3), 253
  interleukin–4 (IL–4), 19, 20, 30
  interleukin–5 (IL–5), 253
  interleukin–6 (IL–6), 105, 254–256, 373
  interleukin–10 (IL–10), 24, 30, 184
  interleukin–12 (IL–12), 19–21
  macrophages, 30, 311
  mast cells, 252
  NK-cells, 18
  T-, B-Lymphocytes, 21, 107, 108
  Th1/Th2-dominance, 17, 23, 43, 253, 291
  TNF-alpha, 21, 30, 40, 60, 61, 84, 88, 105, 184, 252, 254–256, 290, 291, 293, 373
Induced nitric oxide synthase (iNOS), 18, 20, 29, 30, 34, 47–50, 63, 66, 79, 82, 91, 99, 110, 119, 152, 236, 254, 295, 298–300, 307, 311, 326, 330, 332, 337, 371
  chronic pain, 100
  hypoxia, 28, 29, 85, 236
  inflammation, 29, 30, 50, 254
Inflammatory disease, 234, 293, 311
iNOS. *See* Induced nitric oxide synthase (iNOS)
Intermittent hypoxia therapy (IHT), 347–350, 352
Irritable bowel syndrome (IBS)
  bacteria in intestine, 178, 184–185, 198, 199
  D-lactate formation, 148, 180, 181, 184, 200
  gut fermentation, 148, 181, 182, 184–186, 230
  Kefir, 180, 199, 200, 241–244
  leaky gut, 23, 132, 183, 242, 283, 359

malabsorption, 132, 182, 196, 229, 283, 292
neem (*Azadirachta indica*), 201
noxious gases, 181, 378, 380, 381
pancreatic enzymes, 178, 196, 198, 200
prebiotics, 178, 199, 200, 240–241, 243, 292
probiotics, 22, 24, 176, 180, 182, 194, 199, 200, 241–243, 253, 288, 292, 302

K

Krebs-citric acid cycle, 5–7, 11, 48, 51, 57, 81, 87, 94, 129, 134, 154, 160, 161, 227, 245, 257, 259, 272, 284, 316, 324, 328, 340

L

L-(acetyl) carnitine, 56, 82, 123, 155, 158, 161, 191, 298, 325, 327, 341, 374
Lactate/pyruvate ratio, 51, 372, 381
L-arginine, 28, 60, 62, 64, 66, 73, 74, 81, 83, 94, 103
Leptin, 60–61, 172, 255
Linolenic acid (Ω–6 fatty acid), 96
Lipopolysaccharides (LPS), 66, 84, 88
LOGI diet. *See* Low glycemic index (LOGI) diet
Long term potentiation (LTP), 34, 59, 117, 118
Low glycemic index (LOGI) diet, 177, 193, 223, 242, 272, 288, 339

M

Magnesium, 41, 101, 102, 139, 152, 154–160, 195, 198, 200, 202, 218, 220, 247, 253, 254, 277, 278, 286, 294, 298, 299, 307, 309, 310, 323–326, 330, 338, 353–355, 359, 373, 374
Malondialdehyde (MDA), 39, 40, 266, 267, 378
Medium chain fatty acids (MCF), 162–170, 191, 238, 239, 271, 286, 325, 340
Medium-chain triglycerides (MCT), 130, 162–164, 220, 237–240
Melatonin, 58, 59, 61, 63, 91, 135, 146, 169, 187, 258, 270, 299, 300, 329, 334
Metabolic syndrome X (MSX), 86
Methylation cycle, 154, 187–203
Methylmalonic acid (MMA), 371

Mitochondria
  acetyl coenzyme A, 48
  adenosine diphosphate (ADP), 9, 10, 12,
      39–41, 50, 121–124, 134, 135, 154,
      155, 157, 158, 160, 284, 285, 294,
      325, 341, 374, 375
  adenosine monophosphate (AMP), 123,
      128, 285
  adenosine triphosphate (ATP), 5, 9, 11,
      40, 112, 121, 125, 127, 134, 146,
      244, 284
  complex I-V, 5–11, 16, 43, 47, 58, 84, 89,
      92, 95, 98, 105, 106, 122, 124, 127,
      160, 232, 244–246, 294, 314, 325,
      328, 341, 374
  disease, inborn errors, 13
  electronic transport chain (ETC), 5, 7, 11,
      39, 40, 43, 44, 81, 88, 148, 320,
      323, 325, 375
  function, 3–12, 15, 19, 38, 43,
      123, 186
  oxidative phosphorylation, 4, 5, 7–13,
      40, 41, 47, 81, 82, 88, 91, 122,
      127, 128, 144, 154, 157, 161,
      244–246, 269, 284, 309, 320,
      356, 371
  OXPHOS system, 11 ff
  proton gradient, 7–9, 12
  pyruvate dehydrogenase, 11, 48, 51,
      94, 106
  translocator protein (TL), 39, 40, 122,
      124, 154, 155, 157, 158, 309,
      374, 377
Mitochondrial NOS (mtNOS), 29, 30, 47, 107,
    295–296, 307, 311
Mitochondropathy, acquired
  Alzheimer disease (AD)
    Aβ fragments, 110
    BACE activity, 109–110
    coconut oil, 148, 162
    glutamate-NMDA-receptor,
        81, 258
    ketones, 237, 240
    lewy body formation, 111
    new treatment options, 240
    short/medium chain fatty acids
        (MCFA), 238
    tau protein aggregates, 110
  amyotrophic lateral sclerosis (ALS), 107,
      110, 111, 238
  excitotoxicity, 58, 113, 294, 299, 311,
      326, 337
  hypertonus, 71
  Parkinson's disease (PD)
    BH_4-deficit
      L-DOPA, 319, 334, 337
      NADH, 85, 314, 319, 321, 333–338
      tyrosine hydroxylase (TH), 334
    type 2 diabetes, 19, 294
Mitochondropathy, inborn error, 13, 371
Mitoptosis, 347, 349
Multiple chemical sensitivity (MCS), 23, 37,
    38, 42, 59, 62, 81, 98, 111,
    115–121, 250, 266, 293, 297,
    331–333

N
N-acteyl-cysteine (NAC), 78, 254, 266, 267,
    269, 295, 325, 326
Neural sensitization, 59, 102,
    117–121
Neurodegenerative diseases, 37, 106–115,
    316, 373
Neuronal NOS (nNOS), 29, 47, 76, 105, 108,
    152, 236, 296, 334
Niacin (Vit B_3), 161, 294, 298, 321,
    324, 327
Nicotinamide adenine dinucleotide (NAD^+), 5,
    7, 40, 73, 134, 158, 161, 285,
    313–317, 320, 321, 333
Nicotinamide adenine dinucleotide phosphate
    (NADPH), 5, 28, 32, 40, 72, 74, 80,
    175, 237, 307, 313–317, 320, 321,
    324, 328
Nitric oxide (NO)
  biological function, 28–31
  excess formation, 43–50
  inflammation, 43–50, 56–61, 84, 254
Nitrosation, 57, 58, 86
Nitrosative stress, 13–24, 37, 38,
    55–66, 71–202, 249, 251–265,
    272–274, 311, 312, 331, 353,
    371–379
Nitrotyrosine, 49, 81, 85, 103, 249, 371,
    372, 381
N-methyl-D-aspartate (NMDA) receptor, 56,
    59, 83, 101, 152, 296, 335
NO/ONOO-cycle, 99, 297
NO-syn(te)thase, 28, 56, 64, 66, 71, 152, 266,
    268, 327, 330, 332, 337
Nuclear factor kappa-light-chain-enhancer of
    activated B cells (NF-κB), 23, 50,
    58, 60, 64, 83, 99, 105, 106, 109,
    115, 116, 119, 150–153, 236, 246,
    254–256, 307, 310–314, 322, 326,
    329, 330, 337, 338, 340
Nutraceuticals, 302–305, 353

O

β-Oxidation, 11, 48
Oxidative stress, 5, 22, 40, 48, 57, 60, 65, 74, 79, 83, 90, 91, 99, 104, 106, 109, 110, 115, 116, 120, 121, 150, 153, 154, 188, 233, 246, 254, 330, 331, 349, 353

P

Pantothene (Vit B_5), 324
Parkinson, 19, 38, 57, 67, 85, 86, 101, 102, 104–107, 110–112, 138, 174, 190, 219, 230–232, 238, 294, 314, 316, 319, 321, 323–355
Pentose phosphate pathway (PPP), 174, 175, 342–344
Peroxynitrite (ONOO⁻), 29, 31–35, 42, 45, 49, 50, 55–60, 62–66, 71, 79–81, 83, 85–87, 99, 101–107, 109, 110, 115, 116, 119–121, 127, 145, 152, 153, 234, 235, 237, 249–251, 254, 265, 294–296, 307, 311, 312, 324–329, 331–333, 336, 337, 372
Phosphatidylcholine, 154, 164, 165, 217, 219, 220
Poly(ADP-ribose) polymerase (PARP), 73, 83, 107, 237, 321
Polysaccharide A (PSA), 24, 66, 349
Post traumatic stress syndrome (PTSD), 37, 38, 42, 266, 294, 298
Pyridoxin (Vit B_6), 93, 299, 300
Pyrroloquinoline quinone (PQQ), 356, 357

R

Reactive nitric oxide substrates (NOS) species, 31
Reactive oxygen species (ROS), 15, 17, 31, 39, 43, 45, 47, 56, 58, 60, 71, 74, 82, 83, 85, 88, 91, 92, 95, 109–111, 113, 145–147, 152, 158, 236, 244, 245, 254, 256, 268, 273, 293, 294, 308, 310, 311, 314, 325, 327, 328, 347, 349, 371
Recommended daily allowance (RDA), 196, 224–226, 312, 330
Ribofolavin (Vit B_2), 63, 93, 94, 294, 298, 299, 324, 328
Ribonucleic acid (RNA), 174, 187, 188, 295, 329

S

Selenium, 93, 98, 141, 144, 145, 202, 221, 227, 247, 269, 278, 295, 299, 300, 307, 324, 327, 359, 373, 377, 379
Siberian ginseng, 314
S-nitrosoglutathione (S-GSHG), 29
S-100 protein, 373
Stroke, 13, 19, 86, 99, 111, 112, 126, 133, 170, 190, 219, 234, 236, 237, 273, 283, 292
Superoxide (O_2^{+}), 49, 66, 73
Superoxide dismutase (SOD), 49, 56, 82, 104, 144, 146, 199, 249, 258, 270, 295–299, 307, 329, 331, 355, 373, 376

T

Tetrahydrobioterin (BH_4), 29, 58, 61–66, 71, 73–88, 101, 107, 232, 295, 307, 314, 326, 327, 331–337
Thiamine (Vit B_1), 148, 231, 298, 299, 324, 328
Th1/Th2-shift, 19, 22, 43, 44, 253, 254, 291, 292
Tocotrienols, 253, 294, 298, 299, 326
Toxic load
  acidifiers, 96
  analgesics, 89
  antianginals, 89
  antibiotics, 89
  anticonvulsants, 227
  antiepileptics, 89, 231
  antiinflammatory agents (NAIDs), 90
  antituberculostatics, 89
  antivirals, 89
  anxiolytics, 95
  artificial coloring, sweetening, 249
  aspartame, 89
  azo dye stuff yellow-orange colorings, 96
  beta-blockers, 90
  bisphenol A, 143
  cake glaze (alumium and citric acid), 96
  cardiotonic agents, 90
  chemotherapy, 249
  clofibrates, 89
  diuretics, 90
  flavor enhancer, 96
  food emulsifier, stabilizer and thickeners, 96
  food preservatives, 249
  fungicides, 89, 249
  heavy metals (Al, As, Cd, Pb, Ni, Hg), 141, 143–144, 156, 227–229
  herbicides, 89

Toxic load (*cont.*)
    household poisons, 141
    insecticides, 89
    local anesthetics, 89
    nanotechnology, 92
    pesticides (organophosphates), 145, 270
    proton pump inhibitors (PPI), 249
    psychotropics, 89
    silicone, 146
    statins, 89
    steroids, 89, 90
    sugar substitutes, 96
    trans fats, 89, 96, 361, 378
    tricyclic antidepressants, 89, 99
    vasodilators, 28, 254
    volatile anesthetics, 93, 147
    volatile organic compounds (VOCs), 59, 132, 283
Transient receptor potential channel vanilloid (TRPV), 76, 99

U
Urine analysis, 140
Urine drug screening (UDS), 378
Urine dry spot (test), 381

V
Vaccination, 22, 262
Vitamin C, 65, 79, 98, 99, 139–141, 145, 195, 196, 253, 254, 270, 279, 298–300, 307, 314, 326, 329, 330, 332, 359, 362
Vitamin D, 23, 145, 226, 247, 248, 253, 262, 269, 277, 304, 310–313, 340, 358, 362, 372, 373
Vitamin E, 79, 98, 141, 145, 219, 253, 269, 294, 298, 299, 312, 326, 359, 360, 362

X
Xenobiotics, 43, 66, 80, 81, 145, 180, 185, 186, 226–244, 269, 288, 328, 329, 371

Printed by Publishers' Graphics LLC USA
MO20120416-162
2012